Chemical Environmental Pollutants and their Effect on Health

Aikaterini Salavoura

Chemical Environmental Pollutants and their Effect on Health

Springer

Aikaterini Salavoura
Athens, Greece

ISBN 978-3-031-80270-6 ISBN 978-3-031-80271-3 (eBook)
https://doi.org/10.1007/978-3-031-80271-3

This Springer imprint is published by the registered company Springer Nature Switzerland AG
The registered company address is: Gewerbestrasse 11, 6330 Cham, Switzerland

Preface

There is a growing interest of the public on the impacts of the environment on health. However, scientists have not a ready solution. Although fatal bacterial infections have been eradicated with the use of new generation antibiotics, other diseases increase rapidly and the demand for new treatments, such as monoclonal antibodies, is mandatory. Since the decoding of the genes of the human genome, questions regarding multifactorial diseases have not been answered. However, there is not yet a convincing theory to explain the etiopathogenesis of these diseases, not even the effectiveness and safety of new drugs. The increase in the prevalence of diseases, such as allergy, autoimmune diseases, neurodevelopmental and neurogenerative diseases such as Alzheimer disease, Diabetes mellitus type I, and cancer, grows and possesses a serious burden to the health system. Since our genes have not changed in the last 50 years, other factors are implicated to the emergence of these "epidemics."

My clinical work as a pediatrician in Children's Hospital "Agia Sofia" in the last 20 years, which used to be the main hospital for children in Athens, contributed to my decision to analyze the literature. I realized that the emergency visits for asthma attacks increased. In addition, diseases like first diagnosed diabetes mellitus type I, inflammatory bowel disease, and rheumatoid arthritis soared. These chronic diseases apart from the burden in costs for the health system count for a serious deterioration in the life of patients and their families with life-long consequences.

The above observations were points of skepticism. Was this soaring of allergic and autoimmune diseases phenomenal or pragmatic? Going through the literature, research has proved that the increase is pragmatic. In-depth studying showed me the complexity of the issue and it gave me the opportunity to scrutinize new knowledge regarding the etiopathogenesis of multifactorial diseases and the role of environment by the epigenetic regulation of the genes, which is suggested to be associated with the growing impact of the environment on our genes.

Athens, Greece Aikaterini Salavoura

Acknowledgments

I have to thank Professor Neni Nicolopoulou-Stamati for giving the knowledge and the ideas to start writing this book. Additionally, I have to thank her for her pernicious advice after reading the draft of the book.

Contents

1 **The Epigenome** 1
Introduction 1
A Unified Theory of Gene Expression 1
The Epigenotype 3
Epigenetic Mechanisms 4
DNA Methylation 5
Histones 6
RNA Regulatory System 9
Repeated DNA Sequences 11
Transposable Elements (TEs) 11
DNA Transposons 12
Retrotransposons and Endogenous Retroviruses 13
Transcription Factors 13
Genomic Imprinting 14
Epigenetic Reprogramming 15
Methylation of Other Key Elements Within the Genome 18
The Epigenotype Model 20
The Thrifty Phenotype Hypothesis 20
Evolution Models of Development 24
Executive Summary 28
References 29

2 **DNA Methylation and Disease** 31
DNA Hypomethylation and Disease 31
Transcriptional Modifications of Transposable Elements 32
Executive Summary 35
References 35

3 **Oxidative Stress** 37
Reactive Oxygen Species 37
The Central Role of Glutathione in the Pathophysiology of Human Diseases [10] 45

Metals and Their Effects on Health ... 48
Executive Summary ... 52
References ... 52

4 Metals and Disease ... 55
Mercury ... 55
The Metal, Its Forms, and Its Distribution in the Environment ... 55
Accidental Exposures in History ... 57
Effects on Health ... 58
Epigenetic Mechanism of Intoxification ... 59
Chromium ... 61
The Metal, Its Forms, and Its Distribution in the Environment ... 61
Accidental Exposures in History ... 65
Effects on Health ... 66
Epigenetic Mechanism of Intoxification ... 66
Arsenic ... 69
The Metal, Its Forms, and Its Distribution in the Environment ... 69
Accidental Exposures in History ... 72
Effects on Health ... 73
Treatment ... 78
Mechanism of Intoxification ... 78
Epigenetic Mechanism of Intoxification ... 80
Lead ... 84
The Metal, Its Forms, and Its Distribution in the Environment ... 84
Effects on Health ... 86
Blood-Lead Level ... 89
Mechanism of Intoxification ... 90
Epigenetic Mechanism of Intoxification ... 98
SNPs Related to Lead Sensitivity ... 100
Cadmium ... 102
The Metal, Its Forms, and Its Distribution in the Environment ... 102
Accidental Exposures in History ... 105
Effects on Health ... 105
Mechanism of Intoxification ... 107
Epigenetic Mechanism of Intoxification ... 110
Aluminum ... 114
The Metal, Its Forms, and Its Distribution in the Environment ... 114
Effects on Health ... 115
Mechanisms of Intoxification ... 119
Epigenetic Mechanism of Intoxification ... 124
Manganese ... 127
The Metal, Its Forms, and Its Distribution in the Environment ... 127
Accidental Exposures in History ... 128
Effects on Health ... 129
Mechanism of Intoxification ... 131
Epigenetic Mechanism of Intoxification ... 135

Magnesium 138
The Metal, Its Forms, and Its Distribution in the Environment 138
Effects on Health. 138
Mechanism of Action 139
Epigenetic Changes. 140
Nickel 141
The Metal, Its Forms, and Its Distribution in the Environment 141
Accidental Exposures in History 142
Effects on Health. 143
Mechanism of Intoxification. 145
Epigenetic Mechanism of Intoxification. 149
Calcium. 153
Effects on Health. 155
Asbestos 156
The Metal, Its Forms, and Its Distribution in the Environment 156
Accidental Exposures in History 158
Effects on Health. 159
Mechanisms of Intoxicification. 162
Epigenetic Mechanism of Intoxification. 167
Silica 167
The Metal, Its Forms, and Its Distribution in the Environment 167
Effects on Health. 168
Mechanisms of Intoxification 170
Copper. 175
The Metal, Its Forms, and Its Distribution in the Environment 175
Effects on Health. 177
Mechanism of Intoxification. 180
Epigenetic Mechanism of Intoxification. 182
Cobalt 183
The Metal, Its Forms, and Its Distribution in the Environment 183
Effects on Health. 183
Mechanisms of Intoxification 184
Epigenetic Mechanism of Intoxification. 185
Selenium 185
The Metal, Its Forms, and Its Distribution in the Environment 185
Accidental Exposures in History 186
Effects on Health. 186
Mechanism of Intoxification. 187
Iron 189
The Metal, Its Forms, and Distribution in the Environment. 189
Effects on Health. 191
Mechanism of Intoxification. 192
Epigenetic Mechanism of Intoxification. 195

Beryllium 196
The Metal, Its Forms, and Its Distribution in the Environment 196
Effects on Health. 197
Tin Poisoning: An Example of Selective Toxicity of Organic
Metal Complexes 199
Executive Summary 199
References. 207

5 Organic Pollutants and Disease 215
Persistent Organic Pollutants (POPs) 215
The Chemicals, Their Forms, and Their Distribution
in the Environment 215
Effects on Health. 218
Unintentionally Produced Chemical Agents. 219
Polycyclic Aromatic Hydrocarbons (PAHs) 219
Accidental Exposures in History 222
Intentionally Produced Products. 223
Polybrominated Diphenyl Ethers (PBDEs) 223
Polychlorinated Biphenyl Compounds (PCBs) 224
Effects on Health. 225
Chlorinated Compounds 225
Chlorofluorocarbons (CFCs) and Hydrochlorofluorocarbons
(HCFCs) 225
Polychlorinated Naphthalene (PCN). 226
Sodium Cyanide (HxCN) 227
Perfluorinated Compounds (PFC). 228
Trichloroethylene (TCE). 228
Epigenetic Mechanism of Intoxification. 228
Brominated Flame Retardants (BFR) 229
Mechanisms of Intoxification of POPs 229
Epigenetic Mechanisms of Intoxification 231
Epigenetic Mechanisms of Intoxification 236
Pesticides. 236
Effects on Health. 237
Mechanisms of Intoxification 237
Epigenetic Mechanism of Intoxification of POPs. 238
Benzene. 239
Effects on Health. 239
Epigenetic Mechanism of Intoxification. 239
Dioxin and Dioxin-Like Chemicals 240
Mechanism of Intoxification. 240
Epigenetic Mechanisms of Intoxification 242
Air Pollution 244

Indoor Pollution ... 244
The Chemicals, Their Forms, and Distribution in the Environment... 244
Effects on Health... 246
Epigenetic Mechanisms of Intoxification... 247
Outdoors Pollution ... 248
The Chemicals, Their Forms, and Their Distribution in the Environment ... 248
Effects on Health... 250
Mechanism of Intoxification... 252
Epigenetic Mechanism of Intoxification... 253
Alcohol ... 254
Effects on Health... 256
Tobacco Smoking ... 256
Effects on Health... 257
Mechanisms of Intoxification ... 257
Executive Summary ... 258
References... 260

6 Endocrine Disruptors ... 263
Endocrine Disruptors ... 263
The Chemicals, Their Forms, and Their Distribution in the Environment ... 263
Effects on Health... 265
Mechanism of Intoxification... 266
Phthalates ... 272
The Chemicals, Their Forms, and Their Distribution in the Environment ... 272
Effects on Health... 274
Bisphenols (BPs)... 275
The Metals, Their Forms, and Their Distribution in the Environment ... 275
Effects on Health... 276
Mechanism of Intoxification... 277
Epigenetic Mechanism of Intoxification... 279
Parabens ... 280
The Chemicals, Their Forms, and Distribution in the Environment... 280
Effects on Health... 281
Pesticides... 283
The Chemicals, Thier Forms, and Their Distribution in the Environment ... 283
Effects on Health... 283
Thalidomide, α-Phthalimido-Glutarimide ... 283
Diethylstilbestrol (DES) ... 284
Dibutyltin (DBT)... 285

The Chemicals, Their Forms, and Their Distribution in the Environment . . . 287
Effects on Health. . . . 295
Executive Summary . . . 307
References. . . . 308

7 **Mechanisms of Aging and Cancer**. . . . 311
Molecular Models Implicated in Aging and the Development of Cancer. . . . 311
Senescence: The Mechanism of Aging. . . . 312
Senescence as a Mark of the Proinflammatory Profile . . . 316
Executive Summary . . . 327
References. . . . 328

8 **The Food as an Antioxidant System** . . . 329
The Food as a Complex Antioxidant System . . . 329
Inflammation and Oxidative Stress as a Cause of Disease . . . 331
The Microbiome . . . 333
A Balanced Diet . . . 334
Natural Antioxidants. . . . 341
Vitamins . . . 341
The Chemical Structure, Forms, and Distribution in the Environment . . . 341
Effects on Health. . . . 342
Plant Pigments: Carotenoids. . . . 344
The Chemical Structure, Forms, and Distribution in the Environment . . . 344
Effects on Health. . . . 345
Polyphenols. . . . 347
The Chemical Structure, Forms, and Distribution in the Environment . . . 347
Mechanism of Action . . . 358
Soya. . . . 359
The Chemical Structure, Forms, and Distribution in the Environment . . . 359
Effects on Health. . . . 362
Mechanism of Action . . . 364
Beverages . . . 365
Genetically Modified Products . . . 368
Executive Summary . . . 372
References. . . . 374

9 Reprogramming 377
Reprogramming 377
Epigenetic Changes in Behavioral, Neurodevelopmental, Psychiatric, and Neurodegenerative Diseases. 380
Childhood Neurodevelopmental Diseases 384
Psychiatric Disorders 386
Neurodegenerative Disorders 388
Conclusion 389
Executive Summary 392
References. 392

Index. 395

About the Author

Aikaterini Salavoura is a pediatrician with subspeciality in immunology and allergy working in Children's Hospital "Agia Sophia" in Athens, Greece. Her PhD thesis was on the molecular investigation of genetic syndromes and MSc on environmental health, both obtained from the Kapodistrian University of Athens. She currently works as a consultant in a Unit of Pediatric Allergy at the 1st Department of Pediatrics, School of Medicine, National Kapodistrian University of Athens. She works as a Pediatrician with specialization in Immunology and Allergy. Environmental issues and their clinical impact are her main research interest. She has had a teaching role in the training of specializing physicians and nurses in the Children's Hospital "Agia Sophia" in Athens, Greece, and she participated in the publication of local professional and public journals voluntarily.

Chapter 1
The Epigenome

Introduction

In 2001, the sequencing of the human genome project was a reality, and 35,000–50,000 genes were identified and sequenced. Five percent of these genes are transcriptional factors that bind to regions upstream of other genes and regulate their expression. Although there were considerable expectations unmet, the human gene project is a prominent point in human history with significant consequences in the era of biology and medicine. However, one more time, knowledge has been recognized as unlimited and the human genome project has just opened "the sack of Aeolus" to show that many diseases are multifactorial in etiology and there is a considerable influence of the environment related to them. Not only did our expectations to explain, to analyze, to predict disease, and to produce new treatments remain limited, but the human genome project raised new questions to be answered. It was soon recognized that genes could not explain the high diversity of cell function, the differentiation of cells, or even the different phenotypes of monogenic and, moreover, multifactorial diseases [1]. There are a lot of interesting articles in the literature about the link between genes and the environmental marks that influence their expression.

A Unified Theory of Gene Expression

The genes manipulate the function and reproduction of the cells. They consist the template for transcription and production of proteins as well as translation and replication of DNA. The human genome project was completed at the beginning of the twenty-first century, when genes of monogenic diseases were sequenced and their relation to diseases was established. In monogenic diseases, the rule is "one

A. Salavoura, *Chemical Environmental Pollutants and their Effect on Health*,
https://doi.org/10.1007/978-3-031-80271-3_1

gene-one copy-one protein." The number of genes functioning according to this rule is limited, and their function is usually related to rare diseases. Examples of monogenic diseases are cystic fibrosis and thalassemia.

However, epidemiological analysis shows that even monogenic diseases are not equally distributed among different ethnicities, and there is a trend of increase in different geographical areas. Thus, although they are related to disease, some monogenic diseases offer a selective advantage for survival in the specific population due to the resistance to invasive environmental factors. For example, sickle cell anemia in people of African origin offers a selective advantage against malaria.

The advances of technology and the ability to figure out large-scale statistical analysis gave the opportunity to analyze the non-coding areas of the genome and their relation to disease. Numerous numbers of genetic loci were screened to find the ones related to the predisposition to multifactorial diseases by Genome Wide Association Studies (GWAS). The analysis of populations for predisposing genes for these diseases was inconclusive. Different genes, mainly genes linked to the function of the immune system and inflammation, were isolated as candidate regions predisposing to a certain disease. It was found that these regions differ between populations, suggesting heterogeneity in predisposition to inheritance. Research showed that predisposition, for example, to develop Alzheimer disease was detectable in less than 10% of cases. Studies of identical twins detected a predisposition to the disease, which was no more than 40–50% for the 2nd twin [1].

To the best of the published knowledge, the regulation of the cells and, in a more extended degree, specific functions of the body are dependent on Mendelian inheritance ("one gene-one product"). However, more complicated functions are regulated by transcriptional factors and non-coding genetic loci. We name "transcriptional factors" genes that work together and in different combinations, and they produce a spectrum of protein products with different functions. In addition, non-coding loci are areas within the genome that influence the function of the genes in different ways. These areas present in a variety of polymorphisms; they are distributed within the genome and they become functional or non-functional depending on epigenetic factors.

The whole genome is influenced by the regulation of imprinted "stamps or marks" produced during embryogenesis and early development, the so-called "epigenome." The epigenome consists mainly of the histones that pack the DNA, and they permit accessibility to translation. Moreover, more sophisticated regulation of the genes, including methylation of parts of the genome, plays a significant role in the expression of the genes. Hyper or hypomethylated areas predispose differently to disease. The epigenome has become a main target of scientific interest and research, because it seems that it is related to the effects of the environment on the regulation of genes. It offers the opportunity to the developing organism for better adjustment and plasticity to confront environmental stimuli and more chances for reproduction and survival [1].

The Epigenotype

We consider the epigenome as a "collection" of marks on the DNA sequence that are mainly imposed during the early stages of life, and they are highly susceptible to environmental influence. We could imagine these modifications as a "coat" covering the DNA or as "stamps" on it. These specific marks vary between tissues, and they are responsible for the establishment of cell differentiation during fetal life. They are characterized by maintenance through cell division called mitosis, and they could further differentiate during lifespan depending on environmental factors [1].

The epigenotype model postulates that gene expression changes that occur during early life can remain stable during development and adult life, and they create a "memory" of early fetal experiences. These changes, as will later be explained, happen during critical time "windows" and they could or they could not be reversible. Examples showing the existence of such modifications have been reported. They could modify normal responses to cold in plants, allowing adaptation to environmental "winter signals" so that the plant could recognize and react to the stress of low temperatures. They could manipulate the coloring of the furs of animals depending on the habitat geographical region. They create the background of susceptibility of humans to environmental insults like infections or environmental pollution and they are related to disease. In a way, during early development, the organism predicts the environment in which it should adapt and it is organized to adjust and confront insults [1]. The applications of this knowledge are enormous.

The scientific field of epigenetics concerns the mode of alterations of the genome that are induced during early life, their consequences in the regulation of genes, and their relation to disease. Moreover, there are indications that these functional "marks" could influence next generations. Studies report that they could be inherited up to three generations. Thus, environmental factors leave their marks not only on the current generation but also on the descendants (Fig. 1.1).

The epigenome is regulated by modifications of the DNA sequence by the function of histone proteins, by changes in the methylation patterns, and by the regulation of mRNA for the production of proteins in the function ribosomes. This complex regulatory network modulates activation and silencing of genes by modifying the accessibility of genes to translation, and it could interfere with the regulation of translation in the cytoplasm. These changes affect the expression of genes and the properties of their products [2]. Moreover, they are responsible for cell differentiation from the pluripotent zygote to different tissue cells, for the development of the cell, and they are related to multifactorial diseases [2].

To summarize, the mechanism for the regulation of gene expression is based on three models. The first model is Mendelian inheritance, in which the expression of genes to proteins happens by transcription and production of mRNA in the nucleus and translation of the mRNA within ribosomes in the cytoplasm. Finally, the DNA sequence is translated into protein, and any change in the sequencing named

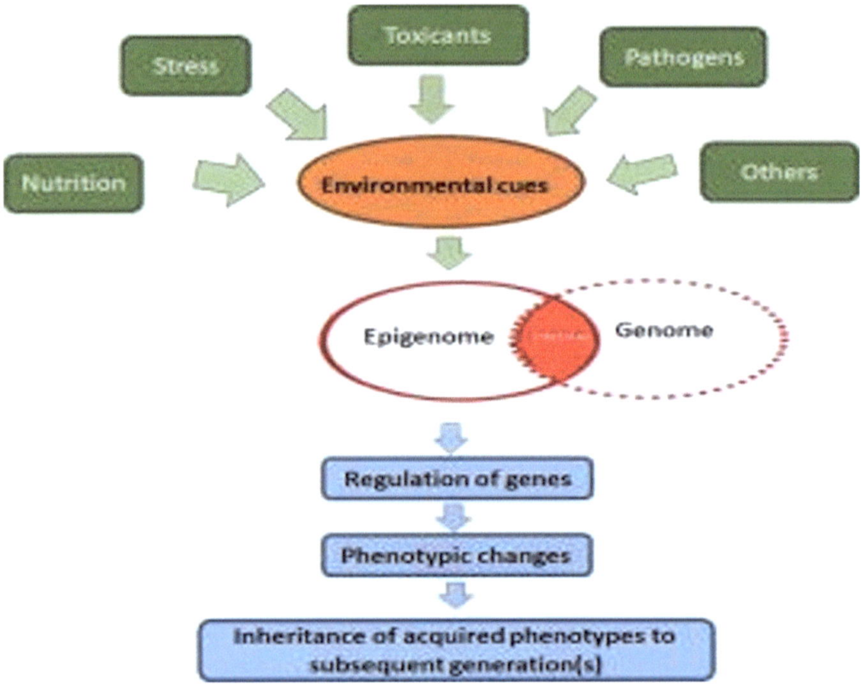

Fig. 1.1 Interaction of genetic predisposition and environmental factors to the development of disease

"mutation" could lead to disease with a milder or more severe phenotype depending on the location of the change within the gene. Mendelian inheritance is the traditional model of genetics. The second model of gene regulation is related to the immediate control of genes by transcriptional activators and suppressors distributed within the genome in introns as well as exons. These transcriptional factors bind to the promoters of genes and regulate their function. The third model of gene regulation is the epigenetic model, which is related to the control of the expression of genes by epigenetic modifications [3] (Fig. 1.2).

Epigenetic Mechanisms

The known epigenetic mechanisms include DNA methylation, modifications of chromatin structure, RNA-associated silencing and changes in protein conformation, as well as imprinting.

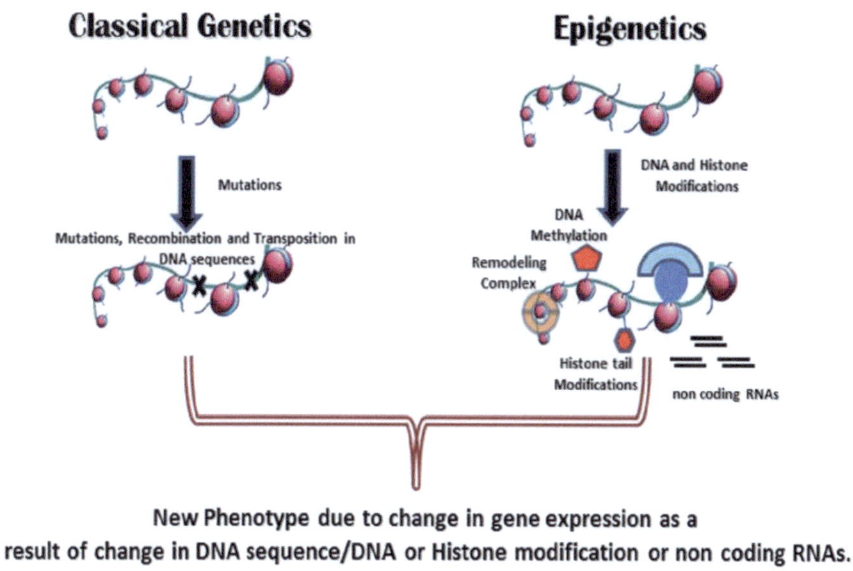

Fig. 1.2 Mendelian inheritance and epigenetic inheritance

DNA Methylation

DNA methylation is a well investigated epigenetic mechanism. In mammals, DNA methylation occurs almost exclusively at cytosines within a cytosine-guanine group of dinucleotides (CpGs) (the "p" denotes the intervening phosphate group), converting cytosine to 5-methylcytosine. The process includes the addition of a methyl group to the 5 position of a cytosine ring, and it is mediated by methyltransferases (DNMT) [4]. Following DNA replication, the DNA methyltransferase 1 (DNMT1), a maintenance methylase, it restores the original pattern of CpG methylation in the daughter strands (Fig. 1.3).

CpG methylation regulates gene expression by interfering with the binding of methylation-sensitive DNA-binding proteins and by interacting with various histone proteins that regulate DNA accessibility. Although CpG methylation is generally associated with gene silencing, methylation-sensitive DNA-binding proteins enable CpG methylation to regulate diverse effects on transcription [2].

There are four major methyltransferases: DNMT 3A and 3B that are related to methylation during mitosis; the DNMT3L, which functions only during early development; and the DNMT1, which maintains methylation patterns through life [2]. The consistency of maintenance of methylation patterns during mitosis is as high as 97–99.9% due to DNMT1. The DNA methyltransferases interact with a variety of gene regulatory proteins through feedback loops that can directly and/or indirectly impact chromatin structure, which in turn may also influence DNA methylation [4]. In other words, the function of methyltransferases is dependent on the structure of chromatin that covers the genome.

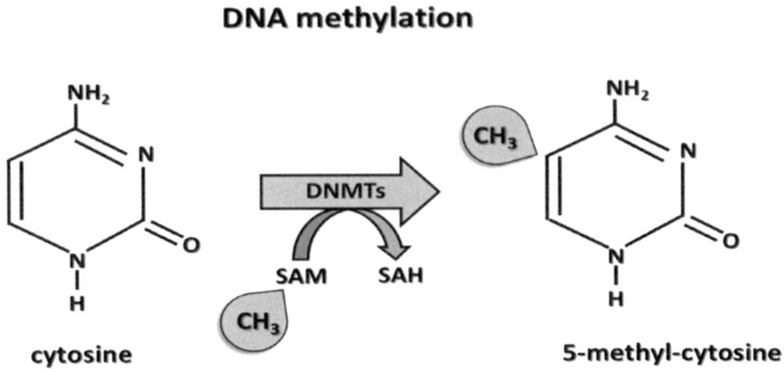

Fig. 1.3 DNA methylation of DNA in CpG loci

CpG regions are dispensed within exons and introns. The genome includes regions poor in CpG and regions dense in CpG sequences, also called CpG islands. CpG islands form 5–10% of the genome. Almost half of these high-density regions are located in the promoter region of genes. CpG islands located close to the genes are normally unmethylated, while the sporadic ones are normally methylated. The pattern of methylation is related to the activity of the gene, and generally speaking, methylation of the promoter is related to silencing of the gene. There is a gradual reversal of this pattern during aging that leads to sporadic methylation of CpG islands and global loss of methylation, but this methylation pattern is particularly pronounced during carcinogenesis [4].

Histones

Histones are small proteins (11–22 kDa) that cover DNA. The four major classes of histones in humans are H1, H2A, and H2B, as well as H3 with H4. DNA is wrapped around octamers of core histones, each consisting of one H3–H3–H4–H4 tetramer and two H2A–H2B dimers, to form the nucleosome core particle. Histone H1 connects nucleosome core particles, which are further stabilized by electrostatic interactions between negatively charged phosphate groups and positively charged q-amino groups. Nuclear DNA is packaged by histone proteins in a highly complex and dynamic way [3] (Fig. 1.4).

Histone proteins are the building blocks of genomic DNA organization. Importantly, histone proteins undergo various post-translational modifications in their tails. The modifications change the local chromatin structure by altering the protein constitution and thus the electrical charge of the histone tails. It is estimated that there are >100 histone modifications and only a few have been studied. Recent mapping of various histones at the whole genome level has provided important insights into the biological function of these proteins [3].

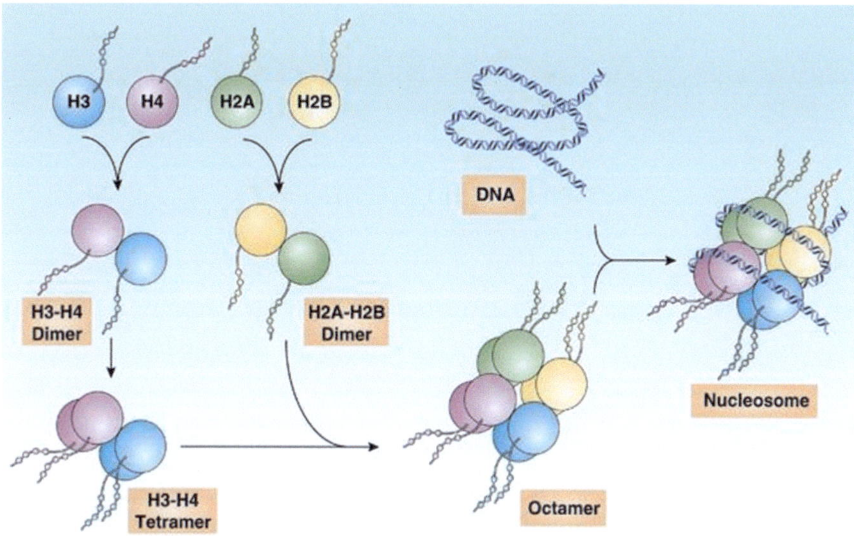

Fig. 1.4 Octamer construction of histones in the nucleus and DNA packaging

Histones, apart from the protection of the genetic code, are characterized by their diversity in different cells of the tissues and they are related to cell differentiation [5]. Inside the cells, histone modifications control the accessibility of chromatin to the transcriptional activities [4]. Thus, certain modifications are related to condensed and inactive chromatin called heterochromatin, which is not translated, while others induce a detachment from DNA, generating an open frame, which is the starting point of DNA replication. These regions of active chromatin are called euchromatin (Fig. 1.5).

Important modifications take place at specific locations of the amino-terminal tails of histone proteins relative to lysine residues. The known ones are acetylation, methylation, phosphorylation, ubiquitination, and sumoylation [4]. Acetylation is associated with activation of gene transcription, while methylation of specific histones such as H3K9 or H3Lys9 and H4Lys20 is related to inactivation of gene expression. Acetylation and methylation show high histone specificity, which is related to the role of histones in regulating gene expression. Direct or indirect interactions between histones and DNMTs as well as other enzymes recruited to specific regions of the genome suggest an interdependence of the chromatin structure to gene DNA strain that determines the epigenetic state [3]. It seems that a histone code exists, similar to the genetic code dictating locus-specific transcriptional competence [3]. Chemical modifications of histone proteins and cytosine bases provide heritable epigenetic information that is not encoded in the nucleotide sequence. The stable maintenance of gene expression patterns is thought to arise by the "tagging" of DNA to histones with chemical attachments related to enzymatic reactions [1].

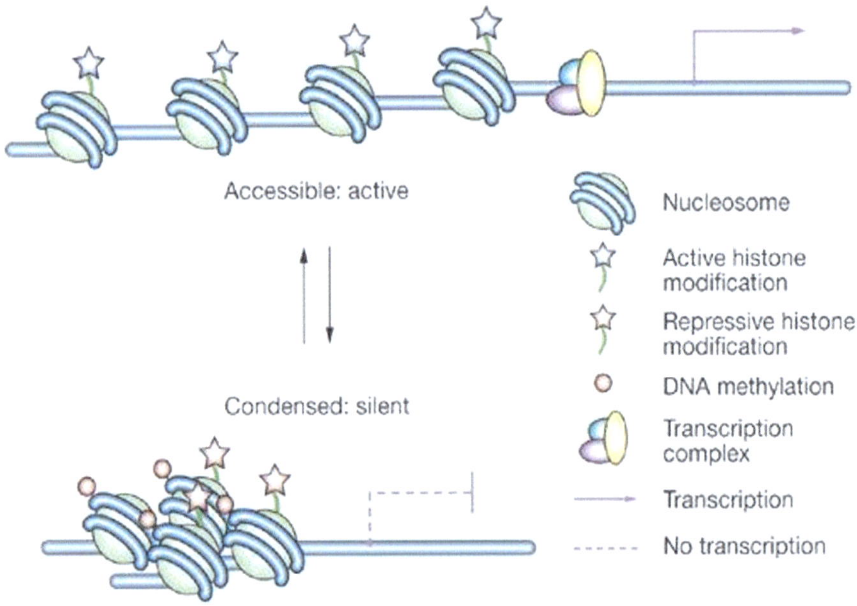

Fig. 1.5 The complex system of epigenetic modifications on the DNA

Unique sets of histone modifications are associated with the regulation of distinct regions and genomic activity. For example, active enhancer regions are marked with histone H3 lysine 27 acetylation and histone H3 lysine 4 monomethylation (H3K4me1), active promoters are marked with histone H3 lysine 4 trimethylation (H3K4me3), and repressive genomic regions are marked with H3K27me3. Importantly, mutations in histone modifiers (enzymes that add or remove various histone marks) are recurrently found in the states of many diseases, including cancer (Fig. 1.6) [3].

Overall, the gene activity of a cell is determined not only by the DNA sequence itself and its regulatory elements, such as transcription factors and methylation status, but also by changes in the way DNA is modified and packaged within the chromatin. Epigenetic mechanisms control the information of the DNA sequence, and thus, the genetic background of an individual is being used. DNA methylation and histone the mechanism that proteins within the nucleus, as well as ncRNA molecules in the nucleus and cytoplasm, are implicated in chromatin structure and function, and they are important forms of epigenetic information [3]. The vast potential information content of these "permutations" suggests that a histone code (analogous to the genetic code) exists in differentiated cells in order to dictate locus-specific transcriptional competence [2].

The exact mechanism that governs the transmission of epigenetic information through mitotic or meiotic cell cycle divisions is not yet completely understood. DNA methylation is mitotically transmitted in a semi-conservative fashion; however, there is no clear evidence that histone modifications are transmitted in the

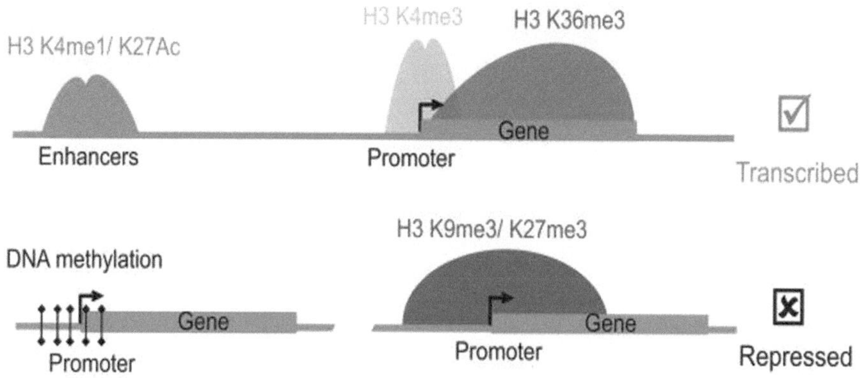

Fig. 1.6 "Marks on DNA" with different histone states act as regulatory mechanisms of genomic activation

same way. It is suggested that an alternative scenario for transmission of histone epigenetic changes exists. Possibly, they remain bound to DNA during DNA replication, and hence, they could reestablish the histone marks on the sister chromatids [3]. The least recognized epigenetic mechanism is feed-forward autoregulation by transcription factors [2].

RNA Regulatory System

RNA has been viewed from the perspective that genes encode proteins by an mRNA intermediate, which is the traditional model of the transportation of signals from the nucleus to the cytoplasm. Recent evidence has shown that non-protein-coding RNAs (ncRNAs) derived from introns of protein-coding genes and the introns and exons of non-protein-coding genes make up a large proportion of the genomic output [4].

NcRNAs could interfere with chromosome architecture, mRNA turnover, protein translation, and timing of translation, but they could also regulate DNA transcription and alternative splicing. They might also play a role in DNA methylation and histone modification in specific loci, determining heritable and stable silencing. Moreover, they seem to play an important role in genetic imprinting in early development [3]. Recent evidence supports the idea that RNA silencing in vertebrates serves as a natural immune response to protect against pathogen invasion [6] (Fig. 1.7).

MicroRNAs (miRNAs) are designated RNAs with 22 or fewer nucleotides that can control translation and thus protein production post-transcriptionally. Precursor miRNA is transported to the cytoplasm with exportin-5, where it finally takes its mature form. miRNAs bind to their target mRNAs either in a complementary way, leading to degradation of the target, or incompletely, leading to the suppression of

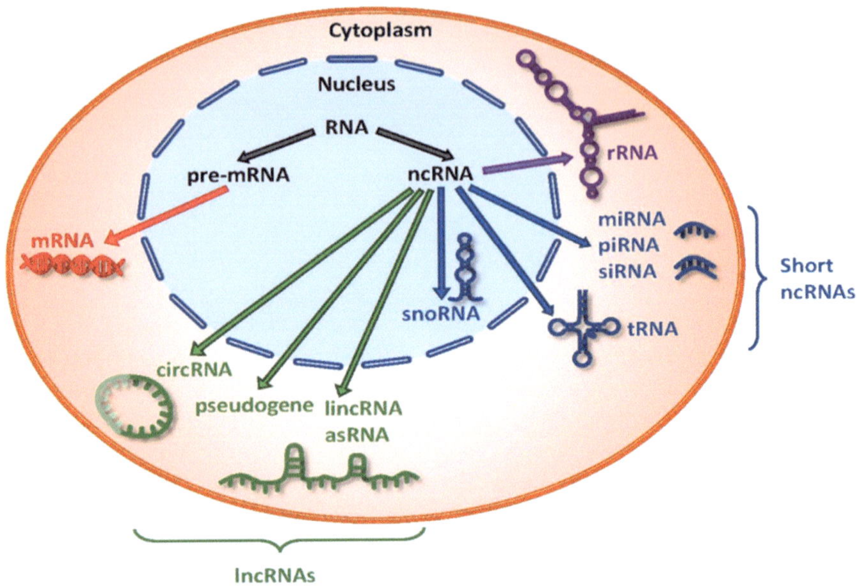

Fig. 1.7 Different types of RNAs in the nucleus and cytoplasm, including miRNA, siRNA, and other regulatory RNAs (NcRNA) and mRNA, which is translated

the translation process. Each miRNA is predicted to have many targets, and each mRNA may be regulated by more than one miRNA. The known miRNAs are approximately 460, and their number is growing every day. They are all related to a specific disease. There is some evidence about how these miRNAs are produced and how their production is regulated [7]. The expansion of our knowledge regarding these regulatory miRNAs is a promising field of new therapeutic interventions for the treatment of neurogenerative diseases such as Alzheimer or, moreover, cancer (Fig. 1.8).

RNA interference (RNAi) is the process by which double-stranded RNA (dsRNA) can induce epigenetic gene-silencing processes by degradation of homologous mRNAs, resulting in loss of protein production [6]. Endogenous and exogenous dsRNA arise after infection by dsRNA viruses, sense-antisense transcription pairs, transcription of inverted repeats, and the delivery of dsRNAs manufactured in vitro.

Small interfering RNAs (siRNAs) are 21–25 bases long, and they are incorporated into an RNA-induced silencing complex (RISC) that also includes members of the Argonaute family of proteins. Activation of siRNAs within the complex results in complementary target RNA destruction, and thus, they control the production of certain proteins [6].

DNA heterochromatin The mode of regulation of euchromatin is highly dependent on regions within heterochromatin. Our knowledge, regarding this DNA "garbage," or the regions in the genome that do not contain genes or they are not directly translated into proteins, is expanding [6].

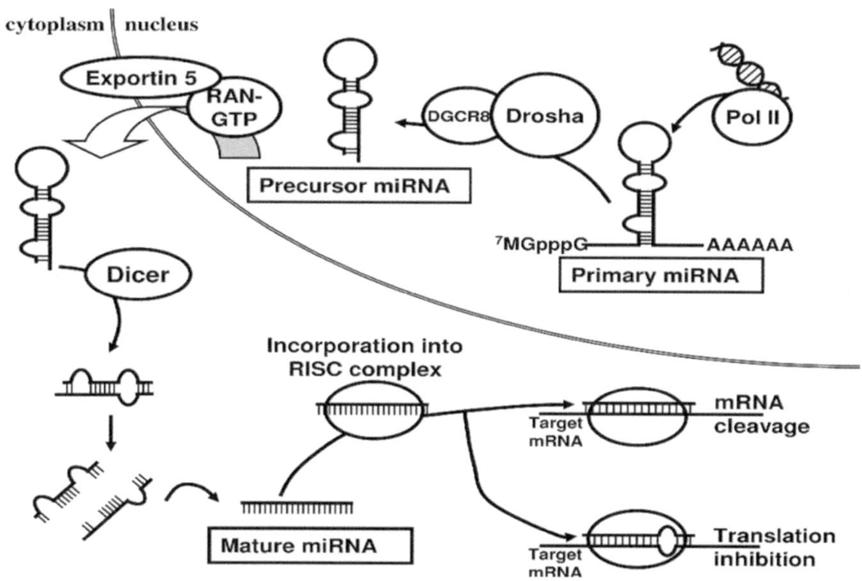

Fig. 1.8 Regulatory role of miRNAs, production, precursor miRNA, transportation from nucleus to cytoplasm by exportin 5, maturation and binding to mRNA

Repeated DNA Sequences

Repeated DNA sequences may be transposable elements found interspersed throughout the genome or simple repeat sequences, such as DNA satellites, that are usually found in pericentromeric or subtelomeric heterochromatin. These are normally methylated within the healthy genome [7].

Transposable Elements (TEs)

Transposable elements (TEs) are DNA sequences that they have the ability to move and to integrate into the genome at different sites. In mammals and humans specifically, they account for almost half of the DNA content and they are located in heterochromatin regions. There are three types of TEs in the human genome: DNA transposons, retrotransposons and endogenous retroviruses [7].

TEs and their methylation patterns are not only different between species, but also between individuals of the same species. They are generally inactivated by methylation and they are activated by demethylation. Their major role as parts of the non-expressed genome is to regulate the activation of neighboring genes. McClintock first suggested that TEs are activated in response to environmental challenges of the genome, and since then, the release of epigenetic silencing of TEs has been described in response to UV exposure, changes of temperature, radiation, during wounding, in cell cultures, during pathogen infection, and in polyploidization [7].

Stress-activated TEs are highly unmethylated, and they have the ability to replicate by an RNA intermediate, to move, and to integrate in different unpredictable regions of the genome. Thus, they might generate the raw diversity of replicates that a species requires over evolution to survive during different environmental stressors. This adaptive response is a long-term strategy to increase variability for selection advantage, and in a way, TEs provide the fuel for evolution [8].

Additionally, they are responsible for the genetic and phenotypic variability of individuals, and they serve as polymorphic locations in the study of the genome [8]. For example, although monozygotic twins are genetically identical, as they grow older, they manifest an increasing discordance due to the epigenetic marking of TE loci, and thus, they develop different expression patterns of the same genes. Phenotypic differences and differences in disease experience are related to the differences of TE distribution and function, and they are tightly correlated to environmental influence. The nature of the environmental signals that are capable of inducing changes in TEs is poorly understood, but it seems likely that a wide range of exposures can have such effects, ranging from differences in diet and lifestyle to stress and behavioral influences [4].

Although they offer selective advantage, the activated TEs are highly mutagenic regions due to their structural sequencing. They consist of loci with a high frequency of aberrations such as chromosomal breakages, sister chromatid recombination, and multiple other rearrangements. TEs are often considered "selfish" or "parasitic" elements, because they tend to predominate in the genome, meaning they tend to increase in number of copies and they constantly change position [9]. To combat the potentially harmful effects of activated TEs, the genome has evolved epigenetic "defense" mechanisms to suppress their activity. Research shows that miRNAs and modifications of histones introduce methylation and silencing of TEs. An epigenetically inactivated TE retains the coding potential to mobilize itself, but it does not produce the necessary proteins in order to reproduce, due to a suppressive chromatin environment. The activated, unmethylated state is negatively correlated with the fitness of the host [7].

Thus, although we consider these heterochromatin regions as inactive with unknown function, often called the "garbage of the genome," we now know that they have a dynamic role in the integrity of the genome.

DNA Transposons

DNA transposons are probably the oldest transposable elements of the human genome, and their origin is probably related to deletions and truncations rendering them fixed within the genome. These 1300–2400 bp elements used to code enzymes, but became ineffective due to nonsense or frameshift mutations. However, some remain active and a potential threat to the genome [7].

Retrotransposons and Endogenous Retroviruses

There are two classes of retrotransposons: LTRs (flanked by long terminal repeats) and non-LTRs that terminate to a polyadenylated tail. These elements are able to move and integrate into other parts of the genome by a "copy and paste" mechanism involving an RNA intermediate. Although many of them degenerate, they still interfere with the function of strong promoters and they drive transcription. LTR retrotransposons (LTRs) and human endogenous retroviruses (HERVs) are similar to other retroviruses, but they lack an envelope. Many of these sequences are not active due to deleterious mutations. However, the active ones, although regulated by a DNA methylation mechanism, they could modify gene expression, and their activation could be related to human disease [7]. Non-LTR retrotransposons are elements that constitute almost 30% of the human genome, and they are classified as Long Interspersed Nuclear Elements (LINEs) and non-autonomous Short Interspersed Nuclear Elements (SINEs) [7].

The LINE Family

Full-length LINEs are 6 kb in length, and they are part of the promoter region of many genes or they encode genes that make them eligible to integrate in any part of the genome. Many of them, the oldest ones, are inactive, because of mutations. The younger ones reach a number of 80–100 families. They are known due to the fact that methylation of LINE sequences in the promoters is related to the activation of genes. They comprise an important regulator of mutagenesis. Hypomethylated LINEs have been observed in numerous cancers [7].

The SINE Family

This subset of non-autonomous elements relies on LINEs for transposition. The well-known Alu family covers 11% of the genome with 1.1 million copies. Alu family is highly methylated and inactive. Some parts are also imprinted, showing different methylation patterns depending on the parental origin [7].

Transcription Factors

The least investigated epigenetic mechanism is autoregulation by transcription factors. The transcription factor proteins regulate gene expression by binding to recognition sequences within gene promoters. Transcription factors that activate their own gene promoter can perpetuate their expression through cell division [6].

Genomic Imprinting

The epigenetic modification that defines the parental origin of the genetic information is called genetic imprinting, and it is acquired in the germ line. "Imprinted" areas are characterized by different methylation patterns of paternal and maternal allele sequence. Thus, genes within an imprinted region acquire functional differences depending on their parental origin. Furthermore, "imprinting" persists through life and it exhibits a functional asymmetry of genes.

Most of the imprinted genes occur in clusters in the genome, and they are key control elements that regulate neighboring genes. They silence genes when methylated or they serve as boundary elements when unmethylated. The result, in all cases, is to ensure monoallelic expression of the relevant genes. Any disruption of these clusters by chromosomal translocations or mutations could cause abnormal expression of many genes within the cluster, accounting for some of the genetic syndromes associated with growth and neurodevelopment. A well-known example of methylation changes in the imprinted area 15q21 is Prader-Willi syndrome, characterized by obesity and short stature due to hypothalamus deficiency, mental retardation and motility problems due to a general central nervous system dysfunction and some distinct dysmorphology features. So far, about 80 imprinted genes have been identified in mice and humans.

The role of imprinting is highly correlated with normal embryonic development. The paternal genome is more important for the development of the extraembryonic tissues, such as the trophectoderm, whereas the maternal genome apparently has a greater influence on proper development of the embryo. Both paternal and maternal alleles are necessary for normal development. The imprinted loci of sperm and oocytes are reciprocal. For example, the paternal gene would accelerate and the maternal gene would inhibit metabolism.

Moreover, the paternal allele comes to "rescue" the embryo from the expression of the maternal alleles and to prevent rejection and abortion. The embryo is a mismatched HLA implant, and there are some considerable changes of the female body in order not to reject embryos. Generally speaking, paternally expressed alleles propagate the development and metabolism of the embryo, while maternally expressed ones protect the mother and her reserves. As an example, it is frequently used in the literature the imprinting of the *Igf2* gene, which encodes the fetal insulin-like growth factor 2, a gene only expressed from the paternal genome. This gene promotes fetal development, and it is antagonistic for maternal resources.

Postnatally, the major function of imprinted genes is the provision and control of resources in mammals. According to this theory, paternally expressed genes and maternally expressed genes have opposing interests in terms of the number of resources. Paternally expressed genes increase resource acquisition to promote growth and postnatal fitness, whereas maternally expressed genes are "interested" in preventing exhaustion of metabolic resources and reducing resource transfer.

A number of imprinted genes have been shown to have an important role in adiposity and glucose-regulation metabolism, and some are associated with human metabolic disorders. Imprinted genes may regulate some of the crucial aspects of mammalian physiology associated with reproduction, placentation, energy homeostasis, lactation, and behavior [9].

Epigenetic Reprogramming

The known patterns of epigenetic regulation and the different ways that epigenetic mechanisms interfere with the expression of genes are discussed up to this point. It is also analyzed how those mechanisms are being established and inherited [9].

Epigenetic marks are established mainly during "critical windows" of development, and they are erased during gametogenesis. Reprogramming in germ lines is crucial to establishing totipotency. They are maintained during mitosis and they remain stable in somatic differentiated cells [9].

Gametogenesis is the first step of erasure of previous epigenetic modifications. A genome-wide demethylation occurs early in the development of primordial germ cells in mice. The demethylation is completed during fetal life before germ cells enter the gonads. Remethylation appears during maturation of spermatozoa and oocytes (Fig. 1.9). Sperm and oocytes are methylated on the CpG sites nearly by 90% and 40%, respectively, suggesting that the sperm is more methylated than the oocyte. During remethylation, a different imprinting pattern is established in order to differentiate the paternal from the maternal allele. The paternal allele is methylated first, and it influences the methylation pattern of the maternal copy, which is methylated after the paternal. The genomes of the mature sperm and egg in mammals are highly methylated, comparable to the methylation of somatic cells, although there occur differences in specific tissues [9].

The second critical period of reprogramming is zygote with a massive demethylation after fertilization. The next step of the remethylation stage is the postzygotically, which is probably linked to the first lineage decisions during mammalian development. Thus, there is a wave of DNA demethylation after fertilization followed by a wave of de novo methylation upon embryo implantation and blastocyst formation (Fig. 1.9). Cloned preimplantation embryos lack some of these processes, and thus, they are prompt in developmental "faults" in nuclear reorganization and cellular differentiation that are responsible for the frequent abortion of these embryos.

After fertilization, the paternal allele undergoes remarkable transformation in the cytoplasm of the oocyte. Sperm chromatin, by removal of protamine and replacement by acetylated histones, undergoes demethylation, which is completed before DNA replication commences. On the contrary, the maternal allele undergoes a replication-dependent, passive DNA demethylation process that is completed by the blastocyst stage. As a result of this major wave of DNA methylation erasure, the genome of the inner cell mass is substantially hypomethylated and only 20% of CpG sites are methylated. As embryonic development progresses from the blastocyst to the epiblast stage, the genome undergoes de novo remethylation, leading to

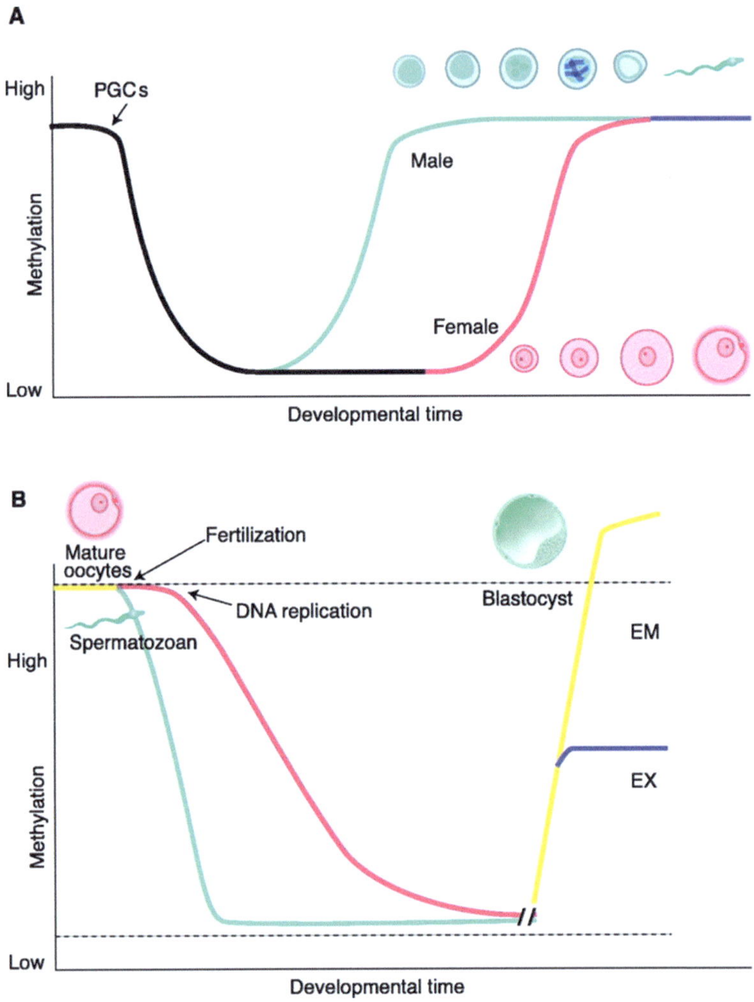

Fig. 1.9 Demethylation and remethylation during maturation of germ cells and after fertilization

nearly 70% methylation of the genome. Around this developmental stage, subsets of epiblast cells follow specific sets of developmental cues [7, 9].

As gonad mesenchymal cells begin to migrate through the developing hindgut to colonize the undifferentiated gonads, the genome undergoes global DNA methylation erasure. During this process, major components of de novo DNA methylation are inhibited, and key players of the DNA demethylation program, such as the recently discovered TET enzymes, are activated [7, 9] (Fig. 1.10).

The imprinting process participates in the differentiation of somatic cells. Although all the cells of the organism have the same set of genes, they differ in their

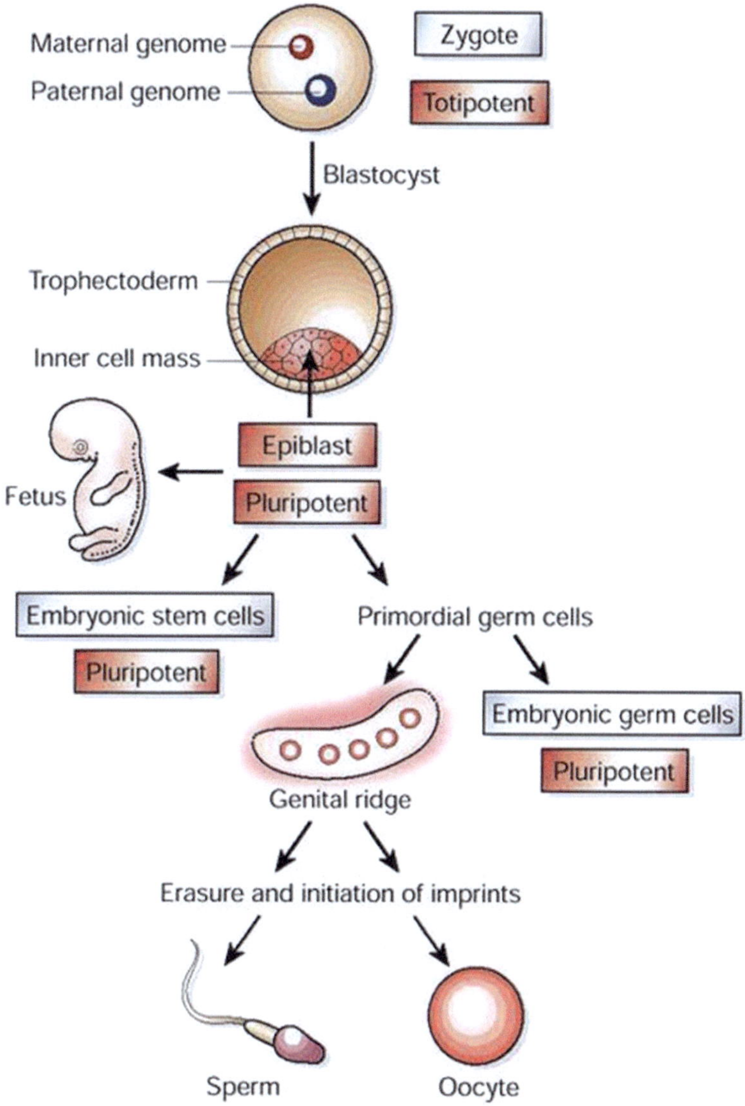

Fig. 1.10 The demethylation process during the formation of gonads

appearance and function. It is the selective expression and suppression of genes that determines the specific properties of individual cells in different tissues, which are established by different methylation patterns. The same pool of genes regulates a number of different functions ranging from neuronal networks and defense mechanisms to gut epithelium related to the absorption of nutrients. The persistence of these epigenetic marks during mitotic cell divisions works as a "molecular memory" that propagates appropriate patterns of gene expression [10]. Nevertheless, even when fully differentiated, any cell can potentially be reprogrammed back to totipotency.

Although the process of imprinting is quite accurate and it maintains epigenetic modifications up to 99% of divisions, it can be erased under specific circumstances. Thus, early marks caused by epigenetic modifications are partly reversible. Programming is related to the plasticity that the genome acquires during early development, producing different phenotypes to confront different environmental conditions [11].

Methylation of Other Key Elements Within the Genome

The methylation mechanism is also important to imply the regulation of transposable elements (TEs) within heterochromatin. Methylation often extends from the regions of chromosomes with a high density of TEs to the neighboring genes. Thus, TEs serve as building blocks of epigenetic phenomena. Taking into account the fact that TEs constitute 50% of the genome, their significance remains prominent. Recent studies have implicated TEs in imprinting and X-chromosome inactivation, as well as in the developmental regulation of gene expression, a role that was first described by McClintock shortly after the discovery of TEs [4].

In addition to methylation of transposable elements, it is also described in literature the methylation of the centromeres of chromosomes and of the telomeres at the end points of chromosomes, which follow the same patterns of regulation.

The centromere constitution is mainly heterochromatin. The centromere in mammals varies in size from several kilobases to megabases, and it contains long tandem repeats or satellite repeats that are surrounded by TEs. Pericentromeric regions include long blocks of LINE and SINE retroposons. Most of these regions are highly methylated and thus inactive. However, they seem to contribute to a large extent to centromeric function [4].

The centromere exhibits an important role during cell division. The condensed structure of the centromere helps to properly traffic the chromosome and to maintain chromatid structure. In addition, tandem repeats such as centromeric satellite repeats are a major source of siRNAs, and it has been suggested that tandem duplications have been retained in the genome to efficiently maintain epigenetic silencing of the region owing to their ability to replenish the siRNA pool [4].

Telomeres are conserved DNA-protein structures of the linear chromosome termini. In mammals, the telomere DNA component consists of hexanucleotide repeats that have the sequence TTAGGG. The length of the repeats varies among species, with an average of 5–15 kb in humans and notably greater in mice [12]. Whereas most of the telomeric DNA is double-stranded, a G-rich single strand forms a terminal 3′ overhang at the very ends of chromosomes [12]. Evidence indicates that the structure of the telomeric end consists of a “T loop,” which is formed by the

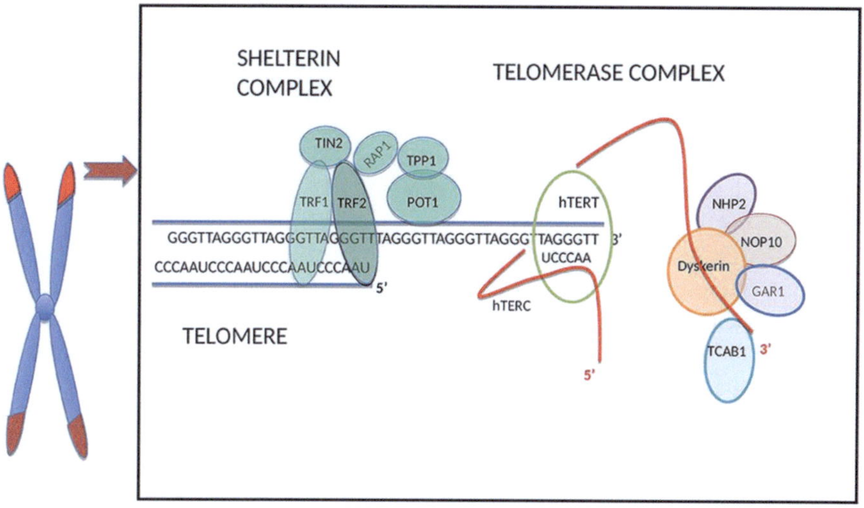

Fig. 1.11 Regulation of heterochromatin at the ends of chromosomes called telomeres

invasion of the single-strand terminus into the double-strand telomeric DNA [13]. Several telomere DNA-binding proteins, collectively named Shelterins, protect the structure. These include telomeric repeat-binding factor 1 and 2 (TRF 1, 2) and TPP1 proteins, which bind to the double-stranded, while POT1 (protection of telomeres 1) binds the single-strand telomeric DNA. In particular, TRF1 regulates telomeric length by inhibiting telomerase, protects telomeric ends, and suppresses checkup points [13]. The proteins TIN2 and RAP1 complete the Shelterin complex [13] (Fig. 1.11).

The progressive shortening of telomeres to a critical limit due to multiple mitosis either reaches a state termed "replicative senescence," in which they are incapable of further cell division (Hayflick limit) or they can provoke apoptosis [14]. The lifespan of a cell is predictable in connection to the telomere length, and it is called "the mitotic clock" of the cell [13]. The telomere hypothesis suggests that telomere length could serve as a biological indicator of previous cell divisions in the history of the cell as well as of the residual replicating capacity of the cell [13].

The presence of telomeres maintains the integrity of chromosomes, prevents illegitimate recombination and fusion, it ensures complete replication in cell divisions, and it is implicated in the localization of chromosomes in the nucleus, the separation of sister chromatids during replication, the regulation of gene expression, and cellular senescence [14]. Telomeres also consist of heterochromatin rich in full length and fragmented TEs within a region of short tandem repeats [4].

The Epigenotype Model

The epigenotype model introduces the concept of how significant the interference of the environment is in modulating the health status and behavioral patterns in humans. Epigenetic and gene expression changes that can occur in early life, as well as during adult life, interfere with gene expression and the appearance of different phenotypes [1].

The possibility that environmentally induced changes in the epigenotype can be transmitted through generations is an interesting scenario to mention, because it suggests that not only the genetic code could be transmitted through generations but also environmental "marks." Several studies suggest that fetal exposure to harmful stimuli (e.g., caloric restriction or exposure to diethylstilbestrol) causes abnormalities in the second (F2) generation. Moreover, the altered DNA methylation was observed in several genes of the F3 generation germ lines, strongly suggesting epigenetic inheritance. It is rather possible that the potential for resetting the normal epigenetic states of the affected genes in subsequent generations is not possible and that the altered epigenetic state is associated with transgenerational inheritance [1].

A cartoon representing the inheritance of the epigenetic "marks" to multiple generations is illustrated. A pregnant woman (P0) carries the epigenetic information up to the F2 generation. Her fetus (F1) and its germ cells (F2) may be affected by environmental exposure. Any changes in the F1 phenotype or in the transgenerational epigenetic state remain stable at least up to the F3 generation (Fig. 1.12) [1].

The Thrifty Phenotype Hypothesis

A common epidemiological and research reference related to epigenetic changes due to environmental cues is the "thrifty phenotype hypothesis." According to these data, it is suggested that poor fetal nutrition, due to either poor maternal nutrition or poor delivery of nutrients to the fetus from other causes (e.g., placental dysfunction), would permanently alter the structure and function of different organs in order to protect the development of the vital organs. Thus, the brain will develop at the expense of other tissues (e.g., the endocrine pancreas). There is a metabolic programming of the fetus to confront a nutrient-disadvantaged environment. This metabolic programming functions in a manner beneficial for survival [1].

However, research shows that adequate or excessive postnatal calorie balance could lead to obesity, which is inconsistent with the fetal programming. This discrepancy is a predisposing factor for the development of diseases such as type 2 diabetes, hypertension, and ischemic heart disease in adult life. In the long term, these first beneficial adaptations for survival will lead to diseases due to the fact that the body functions all the years with an overload of storage materials. In other words, the body functions with the wrong "tools" or "maladaptive programming,"

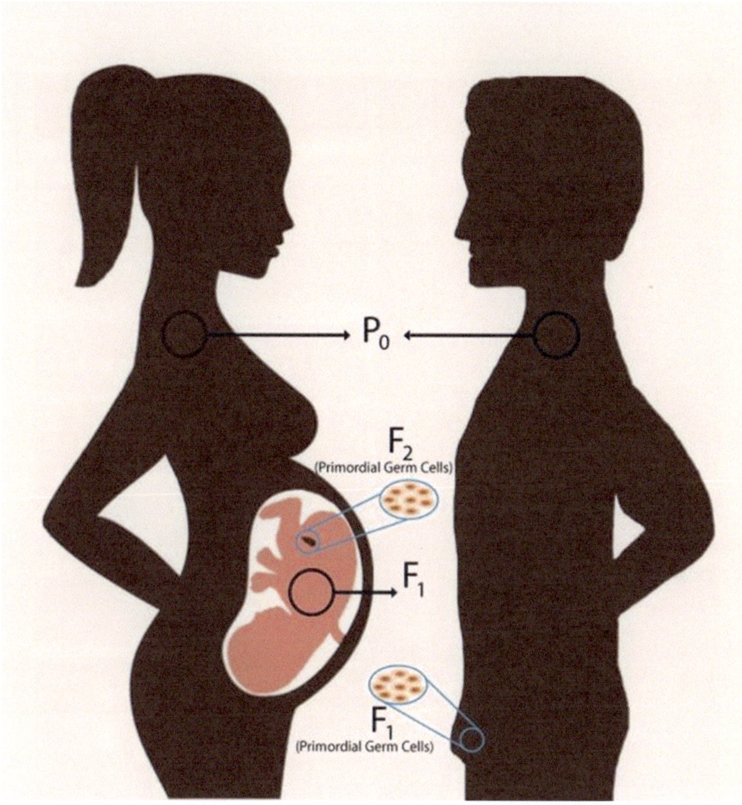

Fig. 1.12 Epigenetic modifications are transmitted up to 3 generations

which is the cause of disease. Earlier death and disability might therefore be the long-term price of the maladaptive programming of early life [1] (Fig. 1.13).

The model was proposed 10 years ago in an attempt to explain the association between poor fetal and infant growth and the increased risk of developing impaired glucose tolerance in adult life [5]. It explains the epidemiological data, which show an association between poor nutrition in early life and the permanent changes in glucose-insulin metabolism leading to type 2 diabetes and metabolic syndrome. These changes include reduced capacity for insulin secretion combined with insulin resistance, which together with other factors such as obesity, ageing, and physical inactivity are the most important factors for type 2 diabetes [5].

Since this hypothesis was proposed, many studies world-wide have confirmed the initial evidence, although the strength of the relationships varied from one study to another. The relationship of fetal sources depreviation to insulin resistance is clear at all ages studied. Less clear is the relationship with insulin secretion [5].

It is proposed that poor development of pancreatic β-cell mass and their function (including islet of Langerhans vasculature and possibly innervation) are key

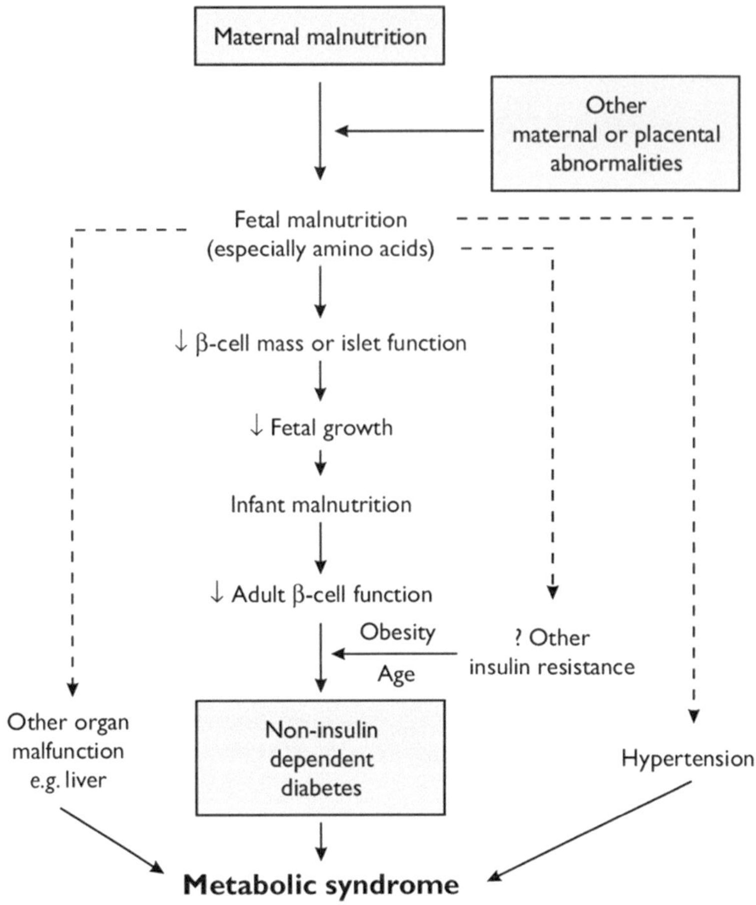

Fig. 1.13 How fetal malnutrition with subsequent adequate calorie uptake could lead to metabolic syndrome in adult life

elements indicating that poor early nutrition is associated to type 2 diabetes and insulin resistance. Fetal nutrition thereby sets "a model programming" based on fetal nutritional status, which has a differential impact on b-cells (Fig. 1.14) [5].

A number of animal models have been used to demonstrate proof of the principle of this epigenetic model. In the rats, for example, maternal protein restriction, calorie restriction, high-fat feeding, intrauterine artery ligation, or overexposure to glucocorticoids results in the development of fetuses with metabolic syndrome. The phenotypic outcomes of these different insults are remarkably similar, suggesting that these insults might manifest by a common metabolic pathway [1].

In rodents, the effects of maternal diet can also be transmitted up to the second-generation offspring. The effects could be transmitted and they manifest as changes in the hormonal milieu (e.g., offspring of women with gestational diabetes are at

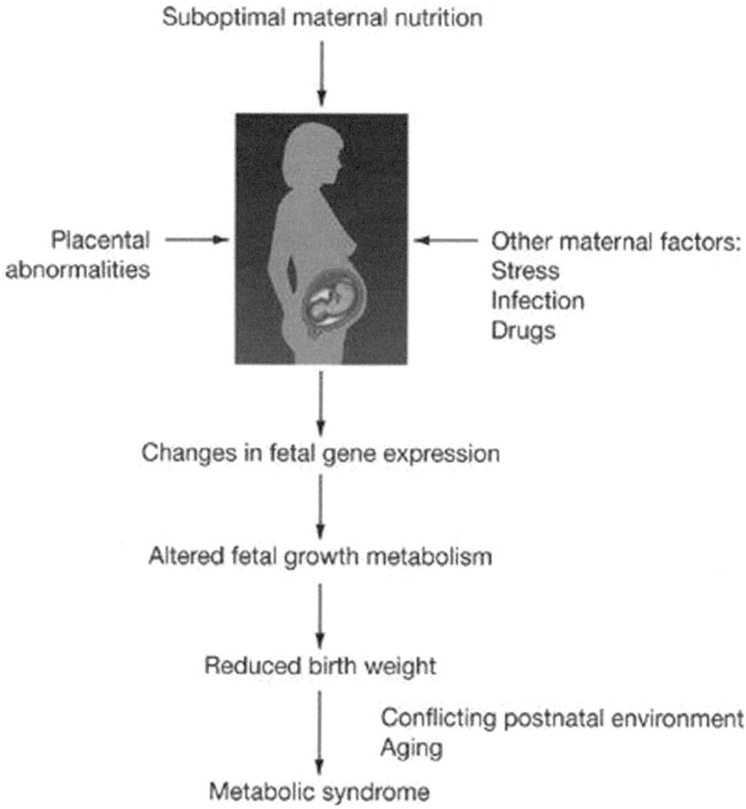

Fig. 1.14 The model of the process of malnutrition in early fetal life and the development of metabolic syndrome and glucose intolerance in adult life

increased risk of type 2 diabetes). Studies in the offspring of glucocorticoid-treated rats show that the effects can also be transmitted paternally [1].

Historical epidemiological studies corroborate the model, examples being the so-called "Dutch Hunger Winter" and the "Wartime famine in Holland and Leningrad." Studies of individuals born during the *Dutch Hunger Winter* express the effects of malnutrition, which are transmitted down to even their grandchildren [1]. The "Dutch famine" is a unique example, because a brief period of intense deprivation of a well-nourished population, was followed by a return to normality with adequate calorie intake after the war. Children exposed to famine in utero were well-nourished in childhood, and thus, they manifested an accelerated weight gain. Similarly, the *famine in western Holland* began abruptly in November 1944 and it ended abruptly with the liberation of Holland in May 1945. Men and women exposed to the famine, while in utero, had positive glucose intolerance testing and higher pro and after fasting insulin concentrations, which suggests that their poor glucose tolerance was partly determined by insulin resistance [5].

In contrast, the "*Siege of Leningrad*" occurred over a prolonged period, 1941–1944 in a previously malnourished population that remained badly nourished after the siege. Children probably did not have accelerated weight gain. The stable environment after the famine explains why the Russian population has not suffered from an altered glucose-insulin metabolism and cardiovascular diseases. Although pointed in the expected direction, the number of people examined was small, making results not statistically significant in the Russian population [5].

The experience on the *Pacific Island of Nauru* gave insights into how improving nutrition may be associated with a rise in type 2 diabetes. The island population was chronically malnourished until the end of the Second World War. The flourishing phosphate mines built after the war drastically changed the economic situation and the nutritional welfare of the population. The immediate consequence was a significant increase in cases of obesity and the emergence of an "epidemic" of type 2 diabetes. However, subsequent studies of individuals born after the war in better nutritional circumstances have shown a substantial reduction in glucose intolerance [5].

Evidence linking environmental stimuli and the epigenetic state of the genome is rapidly growing. For example, normal physiological responses to cold exposure in plants can lead to methylation and chromatin changes of genes involved in the "winter response," thereby allowing the mitotically stable inheritance of a new environmentally induced status. "Behavioral programming" can also exert an effect on the epigenotype. In rats, maternal care, measured by licking and nursing behavior in early life, can affect the offspring's epigenotype in the promoter of the glucocorticoid receptor gene in the hippocampus, which is strongly associated with subsequent life responses to stress [1].

Evolution Models of Development

Darvin's hypothesis that mutations in genes could be either deleterious and lethal or they can offer a selective advantage to a specific environment has been documented in epidemiological studies. Research, as was mentioned previously, shows that TEs, the garbage of the genome, they could replicate independently from the rest of the genome and change their position under environmental stressors. They offer an evolution pool to the organism in order to adapt better to the environmental changes.

The term "*developmental programming hypothesis*" has been used to describe the process by which an environmental stimulus or insult applied at a critical period of development results in long-term effects on the structure and function of the cell by epigenetic changes. This critical period varies for different organ systems, and thus, programming could occur at various times, including conception, fetal life, infancy, or rarely later in adult life [1]. For example, a nutritional insult during the critical period of cell division or differentiation of the pancreas could permanently alter the structure and function as well as the metabolic processes of the organ. These early responses could become maladaptive when the environment changes after birth or simply as a consequence of aging. The outcome of the maladaptation is increased susceptibility to disease [1].

Imprinted genes are strong candidates for being important developmental programming genes. These epigenetic gametic imprints persist after fertilization to adulthood, not an "easy" task, due to the genome-wide epigenetic erasure that occurs soon after fertilization. This perpetuation of the imprint from germ cells to somatic tissues ensures that imprinted alleles retain a memory of their parental origin with consequences in gene activity [1].

The evolution of imprinting is best explained using the *genetic conflict theory*, which argues that a major function of imprinted genes is to control the provision and supply of resources in mammals. According to this theory, paternally expressed genes and maternally expressed genes in utero have opposing interests in regards to the number of resources extracted from the mother. Paternally expressed genes increase resource acquisition to promote fetal growth and postnatal fitness, whereas maternally expressed genes are "interested" in preventing the exhaustion of maternal resources prematurely, so they reduce resource transfer to the embryo [1].

In the mid-1980s, David Barker, an English physician and epidemiologist, first published his observations regarding the Dutch famine, which was originally called the Barker Hypothesis, then the Fetal Origin of Adult Disease, and now the "*Developmental Origins of Health and Disease (DOHaD)*," which suggests that adult disease has its origins during early programming in fetal life. Although the original focus was on how fetal malnutrition contributed to adult hypertension, obesity, and insulin resistance, newer studies have identified other diseases that can result from fetal over or undernutrition including immunological, mental health, and reproductive diseases [6]. It should be noted that although fetal life is a sensitive period for the effects of environmental influences on disease risk, there are likely other periods that are equally important. These include periods during which significant changes in hormones and growth occur, such as the period of infancy to childhood, puberty, perimenopause, pregnancy, and aging. It is not clear how important each of these transition periods is to disease risk or how they might interact with each other [6].

The model of the *human exposome* is another evolution perspective. The human exposome refers to the interference of the exposed surfaces to environmental stimuli during development and lifespan. Part of the exposome is the microbiome, which includes trillions of cells covering the body, including the skin, the gut, and the endothelium. The microorganisms of the so-called huge "second organism" show a high diversity and a complex symbiotic stage with the host. The diversity is influenced by several factors, such as diet, the use of antibiotics, and the ingestion of chemical compounds, and it is vital for the host's physiology, especially during early development. The constitution of the microbiome has been linked to metabolism, energy homeostasis, gut inflammatory diseases, immunologic, and even neurodevelopmental diseases. The westernized lifestyle compromises the diversity of the microbiome, leading to a proinflammatory balance, and it profoundly affects the health of the host. The influence of the exposome on the development of disease is related to the *model of biodiversity*, which includes an additional parameter to the

Barker's hypothesis. According to this model the inclusion of a variety of beneficial microorganisms in the exposome is related to a healthy being. As the Barker hypothesis expanded with new data, it came to include environmental chemical exposures [6].

It is an emerging concept that preliminary exposures are associated with neurodevelopment, including neurobehavioral diseases such as autism and learning disabilities or IQ loss. For example, in 2011, three separate laboratories identified associations between prenatal exposure to organophosphate pesticides, and altered IQ, and cognitive development in separate cohorts of children. Animal studies have also provided important information showing that embryonic exposure to environmental chemicals can induce obesity, cardiovascular diseases, neurodevelopmental and neurogenerative diseases and dysfunctions, reproductive conditions such as endometriosis and uterine fibroids, asthma, and immune dysfunction, as well as different types of cancer, including breast and prostate cancer, in later life [6].

The concept of "*Developmental plasticity*" focuses on the fact that the cell has the ability to change structure and function in response to environmental cues. These responses usually operate during critical time windows before becoming irreversible. Such plasticity permits a wide range of phenotypes to develop from a single genotype. Developmental plasticity sets the template on which continuous postnatal homeostatic and homeorhetic adaptations can occur. The production of different phenotypes that are better or worse adapted in the specific environment is the improvement of the "natural selection" and the "selective advantage" of a phenotype over the others (Fig. 1.15) [7].

The development of disease due to "genome marking" with epigenetic modifications is a multistep process. Firstly, an environmental insult induces changes in the epigenetic status of the genome in the progenitor cells of a given organ. The specific

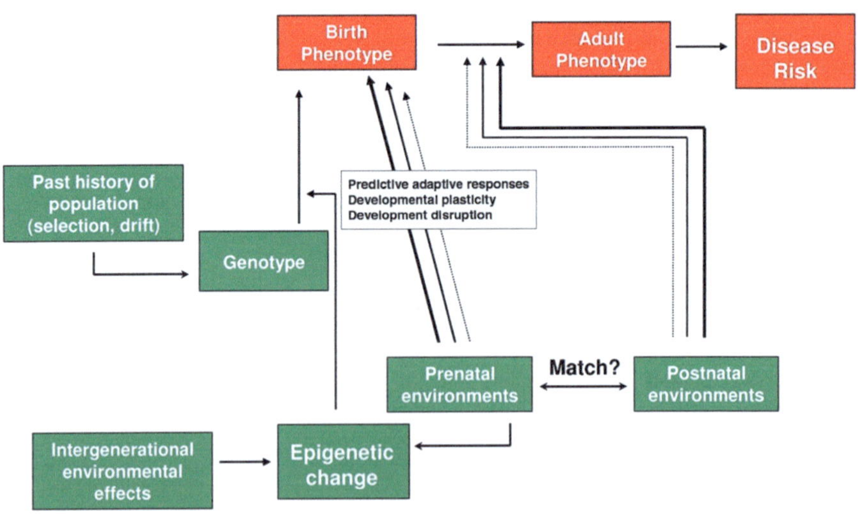

Fig. 1.15 Genotype and Epigenetic changes are related to disease risk or "selective advantage"

change in status leads to or it could be associated with subtle changes in the expression levels of "development programming" genes. The second step of the model is the maintenance of the new epigenetic state and the altered gene expression during adulthood. Adaptive physiological responses might ensue. The third step is the onset of an exhaustion of the new maladaptive pathway and the initiation of a disease process [1].

To summarize, the initial model explaining the etiology of different diseases was the Mendelian mode of inheritance. However, this model could not apply to multifactorial diseases. Interindividual epigenetic variation proved to be a better model to explain variation of disease risk in multifactorial diseases [2]. In addition to genetic and epigenetic inheritance, stochasticity affects the final phenotype. Cumulative errors in maintenance of genetic and epigenetic information contribute to interindividual epigenetic variation with age (Fig. 1.16) [2].

The *predictive adaptive response (PAR)* appears to have a primary value to adaptation. Thus, characteristics interesting for survival can be predicted. For example, the coat of a meadow vole offspring is thicker if the offspring is born at a season with decreased day length than if it is born in spring, even though the perinatal thermal environments are similar. It is proposed that PARs allow the developing organism to use developmental plastic processes to set its phenotype to the one that is predicted to give an optimal chance for survival and reproduction [7]. Epigenetic differences among MZ twins provide support for the hypothesis that stochastic differences establish different gene regulation and different risk for the development of a specific disease. The stochasticity of mammalian development enables one genotype to result in a wide range of phenotypes [2].

Overcoming the daunting challenges presented by epigenetic epidemiology will enable crucial advancements in our understanding and the possibility to interfere with these inherited marks and their long-term effects in order to produce a new generation of medication [2]. The siRNA therapeutics are a promising perspective for treating certain neurological disorders. For example, siRNAs targeting the

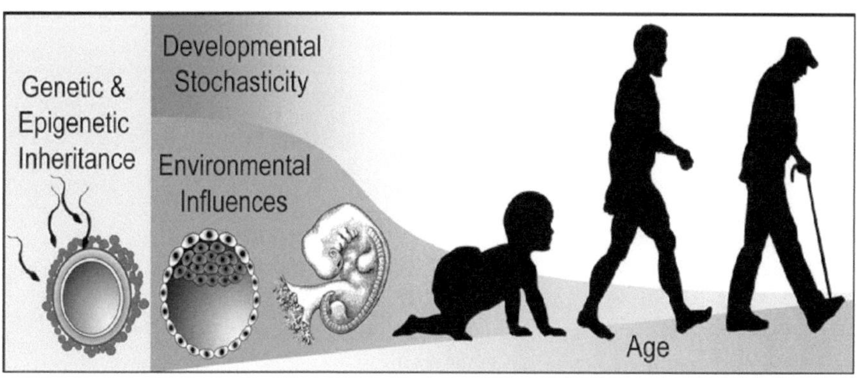

Fig. 1.16 Genetic background, epigenetic changes and Developmental Stochasticity is a more recent model used to explain predisposition to disease

BACE1 molecule have been used in the therapeutics of Huntington's and Alzheimer's diseases. The secretase (BACE1) is upregulated in Alzheimer's disease and plays a role in the cleavage of amyloid precursor protein. Other siRNA targets to treat neurologic diseases are being studied, such as SCA1 for spinocerebellar ataxias and superoxide dismutase 1 (SOD1) for amyotrophic lateral sclerosis. In addition, siRNA therapeutics may be beneficial in cancer therapeutics, targeting oncogenes, oncogenic fusion proteins, apoptotic pathway genes, and as antiviral targets such as for hepatitis C [4].

It is defined as *epigenetic epidemiology*, the study of the associations between epigenetic variation and the risk for disease. Because of genetic-epigenetic interactions, epigenetic epidemiology cannot be completely separated from genetic epidemiology [1].

The environmental influences leading to disease have many features in common:

1. Specific sensitive windows of exposure leading to time-specific and tissue-specific effects. In general, increased sensitivity is noted during the time a specific tissue is forming, which can occur in utero or extend to the first years of life, depending on the tissue.
2. The insults do not result in birth defects or malformations, which are apparent at birth. Instead, subtle functional changes occur, such as altered metabolism.
3. These functional changes increase the risk of disease long after the exposure, with a latency period of months, years, or decades.
4. The result of these functional changes is associated with an increased incidence, earlier onset, and increased severity of a disease.
5. The effect may depend on genetic background and can be sex-specific. Some effects can be transmitted by the germ line to subsequent generations, resulting in multigenerational and transgenerational effects [6]. Changes in the F3 generation sperm epigenome have been identified after exposure to the fungicides vinclozolin, dioxin, DDT, hydrocarbons, and DEET. In addition, BPA is implicated in social behavioral changes in the F4 generation, tributyltin on adipose depot weight, adipocyte size, mesenchymal cell fate, and fatty liver in the F1, F2, and F3 generations [6].

Executive Summary

- 2001, human genome project.
- Mendelian inheritance: One gene, one copy, one protein.
- Genome Wide Association Studies (GWAS): Analysis of populations for predisposing genes in multifactorial diseases.
- Epigenotype: More complex regulation of phenotype.
- Epigenetic mechanisms: DNA methylation, changes in chromatin structure, imprinting, changes in protein confirmation, and RNA-associated silencing.
- CpG methylation is generally associated with gene silencing .

- Four major classes of histones in humans are H1, H2A, and H2B, as well as H3 with H4. DNA is packed around histones that protect the genetic code and additionally regulate the active and silenced regions.
- NcRNAs are regulatory RNAs. Interfere with chromosome architecture, mRNA turnover, protein translation, and timing of translation, but they could also regulate DNA transcription and alternative splicing.
- MicroRNAs (miRNAs) are RNAs with 22 or fewer nucleotides that can control translation and thus protein production post-transcriptionally.
- Transposable elements (TEs): DNA sequences with the ability to move and integrate into the genome at different sites.
- Heterochromatin consists of these repeatable elements that control the function of neighboring genes.
- Imprinting is the pattern of asymmetrical methylation of genes depending on the parental inheritance associated with reproduction, placentation, energy homeostasis, lactation, and behavior.
- Reprogramming in germ lines is crucial to establishing totipotency.
- The epigenotype model: The significance of the environment in modulating the health status and behavioral patterns in humans.
- Environmental "marks" are transmitted up to three generations.
- The thrifty phenotype model as a model leading to metabolic syndrome diseases.
- TEs and their methylation status offer an evolutionary pool for more adaptive phenotypes.
- The Barkers hypothesis about the Developmental Origins of Health and Disease (DOHaD).
- The influence of the exposome on the development of disease consists of the model of biodiversity.
- Epigenetic epidemiology is the study of the associations between epigenetic variation and the risk for disease.

References

1. Orphanides G, Reinberg D. A unified theory of gene expression. Cell. 2002;108(4):439–51.
2. Ozanne SE, Constância M. Mechanisms of disease: the developmental origins of disease and the role of the epigenotype. Nat Clin Pract Endocrinol Metab. 2007;3(7):539–46.
3. Waggoner D. Mechanisms of disease: Epigenesis. Semin Pediatr Neurol. 2007;14(1):7–14.
4. Chuang JC, Jones PA. Epigenetics and MicroRNAs. Pediatr Res. 2007;61(5 Pt 2):24R–9R.
5. Oommen AM, Griffin JB, Sarath G, et al. Roles for nutrients in epigenetic events. J Nutr Biochem. 2005;16(2):74–7.
6. Wilson AS, Power BE, Molloy PL. DNA hypomethylation and human. Biochim Biophys Acta. 2007;1775(1):138–62.
7. Keith Slotkin R, Martienssen R. Transposable elements and the epigenetic regulation of the genome. Nat Rev Genet. 2007;8(4):272–85.
8. Reik W, Dean W, Walter J. Epigenetic reprogramming in mammalian development. Science. 2001;293(5532):1089–93.
9. Zeng Y, Chen T. DNA methylation reprogramming during mammalian development. Genes (Basel). 2019;10(4):257.

10. Surani MA. Reprogramming of genome function through epigenetic inheritance. Nature. 2001;414(6859):122–8.
11. Mathers JC. Early nutrition: impact on epigenetics. Forum Nutr. 2007;60:42–8.
12. Wong J, Collins K. Telomere maintenance and disease. Lancet. 2003;362:983–8.
13. Weng N, Hodes R. The role of telomerase expression and telomere length maintenance in human and mouse. J Clin Immunol. 2000;20:257–67.
14. Weng N, Granger L, Hodes R. Telomere lengthening and telomerase activation during human B cell differentiation. Proc Natl Acad Sci. 1997;94:10827–32.

Chapter 2
DNA Methylation and Disease

DNA Hypomethylation and Disease

It was previously mentioned how methylation patterns are related to multifactorial diseases [1].

Different epidemiological models of *cancer* have been proposed in order to explain the mechanisms of development of cancer. Early studies demonstrated an effect of both carcinogen exposures and methyl-deficient diets on global methylation levels. The stochastic accumulation of carcinogen exposure increases the risk for cancer (Fig. 2.1) [2].

A hallmark of tumorigenesis is the coexistence of genome-wide hypomethylation and hypermethylation of CpG islands in the promoters of specific genes. Promoter-region CpG island hypermethylation, associated with inappropriate transcriptional silencing of tumor suppressor genes, is the most frequently documented epigenetic alteration. Nearly half of the tumor suppressor genes that cause familial cancer, when mutated, also undergo epigenetic silencing in sporadic forms of cancer [3].

Many different pathways can lead to the initiation of transformation of a normal cell into a carcinogenic one [4]. However, an important change is generalized hypomethylation of the genome, including transponsable elements of the genome. Global hypomethylation could lead to the activation of the promoters of oncogenes, the expression of retroviruses widely distributed within the genome, an increased mutagenic activity within genes, or gross aberrations such as insertions or deletions and chromosomal rearrangements (Fig. 2.2).

Stressors and exogenous stimuli could initiate global hypomethylation, transformation of normal cells, and cancer. Some factors that can contribute to the global hypomethylation are diet, due to methyl donor deficiency diet, due to exposure to

A. Salavoura, *Chemical Environmental Pollutants and their Effect on Health*,
https://doi.org/10.1007/978-3-031-80271-3_2

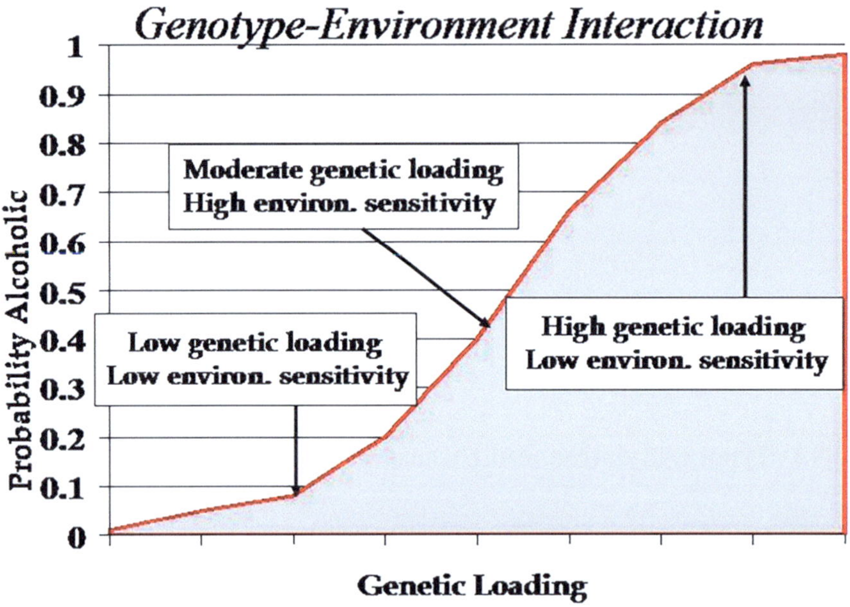

Fig. 2.1 Both epigenetic programming and stochastic accumulation of environmental insults generate risk for a specific disease

harmful substances and metabolites and the deficiency of antioxidants, infections, environmental UV and chemicals and the production of harmful metabolites, infections, environmental UV, and chemicals [5].

Transcriptional Modifications of Transposable Elements

The hypomethylation of repeated DNA sequences is expected to lead to the transcriptional activation of the promoters of genes. An elevated expression of endogenous retroviral sequences, especially HERV-K, has been reported in a number of cancers as well as other diseases.

In the event of global hypomethylation, transposable elements are prone to high mutagenic activity. These mutations can have variable forms, such as insertions, deletions, or disruptions of the coding sequence. Recently, in an investigation on 50 families with inherited breast cancer, three families were found to have heritable Alu insertions within the BRCA1 or BRCA2 genes. Interestingly, one of the insertions was created by an older Alu S element, and the insertion within the BRCA2 gene resulted in the removal of the targeted exon from its corresponding mRNA.

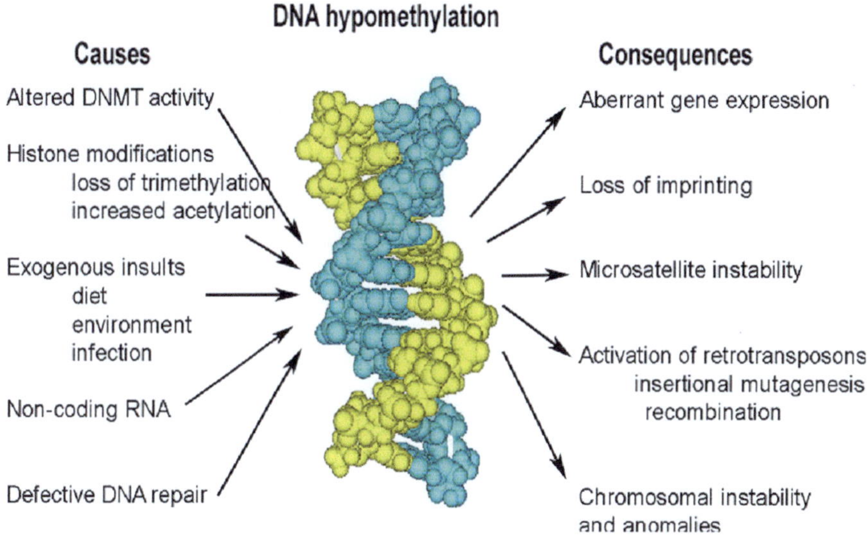

Fig. 2.2 Genome hypomethylation. Causes and consequences

Repeated DNA sequences represent a significant threat to the structural integrity of the genome. These elements can potentially mediate the recombination between non-allelic repeats, causing chromosomal rearrangements or translocations. The relationship between karyotype abnormalities and hypomethylation has been analyzed in a number of cancer types.

Hypermethylation of specific genes has been detected in the tissues of aging individuals. Thus, it has been proposed that the increased risk of cancer in some individuals might result from an age-related predisposition to hypermethylation of CpG islands that silence tumor suppressor genes. Such epimutations (i.e., abnormal alterations to the normal epigenotype) may apply to early events of carcinogenesis at the level of stem cells or progenitor cells in nonneoplastic cells [4].

The recognition of promoter CpG island methylation as a potential mediator of epigenetic alterations in neoplastic cells prompted more recent studies [2]. Not only does the silencing of tumor suppressor genes but also the silencing of functional genes could generate cancer. For example, the estrogen receptor (ER) gene is essentially unmethylated in normal breast epithelium. Promoter methylation of the estrogen receptor (ER) gene is associated with methylation-related gene silencing and the development of breast cancer. Contrary, the estrogen receptor of the tumor cells of the colon is methylated. Research shows that in normal colon tissue is frequently detected a partial methylation, which increases linearly with age [6].

An epigenetic progenitor model of cancer highlighted the potential importance of polyclonal epigenetic disruption of stem cells due to alterations of tumor progenitor genes [3]. For example, the insulin-like growth factor 2 (*IGF2*) gene is dysregulated in colon cancer. *IGF2* is a genomically imprinted gene usually expressed only from the paternally inherited allele, while in malignant cells both copies are expressed [3].

A population-based study of peripheral blood DNA from 262 Japanese individuals demonstrated that epigenetic changes in the gene *IGF2* occur in 10% of normal adults. Compared to patients with no personal or family history of colorectal cancer, those with a family history of colorectal cancer had a fivefold elevated odds ratio for dysregulated *IGF2* in peripheral blood, and those diagnosed with colorectal cancer had a 20-fold elevated odds ratio. *IGF2* is a gene with biallelic expression in colorectal cancer due to demethylation processes. *IGF2* changes also occur in several childhood cancers, including a large proportion of Wilms' tumors [3].

Ageing is associated to incomplete methylation patterns of the epithelium of the colon. These patterns either increase uniformly with age or develop a patchy methylation pattern. The result is that in some crypts the cells of the epithelium are unmethylated, with associated methylation silencing, and some show a methylated pattern. These observations demonstrate that aging in human epithelium is accompanied by a progressively increasing epigenetic mosaicism in colonic stem cells to such a level that there should be considerable crypt-to-crypt variability in gene expression patterns [6].

The epigenetic mosaicism model has potential implications in the pathogenesis of several other age-related diseases. Atherosclerotic vascular disease, for example, is characterized by discrete foci of abnormal proliferation in the arteries. Thus, there should be a variability in the profile of gene expression of the epithelium of the arteries associated with different patterns of proliferation. Similar dynamics could be relevant for other aging diseases, including neurodegenerative diseases [3].

The most discussed link between epigenetics and cardiovascular disease is hyperhomocysteinemia [2]. A population-based study of Italian adults explored the combined effects of both plasma folate levels and a common polymorphism (C677T) in 5,10-methylenetetrahydrofolate reductase on global DNA methylation. Plasma homocysteine was also related to epigenetic alterations and an informative polymorphism in the imprinted *H19* gene [3].

Data from more than 12 million Medline records identified epigenetic factors among the strongest statistical associations to type 2 diabetes. The most direct evidence implicating epigenetic dysregulation in human diabetes was from studies of the transient neonatal diabetes (TND). Recent studies show that several imprinted genes of peripheral blood leukocytes of infants with sporadic TND have an aberrant methylation [3].

Overall, developmental responses to environmental stimuli may be either disruptive or adaptive. Disruptive responses have no evolutionary significance, while adaptive ones should lead to changes in the expectations of the future environment. Such changes happen during the early stages of developmental plasticity to optimize the phenotype for the best adjustment in the expected environment to live as a mature organism. Where a match exists between the predicted and actual mature environment, these changes are appropriate and assist survival. Conversely, inappropriate predictions increase the risk of disease [7].

Executive Summary

- Epigenetic marks control gene activity in different and multiple ways.
- Development of diseases due to epigenetic modifications is a multi-step process.
- Disruption of methylated patterns is detected in many cancers including breast and colon cancer as well as in other diseases such as atherosclerosis and neurodegenerative diseases.

References

1. Hales CN, Barker DJ. The thrifty phenotype hypothesis. Br Med Bull. 2001;60:5–20.
2. Issa J-P. Epigenetic variation and human disease. J Nutr. 2002;132(8 Suppl):2388S–92S.
3. Carpinello OJ, DeCherney AH, Hill MJ. Epigenetic epidemiology of the developmental origins hypothesis. Semin Reprod Med. 2018;36(3–04):177–82.
4. Ozanne SE, Constância M. Mechanisms of disease: the developmental origins of disease and the role of the epigenotype. Nat Clin Pract Endocrinol Metab. 2007;3(7):539–46.
5. Heindel JJ, Vandenberg LN. Developmental origins of health and disease: a paradigm for understanding disease cause and prevention. Curr Opin Pediatr. 2015;27(2):248–53.
6. Gluckman PD, Hanson MA. Living with the past: evolution, development, and patterns of disease. Science. 2004;305(5691):1733–6.
7. Chuang JC, Jones PA. Epigenetics and MicroRNAs. Pediatr Res. 2007;61(5 Pt 2):24R–9R.

Chapter 3
Oxidative Stress

Reactive Oxygen Species

Oxygen is released into the atmosphere by photosynthesis happening in the leaves of the plants. Even though atmospheric oxygen is essential for aerobic life, some of its derivatives have detrimental effects. Atmospheric molecular oxygen, a free radical itself, is relatively unreactive because of spin restrictions. This property prevents materials from undergoing spontaneous combustion in an ambient oxygen-rich atmosphere.

Aerobic life is highly dependent on molecular oxygen. Oxygen is essential for the production of energy in mitochondria by the oxidative phosphorylation process. Homeostasis in the cell is maintained by optimal cellular and tissue oxygen status through complex oxygen-sensing mechanisms, signaling cascades, and transport processes. Oxygen levels fluctuate due to metabolic processes, and they either increase (hyperoxia) or decrease (hypoxia) [1].

Most of the oxygen entering the living cell is used to phosphorylation processes in the mitochondria in order to generate the currency of energy, ATP. A proportion of the cellular oxygen is an important substrate for several non-mitochondrial reactions that are catalyzed by oxygenases (monooxygenases and dioxygenases). Iron (Fe) plays a pivotal role in the catalysis of oxygenation [2].

Molecular oxygen (2–3% of the total oxygen) in the mitochondria undergoes four-electron reductions to finally produce water. This four-step reduction generates intermediates that are commonly and collectively called reactive oxygen species (ROS). ROS is responsible for various types of stress in the cell. The redox status and the presence of transition metals also contribute to the mitochondrial oxidative stress, which ultimately determines the homeostasis, survival, or death (apoptosis) of the cell. Therefore, mitochondria play a pivotal role in the survival and death of the cell, and molecular oxygen is a critical player [2].

A. Salavoura, *Chemical Environmental Pollutants and their Effect on Health*,
https://doi.org/10.1007/978-3-031-80271-3_3

In addition, molecular oxygen is activated by both enzymatic and nonenzymatic mechanisms into highly reactive oxygen species (ROS). Typical oxygen-free radicals are the peroxide ion (O_2^{-2}), the superoxide ions ($O_2OH\bullet$), the peroxyl radical (RO_2), and the hydroperoxyl radical ($HO_2\bullet$). The non-radical ROS are the hydrogen peroxide (H_2O_2), the hypochlorous acid (HOCl), the singlet oxygen ($^1O_2$), and the ozone (O_3). Reactive nitrogen species (RNS) such as the peroxynitrite (ONOO) and the nitric oxide (NO) are also sometimes included among the ROS. The members of the ROS family have different specificities as reactants, a different degree of chemical reactivity, and a wide range of lifetime [3] (Fig. 3.1).

The first step of oxygen activation is the production of the peroxide ion by a nonenzymatic process (ionizing radiation) or by enzymatic catalysis. Further catalysis by dismutase produces the hydrogen peroxide and by-products such as HOCl (hypochlorous acid) or NO_2 (nitrogen dioxide) [3] (Fig. 3.2).

A major catalytic reaction for the production of free radicals is the Fenton reaction. It is catalyzed by enzymes supported by different metals, mainly iron and copper. The typical reaction is metal $+H_2O_2$ = metal(+electron) + (OH^-) + (OH^+) (Fenton reaction). The redox cycling is maintained by an electron reduction to (O_2^-) by the reaction: Metal (+electron) + O_2 = $Metal^+ + O_2^-$ to be balanced by the reaction $H_2O_2 + (O_2^-) = O_2 + OH^- + OH\bullet$(Haber-Weiss reaction).

ROS-mediated signaling cascades from mitochondria to nucleus are considered one of the most important regulators of aging and disease in humans. They are tightly balanced by antioxidants, which scavenge the highly reactive oxygen metabolites produced in mitochondria [2]. Although ROS are also required for many normal physiologic functions of the aerobes, overwhelming production of ROS coupled

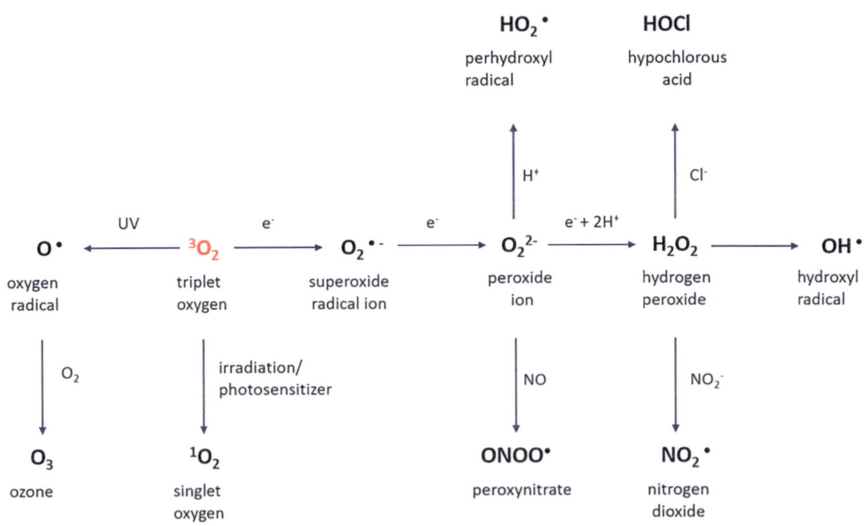

Fig. 3.1 The fate of molecular oxygen in the cell. Enzymatic processes produce hydrogen hyperoxide and nonenzymatic ones produce the superoxide radical ion

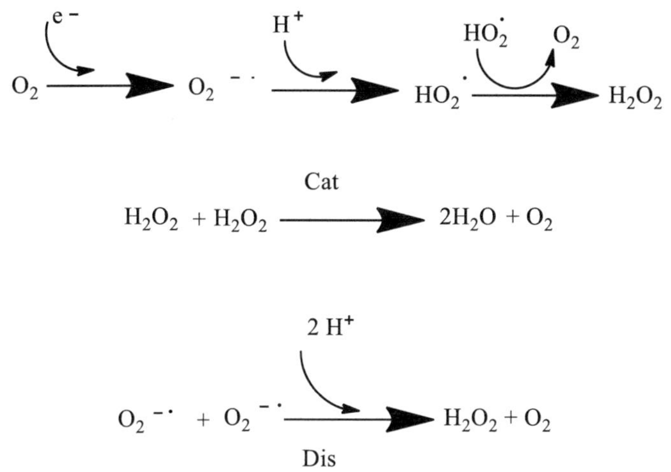

(2.1)

(2.2)

(2.3)

Fig. 3.2 Production of and degeneration of H_2O_2 and the safety valve of production of water from H_20_2

with their insufficient scavenging by endogenous and exogenous antioxidants could lead to detrimental oxidative stress [3].

One of the surprises in redox biology was the relatively recent concept of hydrogen peroxide functioning as a messenger molecule. It is now widely accepted that this low-weight molecule is utilized in many metabolic regulation pathways. H_2O_2 diffuses through tissues, and it initiates immediate cellular reactions, such as cell shape changes, the formation of functional actomyosin structures, and the recruitment of immune cells [3].

A major source of hydrogen peroxide is the dismutation of the peroxide anion radical, formed by 1-electron reduction from oxygen. Other sources of hydrogen peroxide include the NADPH cycle and the respiratory chain in mitochondria. Although there is also a spontaneous dismutation, superoxide dismutases catalyze the production of hydrogen peroxide from oxygen. NAD(P)H oxidases are ubiquitous, operating under the control of growth factors and cytokines.

Respiratory chain-linked H_2O_2 production in the mitochondria involves the cytochrome *bc1* complex (Complex III). It is noteworthy that Complex I is another major source of mitochondrial peroxide production, which is released directly to the mitochondrial matrix space, whereas Complex III produces peroxide released to the intermembrane space. Mitochondrial Complex II is another independent source of mitochondrial reactive oxygen species production. The catalytic reaction of H_2O_2 to H_2O and O_2, may be regarded as a safety valve, occurring at high ranges of H_2O_2 concentration, for example, under toxic conditions [2] (Fig. 3.2).

Antioxidants, by definition, either slow or inhibit a free radical-mediated oxidative reaction by different mechanisms. Their primary function is to protect from the harmful effects of ROS and to maintain redox homeostasis [3].

"Oxidative stress is an imbalance between oxidants and antioxidants in favor of the oxidants, leading to a disruption of redox balance and control and/or molecular damage." Moreover, oxidative stress leads to the damage of the biologic system at the molecular, cellular, organ, and organism level [1]. Oxidative stress also occurs when there is an imbalance between the formation and the removal of RONs because of an overproduction and/or an impaired ability to neutralize them or to repair the resulting damage [4].

As a consequence of oxidative stress, there is considerable damage to biomolecules. ROS are highly reactive and capable of reacting with almost every molecule in the body. The reactivity of ROS with biomolecules leads to the damage of lipids, carbohydrates, proteins, and DNA. Lipid peroxidation of the membrane phospholipids by ROS leads to dysfunction of the cell. Cellular carbohydrates are also attacked by ROS, leading to abnormal molecular processes. Proteins, upon oxidation by ROS, inactivate enzymes and give rise to a number of by-products. DNA damage is also induced by ROS, especially by the highly reactive hydroxyl radical, which leads to base damage, single-strand breaks, and the formation of crosslinks. These aberrations have been detected in chemically induced mutations and in carcinogenesis [2].

Redox homeostasis and balance fluctuate due to the metabolic processes of the cell (Fig. 3.3). Conditions related to deficiency or excess of oxygen metabolites are infection, inflammation and toxic insults, which influence the use of oxygen in the mitochondria and they are related to disruption of the oxygen balance within the

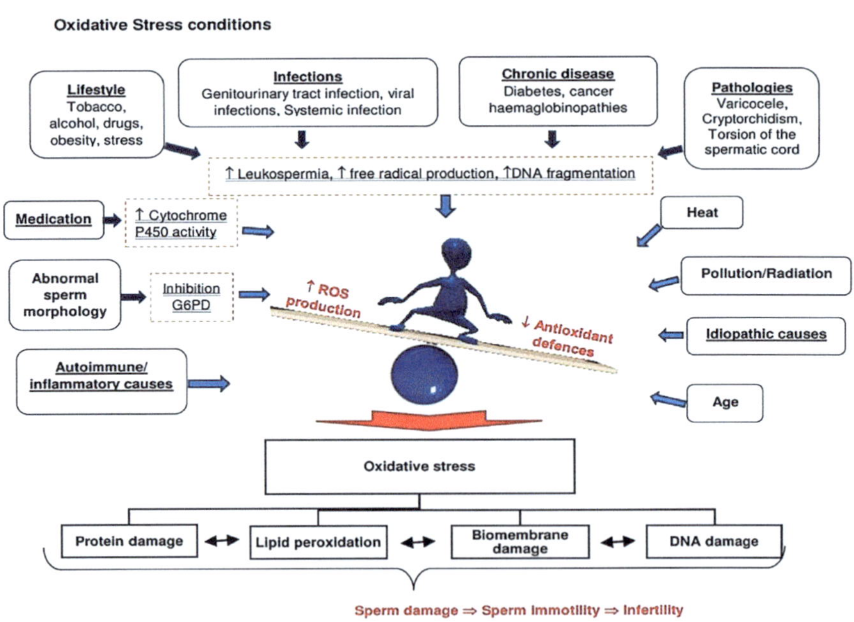

Fig. 3.3 Redox homeostasis insults

cell. This starts a cascade of responses, including the induction of transcription factors [3] (Fig. 3.3).

Under certain conditions, although the tissue/organ uses repair strategies to restore the altered redox homeostasis are activated, the balance is not restored, leading to pathology [1].

The brain is especially susceptible to the alteration of the redox status. First, it is well known that the brain has a high demand for oxygen, and thus, it is more vulnerable to hypoxia. Additionally, the brain has a low antioxidant ability; for example, it manifests ineffective reactions dependent on catalase. Thirdly, the brain has higher levels of membrane polyunsaturated fatty acids, which are susceptible to the free radical-mediated peroxidative attack. Fourth, the cerebrospinal fluid has low-molecular-weight iron and copper complexes, and neurons have non-heme iron to bind oxygen [5]. Oxidative stress in the brain is involved in most of the neurodegenerative disorders such as Alzheimer's disease, Parkinson's disease, and amyotrophic lateral sclerosis (ALS) [5].

ROS are also involved in vascular diseases and disorders including atherosclerosis, peripheral vascular disease, ischemia, ischemia-perfusion injury, congestive heart failure, and stroke [1]. During the process of atherosclerosis, enzymes present in the vessel wall, including NAD(P)H oxidase, myeloperoxidase, and lipoxygenases, generate ROS, leading to lipid peroxidation. Although the cause-effect relation between lipid oxidation and atherosclerosis is not yet clearly established, it is detected that the initiation of atherosclerotic lesions could start with the interaction of lipids with oxygen-derived species [1].

Vascular infraction causes cell death. The two most well-known conditions of infraction are myocardial ischemia and ischemic stroke. Myocardial ischemia (oxygen starvation of the heart tissue) arises from either a disproportionate increase in oxygen demand or the interruption of blood flow to the cardiac tissue. Myocardial dysfunction leads to different cardiac disorders, including cardiomyopathy and congestive heart failure, which are the leading causes of death in the United States. Ischemic stroke is the response of the brain to oxygen deprivation. Stroke results in neuronal death in the ischemic region, and depending on the nature and extent of the damage, cerebral functions are affected [1].

However, the excess of oxygen could also cause disease. For example, retinopathy of prematurity is caused by high oxygen supplementation (hyperoxia) that causes retinal vascularization. It has also been suggested that other factors, including low vitamin E concentration in premature infants, birth weight, gestational age, and exposure to light, contribute to the development of retinopathy. Clinical trials have suggested that the supply of lower oxygen saturation levels (SpO2 85–89%) is safer compared with the currently used high-end SpO2 levels (91–95%) to reduce the severity of premature retinopathy. Conversely, retinal hypoxia has been suggested as a contributing factor in retinal neovascularization in the ischemic proliferative retinopathies, with special emphasis on diabetic eye diseases [1].

Pulmonary oxygen toxicity is associated with the influx of phagocytotic inflammatory cells in the lung and an enhanced production of cytokines. Signaling mechanisms of the damaged lung epithelium play a crucial role in the oxidant-mediated lung diseases, including asthma, cystic fibrosis, bronchitis, and chronic obstructive pulmonary disease (COPD). Mitogen-activated protein kinases (MAPKs) are the main signaling molecules activated by ROS [2].

The environmental chemicals with detrimental health effects due to the production of free radicals are polychlorinated biphenyls (PCBs, including dioxins), polybrominated diphenyl ethers (PBDEs), perfluoro-carboxylic acids (PFCAs), perfluoro-octanesulfonate (PFOS), benzene, and bisphenol A (BPA) that originate from industrial processes (e.g., material additives or combustion products) and pesticides (used to optimize crop yields by prevention of pest infestation). Metal radicals have an important role to the development of disease (Fig. 3.4) [1].

As far as the immune system is concerned, ROS activate the production of cytokines that mediate and control immune and inflammatory responses. The inflammatory reactions result to the stimulation of four major programs: the acute-phase reaction,

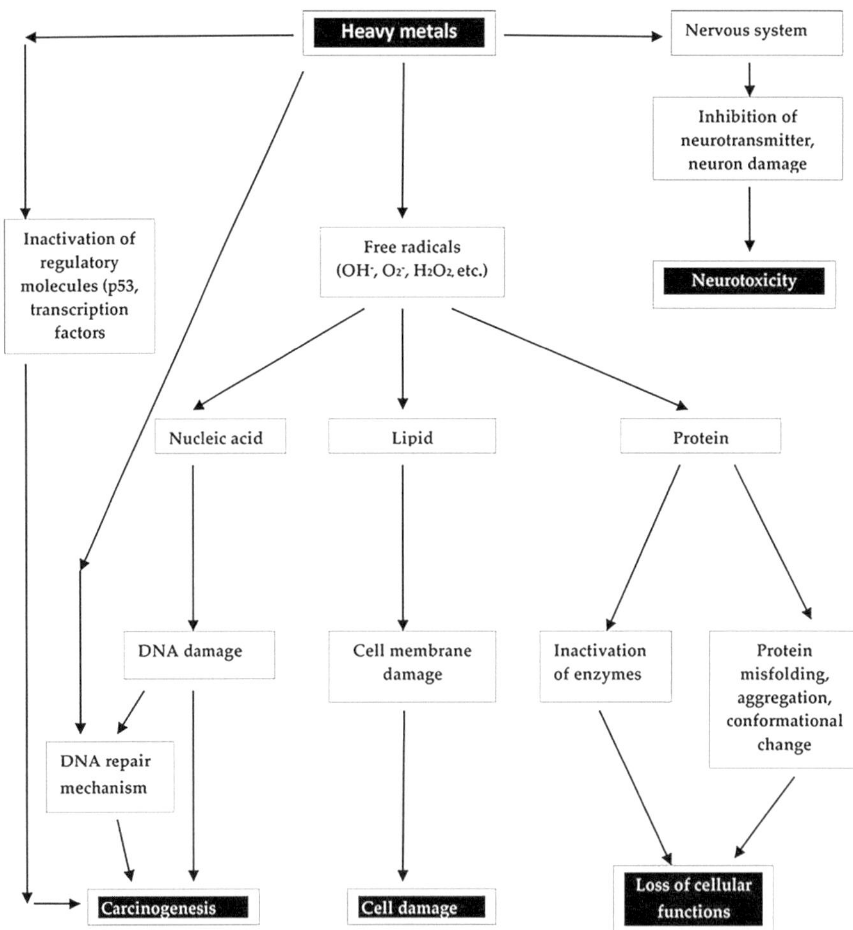

Fig. 3.4 A schematic illustration of how metals produce free radicals and disease

the sickness syndrome, the pain program, and the stress response. Complex interactions exist between cytokines, inflammation, and the adaptive responses in maintaining homeostasis, health, and well-being. For example, during the stress response, the inflammatory reaction is crucial for survival [5].

The activation of the stress system, through induction of a TH2 shift, protects the organism from systemic "overshooting" with TH1 proinflammatory cytokines. When a harmful signal starts the inflammation process, the stress cascade is activated and it facilitates the production of a number of proinflammatory cytokines, including interleukin IL-1, IL-6, IL-8, IL-18 and tumor necrosis factor-alpha through activation of the corticotropin-releasing hormone/substance P-histamine axis [6]. However, at the same time, the overproduction of IL-6 mediates the activation of the TH2 response with anti-inflammatory properties. The dysfunctional neuroendocrine-immune interface is associated with abnormalities of the "systemic anti-inflammatory feedback".

The prolonged inflammation is related to the pathogenesis of atopic/allergic and autoimmune diseases, obesity and atherosclerosis. The failure of the adaptive systems to resolve inflammation effectively is implicated in behavioral changes including quality of life, sleep and depression [6]. The interplay of the immune and endocrine system tends to lead to a homeostasis status under normal conditions which is connected with the well-being and the non-disease state. When the above balance is dysregulated, the new status is called cacostasis, which if extended, leads to disease [6].

The stress system also interacts with other major central nervous system (CNS) elements, primarily with the mesocorticolimbic dopaminergic system ("reward" system), the amygdala/hippocampus complex and the arcuate nucleus proopiomelanocortin (POMC) neuronal system [7]. Conversely, these systems after their activation by a stressor signal act by specific neuronal pathways to modify the activity of the individual. It is of note that well established interactions exist between the stress system and centers of the CNS that are crucial for survival, such as the thermoregulatory and the appetite-satiety centers [7].

Stress occurs when homeostasis is threatened or perceived to be so; homeostasis is re-established by various physiological and behavioral adaptive responses. Neuroendocrine hormones have major roles in the regulation of both basal homeostasis and responses to threats, and they are involved in the pathogenesis of diseases characterized by dyshomeostasis or cacostasis. The stress response is mediated by the stress system, partly located in the central nervous system and partly in peripheral organs [7].

The benefits of the antioxidant medication which decrease the reactive oxygen radicals within the cell, has been investigated. Such drugs could be used for the prevention of aging, cardiovascular disease, and cancer [1].

The antioxidant defense mechanisms include endogenous antioxidants, enzymatic and non-enzymatic, as well as exogenous antioxidants, that people acquire by their diet [3]. The primary endogenous antioxidant enzymes are SOD, catalase (CAT) and glutathione peroxidase (GSH-Px). They can be produced by the body and they contribute to redox equilibrium within the cell. As mentioned above, O_2·is converted by SOD to H_2O_2, which is decomposed to water and oxygen by CATalase, preventing hydroxyl radicals' overproduction. In parallel, GSH-Px converts peroxides and hydroxyl radicals into nontoxic forms by oxidating glutathione (GSH) into glutathione disulfide by glutathione reductase. Other related antioxidant enzymes are glutathione-S-transferase and glucose-6-phosphate dehydrogenase (NADPH) [4] (Fig. 3.5).

Glutathione peroxidases (Sec-GPx) in various subcellular compartments and cell types have a major function in the control of H_2O_2 and of other hydroperoxides [2]. They typically contain a peroxidatic Sec residue or 1-Cys Prx-6. They produce water from peroxides based on the reaction: $H_2O_2 + 2GSH \rightarrow Sec - GPxsPrx{-}6{-}2$ H_2O + GSSG. The two-electron reduction from H_2O_2 occurs by the oxidation of thiols (thiolate anion, RS^-): $H_2O_2 + RS^- \rightarrow OH^- + RSOH$. This reaction occurs spontaneously, and the energy produced is too high. When O_2·radical ion is generated outside the cells, for example by the NADPH oxidase isoform or NOX2 in endothelial cells, the radical could enter the cell, where it functions as a signal with a short duration of action [1].

The non-enzymatic antioxidants are molecules that interact with ROS and they terminate the free radical chain reactions: bilirubin, α-tocopherol (vitamin E), and β-carotene are present in blood while albumin and uric acid account for 85% of antioxidant capacity in plasma [8]. Exogenous antioxidants are essentially acquired from the diet and they include ascorbic acid (vitamin C), which scavenges hydroxyl and superoxide radical anion, α-tocopherol (vitamin E), which is involved against lipid peroxidation of cell membranes, and phenolic antioxidants, which include stilbene derivatives (resveratrol, phenolic acids, and flavonoids), oil lecithins,

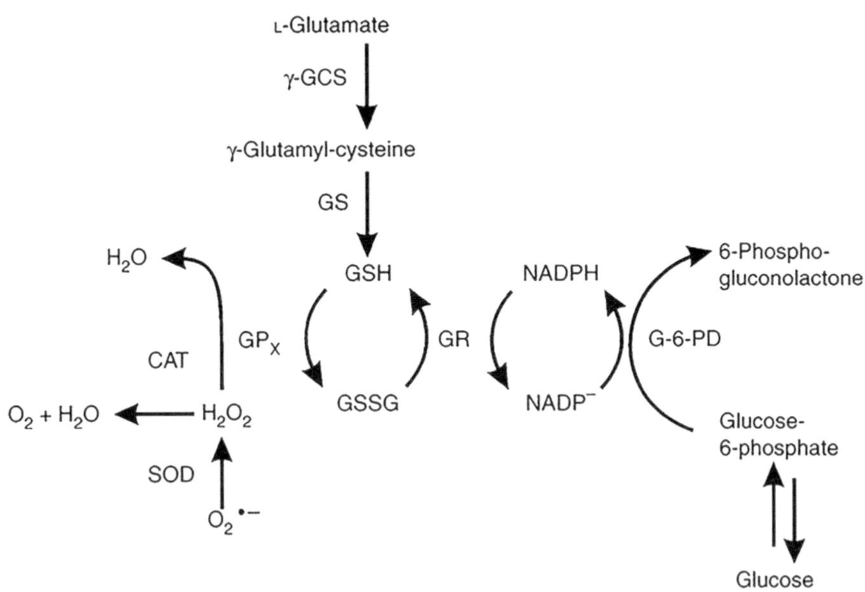

Fig. 3.5 Glutathione enzymatic antioxidant mechanism

selenium, zinc, and drugs such as acetylcysteine [9]. The various components of the antioxidant defense system are interrelated and function together to ensure adequate protection of the cell. Antioxidant defense mechanisms range from very simple reactions to a complex network of redox switches that can sense redox imbalance in the environment and they could immediately be immobilized to regain equilibrium [4].

The Central Role of Glutathione in the Pathophysiology of Human Diseases [10]

It would be discussed in more detail, the central role of glutathione (GSH) in the antioxidant network. Glutathione is a water-soluble molecule located in the cytoplasm. It acts either as a proton donor or as a cofactor of nucleophilic conjugates. GSH can scavenge superoxide anion ($O_2\cdot$), hydroxyl radical (−OH), and singlet oxygen ($O_3\cdot$) by donating electrons and becoming oxidized to thyil radical (GS). It is important to mention that the reductive ability of GSH is associated with the formation of thiyl radicals (GS).

The metabolic pathway of production of glutathione is the following: it is formed in a two-step enzymatic process including, first, the formation of g-glutamyl cysteine from glutamate and cysteine by the activity of the g-glutamyl cysteine synthetase; and second, the formation of GSH by the activity of the GSH synthetase, which uses g-glutamyl cysteine and glycine as substrates. While its synthesis and metabolism occur intracellularly, its catabolism occurs extracellularly by a series of enzymatic and plasma membrane transport steps.

Glutathione metabolism and transport participate in many cellular reactions, including antioxidant defense of the cell, drug detoxification, and cell signaling (involved in the regulation of gene expression, apoptosis, and cell proliferation). Glutathione metabolism and degradation are mediated by the g-glutamyl transpeptidase (g-GT), expressed mainly on the apical surface of cells (Fig. 3.6).

Reactive oxygen is connected with two glutathione molecules with bisulfide bonds to produce glutathione conjugates (GSSG). This pathway is enzyme-dependent and it serves as a detoxification mechanism to scavenge protein or DNA radicals. Glutathione reductase removes the g-glutamyl moiety from GSH and GSH conjugated compounds by its transfer to other acceptors, thus producing cysteinyl-glycine (Cys-Gly) or Cys-Gly-conjugates and g-glutamyl-amino acid (g-Glu-aa) products (Fig. 3.6). In order to have an adequate concentration of glutathione in the cytoplasm, glutathione is actively transported by GSH transporter families. One such family is the multidrug-resistant protein family (MRP) pump. GSH is actively transported within the mitochondria.

A nonenzymatic reaction between GSH and nitric oxide forms the S-nitroso glutathione (GSNO) to scavenge proteins also. Glutathione peroxidases with catalytic action are selenoproteins that vary in their peroxide substrate specificity.

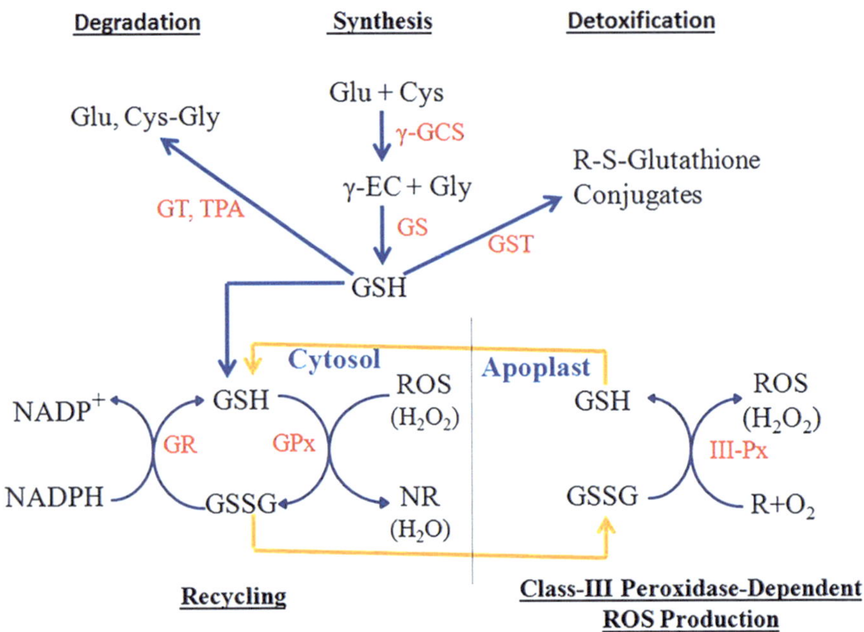

Fig. 3.6 Glutathione as a ROS scavenger within the cell

Glutathione is a ubiquitous molecule that is produced intracellularly in all organs and cell types, but it is more abundant in liver and lung tissues. It is a linear tripeptide formed from the amino acids glycine, cysteine, and glutamate. Intracellular GSH can exist as a monomer in its reduced form or as a disulfide dimer formed due to its oxidation (GSSG), which usually accounts for less than 1% of the total intracellular glutathione. Glutathione is 85–90% freely distributed in the cytosol, although it can be in different organelles, including mitochondria, peroxisomes, the nuclear matrix, and the endoplasmic reticulum, after its cytosolic synthesis. Thiol-containing compounds participate in many biochemical and pharmacological reactions (Fig. 3.7).

In its reduced form, glutathione (L-g-glutamyl-L-cysteinyl-glycine, GSH) is the most abundant non-protein thiol in mammalian cells. It was initially described as a potent reducing agent; however, many other cellular functions have been recently ascribed to GSH, focusing the scientific interest in its biological properties. Today, it is widely accepted that GSH not only acts as a reducing agent and a major antioxidant within the cells, maintaining control of the redox status, but it also acts as a mediator of many other physiological reactions, including cellular signaling, metabolism of xenobiotics, thiol disulfide exchange reactions, and also as an important reservoir of cysteine (Fig. 3.7) [6].

Changes in the intracellular thiol-disulfide (GSH/GSSG) balance within the cell are used as an index of the redox status of the cell. As mentioned above, under conditions of oxidative stress, GSH is consumed by scavenging free radicals, while

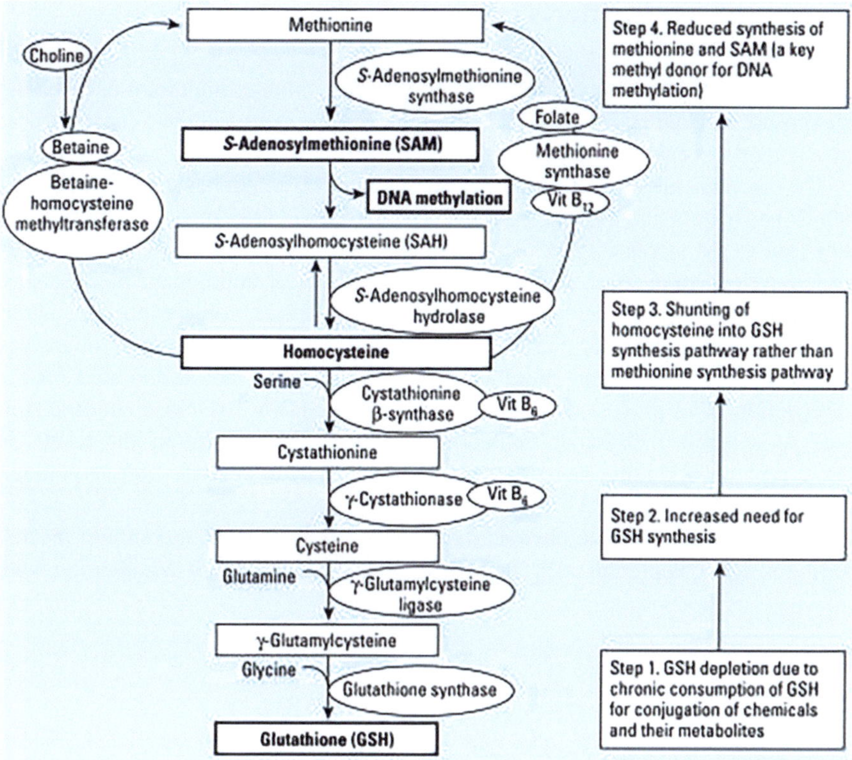

Fig. 3.7 The products produced in the synthesis of glutathione participate in many biological reactions

GSSG levels increase. The ratio of [GSH/GSSG] and the reduction potential (mV) are two ways to express the intracellular redox environment of the cell. The GSH/GSSG couple is considered to be a cellular redox buffer because it provides a very large pool of reducing equivalents.

Without a process to reduce protein disulfides, vulnerable cysteinyl residues of essential enzymes might remain oxidized, leading to changes in their catalytic activity. Higher-order thiol cell systems like metallothioneins and thioredoxins are regulated by the GSH/GSSG balance. The antioxidant function is additive because it participates in other antioxidant systems, such as the oxidized ascorbate (dehydroascorbate).

Metals and Their Effects on Health

Apart from diet as a lifestyle determinant of health, another important determinant is exposure to chemicals. Chemicals fluid the environment and their consequences have recently been recognized [11].

The mechanism of the lifelong impacts of metals on health has started to be clarified. First, metals and organic pollutants increase oxidative stress in the cell, and they lead to the increased production of reactive oxygen species (ROS). Secondly, prolonged exposure to chemicals leads to depletion of antioxidant mechanisms, especially the intracellular glutathione network. Under chronic exposure, metals lead to the consumption of glutathione and the production of homocysteine instead of the production of methionine. Methionine is an important amino acid for the methylation DNA process since it is the substrate of DNMT (DNA methyltransferase). As a result, methionine depletion inhibits DNA methylation and results in DNA hypomethylation. Depletion of glutathione leads to genome-wide hypomethylation (Fig. 3.8) [12].

In populations living in chemically contaminated societies, a common mechanism for GSH depletion may be GSH consumption through conjugation with

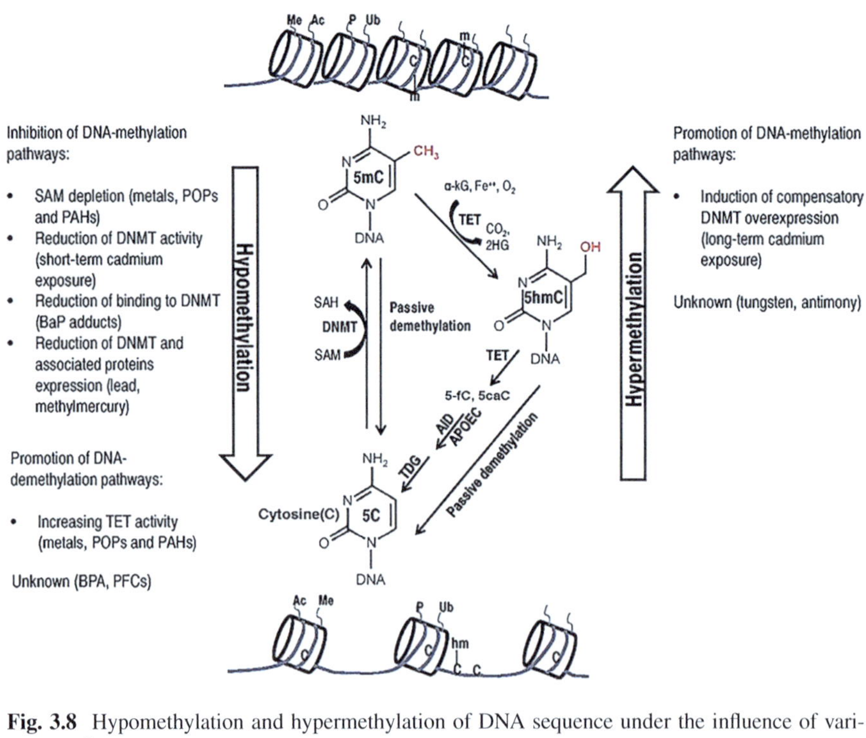

Fig. 3.8 Hypomethylation and hypermethylation of DNA sequence under the influence of various metals

chemicals or their metabolites. A great concern is the fact that some of the chemicals bioaccumulate in adipose tissue during life, and although everyday exposure is within normal limits, the lifelong exposure is hazardous [13]. Third, some metals like cadmium, lead, and mercury can directly reduce the enzymatic activity and concentrations of DNMT. Cadmium in high concentrations could have the opposite effects, and it is related to DNMT overexpression. Fourth, some chemicals can interfere with histone modifications, such as acetylation, changing the methylation status. For example, lead has been related to transcription modifications leading to hypomethylation; methylmercury and nickel lead to hypermethylation; and environmental toxicants and BPA or PFCs lead to hypomethylation.

Apart from exhaustion of the glutathione antioxidant system, another pathway leading to DNA hypomethylation is the production of a-ketoglutarate due to oxidative stress. A-ketoglutate is a compound that acts in demethylation pathways involving DNA repair enzymes (Fig. 3.8) [11].

How metals interfere with the epigenomic mechanism to induce disease has recently been described in literature. Epigenetic marks have been established in intrauterine life and predispose to disease in adult life. Additionally, in general, normal cells are characterized by gene-specific hypomethylation and global hypermethylation, while cancer cells express a reverse profile to this norm. DNA methylation and histone acetylation are partly reversible, and they are related to environmental exposures during adult life. Chemicals are known to participate in these changes and thus predispose to cancer. Some of the well-known metals that predispose to cancer are nickel, arsenic, cadmium, and chromium [14].

Another effect of metals is the methylation status of transposable elements. Methylation of these elements is related to genomic stability, while the unmethylated status transforms these parts of the genome to a mobilized status, and thus they can insert into parts of the genome where regulatory regions are located. Differential methylation located within or near promoters of genes can also interfere with normal gene expression (Fig. 3.9).

People are continually exposed to chemicals in their everyday lives and much more since the lifestyle was "westernized." Exposure occurs through contaminated food, water, and air, and there could be described a range of deliberate and undeliberate forms of exposure. Fish contaminated with mercury, meat contaminated with antibiotics and pesticides, fruits and vegetables with detectable residues of pesticides, air pollution, and water with a lot of soluble metals, especially water mined close to industrial areas. Apart from exogenous exposure, endogenous production of chemicals that are part of the metabolism processes during stress and inflammation is considered significant. Many of those exogenous and endogenous chemical compounds are carcinogenic, and they interfere with the redox balance and epigenetic marks of the cell [15].

Cumulative evidence is published regarding the toxicity of metals and their relation with oxidative stress. Different metals express their toxicity in different tissues. Many of them are not harmful by themselves, but they could become due to the

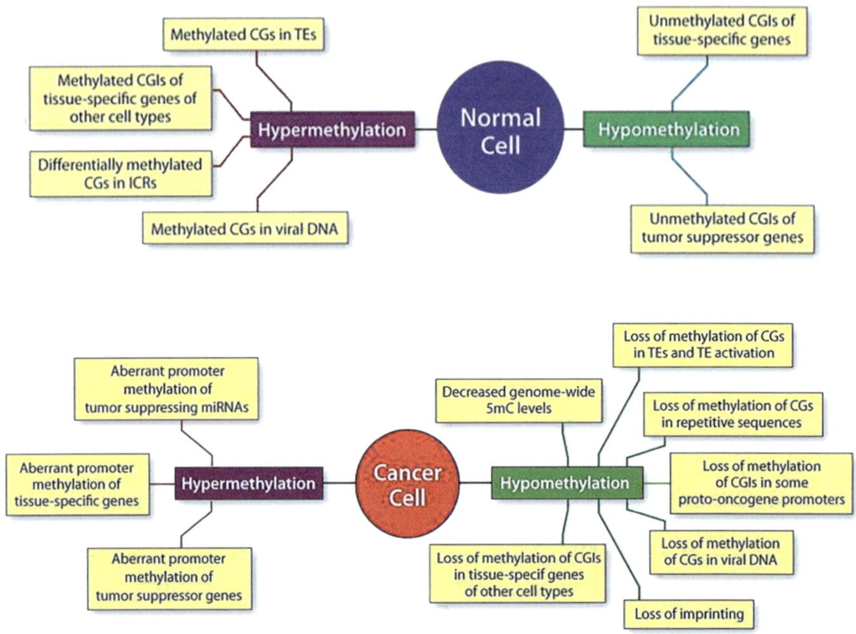

Fig. 3.9 Mechanisms of hypermethylation and hypomethylation in normal and cancer cells

accumulation of their metabolites. However, the underlying mechanism in all cases is increased oxidative stress in the cell. The examples of accidents that released a massive amount of chemicals to the environment and the health impact on the exposed population are described.

The contribution of specific tobacco carcinogens (e.g., PAH and NNK) to tobacco-induced lung cancer can be evaluated by a weight of evidence. Heterocyclic amines produced when meat is heated above 180 °C for a long period have been detected in meat meals in the USA, and they represent an important class of carcinogens in foods [14]. There is a growing list of carcinogens, and until now 88 agents (drugs, environmental chemicals, radiation, viruses, and mixtures) are characterized as highly carcinogenic (http://monographs.iarc.fr) [15].

A new tool for the investigation of the susceptibility of different individuals to the carcinogenic properties of different chemicals is molecular biomarkers. Research contributed to the identification of a number of genetic polymorphisms associated with different cancers. For example, our knowledge regarding the function of detoxification enzymatic systems like cytochrome P450 is expanding. P450 enzymes in the liver and their metabolites, glucuronyl transferase, glutathione *S*-transferase, and others, metabolize chemicals by inactivation pathways, and the relative amounts of activation and detoxification will determine whether a chemical is carcinogenic.

An interesting field of research is the effects of intrauterine exposure to heavy metals. For example, in utero exposures to metals, such as arsenic (As), mercury (Hg), lead (Pb), and cadmium (Cd), as well as organic pollutants, such as polychlorinated biphenyls (PCB), polybrominated diphenyl ethers (PBDE), and dichlorodiphenyldichloroethylene (DDE), have been associated with developmental delay. Early life exposure to heavy metals has long been linked to deleterious outcomes for children [16].

We name environmental pathology the study of the mechanisms by which the environment participates in the initiation of disease. Environmental medicine is a multidisciplinary field involving medicine, environmental science, chemistry, and others, overlapping with environmental pathology. The scope of this field involves studying the interactions between the environment and human health and the role of the environment in causing or mediating disease. This special field of study has developed after the realization that health is more widely and dramatically affected by environmental factors than previously recognized. Environmental factors in the causation of environmental diseases can be classified into: physical, chemical, biological, social (including psychological and culture variables), ergonomic and safety or any combination of the above [17] (Fig. 3.10).

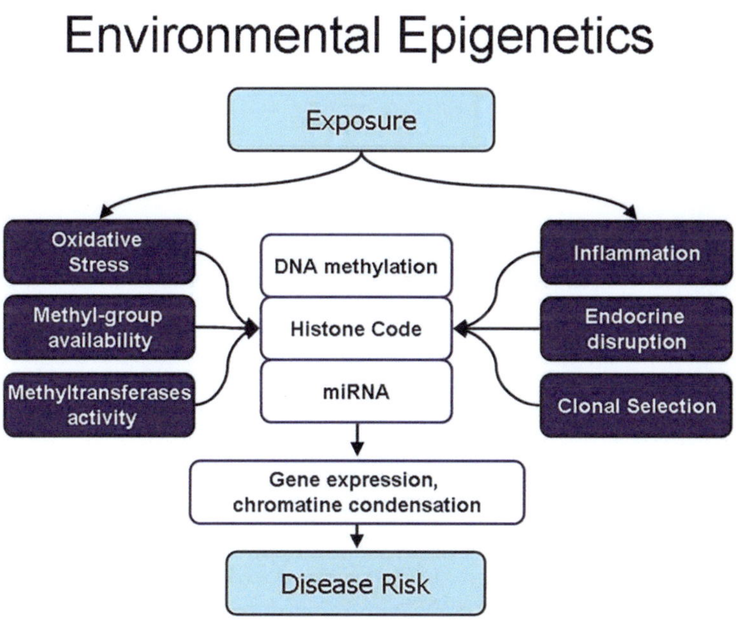

Fig. 3.10 Disease risk due to environmental influence

Executive Summary

- Molecular oxygen is activated by both enzymatic and nonenzymatic mechanisms.
- ROS (reactive oxygen species) are by-products of oxygen use in mitochondria and non-mitochondria processes.
- The most important of ROS is hydrogen peroxide.
- ROS have detrimental effects when in excess, and together with antioxidant mechanisms, they constitute the redox balance of the cell.
- Oxidative stress is an imbalance between oxidants and antioxidants in favor of the oxidants, leading to a disruption of redox signaling and control and/or molecular damage.
- The reactivity of ROS with biomolecules leads to the damage of lipids, carbohydrates, proteins, and DNA and the induction of different diseases (neurogenerative, cardiovascular, retinopathy, etc.).
- ROS lead to the production of cytokines that mediate and control immune and inflammatory responses. Complex interactions exist between cytokines, inflammation, and the adaptive responses in maintaining homeostasis, health, and well-being.
- The dysfunctional neuroendocrine-immune interface contributes to disease.
- The glutathione antioxidant mechanism has a central role in the pathophysiology of human disease.
- People are continually exposed to chemicals in their everyday lives.
- The toxicity of metals is related mainly to oxidative stress and epigenetic changes.

References

1. Kulkarni AC, Kuppusamy P, Parinandi N. Oxygen, the Lead actor in the pathophysiologic drama: enactment of the trinity of normoxia, hypoxia, and hyperoxia in disease and therapy. Antioxid Redox Signal. 2007;9(10):1717–30.
2. Forman HJ, Maiorino M, Ursini F. Signaling functions of reactive oxygen species. Biochemistry. 2010;49(5):835–42.
3. Sies H. Role of metabolic H2O2 generation. Redox signaling and oxidative stress. J Biol Chem. 2014;289(13):8735–41.
4. Minich DM, Brown BI. A review of dietary (Phyto)nutrients for glutathione support. Nutrients. 2019;11(9):2073. Heavy metals effects on oxidative stress and cardiovascular health/disease
5. Elenkov IJ, Iezzoni DG, Daly A, Harris AG, Chrousos GP. Cytokine dysregulation, inflammation and well-being. Neuroimmunomodulation. 2005;12(5):255–69.
6. Rahal A, Kumar A, Singh V, et al. Oxidative stress, prooxidants, and antioxidants: the interplay. Biomed Res Int. 2014;2014:761264.
7. Chrousos GP. Stress and disorders of the stress system. Nat Rev Endocrinol. 2009;5(7):374–81.
8. Gould RL, Pazdro R, et al. Impact of supplementary amino acids, micronutrients, and overall diet on glutathione homeostasis. Nutrients. 2019;11(5):1056.
9. Hemilä H, Chalker E. Vitamin C for preventing and treating the common cold (review). Cochrane Libr. 2013;2013(1):CD000980.

10. Franco R, Schoneveld OJ, Pappa A, et al. The central role of glutathione in the pathophysiology of human diseases. Arch Physiol Biochem. 2007;113(4–5):234–58.
11. Ruiz-Hernandez A, Kuo C-C, Rentero-Garrido P, et al. Environmental chemicals and DNA methylation in adults: a systematic review of the epidemiologic evidence. Clin Epigenetics. 2015;7(1):55.
12. Lee D-H, Jacobs Jr DR, Porta M. Hypothesis: a unifying mechanism for nutrition and chemicals as lifelong modulators of DNA Hypomethylation. Environ Health Perspect. 2009;117(12):1799–802.
13. Brocato J, Costa M. Basic mechanics of DNA methylation and the unique landscape of the DNA Methylome in metal-induced carcinogenesis. Crit Rev Toxicol. 2013;43(6):493–514.
14. Wogan GN, Hecht SS, Felton JS, et al. Environmental and chemical carcinogenesis. Semin Cancer Biol. 2004;14:473–86.
15. Lee D-H, Jacobs Jr DR, Porta M. Hypothesis: a unifying mechanism for nutrition and chemicals as lifelong modulators of DNA Hypomethylation. Environ Health Perspect. 2009;117(12):1799–80.
16. Li Q, Kappil MA, Li A, et al. Exploring the associations between microRNA expression profiles and environmental pollutants in human placenta from the National Children's Study (NCS). Epigenetics. 2015;10(9):793–802.
17. Baccarelli A, Bollati V. Epigenetics and environmental chemicals. Curr Opin Pediatr. 2009;21(2):243–51.

Chapter 4
Metals and Disease

Mercury

The Metal, Its Forms, and Its Distribution in the Environment

Mercury (Hg) is a chemical element presenting in two oxidative forms Hg I and II. Most mercury I compounds are diamagnetic and they feature the dimeric cation, Hg_2^{+2}. Stable derivatives include the chloride and nitrate. Mercury II oxide, the main oxide of mercury, arises when the metal is exposed to air for long periods at elevated temperatures. The bioavailable form is methylmercury (MeHg) which is the toxic form (Fig. 4.1).

Mercury (Hg) is a heavy metal with a known impact on the environment and on human health. It is distributed differently in geographic regions and it is released to the atmosphere from a variety of point and diffuse sources both natural and anthropogenic. Air disperses the metal in a long distance from the original source. Due to precipitation processes, it contaminates the earth's surface and water [1] (Fig. 4.2).

Mercury exists naturally in the environment. Additionally, it is produced in high levels from hydroelectric, mining, pulp and paper industries, incineration of municipal and medical waste and from power plants, especially those using coal [1]. While the main source of human exposure is inorganic mercury due to occupation (dentistry, mining, artisans manipulating mercury-containing materials), the general population is mainly exposed to organic mercury by consumption of fish and to a lesser degree shellfish and other marine animals as well as dental amalgams.

Since the industrial revolution, the concentration of organic mercury has increased by at least three-fold in the ocean. Exposure is widespread in the US and Mexican populations [2]. It has been found in potable groundwater in the United States. It has also been reported in polluted surface waters in Canada, China, Ghana, India, Japan, Romania, and Sweden at levels that could cause adverse health effects. The largest anthropogenic source of mercury emissions in the USA comes from

A. Salavoura, *Chemical Environmental Pollutants and their Effect on Health*,
https://doi.org/10.1007/978-3-031-80271-3_4

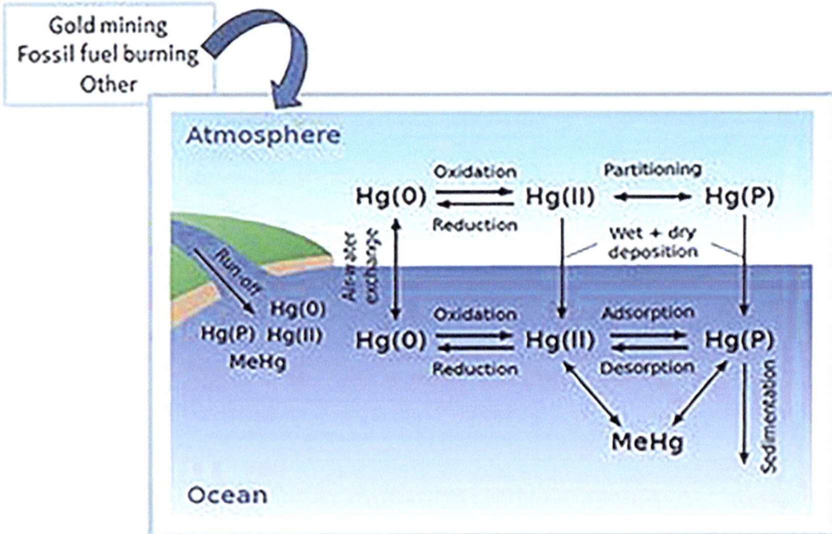

Fig. 4.1 Transformation of mercury in its three stable forms, elemental, mercury II oxide, and methylmercury

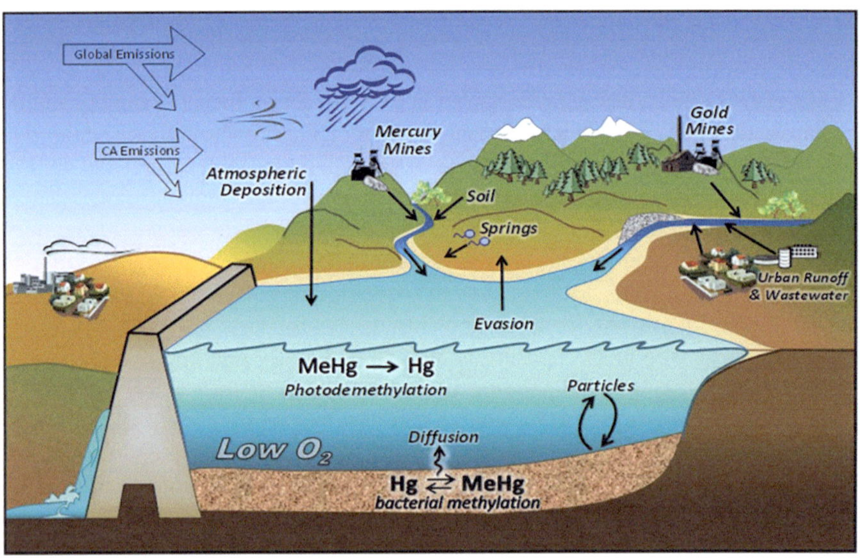

Fig. 4.2 The life-cycle of mercury in the environment

coal-burning power plants; other sources include byproducts of chlorine production, dental fillings using amalgams, thermometers, broken fluorescent bulbs, and batteries [3].

Forest ecosystems serve as important reservoirs of Hg. The metal enters the aquatic systems showing a high mobility in the ecosystem. Vegetation plays an important role in the capture of atmospheric Hg within the leaves of the trees which after foliage and they are deposited on organic layers on the forest floor. Mercury (Hg) is a persistent element not degradable in the environment, being bioaccumulated and biomagnified in biota through the food chain. Microbes in soils and in river and lake sediments convert elemental mercury into organic methylmercury, which is more bioavailable form. The biological transformation of mercury in the soil to methylmercury happens by saprophagous detritivores. Methylmercury concentrations accumulate in different food chains, being highest in fish species at the top of the food chain, such as pike, walleye, and bass in freshwater, and tuna, swordfish, and shark in the sea water [4].

Millipedes in the landfills are used for the detection of the concentration of the metal in litter due to a tendency to mercury toxicity [6]. Litter is an environmental compartment of great importance for ecologists and ecotoxicologists, because it represents the main food source for saprophagous detritivores that are essential in organic matter fragmentation and decomposition. They are also an important food source for a large number of predators in the soil food web [5].

When ingested, almost all of the methylmercury is absorbed. Most methylmercury is converted to inorganic mercury and it is excreted to the feces. In humans, the half-life of non-organic mercury is about 44 days. When inhaled, Hg is oxidized in the red blood cells and lungs to divalent inorganic cation (Hg^{+2}), while inhalation of Hg vapor mainly deposits in the kidney and brain [1].

Hg concentration in urine is the best biomarker of chronic exposure to organic and inorganic Hg, as urinary Hg levels derive directly from deposits in the kidney tissue. In addition, testing the mother's hair or blood for mercury levels may be indicated if she had consumed fish with high mercury levels.

Accidental Exposures in History

The prominent example of a serious poisoning with methylmercury was described in the victims of the Minamata in Japan. In the small town of Minamata, on the southern area of the island of Kyushu, a large chemical factory employed a significant amount of inorganic mercury as a catalyst for the production of the raw ingredients of plastic. Some of this mercury was discharged into Minamata Bay. The mercury was deposited in the sediment of the bay and subsequently entered the food chain. The victims were childbearing women, because the metal crosses the placenta. Newborns developed serious nervous abnormalities. After the accident, 1422 infants, who appeared healthy at birth or with minor problems in sucking and swallowing, were later found to develop mental retardation, disturbances of gait, speech, and abnormal reflexes. Mothers were largely asymptomatic [4].

The hazards of inorganic mercury poisoning have been known since Roman days, and the dangers of working in the mercury mines of Almedon, Spain, have been described in literature. The employees suffered from a disease called "mad as a hatter," a name given previously in people working in the tanning industry as hatters in Victorian England, who also suffered from neurological symptoms. The symptoms included tremor, irritability and behavioral disturbances. The symptoms were reversible upon removal from the exposure source [4].

In 2006, WHO changed the baseline safety limit in drinking water for mercury exposure from 1 μg/L of total mercury to 6 μg/L of inorganic mercury. The guideline was based on the toxicity of methylmercury. Regarding methylmercury, WHO stated "it is unlikely that there is any direct risk of the intake of organic mercury compounds, and especially of alkyl mercurial, such as methyl- and ethyl mercury compounds, as a result of the ingestion of drinking-water" [3].

Effects on Health

Methylmercury crosses the blood–brain barrier (BBB) causing a dramatic presentation from the nervous system. Patients initially experience a constricted visual field, paresthesia, and gait ataxia. Clinical examination shows impairment of the vibration sense (stereognosis) and the two-point discrimination. Subsequently, patients experience a wide range of visual loss progressing to total blindness, ataxia, and sensory-motor dysfunction symptoms.

Cytology shows neuronal destruction. In cytopathology, lesions of methylmercury poisoning are quite distinct, with acute necrosis of the primary visual center called calcarine, pre-central cortex, and the cerebellum. Nerve cell loss is severe with a profound glial response. In the cerebellum, the neuronal loss predominates in the internal granular layer sparing the Purkinje cells.

Although blindness is a common complication of the intoxification with methylmercury, the retina appears to be intact, and the loss of vision is thought to be directly related to destruction of the primary visual cortex. In patients with prolonged survival after exposure to methylmercury, dramatic atrophy and glial scarring of the above-mentioned regions is observed. The above findings are observed in people exposed to extremely small doses of methylmercury also [1].

Another complication of the high levels of mercury in blood is dilated cardiomyopathy. High levels of mercury in blood reach a thousand elevation in the heart tissue and they are detected in sporadic cases of cardiomyopathy. Dilated cardiomyopathy has been described in Greenlanders with high mercury levels in their blood, presumably from eating contaminated seal meat. It is believed that mercury ions may act as calcium antagonists at the actin–myosin junction, inhibiting sarcomere contraction, or that they may disrupt microtubule structure. This latter observation appears to be duplicated, but the mechanism of potential mercury-induced cell damage has yet to be clarified [1].

In addition, recent research has shown that chronic exposure to low concentrations of mercury causes renal, visual, auditory problems in addition to endocrine, reproductive, and immune problems. There is considerable genetic variation in susceptibility to mercury intoxication.

Evidence that mercury promotes autoimmunity appears to be more straightforward in mouse models, where it stimulates autoantibody nuclear antigen (ANAs) and it induces immune complex-mediated lupus nephritis. Developmental exposure to subclinical doses of $HgCl_2$ in maternal drinking water in mice has been shown to cause adverse immune effects in offspring. This includes increased number of activated $CD4^+$ T cells and increased levels of brain-reactive antibodies [6].

The toxic-form organic methylmercury easily crosses the blood–brain barrier and the placental barrier [4]. Environmental contaminants are associated with numerous adverse pregnancy outcomes including infertility, pregnancy loss, and preterm delivery. Regarding exposure to mercury, maternal exposure may contribute to increased risk of spontaneous abortion and preterm delivery. Through epigenetic changes, tissue-specific pathophysiology in the placenta and uterus, including the cervix could contribute to the onset of labor by regulation of uterine contraction and quiescence [7].

Epidemiological studies have found that prenatal exposure to mercury in utero is associated with poor cognitive development, learning, language development, attention deficit problems, sensory/motor dysfunction, low IQ, and motivation behavioral disorders in children, data which have replicated in mice [2]. The presentation is within the autism spectrum disorder. Additionally, it is responsible for cardiovascular problems and possibly affects fetal growth [2].

Epigenetic Mechanism of Intoxification

The most significant mechanism of the effects due to exposure to organic and inorganic mercury in the cell is oxidative stress. Oxidative stress is mainly due to the pathology of the metabolism of glutathione in addition to the increase or decrease of methyl donors [6]. It has been shown that mercury exposure also inhibits other antioxidant enzymes apart from glutathione-S-transferase (GST), which include glutathione peroxidase (GPx), superoxide dismutase (SOD), and catalase (CAT) contributing to the increase of oxidative stress in the cell [1].

Additionally, in vitro studies in human neurons have shown the direct toxicity of methylmercury to the brain. However, the exact mechanism of neurotoxicity in humans is not defined. There are detected direct effects on sulfhydryl groups in pericytes and cerebral endothelial cells, accumulation of the metal within astrocytes, microglial activation, induction of chronic oxidative stress, activation of immune-inflammatory pathways, and impairment of mitochondrial functioning [8].

Moreover, prenatal MeHg exposure disrupts calcium homeostasis, apart from the generation of oxidative stress, due to altered glutamate homeostasis. Oxidative DNA damage is the result of reduced DNA methyltransferase activity, which results in altered methylation patterns of cytosine residues at CpG sites. Both As and Hg can alter DNA methyltransferase activity suggesting the interference in the methylation patterns as a potential mechanism for epigenetic disruption. Mercury has been shown to have strong inhibitory action in the methionine synthase enzyme crucial for SAM regeneration. Thus, the oxidative stress action of both As and Hg, along with the possibility of mercury exposure depleting levels of SAM, suggests that co-exposure to these two common environmental contaminants can disrupt the epigenome synergistically [2].

Genetic factors may increase susceptibility to mercury, for example, polymorphisms in glutathione transferases, other glutathione-related genes, as well as in selenoproteins. Studies detected that mercury-containing adjuvants in vaccines generate inflammatory and immune responses which are strongly influenced by polymorphisms in the human leukocyte antigen (HLA) region and by genes encoding effector proteins, such as cytokines and pattern recognition receptors. However, results should be further confirmed [8].

Prenatal exposure to mercury detected in infant toenails correlated with hypermethylation of CpG islands of cord blood leukocytes [8]. In the literature, methylation patterns after exposure to mercury have been investigated in gold miners, who are highly exposed to mercury, due to the processes to separate gold. It was found a higher Hg content in different tissues from artisanal gold miners when compared to control subjects [9]. Developmental exposure to low levels of methylmercury induces epigenetic suppression of BDNF (Brain Derived Neurotrophic Factor) gene expression in the hippocampus and predisposes mice to depression [6].

Regarding the global methylation profile, a significant decrease in DNA methylation of *Alu (Yb8)* in peripheral blood was found with *Yb-lineage* being one of the young groups of evolutionarily-related *Alu* subfamilies in humans. *Alu (Yb8)* hypomethylation has been associated with tumorigenic events. In contrast, and contrary to expectations, it was found that *LINE1* methylation areas rich in AT elements had significantly increased methylation levels among the miners, and methylation was correlated with *8-OHdG*, which is one of the major products of DNA oxidation and Hg levels. Given the nature of *Alu* and *LINE1* elements, methylation may be differentially regulated in these regions after exposure to the metal. Additionally, it has been shown that different transposable elements could be differentially methylated in response to exposure to the metal within the same experimental system [9].

Regarding the change of methylation patterns of genes, DNA methylation changes were found in genes of the antioxidant response (*SEPP1* and *SOD2)* and base excision repair (BER) pathway (*XRCC1* and *OGG1*) genes. It has been documented that hypermethylation both in *8-Oxoguanine glycosylase (OGG1)* and *in X-ray repair cross complementing 1 (XRCC1)* CpG islands may be associated with oxidative stress conditions.

Other studies show that the decrease of *OGG1* mRNA and protein was linked with hypermethylation of the first exon CpG island of the gene in age-related cataract patients, where oxidative stress and ROS play a critical role in pathogenesis. Suppressed expression of *XRCC1* and *OGG1 mRNA* and proteins was detected during cadmium-induced malignant transformation of human bronchial cells in association with promoter CpG island hypermethylation. Similar changes regarding *XRCC1* methylation were positively correlated with Hg exposure also [9].

Hanna and collaborators have shown an increased methylation in the *promoter of GSTM1 (Glutathione S-transferase* Mu 1) gene, when measured in women with high blood Hg levels. Another study correlated DNA hypomethylation of the *selenoprotein P plasma 1 (SEPP1)* promoter, an important enzyme involved in the antioxidant response of the cell as the previous study in gold miners. *SEPP1* promoter hypomethylation was found in male dental professionals associated with *SOD*2 hypermethylation. *SEPP1* hypomethylation induces SOD2 downregulation impairing redox signaling and creating a proliferative, apoptosis-resistant state as the one seen in carcinogenesis [10].

Moreover, another study showed that chronic exposure to MeHg (10 μM) causes epigenetic landscape modifications of histone H3K4 trimethylation (H3K4me3) marks in *Caenorhabditis elegans*. The modifications correspond to 1467 loci of genes with enhanced and 508 loci of genes with reduced signals. Among enhanced genes are those encoding glutathione-S-transferases, lipocalin-related protein, and a cuticular collagen. In utero, exposure to Hg in mice enhanced marks without alterations in mRNA expression except for the Lrp-5 gene. Biomarkers and data from the above studies could help prevention as well as risk assessment for the exposure to different levels of mercury [11, 12].

Chromium

The Metal, Its Forms, and Its Distribution in the Environment

Chromium is one of the most abundant elements in the Earth's crust, with an average concentration of 125 mg/kg. It has multiple oxidation states, ranging from Cr (0) (elemental chromium) to Cr (VI) (hexavalent chromium). The most common as well as stable form of chromium is the trivalent form, Cr (III), which occurs naturally in chromite ore and it is used in multiple manufacturing processes. The next most stable form is hexavalent chromium (CrVI). Cr (III) and Cr (VI) differ not only in their oxidation state, but also in their chemical properties and their toxicity [13].

According to the US Environmental Protection Agency (EPA), chromium is released into the environment from industrial sources such as electrochromic plating facilities, leather tanning facilities, and coal-fired power plants. Any type of stainless steel or chrome plated object, such as cookware, sink faucets, or automobile parts, are coated with chromium to improve the appearance and add a layer of protection from wear and tear [14].

Environmental pollution with chromium happens by natural and anthropogenic sources. Human activities that are related with various forms of Cr exposure include its numerous uses in the chemical industry, production of dyes, wood preservation, leather tanning, chrome plating, manufacturing of various alloys, and many other applications and products. Stainless steel contains up to 20% Cr by weight and it is the main industrial product that contains the metal [15].

The magnitude of Cr utilization is also evident from the estimated occupational exposure by more than 500,000 workers in the USA alone. Incineration and emissions from cars create an ambient pollution with small Cr (VI) and Cr (III) containing particles, which leads to a low-level inhalation exposure by a significant part of the general population and it increases Cr levels in surface waters [15].

Serious cases of anthropogenic contamination in the USA came from the discharges of toxic Cr (VI) by cooling towers. Large-scale environmental pollution with Cr (VI) involves improper disposal of millions of tons of incompletely processed chromite ore. Hundreds of the largest toxic waste sites in the USA known as superfund sites contain Cr as a major contaminant. Efforts by the chromium-industry have been made to suggest that contrary to research, Cr (VI) is not mutagenic. However, numerous epidemiological studies have reported a high incidence of lung cancer among workers exposed occupationally to Cr (VI) by inhalation [14]. In the public's interest, communication of the side effects of these processes should be forwarded [16].

Trivalent chromium Cr (III) is part of human diet by either supplementation-containing compounds and contaminated drinking water. Cr (III) is widely used in Western societies as a component of many multivitamins, nutritional supplements, and it is even present in foods. Predominant forms of Cr (III) that are taken as supplementation include chromium picolinate, chromium-histidinate, chromium dinicocysteinate, and niacin-bound chromium. Chromium picolinate is the most prominent form of Cr (III) in nutritional supplementation, because this form allows optimal absorption [16].

The Council for Responsible Nutrition (CRN) published data in 2017 revealing that 170 million adults in the USA take some form of nutritional supplementation as a multivitamin, a specialty supplement, herbals or botanicals, sports nutrition supplements, or weight management supplements. In 2016, chromium (III) supplements ranked as the fourth highest selling supplement in the USA, grossing 110 million dollars that year, just behind calcium, magnesium, and iron [16].

Dietary intake of Cr (III) has been estimated to be within the range of 23–29 and 39–54 g/day for women and men, respectively. These limits fall within and sometimes exceed the adequate intake daily values (25 g/day for women and 35 g/day for men) established by the Food and Nutrition Board of the National Academics for Sciences Engineering and Medicine. Thus, additional uptake of Cr (III) with supplements supersedes the daily established limits.

According to some studies, nutritional supplementation is beneficial for diabetics and both in the process of weight loss and muscle anabolism. In particular, supplementation of chromium has been described to have positive effects in regards to the potentiation of insulin action, and to increase the beneficial results from exercise including higher percentage of lean muscle mass and loss of fat. However, the evidence is weak, conflicting, and some studies even suggest that excessive intake of Cr (III) formulations is carcinogenic. Altogether, Cr (III) is efficiently gained by normal diet and supplementation has no nutritional benefit. On the contrary, supplementation could be a risk factor for the development of cancer [16].

Public health concerns are centered on the contamination with Cr VI that is classified as a known human carcinogen mainly by inhalation. In addition to inhaled chromium from natural or anthropogenic sources, ingestion of Cr VI could be also harmful. The main source of intoxication by ingestion is the contaminated drinking water [15].

Drinking water supplies in many geographic areas contain chromium in both forms III and VI. Cr (VI) has a high environmental mobility. Acidic environments with high organic content promote the reduction of Cr (VI) to nontoxic Cr (III). The opposite process of Cr (VI) formation from Cr (III) also occurs, particularly in the presence of common minerals containing Mn (IV) oxides [15].

The presence of Cr (VI) in drinking water can also result from the oxidation of naturally occurring Cr (III) by Mn (III/IV) oxides in birnessite, a common mineral that coats weathered grains and fractures in Cr-rich ultramafic rocks and serpentinites, which are frequently enriched by chromite [FeCr (III)2O4]. In addition to birnessite, the presence of two other Mn (IV) oxide-containing minerals, asbolane and lithiophorite, has also been associated with the formation of Cr (VI) from natural Cr (III). Examination of four minerals made of Mn oxides (birnessite, cryptomelane, todorokite, and hausmannite), showed that birnessite had the highest ability to oxidize Cr (III) under laboratory conditions [15].

Animal waste contains heavy metals such as Cu, Zn, Mn, As, Cd, Pb, Cr, and Ni and halogens such as Br. Thus, soils submitted to successive applications of waste over several years may cause excess accumulation of these elements. Additionally, Cr in the wheat crop showed higher translocation to the grains, which represented 71% of the total content of the plant [19].

Cr (III) could serve as a nutritional supplement, whereas Cr (VI) is a very toxic agent inducing a wide variety of injuries in the cells, such as: DNA damage, chromosomal aberrations, alterations in the epigenome, and microsatellite instability [13]. Another difference between Cr (III) and Cr (VI) is their ability to enter cells, which is the cause of their difference in toxicity. Cr (III) compounds cannot enter cells by any transport mechanism, but only by gradual diffusion. On the contrary, Cr (IV) has the ability to enter the cell wall by anion transporters giving a rate of transportation which is 1000 times higher compared to Cr (III) [13].

Cr (VI) enters the cell by anion transporters and it undergoes a series of metabolic reductions to form intermediate Cr species, Cr (V) and Cr (IV) [13]. Once inside the cell, Cr (VI) undergoes a rapid metabolic reduction to convert ultimately to Cr (III) which is the primary mechanism of detoxification [13]. Hexavalent chromium in aqueous environments predominantly exists as the chromate anion ($CrO4^{-2}$). Chromate closely resembles the sulfate and phosphate anions ($SO4^{-2}$, $PO4^{-2}$, respectively) in structure, and therefore it can readily pass through the cell membrane by anion transporters. Chromate, the predominant form of Cr (VI) at neutral pH, is taken up by all cells by sulfate channels [14].

Intracellular reduction of Cr (VI) is facilitated by a number of non-enzymatic and enzymatic antioxidants. Ascorbate (Asc), glutathione (GSH), and cysteine (Cys) are three reducing agents for Cr (VI), but the major pathway of inactivation is the ubiquitously present ascorbate and small thiols [13].

Unlike Cr (VI) uptake, ligand-bound Cr (III) is postulated to enter the cell by phagocytic mechanisms or by non-specific mechanisms of diffusion. It is estimated that Cr (III) diffusion is about 1% and most ingested chromium Cr (III) is being excreted in the feces. Urinary excretion of Cr (III) is inversely related to Cr (III) intake and the excretion rate increases under different physiological conditions, such as stress and during periods of exercise [14].

The intracellular reduction of Cr VI depends on the cellular availability of different antioxidant molecules as well as their reaction rate. Although ascorbic acid and glutathione (GSH) have similar concentrations in vivo (approximately 1–3 mM), early studies show that ascorbic acid is the kinetically favored reducing agent and it accounts for 80–90% of Cr (VI) reduction in vivo. Depending on the nature of the reducing agents in the cell, Cr (VI) undergoes either one- or two-electron reductions. Ascorbic acid reduces Cr (VI) by a two-electron reaction forming the reduction intermediate, Cr (IV). Reduction of Cr (VI) by GSH can be either by one- or two-electron reactions that produce Cr (V) or Cr (IV). Reduction by cysteine (Cys) is almost exclusively a one-electron reaction. A combined activity of ascorbic acid, glutathione, and cysteine in cells reduces more than 95% of Cr (VI) into Cr (III) [13].

Interestingly, in vivo studies have found that ascorbic acid is the predominant scavenger to reduce Cr (VI), while in contrast, in vitro studies demonstrate that glutathione is the main interactor [15]. However, adding ascorbate into the cell culture medium in vitro, although it decreases overall oxidative stress, it induces DNA double-strand breaks by the formation of ternary DNA adducts containing Cr (III) crosslinked with histidine, cysteine, ascorbate, or glutathione. These different Cr-containing DNA adducts are not easily repaired. They are the main contributor to Cr-induced malignant cellular transformation [14].

Apart from the intervention of antioxidant mechanisms to reduce the toxicity of Cr (VI) and derivatives, other intracellular reducing agents including cytochrome P450 reductase in mitochondria, mitochondrial electron transport complexes, glutathione reductase, aldehyde oxidase also contribute to detoxification. However, these enzymes are only minor players in the intracellular Cr (VI) reduction [13] (Fig. 4.3).

The major extracellular reduction of Cr (VI) occurs in the gastrointestinal system. After ingestion, Cr (VI) can be reduced to Cr (III) by body fluids including saliva and gastric juice and further sequestered by intestinal bacteria. The main reducing agent, gastric juice, has a relatively high reducing capacity, ranging around 8 mg/L (fasting) to 31 mg/L (fed). Considering the large number of fluids secreted in the stomach daily, about 1000–1500 mL/day (fasting) plus 800 mL/meal (fed), the body has a large capacity to convert Cr (VI) to Cr (III) (around 80 mg/day). Thus, extracellular Cr (VI) reduction, primarily in the stomach, has been considered a protective mechanism accounting for the low genotoxicity and carcinogenicity in animals exposed to Cr (VI) by drinking water. However, recent animal studies indicate that a portion of Cr (VI) is able to escape filtering by the gastrointestinal system [13].

Unlike Cr (VI) uptake, ligand-bound Cr (III) is postulated to enter the cell via phagocytic mechanisms or through non-specific mechanisms of diffusion. It is estimated that Cr (III) diffusion is about 1% and most ingested chromium Cr (III) is being excreted in the feces. Urinary excretion of Cr (III) is inversely related to Cr (III) intake and the excretion rate increases under different physiological conditions, such as stress and during periods of exercise [14].

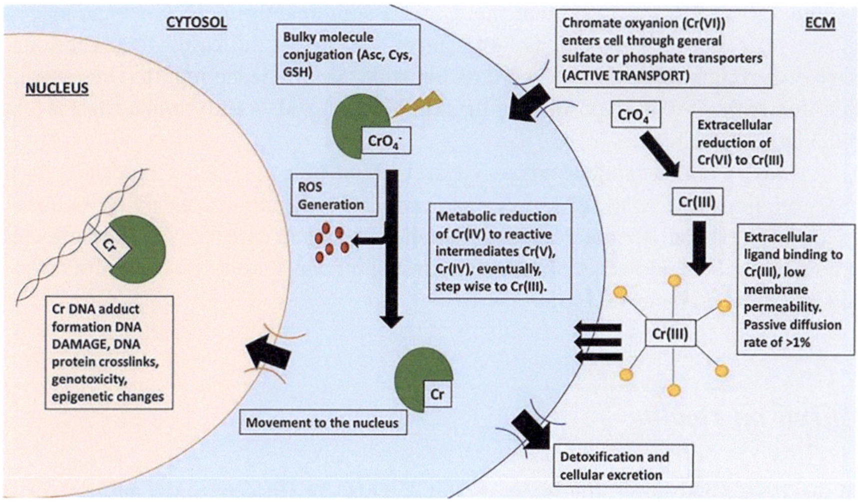

Fig. 4.3 Oxidative state products of chromium and the distribution in extracellular, cytosol, and nucleus space

Accidental Exposures in History

A well-known example of intoxication with chromium from drinking water was in Jersey City, NJ. This area once had large chromate refineries to produce Cr VI from ore that was shipped into Jersey City. The chromate ore residue after refining was transformed to attractive brown rocks that were used for landfill throughout Jersey City and the surrounding areas. A substantial portion of Jersey City contained many feet of landfill from these brown rocks. Water passing over the rocks extracted yellow Cr VI and spread it throughout Jersey City resulting in substantial human exposure to the carcinogenic metal. The contaminated water seeped into basements and destroyed brick and cement walls. Buildings constructed on these mine tailings sometimes collapsed. In the last several decades, many millions of dollars were spent to clean the chromate-contaminated soil and water from Jersey City by digging up and moving the toxic metal to toxic waste dump sites. It was fortuitous that the hexavalent chromate was converted to the nontoxic trivalent chromium in organisms and plants, because it was almost safe to eat vegetables, fish, and games [16].

The most known pollution of chromium leading to disease of the population has been reported in China. Even though inhaled Cr VI is incriminated for lung cancer, a study of the residents of small villages in China, in the Liaoning province, reported an increase of stomach cancer mortality in the area due to elevated Cr (VI) concentrations in drinking water [13].

Several areas of Greece have also suffered the consequences of drinking Cr (VI)-contaminated water. An ecological mortality study in the Oinofyta region of Greece, where the water was contaminated with Cr (VI) (maximum levels ranging between

41 and 156 μg/L), indicated that there was a significantly increased incidence of liver cancer mortality as well as lung cancer, cancers of the kidney and genitourinary organs among women. A more recent study from Greece reported an association between Cr VI exposure in drinking water and some hematological and biochemical parameters [13].

A study in India evaluated adverse health effects on a population exposed to high concentrations of Cr (VI) (20 mg/L) and reported a slightly increased incidence of gastrointestinal and dermatological complaints in an area with Cr (VI)-contaminated groundwater [13]. However, there were limited human studies on the health effects of Cr (VI) when ingested [13].

Effects on Health

In the 1989 assessment, the International Agency for Research on Cancer (IARC) has classified Cr (VI) as a human carcinogen when the exposure is by the inhalation route. It was considered as one of the most active agents in mutation assays and in other assays of genotoxicity [13]. Chronic Cr (VI) exposure causes tumors of the lung, gastrointestinal, and central nervous systems [17].

In the workplace, both OSHA (Occupational Safety and Health Administration) and NIOSH (National Institute for Occupational Safety and Health) have set standards for workday exposure by the air to the three predominant forms of chromium. OSHA standards are Cr (0): 1.0 mg/m^3, Cr (III): 0.5 mg/m^3, and Cr (VI): 0.005 mg/m^3. NIOSH standards are 0.5 mg/m^3 for chromium metal and Cr (III). For Cr (VI) the standard is 0.001 mg/m^3. In 2008, the National Toxicology Program (NTP) released a 2-year rodent study on oral exposure of Cr (VI) in drinking water, indicating that Cr (VI) was a carcinogen when ingested with an estimate of the cancer potency to humans equal to 0.5 (mg/kg/day). The United States EPA has set a water standard for chromium species of 0.1 mg/L for total chromium [16].

Epigenetic Mechanism of Intoxification

Regarding direct toxicity, Cr (VI) itself does not bind to DNA or other macromolecules in cells. Instead, its metabolic intermediates Cr (V), Cr (IV), and the final product Cr (III) are highly reactive and readily form Cr-DNA adducts. Cr (III)-induced DNA adducts can be repaired rather efficiently by nucleotide excision repair mechanisms within minutes after exposure. However, bulky Cr-DNA lesions are not well repaired, they inhibit the repair machinery, and they contribute to the development of persistent mutations [13].

Cr-DNA adducts are a heterogeneous group that includes binary [Cr (III)-DNA] and several ternaries [ligand-Cr (III)-DNA] adducts, where the ligand can be ascorbic acid, glutathione, cysteine, or histidine. All four ternary adducts have been detected in Cr (VI)-treated cells and they are readily formed during in vitro Cr (VI) metabolism. DNA cross-linking of glutathione, cysteine, and ascorbic acid was observed in their reactions with Cr (VI), while DNA attachment of nonreducing amino acid histidine occurred efficiently during the reduction of Cr (VI) with ascorbic acid [13]. Cr (III) concentration diffused within the cell is responsible for DNA damage and it is related with the genotoxicity of the metal. Other side effects from the metal in the long term are chromosomal aberrations, alterations in the epigenome and microsatellite instability [14]. Oxidative stress and free radical species, a prominent mechanism of the intoxication from other metals, it is not the main mechanism of intoxication with chromium [15].

In addition, studies have confirmed that Cr VI is responsible for lung cancer by epigenetic changes. Wang et al. reported global hypomethylation in chromate workers that have not developed tumors. Kondo et al. investigated *p16* hypermethylation using a methylation-specific PCR method in 30 lung cancer cases associated with chromate exposure and 38 non-chromate lung cancers. *p16* (also known as multiple tumor suppressor 1) is a protein that slows cell division by delaying the progression of the cell cycle from the G1 phase to the S phase. He showed that chromate exposure influenced *p16* hypermethylation measured in lung cancer tissues leading to gene silencing. Tissues from non-chromate lung cancer have not this methylation pattern.

Hypermethylation and subsequent silencing of three genes *WIF1* (WNT protein), *APC* (Adenomatous polyposis coli), and *MLH1* (mismatch repair enzyme) that are known tumor suppressor genes has been demonstrated. Silencing of the mismatch repair gene *MLH1* in lung cancers of the chromate workers, was associated with high replication error and microsatellite instability. It was reported also hypermethylation of the *APC* and *MGMT* gene promoters in chromate lung cancers compared with those in non-chromate lung cancers [13].

Inhaled exposure to Cr (VI) is associated with changes in various histone marks including decreased H3K27Me3 and increased levels of H3K4Me3 as well as H3K9Me2 and H3K9Me3. Chromium has been shown to reduce in vitro H3 phosphorylation and trimethylation, as well as various acetylation marks in H3 and H4 [13]. Interestingly, Cr appears to be linked with a positive feedback loop to a few interacting molecules that result in changes in the expression of the mismatch repair gene *MLH1*. The result is a heterochromatin state which decreases the expression of the *MLH1* gene limiting the DNA repair capacity of the cell. Exposure to Cr induces the expression of the histone methyltransferase G9a, which is responsible for adding posttranslational methylation to H3K9 [18].

Limited epidemiological evidence for Cr (VI) ingestion is suggestive of elevated risks for stomach cancers. Exposure of animals to Cr (VI) in drinking water induced tumors in the alimentary tract, with linear and supralinear responses in the small intestine. The most abundant form of DNA damage induced by Cr (VI) is by the formation of Cr-DNA adducts, which cause mutations and chromosomal breaks similarly with lung cancer precursors.

Colorectal cancer in association with carcinogenic metals is a matter of concern. The disease is one of the most common neoplasias in western countries and the second leading cause of cancer-related death. Data from the United States Cancer Statistics show that Kentucky has the second highest colorectal cancer incidence in the USA compared to other states, especially in the Appalachian region (USCS 2011). In this area, the concentrations of carcinogenic metals reported in drinking water are relatively high compared with EPA standards [17].

Although As (III) and Cr (VI) are believed to act by very different mechanisms, they share several properties in regard to their carcinogenicity. Both can activate NADPH oxidase, which is a major source of cellular hydrogen hyperoxide and which is able to induce oxidative stress. Oxidative stress plays an important role in both the initiation and the progression of various types of cancer [17].

NOX1, a member of the NADPH oxidase family, that is highly expressed in colonic epithelial cells, plays a pivotal role in cell signaling, cell growth, angiogenesis, and motility that integrates Wnt/β-catenin signals. It has been reported that NOX1 overexpression in colon tumors could contribute to the development of colorectal cancer.

In addition, the expression of several important antioxidant enzymes, such as SOD and catalase, were decreased in mice exposed to carcinogenic metals in drinking water. The level of *8-OHdG*, which is an oxidative DNA damage marker, was also increased, indicating that oxidative stress was involved in carcinogenic metals-induced tumor promotion effect [17].

In experimental models, the expression of β-catenin and phospho-GSK was increased after treatment with arsenic and chromium in combination or alone. Other carcinogenic metals have the same implications to that part of the antioxidant machinery [17].

Beta-catenin, a central molecule of the Wnt-signaling pathway, is known to be involved in the tumorigenesis of various gastrointestinal cancers, such as gastric cancer and colon cancer. In the absence of Wnt stimulation, this pathway is regulated at least in part by a multiprotein complex consisting of axin, GSK3β, and the tumor suppressor protein APC. Formation of this complex facilitates the phosphorylation of critical residues by GSK3β in the NH2 terminus of β-catenin, leading to degradation by the ubiquitin/proteasome pathway. Wnt stimulation shuts off β-catenin degradation by inhibiting GSK3β in the axin complex. This inhibition is believed to be the key event in the activation of the Wnt/β-catenin signaling pathway [17] (Fig. 4.4).

The Wnt signaling pathway is a group of signal transduction pathways, which begins with proteins that pass signals into a cell by cell surface receptors. Wnt signaling was first identified for its role in carcinogenesis, then for its function in embryonic development. The embryonic processes which it controls include body axis patterning, cell fate specification, cell proliferation, and cell migration. These processes are necessary for proper formation of important tissues including bone, heart, and muscle. As we have mentioned, the role of Cr in this process has not been clarified.

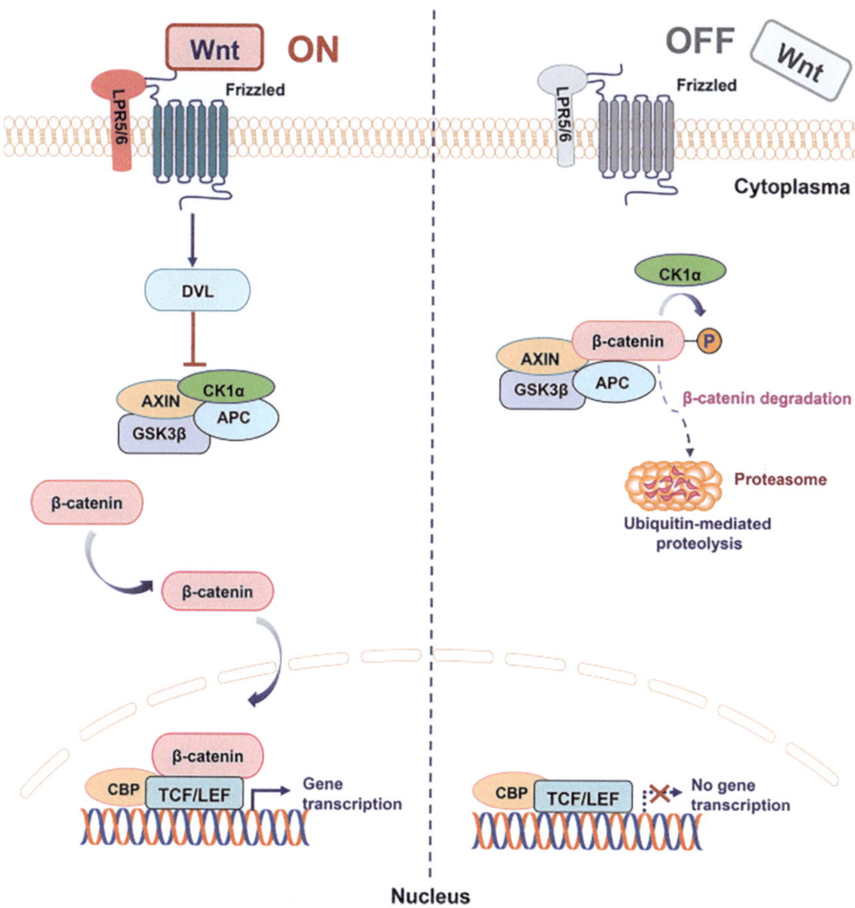

Fig. 4.4 The transcriptional inhibition by the Wnt/β-catenin signaling pathway

Arsenic

The Metal, Its Forms, and Its Distribution in the Environment

Arsenic (As) displays features of both a metal and a nonmetal, and it is called a metalloid. However, it is considered a heavy metal from a toxicological point of view. Arsenic is a chemical known as a poisoning agent since ancient ages. It is odorless and tasteless. Organic arsenic is less deleterious than its inorganic forms due to the fact that it is soluble in water and it quickly excreted with urine.

Arsenic is the twentieth most abundant element in the Earth's crust. In nature, arsenic can occur as a metalloid alloy as well as a variety of chemical compounds. In geological media such as rocks, arsenic is commonly found as a sulfide, such as orpiment (As_2S_3) or realgar (As_2S_2), in the form of arsenopyrite or mixed sulfides

(AsFeS). Significant amounts of arsenic may also be found bound to gold, silver, copper, lead, zinc, and cobalt ores. Mining of these minerals may result in the mobilization and/or transportation of arsenic into drinking water. Deliberate exposure to arsenic or exposure related to occupational or accidental exposure is possible [23].

However, the most frequent source of toxic exposure to arsenic is the drinking water. Water contamination happens in agricultural and mining areas. Arsenic is an active ingredient of herbicides, insecticides, and other pesticides. Arsenic is also used in a variety of industrial (e.g., manufacturing of solid-state detectors, glass, and tinning of woods) and medical applications (e.g., drugs and medical treatments) [20].

Exposure to toxic metals, such as inorganic arsenic compounds (iAs), is reported in traditional Chinese and Western medicine for over 2500 years. In 1909, Ehrlich used arsenic for the treatment of syphilis. Arsenic compounds such as Arsphenamine (Salvarsan) were used as antibacterial agents, before the discovery of sulfa drugs and penicillin. Combined forms of phenylarsonic acid are still used for the treatment of parasitic infections. Arsenic trioxide (As_2O_3) was used in the 1970s for the treatment of acute promyelocytic leukemia (APL) with striking efficacy and a good safety profile. Furthermore, As_2O_3 has also shown promising results for the treatment of other malignancies such as adult T-cell leukemia/lymphoma (ATL) and NPM1 mutant acute myeloid leukemia [20].

Exposure to traces of the chemical could also occur from the consumption of certain foods. The contents of As, Cd, Cr, Cu, Ni, Pb, and Zn in soil after 10 years of cultivation and 14 applications of organic and mineral nutrient sources are below the environmental limits established by regulatory agencies around the world. However, the content of arsenic in wheat, black oat, and corn is approximately three times higher, while in rice, mustard, and amaranth, it is four to five times higher than other crops, when the soil is enriched with nutrients and metals. Contaminated water with arsenic could be a reason for exposure to the metal by consumption of these grains (Fig. 4.5) [21].

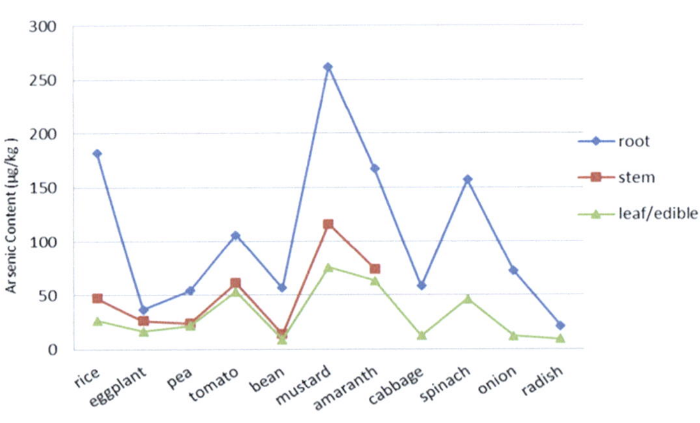

Fig. 4.5 Content of As in various plants when the soil is contaminated with arsenic

In the crops, arsenic may be an important factor of the enrichment, spread, and/or maintenance of antibiotic-resistant genes, and further investigation is required to find if metal toxicants contribute to the development of antibiotic resistance [21].

Naturally occurring inorganic arsenic (iAs) is present mainly in its sulfide form within complex minerals containing silver, lead, copper, nickel, antimony, cobalt, and iron. The greatest concern of arsenic pollution is still mostly present in developing countries, since industrialized Western governments have already adopted fundamental measures to prevent and to address arsenic pollution [22].

Arsenic is a well-known environmental pollutant in many countries, including Argentina, Chile, Bangladesh, India, Hungary, China, Taiwan, Nepal, Myanmar, Cambodia, Taiwan, Mongo, Vietnam, Pakistan, China, Afghanistan, Argentina, Mexico, Chile, and the USA. However, the Asian region is considerably more affected by arsenic pollution compared to other parts of the world [22, 23]. It affects roughly 140 million people in over 70 countries who are chronically exposed to arsenic by the consumption of drinking water containing high levels of the metal (World Bank 2005a, b) [24].

Since arsenic is easily soluble in water, it can contaminate local wells and public water supplies (NRC 2001). According to the World Health Organization (WHO) and US standards for drinking water, As level should be lower than 10 μg/L (WHO 2006; Pesola et al. 2012). Ingestion of inorganic As (iAs) by contaminated drinking water is associated with chronic diseases [22].

Arsenic can be oxidized, reduced, methylated, and demethylated by soil processes in which microorganisms play an important role. The methylated form of arsenic is widespread in nature. In soil, arsenic of herbicides and pesticides is methylated by microbes. The entry of arsenic-containing wastewater in the aquifer is the major source of water contamination [25].

The inorganic forms of arsenic both inorganic arsenite (As3+) and arsenate (As5+) are the predominant forms of arsenic, which transform to As $(OH)_3$ in an aqueous solution. After entering the cells through the aquaglyceroporin 9 (AQP9) transmembrane, arsenic interacts with the thiol (or sulfhydryl) groups (-SH) of proteins with a high cysteine content. The reaction constitutes the basic biochemical reaction of As (III) intracellularly, and it alters the conformation of the thiol protein leading to loss of function. Apart from the direct arsenic-protein binding, recent studies revealed that many key proteins are modulated by arsenic through more complicated post-translational stepwise regulations (Fig. 4.6) [23].

Although arsenic exhibits in both organic and inorganic forms, the inorganic trivalent arsenic compounds are considered to have the greatest toxicity [23]. Biomethylation and the enzymatic conversion of inorganic arsenic to mono-(MMAs) and demethylated (DMAs) species is considered the primary detoxification mechanism of inorganic arsenic, these products also manifest toxicity. Compelling experimental evidence suggests that biomethylation, particularly the production of methylated metabolites from the trivalent arsenic, is a process that could activate arsenic as a toxin and a carcinogen (Fig. 4.6) [20].

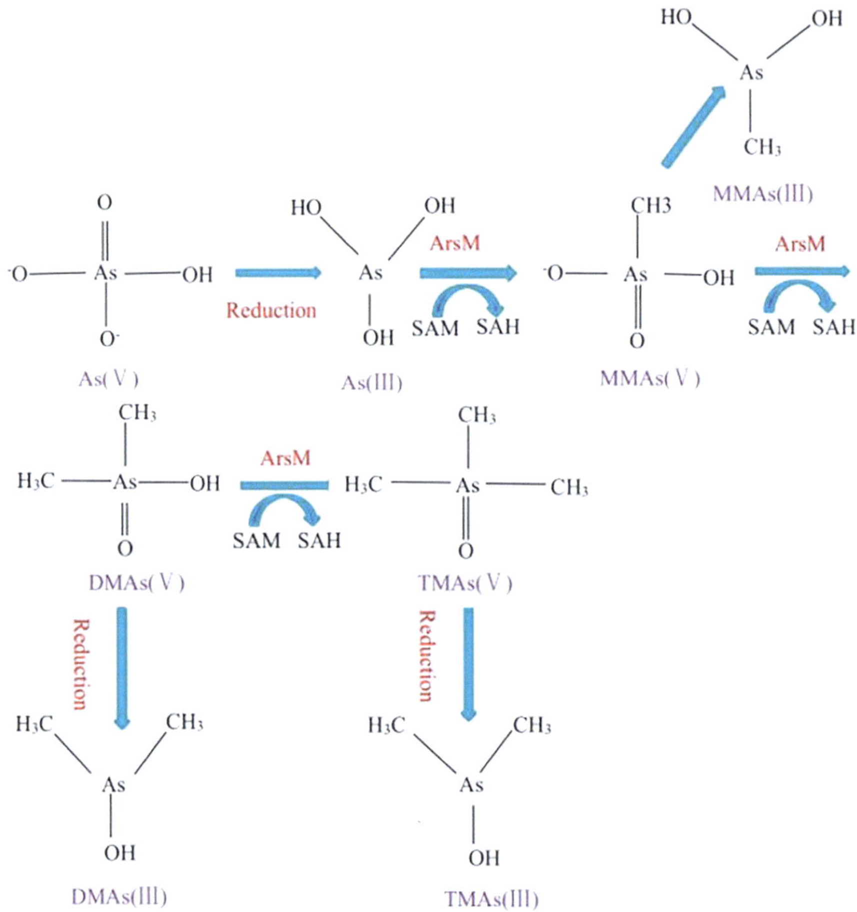

Fig. 4.6 Trivalent and pentavalent arsenic are metabolized to MMA (monomethylarsonate), DMA (dimethylarsinate), and TMA (trimethylarsenate) within the cell [22]

Accidental Exposures in History

Among the countries with higher pollution levels from arsenic, most studies have been conducted in Bangladesh. Survey data from 2000 to 2010 has shown that an estimated number of 35–77 million people in the country are being chronically exposed to As-contaminated drinking water, a fact described as the largest mass poisoning in history [24]. Groundwater containing arsenic is also responsible for the contamination of several food items, such as rice and vegetables, which contain the elemental form of the metal up to 96% of the total content in arsenic. Moreover, in West Bengal, India, it has been reported that cooked rice may contain high arsenic levels (200 μg/kg) that may be even related to genotoxicity [22].

Mazumder et al. conducted an epidemiological study in As-contaminated villages of the Nadia district, West Bengal, India, and they found a high prevalence of arsenicosis in approximately 15.4% in the population. It was estimated that around 84 million people were presumed to be exposed to arsenic in the district, and they projected that 14 million people might suffer from arsenicosis. The highest level of arsenic exposure, up to 1362 μg/L, was found in drinking water sources, and around 23% of measurements of the arsenic level in drinking water was more than 100 μg/L [23].

More than 20 million people are at risk of high arsenic exposure in areas such as the Tarim Basin in Xinjiang, the Ejina region in Inner Mongolia, the Heihe River in Gansu Province, the plains in northern China, Henan and Shandong provinces [24].

The inhabitants of Kamisu city in Japan were exposed to diphenylarsenic acid (DPAA), an organic arsenic compound, by drinking contaminated well water. The symptoms of DPAA poisoning were associated with a cerebellum and brainstem destruction that induced nystagmus, myoclonus, tremors, and cerebellar ataxia, as well as symptoms such as memory loss, sleep disorders, and visual impairment which suggests destruction of the temporal and occipital lobes of the brain. Some of the affected children also exhibited mental retardation [23].

Effects on Health

Arsenic is a naturally occurring ubiquitous metalloid and a Class I human Carcinogen. Approximately 200 million people around the world are exposed to unsafe levels. Arsenic levels can be measured in the blood, urine, hair, and nails. Urine is the most sensitive method for the detection of the metal, but only if measurements are done within the first days of exposure. Measurements of arsenic in the hair and nails can show persistent exposure during the last few months. To evaluate the risk of intake of a given chemical during the course of an individual's life, it is necessary to consider the period of intake, in addition to the daily limits considered safe.

Acute toxicity from arsenic could happen as a result of homicide or suicide; exposure to insecticides, herbicides or rodenticides; industrial exposure; exposure to naturally contaminated materials; or from treatment with arsenical medications. Acute arsenic and mercury intoxication could also result from overdose of Indian ethnic remedies and Chinese herbals.

The clinical features of acute poisoning include severe abdominal pain with muscle cramps, nausea, vomiting, diarrhea, a metallic taste, and extreme thirst, followed by stupor, coma, cardiovascular collapse and eventually death. Death could occur within 24 h of exposure, but with sublethal doses, survival is possible. Liver injury may arise 24–48 h after ingestion. The characteristics of the liver injury have not been well defined, but are likely similar to those related to acute iron poisoning. The characteristics are acute hepatic necrosis, marked elevations in serum aminotransferase levels, early onset of hepatic failure, and rapid recovery after clearance from the metal.

A severe progress of acute hepatic failure is responsible for early fatalities [24]. Other symptoms include conjunctival and respiratory tract inflammation, epistaxis, rashes, renal insufficiency, and painful neuropathy. Progressive peripheral neuropathy develops rapidly after 2–3 weeks. Damage is relatively acute following exposure and the extent of damage is proportional to the exposure dose. The acute arsenic toxicity may also result in toxic myocarditis [20].

The main symptoms of *chronic exposure* to arsenic occur on the skin. Arsenic tends to concentrate in ectodermal tissues, including the skin, hair, and nails. *Arsenical keratosis* is a well-established clinical entity. The lesions are usually most pronounced on the palms and soles, although they can occur on the trunk and on parts of the extremities. Arsenical keratoses are characterized by several specific pathologic features, including hyperkeratosis, parakeratosis, and acanthosis. They histologically show an absence of dermal solar elastosis. Within the spectrum of keratotic lesions, arsenical keratosis may be differentiated from the more commonly diagnosed actinic keratosis due to the absence of epidermal atrophy and basophilic degeneration of the upper dermis. All arsenical skin changes, including keratoses, tend to occur in non-exposed sites to the sun [20] (Fig. 4.7).

Dermal toxicity results from local tissue responses to direct contact of the metal with the skin or, alternatively, it may represent a manifestation of systemic toxicity following ingestion or inhalation. The occurrence of a "raindrop" pattern of hyper- and hypopigmentation has been reported in populations chronically exposed to arsenic [25].

Epidemiological studies have confirmed the role of arsenic in the induction of cancers of the skin. In mild cases, nuclear atypia is sometimes present. In severe atypia, squamous cells exhibit hyperchromatic nuclei and they keep a disordered arrangement within the epithelium. Squamous cell carcinomas in situ (Bowen's disease) and basal cell carcinomas of the skin have been associated with chronic inorganic arsenic ingestion with a latency period of 2–20 years after exposure [22].

Lip cosmetics have become one of the most commonly used cosmetics worldwide among modern women who want to moisturize their lips in order to appear attractive and charming. Nevertheless, some chemical additives in the lipsticks such

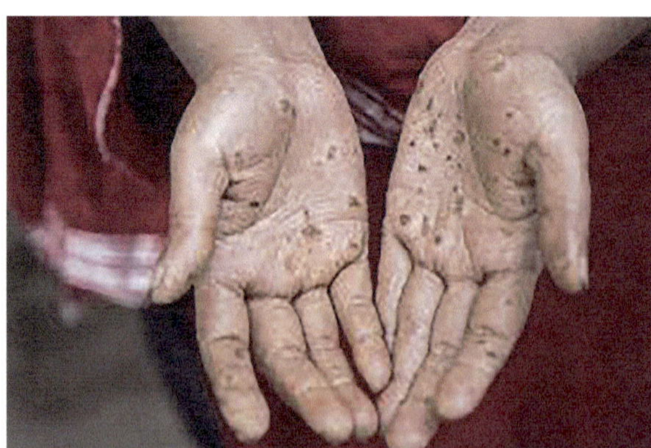

Fig. 4.7 Arsenic keratosis

as dyes, shine, natural mica minerals, and other colorants are incorporated during production and formulation of these products to increase the brightness of lip cosmetics and to booster effects. The usual additives are antimony (Sb), arsenic (As), cadmium (Cd), chromium (Cr), cobalt (Co), copper (Cu), nickel (Ni), manganese (Mn), and lead (Pb), which are considered toxic, when chronically applied. Arsenic is also detected in eye-shadows, eye pencils, shampoos, hair gels, and conditioners as well as showering products. Arsenic exhibits a high affinity for the skin, and therefore, the long-term contact could result in a variability of skin eruptions and in the long-term skin cancer [26].

Arsenicosis is a syndrome that develops after chronic marked exposure to arsenic and it presents as fatigue accompanied by nausea and vomiting, abdominal crampy pain, weakness, stupor, seizures, and neuropathy [26]. Chronic arsenic exposure may also result in pulmonary fibrosis, hepatic fibrosis, and bone marrow suppression (leukopenia and anemia) [20].

In addition, epidemiological studies have provided suggestive evidence that links chronic exposure to arsenic with tumors of the skin, bladder, lung, prostate, and the colon. Chronic Cr (VI) exposure causes tumors of the lung, gastrointestinal, and central nervous systems. As described, these two chemicals function synergically [23]. In the majority of cases, in which the internal cancer is ascribed to arsenic exposure, some dermatologic hallmarks of arsenic poisoning (such as hyper- or hypopigmentation) are identified [20].

Existing experimental data indicate that chronic arsenic exposure also affects the *nervous system* by impairing the nerve and brain tissues of exposed animals. Clinical studies indicate that chronic arsenic exposure leads to both central nervous system (CNS) and peripheral nervous system impairment. It also causes depression, memory impairment, and difficulty in problem solving, body coordination, etc. [21]. It has been stated that arsenic exposure causes numerous nervous system disorders, i.e., polyneuropathy, electroencephalographic (EEG) abnormalities, cognitive impairment, and, in extreme cases, disorientation, hallucinations, and agitation [21]. Apart from gradual impairment of the central and peripheral nervous system, related symptoms are a sense of burning and itching in the hands and feet, as well as hyper-pigmentation after exposure to the sun, hyperkeratosis in the hands and soles, white lines on the nails, liver damage, and cancer.

A prospective study, conducted in Bangladesh, found an increased risk of mortality from *ischemic heart disease* and cerebrovascular disease in populations exposed to high concentrations of inorganic arsenic (iAs) from drinking well water. Similarly, in other studies, increased risk of hypertension, stroke, and changes in cardiovascular (CVD) biomarker levels was observed. Among all the adverse health risks from arsenic exposure, CVD is regarded as the most serious health outcome [27]. The chronic exposure, such as occupational exposure, may cause chronic cardiomyopathy characterized by small vessel disease and an interstitial inflammation with fibrosis [20].

Chronic arsenic exposure is also related to *diabetes mellitus*. The mechanism underlying arsenic-associated diabetes is inhibition of insulin-dependent glucose uptake from tissues by inhibition of GLUT4 localized on membranes. Thus, the metal influences the glucose transport system in the cell. Moreover, it inhibits the pathways of Akt gene expression and antioxidant pathways. In addition, β cell damage due to apoptosis and to induction of chronic inflammation is another mechanism that could induce diabetes. Finally, stimulation of gluconeogenesis due to overexpression of phosphoenolpyruvate carboxykinase is another mechanism of development of diabetes due to chronic arsenic exposure (Fig. 4.8).

Furthermore, inorganic and methylated arsenic concentrations in urine, especially monomethylarsonate (MMA) and dimethylarsinate (DMA), were positively associated with the incidence of *nephrotoxicity*. Among the three arsenicals (iAs, MMA, and DMA), MMA showed a more potent cytotoxic effect on human urothelial cells in vitro. Moreover, recent evidence also indicates that developmental arsenic exposure has an impact on stem cells leading to renal carcinogenesis [25].

Investigations show that trivalent arsenic (As (III)) works as an immunomodulatory agent. Recent reports on animal models revealed the relation of As (III) to several *autoimmune and inflammatory diseases* [24]. As (III) interferes with many pathways of innate immunity. In macrophages, for example, nontoxic

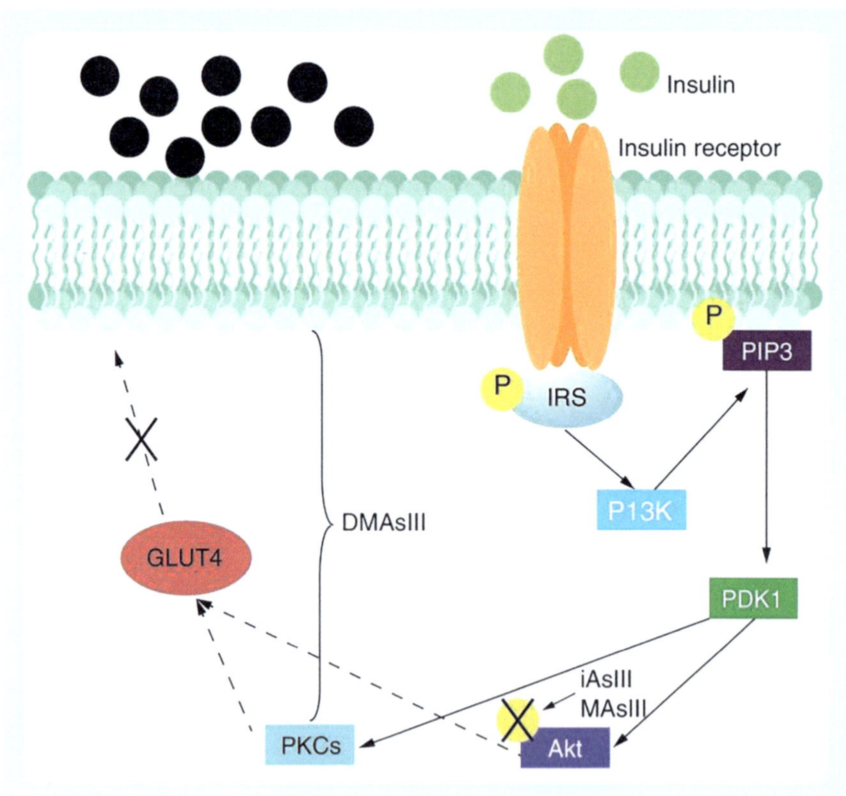

Fig. 4.8 Inhibition of Akt signaling due to mobilization of GLUT4 after exposure to arsenic

concentrations of As_2O_3 induce an up-regulation of 32 and a depression of 91 genes indicating a global multidirectional change of the cell-signaling. It should not be ignored that As (III) may inactivate up to 200 enzymes, many of which are crucial regulators of important signaling pathways. For example, As (III) inhibits IκB kinase (IKK), which is a key enzyme to the activation of the NF-κB pathway. Other affected pathways include the Rho Kinase/p38-kinase pathway, c-jun NH2-terminal MAPK, and others [24].

In neutrophils, high concentrations of As_2O_3 induce apoptosis of the cells by the generation of H_2O_2, de novo protein synthesis, caspase activation, and Syk kinase activation. The As_2O_3-induced de novo synthesized proteins include annexin 1 and heat shock proteins. Moreover, As_2O_3 induces endoplasmic reticulum (ER) stress, which elicits either self-protective mechanisms or cell apoptosis [20]. Additional action is authorized to eosinophils with an impact on the recruitment and chemotaxis. The impact on basophils and mast cells is decreased production of IgE and on monocytes is a downregulation of the Nf-κB-related pathway [26].

Regarding specific immunity, studies in T lymphocytes show that among 44 CpG islands, 75% of them exhibited higher methylation levels in the group exposed to high As compared to the group exposed to low As. DNA methylation on As exposure exhibits a linear, dose-dependent relationship [24].

The multiple consequences of arsenic exposure are illustrated in the following diagram (Fig. 4.9).

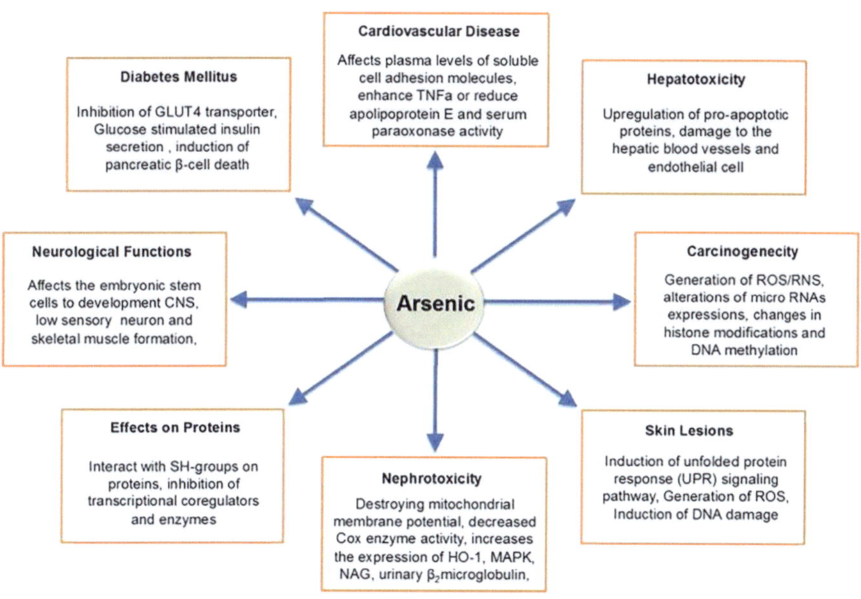

Fig. 4.9 The multiple interference of arsenic in functions of the body

Treatment

In the 1940s, dimercaprol, also called the British anti-Lewisite (BAL), was developed, and it is still used occasionally as antidote for severe arsenic toxicity. Dimercaprol and 2,3-Dimercaptosuccinic acid (DMSA) are the two most common chelators available in the USA. DMPS (2,3 Dimercapto propane sulfonate) may be the chelating agent of choice for arsenic according to several studies, but it has not been approved by the US FDA. DMPS and DMSA have a higher therapeutic index than BAL and they offer therapeutic advantages, as they do not redistribute arsenic to the brain. Chelation therapy is rarely indicated in less severe acute or chronic toxicity [28].

Mechanism of Intoxification

As (III) induces intracellular ROS accumulation by the inhibition of antioxidants, such as glutathione (GSH), glutathione peroxidase (Gpx), and glutathione reductase (GR), and the activation of ROS-generating enzymes, such as NADPH oxidase. As a consequence, expression of redox-sensitive genes, such as *HMOX1* (Heme Oxygenase 1), *NQO1* (NAD(P)H Quinone Dehydrogenase 1), and *GCLM* (Glutamate-cysteine ligase regulatory subunit), is up-regulated. Especially, nuclear factor erythroid 2-related factor 2 (Nrf2), a stress-activated transcription factor, responsible for inducing a battery of cytoprotective effects, is found to be activated by As (III) [24].

However, impairment of cellular respiration by inhibition of various mitochondrial enzymes and uncoupling of oxidative phosphorylation is one of the major mechanisms by which arsenic exerts its toxic effects. At the molecular level, the toxicity of arsenic results from its ability to interact with sulfhydryl groups of proteins and enzymes and to substitute phosphorus in a variety of biochemical reactions. In vitro experiments have demonstrated that arsenic reacts with protein sulfhydryl groups to inactivate enzymes, such as dihydrolipoyl dehydrogenase and thiolase, thereby inhibiting oxidation of pyruvate and beta-oxidation of fatty acids [20].

Several studies have shown that As exposure during gestation leads to future *congenital malformations* and long-term disease consequences including cancers. Exposure to high levels of arsenic in water, meaning arsenic levels 50 mg/L or greater, showed an increased risk of spontaneous abortions and stillbirth and a moderate increased risk of neonatal and infant mortality. Children exposed to inorganic arsenic early in life (in utero or postnatally) had an increased risk of bronchiectasis and lung cancer. Adults with exposure to elevated levels of arsenic in utero, developed bladder and lung cancer in higher rates as adults, despite their arsenic exposure ending as much as 40 years ago [28] (Fig. 4.10).

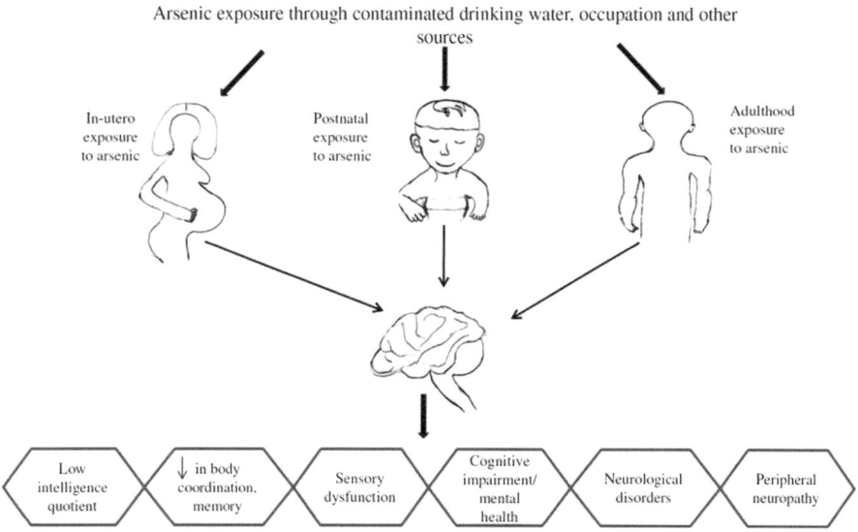

Fig. 4.10 Intrauterine, infant, and adult exposure to arsenic and its consequences (18)

In fact, it has been shown that exposure to even low arsenic levels may affect the fetal epigenome. The authors have investigated DNA methylation of cord blood-derived DNA in 134 infants from a birth cohort in New Hampshire, England, and they observed a 10% increase in the risk of gestational diabetes mellitus relatively to each 5 mg/L increase in arsenic concentration in local well water [29]. They also observed methylation at individual CpG loci [28].

A growing body of evidence indicates that prenatal exposure to arsenic may have neurotoxic effects and it is most severe during brain development and fetal growth. A recent study (Rager et al. 2014) in a Northern Mexico's population, where 450,000 people drink water with inorganic As levels exceeding 50 mg/L, has found that prenatal arsenic exposure is associated with altered miRNA expression levels in newborn cord blood (111).

Arsenic is an established *carcinogen*. Arsenic-related cancer usually takes more than 10 years to develop [28]. The International Agency for Research on Cancer (IARC) and the US EPA have classified arsenic in drinking water as a human carcinogen leading to cancers of the skin, bladder, and lungs [28].

Although the model, by which arsenic induces genotoxicity and carcinogenesis, is not completely clear, there are some indications that arsenite-generated free radicals and arsenite-generated reactive oxygen species lead to genotoxic damage. Arsenic exposure has been associated with DNA methylation alterations in nonpathologic as well as tumor tissues and there is some data suggesting that changes in miRNA expression and histone tail modifications in tumor cells are also associated with exposure to arsenic. The mechanism of the cancer process due to arsenic is the methylation of inorganic arsenic by S-adenosyl-methionine (SAM) which is actually a mechanism of detoxification. DNA methyltransferases require SAM as their methyl donor which further suggests a role of DNA methylation in arsenic carcinogenesis and other arsenic-related effects [27].

Moreover, arsenic's role in carcinogenesis perhaps resides in its ability to compromise the repair of DNA lesions induced by DNA damaging agents. These lesions are formed from endogenous metabolism byproducts or they arise from different environmental exposures. Compromised DNA repair is due to dysfunction of multiple mechanisms including the diminished expression of DNA repair genes, the functional disruption of DNA repair proteins, the induction of a chromatin environment that is not conducive for DNA repair, and the aberrant cell cycle regulation. Hence, chronic arsenic exposure can result in progressive decline in DNA repair capacity, which may represent a crucial pathway by which arsenic contributes to carcinogenesis. In addition, elevated production of ROS/RNS from arsenic exposure contributes to the accumulation of DNA lesions and carcinogenesis [29].

In a multi-step process of colon carcinogenesis in experimental models, ROS were found to enhance colon carcinogenesis at all stages: initiation, promotion, and progression. NADPH oxidase is one of the major sources of cellular ROS. NOX1 (NADPH oxidase 1), a member of the NADPH oxidase family that is highly expressed in colonic epithelial cells, plays a pivotal role in cell signaling, cell growth, angiogenesis, and motility and it integrates Wnt/β-catenin signals. It has been reported that NOX1 overexpression in colon tumors could contribute to the development of colorectal cancer [31].

Christensen C et al. showed that As (III) or Cr (VI), alone or in combination, increased NOX1 expression in vivo. In addition, the expression of several important antioxidant enzymes, such as SOD and catalase, was decreased in mice exposed to the carcinogenic metals in drinking water. The level of *8-OHdG*, an oxidative DNA damage marker, was also increased, indicating that oxidative stress was involved in a carcinogenic metal-induced tumor effect. The expression of β-catenin and phospho-GSK3β, as well as the level of ROS generation, was also increased by As (III) or Cr (VI), alone or in combination. ROS is important in causing β-catenin activation induced by carcinogenic metals. SOD and catalase decreased metal-induced β-catenin expression and activity as already described [31].

Epigenetic Mechanism of Intoxification

More recently, investigations showed that exposures to 250–500 μg/L of arsenic in drinking water results in global hypermethylation due to increased ^{3}H methyl group uptake, but that exposure >500 μg/L results in global hypomethylation [29]. Furthermore, DNA methylation patterns in the promoter regions of tumor suppressor *p16* and *DAPK* genes lead to As-induced precancerous and cancerous skin lesions [32].

In rat=liver epithelial cell lines treated with chronic low arsenic doses, Zhao et al. showed malignant transformation associated with depressed SAM levels, global DNA hypomethylation, and decreased DNA methyltransferase activity. The global, dose-dependent hypermethylation in Bangladeshi adults with chronic arsenic exposure was an unexpected finding, recently reported in literature. This effect was modified by folate, which suggests that arsenic-induced changes were dependent on methyl availability [32].

Higher levels of As exposure were positively associated with DNA methylation in *LINE-1* repeated elements, and to a lesser degree at CpG sites within the promoter region of the tumor suppressor gene *p16*. Another study was conducted by Kile et al. (2014) on pregnant women. This study in a prospective birth cohort recruited in Bangladesh to evaluate the effects of exposure to low arsenic levels on reproductive health outcomes. The study showed that prenatal arsenic exposure was associated with altered DNA methylation at the 473844 CpG region in the cord blood of newborns, after adjusting for leukocyte distribution [22].

After the publication of these findings, several studies have shown that arsenic is associated with gene-specific hypermethylation. In a human study in India, significant DNA hypermethylation of *p53* and *p16* promoter regions was observed in blood DNA of subjects exposed to toxic levels of arsenic compared to controls. In this study, hypermethylation showed a dose–response relationship to the arsenic measured in drinking water [30]. Other studies also confirm a dose-dependent DNA hypermethylation of the promoters of these genes. It was shown that arsenic exposure can result in *CDKN2A* (cyclin-dependent kinase inhibitor 2A) hypermethylation in human blood DNA and a subgroup of cases had hypomethylation with high arsenic exposure [19]. *CDKN2A* is a gene, ubiquitously expressed in many tissues and cell types. The gene codes for two proteins, including the INK4 family member p16 (or *p16INK4a*) and the p14arf (*CDKN2A locus*). Both act as tumor suppressors by regulating the cell cycle.

In human bladder cancer, relatively low levels of inorganic arsenic exposure have been associated with hypermethylation of *RASSF1A*, which suppresses the proliferation of tumor cells, and *PRSS3*, within the locus of T cell receptor beta variable orphans, but not *CDKN2A*. A recent study of a large number of human bladder tumors, which examined hundreds of genes, demonstrated a relationship in promoter methylation profiles with arsenic exposure, and it concluded that highly methylated tumors came from individuals with larger drinking water arsenic exposure, and that these tumors were more aggressive [31]. In the same time, emerging evidence indicates As (III) as a promising immunoadjuvant for hematological malignancies and solid tumors in the era of cancer immunotherapy [20].

Changes in histone marks are mediated by histone kinases, such as the nuclear mitogen and the stress-activated protein kinase 1 (MSK1), which are both activated upon arsenic exposure. In response to stress, MSK1 activation can lead to both the demethylation of H3K9, which increases transcriptional activity, and an increase of phosphorylation of H3S10, which induces activation of proto-oncogenes such as *c-fos* and *c-jun*. In addition, emerging evidence of modifications on highly conserved histone variants has provided new insights. Examples of altered histone variants after arsenic exposure include H2B1K, H2B1C, H2B1D, and H2B1B. Furthermore, arsenic has also been found to induce histone H3.1 polyadenylation by degradation of a stem-loop binding protein (SLBP). The presence of a poly(A) tail increases mRNA stability by preventing degradation and by subsequently promoting the accumulation of higher levels of polyadenylated H3.1 mRNA outside the S phase. Polyadenylated H3.1 alone can induce genomic instability, mitotic blockade, and cell transformation. Increased SLBP degradation and subsequent polyadenylation of H3.1 can interrupt and block histone H3.3 function, which in turn promotes overall genomic instability and subsequent aberrant gene expression [20].

Generation of open chromatin involves the actions of multi-subunit chromatin-remodeling complexes and post-translational modifications of core histone proteins. In regards to the latter, acetylation of lysine 16 in histone H4 (H4K16Ac) and mono-ubiquitination of lysine 120 in histone H2B (H2BK120ub) represent the histone epigenetic marks that promote the formation of biochemically accessible chromatin at, or near DNA DSB (Double-Strand Break) sites. It is well known that DSBs are among the most deleterious types of DNA lesions, which can lead to mutations, loss of heterozygosity, and chromosomal rearrangement; if not properly repaired, they can lead to cell death and cancer [33].

H3K9me2, mediated by increased mRNA and protein levels of histone methyltransferase G9a, is correlated with transcriptional repression, which has been shown to be involved in the silencing of tumor suppressor genes in cultured cancer cells. H3K27me3 is frequently accompanied with inactive promoters and gene silencing, and it labels chromatin by polycomb repressive complex 1 (PRC1) by H2AK119 ubiquitination in order to facilitate chromatin compaction [33].

Recent studies have also demonstrated that arsenic-induced aberrant alternative splicing may be associated with cancer development. Alternative splicing of pre-mRNA is important for the biodiversity of proteins and it occurs in 95% of multi-exon genes. Dysregulation of normal splicing events is known to occur in various human cancers and it is correlated with angiogenesis, carcinogenesis, and mesenchymal transformation (EMT) [20].

Arsenic toxicity has also been recently related to changes in miRNA expression. Marsit et al. showed alterations in the miRNA profiles of human lymphoblastoid cells grown under sodium arsenite treatment. Interestingly, the expression of specific miRNAs altered by arsenic are involved in one-carbon metabolism, relating the toxicity of arsenic to SAM [34]. An initial study identified five miRNAs (miR-210, miR-22, miR-34a, miR-221, and miR-222) that were significantly modulated by As (III) exposure in human lymphoblasts after 6 days of exposure. These changes in miRNA levels were not accompanied by any changes in the overall DNA methylation status [34].

Arsenic has also been found to alter the regulation of other non-coding RNAs (ncRNAs). Evidence suggests that arsenic is able to stimulate global changes in miRNA expression. In fact, one particular study showed that arsenic exposure could lead to the deregulation of 36 different miRNAs. Likewise, the piRNAs are known epigenetic regulators and they are implicated in various tumors due to their ability to regulate gene expression and genome instability [20].

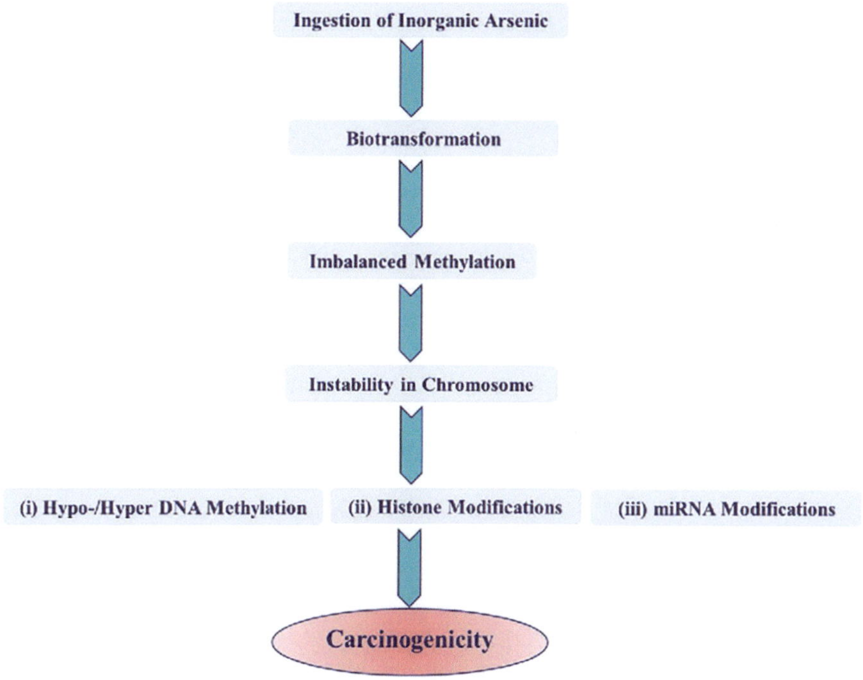

Fig. 4.11 The multistep pathway of carcinogenesis due to arsenic exposure

These alterations have the potential of a great impact on cellular homeostasis as each of these epigenetic components play a crucial role in regulating gene expression [24] (Fig. 4.11).

Many developed countries have taken measures focusing on reducing arsenic exposure and on early diagnosis and treatment of arsenic-induced cancers. Arsenic toxicity has become a public health and environmental problem, which is a serious issue in certain parts of the world [21].

Many factors, such as diet and nutrition, can influence the bio-accessibility, metabolism, and toxicity of arsenic. Several studies have reported that intakes of some vitamins, jaggery, fruit, tea, or N-acetylcysteine, as well as elevated levels of glutathione, zinc, and selenium may reduce arsenic-induced toxicity, presumably by reducing availability or formation of toxic mono-methylated species [22, 35].

Lead

The Metal, Its Forms, and Its Distribution in the Environment

Lead is a chemical element with the symbol Pb (from the Latin *plumbum*). It is a dense, heavy, relatively unreactive post-transition metal. Its weak metallic character is illustrated by its amphoteric nature and frequently lead and lead oxides react with acids and bases, and they tend to form covalent bonds. The compounds of lead are usually found in the +2 oxidation state rather than the +4 state and they usually associated with lighter members of the carbon group.

Lead is a toxic heavy metal known for causing extensive environmental pollution due to its wide range of use, environmental persistence, and transportability. The non-biodegradable nature of lead is the primary reason for its persistence in the environment. It finds its use in various industrial products and mining activities due to its distinctive physical and chemical properties. Widespread usage of the metal results in an increase of human exposure, which mainly affects the developing countries [36]. In Mexico, lead exposure occurs mainly due to the production of glazed pottery, lead-based paints, and air pollution [37].

Exposure to lead could come from the air, in the form of lead fumes (e.g., firing range operators, stained glass workers, solderers), from water contaminated by lead plumbing components (particularly in acid conditions), the usage of lead-containing vessels (lead crystal, lead glazes of pottery in which acid liquids are stored), and airborne particulates, found around old homes and bridge constructions. Lead in drinking water is known to be a significant source of lead deliberate exposure [38]. Workers particularly prone to high lead exposure are battery production workers and bronze workers [39].

The toxicity of lead is correlated with its chemical properties and its affinity to other metals. Lead absorption is inversely associated with the body's iron burden [36]. However, some individuals show increased susceptibility to the detrimental effects of lead which is associated with a predisposition genetic background [40] (Table 4.1).

Table 4.1 Examples of common sources of lead exposure

Occupational	Lead mining and refining, plumbing and pipe fitting, auto repair, glass manufacturing, printing, battery manufacturing and recycling, construction work, firing-range instruction, plastic manufacturing, gas station attendant
Environmental	Air pollution and contaminated drinking water, lead paint, soil or dust near roadways or lead-painted homes, plastic window blinds, plumbing leachate (from pipes or solder), ceramic ware, lead-core candle wicks
Hobbies	Glazed-pottery making, target shooting, lead soldering, preparing lead shot or fishing sinkers, stained-glass making, painting, car or boat repair
Other	Folk remedies, gasoline sniffing, costume jewelry, cosmetics [41]

However, the more discussed source of exposure to lead is air pollution. The widespread use of lead until the early 1990s in gasoline to improve performance was associated with variable diseases, only noticed at that time. Lead was added to gasoline in 1921. However, the increasing number of cars, which became the preferred mode of transportation, especially in big cities in the mid-1970s, resulted in high exposure of the population to the chemical. In 1976, the US medical community noticed an increase of diseases that could probably be related to the fact that American's blood-lead level (BLL) was three times higher than the current reference value. Individuals, and children in particular, living in highly urban areas, near busy roads, or near lead-emitting industries, had the highest exposure. The most discussed was Alzheimer's disease [40].

In the mid-1970s, platinum catalytic converters were introduced in the car industry due to the fact that lead ruins converters. Moreover, there were warnings from the US Environmental Protection Agency regulations responding to public health concerns [40]. Scientific research introduced the finding that increased risk for neurodegenerative diseases and especially Alzheimer's disease was associated with the levels of lead in blood testing. After 1976, a routine population surveillance lead testing was added to the second National Health and Nutrition Examination Survey (NHANES-II). Results of testing showed that children under the age of 5 years had acceptable levels of lead in their blood (on average 3.75 μg/dL), which confirms that the protection of the population from increased lead exposure was effective [40].

Since 1970, both child and adult lead exposure have been steadily dropping in the USA following the removal of lead from gasoline and food cans, the general deindustrialization of the economy, and the gradual strengthening of air quality standards and pollution preventing technology. Indeed, the prevalence of American adults with elevated blood-lead levels ($\geq$25 μg/dL) has declined nearly three-fold over the last years. Lower adult lead exposures could have plausibly contributed to recent declines in national dementia rates. However, the intervention measures were not applied to adults born in the 50s–60s who still suffer from the disease [39]. A multi-decadal Swiss study detected lower rates of brain amyloid-β burden in elders autopsied across the years of 1972–2006. Declines in blood-lead levels in the Swiss population over the past three decades mirror those in the USA and the UK [36].

In 1997, data from the NHANES showed that 4.4% of children in the USA still had elevated blood-lead levels [41]. Whether children highly exposed to lead will reverse with aging, it remains to be seen [39]. The ban on lead petrol has been very successful in lowering blood-lead levels of children, which clearly indicates a reduced exposure has been achieved [36]. More data on lead exposure of European citizens is necessary and it is currently being collected.

Lead in drinking water is considered a significant source of lead exposure. Candy, folk and traditional medicines, ceramic dinnerware, children's jewelry and toys, clothing ornaments, key chains, and other metallic or painted objects as well as products including vinyl, plastic, and rubber are other potential sources of lead. Since the banning of production of leaded gasoline, lead-rich paints are the main source of human exposure globally [36]. The use of lead in paints could poison children who eat the lead-containing paint chips, as well as adults who restore old furniture and houses painted with lead-containing paint. Even nowadays, we keep producing lead crystals for decanters and glasses to hold wine, resulting in the leaching of the metal from the container into the wine [41]. Inorganic lead is found in progressive hair dyes (as lead acetate) as well as in lipsticks [36].

Environmental lead is ubiquitous, which means that most people in the world absorb lead in trace amounts [36]. After inhalation or ingestion, lead is absorbed into the body through the mucous membranes. Lead is available in an organic and inorganic form and its absorption rate depends on its physicochemical form and the condition of the exposed individual. Inorganic lead is not metabolized, but it is distributed and deposited in soft tissues and bones from where it is secreted in blood. Organic lead in the air is easily absorbed by the skin and the respiratory tract. Inhaled lead is almost completely absorbed, while absorption of ingested lead is about 10–15%. The rate of absorption is higher in pregnant women and children and it may further increase during fasting. Studies have shown that dermal absorption of inorganic lead happens at a rate of 2.9 ng/cm [36].

The half-life of lead in the blood of adults is approximately 40 days. However, blood lead has a longer half-life in children and pregnant women, as their bones are rapidly remodeled causing an increased release of lead in the blood. Lead deposits in the bones are eliminated very slowly by urine. Additionally, lead is excreted by the stool, the nail deposits, and sweat. The estimated half-life of lead deposits of the bones is 20–30 years. Thus, bones keep releasing the metal continuously into the bloodstream for years after exposure [36].

The World Health Organization and the US Centers for Disease Control and Prevention (CDC) announced that blood-lead levels >10 μg/dL may affect health. However, the accurate level of safe exposure to lead has not yet been in CDC guidelines. It seems that even blood-lead levels <10 μg/dL could also cause disease. As per CDC blood-lead level (BLL) 5 μg/dL is considered high for adults and children. That is to say that there is not an accepted safe limit for blood-lead levels [36].

In 1998, the CDC issued new guidelines, adopted by the American Academy of Pediatrics, that endorse universal screening in areas where 27% or more of the housing was built before 1950 and in populations with elevated blood-lead levels ≥12 μg/dL or higher, especially in 1 to 2-year-old children [41].

Effects on Health

Lead is an established *neurodevelopmental toxicant* to humans and it is probably the most important neurotoxin [36]. Adults are more resistant to lead poisoning and they develop symptoms of peripheral neuropathy only in high doses of exposure. In children, lower doses could be responsible for severe symptoms [36].

Acute poisoning with lead manifests with serious symptoms. High doses in children can cause acute lead encephalopathy, a life-threatening condition characterized by generalized cerebral edema with increased intracranial pressure, which leads to trans tentorial and cerebellar tonsillar herniation. However, lower doses could also be harmful. Clinical features of acute lead encephalopathy may include ataxia, seizures, stupor, coma, and often death. In adults, high doses due to acute poisoning cause motor peripheral neuropathy [39].

Autopsy shows that the brain is markedly swollen with compressed gyri, obliterated sulci, and collapsed lateral ventricles. Uncal and cerebellar tonsillar herniation are commonly encountered. Microscopically, there is a breakdown of the blood–brain barrier with transudation of fluid into the pericapillary space and an ischemic damage to cerebral cortical neurons [39].

On the other hand, *chronic exposure* to lead causes impairment of the central nervous system expressed differently in children and adults. Some of the characteristic clinical features of chronic lead exposure *in children* are distractibility, attention deficit, and memory problems. The symptoms are consistent with the deposition of the metal to specific areas of the brain [41].

The earliest evidence showing an association between childhood lead exposure and neurodevelopmental disorders comes from a case of a lead-poisoned Cincinnati child in 1931. Hospitalized for lead-related encephalopathy after ingesting leaded paint at the age of two, the child experienced progressive mental deterioration until his death at age 44 years. A brain autopsy following his death revealed extensive atrophy similar to Alzheimer's disease (AD) in cortical, temporal, and hippocampal brain areas. Neural tissues of these regions were clouded by neurofibrillary tangles. At that time, at least two other cases of individuals highly exposed to lead "from a very early age" were found to experience AD-type pathology. Following these cases, researchers at Indiana University observed the formation of neurofibrillary tangles in cortical and hippocampal neurons of rabbits within hours of injecting the animals with tetraethyl lead [40].

Early signs of lead chronic poisoning in children manifest as subtle neurobehavioral symptoms that affect social interaction that could be frequently disregarded [39]. Classroom behavior gradually changes and it could be disturbed. Follow-up studies on children with elevated lead levels showed a markedly high dropout level from high school, an increased absenteeism and an impaired eye–hand coordination [39]. Cognitive, affective, and physiological changes related to the function of the nervous system are observed depending on the dosage and duration of exposure. Concentration problems and impulsive and unpredictable behavior is established after long-lasting exposure. There is evidence for effects on motor coordination, spatio-visual capacity, and verbal capacity [39]. A number of studies showed that even low levels of blood lead are harmful. Lead is considered a risk factor for the development of the well-known attention deficit hyperactivity disorder (ADHD). Childhood lead exposure is estimated to contribute up to 600,000 new cases of intellectual disabilities every year with maximum occurrence in developing countries [38].

Apart from the harm of the developing child's brain, the risk for diseases emerging later in life is another question and active area of research. Accumulating evidence from experimental and observational studies suggests that childhood lead exposure may result in lasting neural, epigenetic neural and behavioral changes and they could possess a significant risk for the development of neurodegenerative diseases in an older age [40].

In adults, the first report of detrimental consequences of chronic lead exposure to the Central Nervous System was reported in 1990 by Needleman et al., who conducted a 11-year follow-up study of 132 young adults who had studied in the same primary school. The subjects with high lead levels were more likely than those with low levels to have attention problems, aggressiveness, antisocial and delinquent behavior [41]. Lead exposure was estimated by the calculation of lead levels in the dentine of deciduous teeth. These studies suggested that exposure to relatively low levels of lead in childhood was associated with a slight, but significant, drop in IQ level [39].

Chronic lead exposure in adults results in peripheral neuropathy associated with weakness of the distal muscles as well as marginal sensory involvement due to axonal degeneration. Long-term inorganic lead exposure has been associated with features of mild sensory and autonomic neuropathy [38].

Studies in adults exposed to lead due to their home environment or employment have indicated a link between exposure and neurodegenerative diseases. Lead is hypothesized to increase disease risk in exposed adults by directly causing inflammatory and pro-degenerative intracellular oxidative damage. It is not yet clear if the mechanisms of brain damage are similar to those in children. Nevertheless, studies of adults exposed to lead provide confirmation of pro-degenerative neurotoxic activity [40].

The best evidence of the risks for the development of neurodegenerative disorders is expected to be published from follow-up ongoing studies in the general population in the USA which analyze the cognitive function of adults who were environmentally exposed to lead as children in the 1960s, 1970s, and 1980s. The few such studies conducted to date suggest that the neurologic harm and cognitive deficits associated with childhood lead exposure persist into adulthood and middle age and they are accompanied by changes in brain morphometry, white matter integrity, and brain metabolism [40].

Research studies on the potential neurodegenerative consequences of lead exposure have primarily focused on the risk of the two most common diseases: Alzheimer's and Parkinson's diseases. Although these specific pathology syndromes result in selective degradation of different brain networks, there is considerable pathological overlap between them and some related diseases (e.g., Lewy-body dementia, amyotrophic lateral sclerosis, and the so-called "mixed pathologies"). Lead exposure is a common, although non-specific, risk factor for these disorders [40].

In a US case-control study with elderly participants, individuals with the highest lifetime exposure, calculated by bone and blood-lead measurements and by analysis of probable occupational exposure, were found to have a two-fold increased risk for

Parkinson's disease (PD) than those with the lowest lifetime exposure. A different case control study, using US elders from the general population and bone-lead measurements for cumulative lead exposure, reported a three-fold higher risk for those in the highest percentile of exposure compared to those in the first percentile curve [40].

Occupational studies based on self-reported measurements of lead exposure have generally failed to find associations between lead exposure and Parkinson's disease. Only one such study reported a considerably elevated risk for workers possibly exposed to lead, in addition to other metals (e.g., lead and copper or lead and iron) for more than 20 years. In the USA, prospective studies linking industry lead emissions data to Parkinson's disease risk failed to identify any association, suggesting that occupational levels may not alter risk or, conversely, that measures for industrial lead emissions probably decrease actual lead exposure [40].

Apart from implication in central and peripheral nervous system, acute poisoning as well as cumulative lead exposure could affect cardiovascular, renal, skeletal, hematological, immune, respiratory, gastrointestinal, reproductive, and endocrine systems [38].

Physicians should consider *screening* for lead exposure in children who present with growth failure, behavioral disorders, hearing loss, speech, language or attention deficits, developmental delay, microcytic anemia, or sleep problems. Adults presenting with unexplained neurological symptoms such as tremor, attention deficit disorder, unexplained arthralgias or headaches, glycosuria, or a history of occupational or hobby exposure may also require screening. The first question to ask is "the age of the housing" which has been nominated as the single most important environmental marker of lead poisoning in a community [41].

Recommended actions to be taken by primary care physicians depend on the levels of blood lead, especially in children.

Blood-Lead Level

0.30 to <0.50 μmol/L. Take an exposure history.

0.50 to 0.75 μmol/L. Repeat blood test and reexamine in 3–6 months; take full exposure history; contact local public health department to obtain advice on home remediation; counsel parents on exposure reduction and home remediation.

>0.75 to 1.00 μmol/L. Refer patient to a specialist for a lead poisoning assessment.

>1.00 to 2.10 μmol/L. Refer immediately the patient to a specialist for treatment; consider chelation therapy.

>2.10 μmol/L. Urgent referral and individualized case management are required; admission to hospital may be necessary [39].

In the case of increased lead blood levels, the parents should be advised to remove the children from the house immediately and proceed to thoroughly clean it up. All surfaces should be scrubbed with a general all-purpose cleaner twice at first and then, on a weekly basis and the house should be vacuumed frequently. Trisodium phosphate [TSP] is no longer recommended for cleaning; more recent advice on lead remediation suggests that a general all-purpose cleaner is just as effective as TSP in cleaning up lead dust at home [39]. An accurate assessment of lead poisoning requires knowledge of the source of contamination, the identification of high-risk groups, and the application of relevant laboratory tests. Most incidents could be prevented and treated accordingly. The diagnosis of lead poisoning is relatively simple compared to the diagnosis of exposure to other environmental toxins [41].

When blood-lead levels remains high after 2 years, a continuous exposure, probably from drinking water, is possible. The physician should advise parents to test their drinking water for lead and to replace the plumbing system or to use a reverse osmosis filter [41].

There are two high-risk populations for lead intoxification: children aged from 9 months to 3 years and people living near point sources of lead pollution. Physicians should be aware of these point sources and information of such "hot spots" should be available from the local public health department. Well-known "hot spots" are Trail, BC, and the South Riverdale area of Toronto [41].

Mechanism of Intoxification

Although the multiple, deleterious effects of lead intoxication have been described, the *neurotoxic effects* are the most known and better studied. Lead disrupts the blood–brain barrier (BBB) mainly by mimicking calcium transporters and it induces a direct injury to the main structural components of the endothelial microvasculature. In the Central Nervous System (CNS), the metal causes injury of astrocytes preferentially in the prefrontal cerebral cortex, hippocampus, and cerebellum [41].

The *Blood–Brain Barrier (BBB)* maintains the integrity of the brain tissue by protecting the entrance of toxic insults, such as chemicals. However, some metals, including lead, mimic the behavior of essential nutrients of the brain and they utilize physiological ionic transporters. It is well known that calcium is a critical element for neuronal signaling, neurogenesis, myelination, synaptic plasticity, and functioning of glial cells [38].

In particular, lead modulates the calcium-dependent protein kinase C (PKC) pathway and it alters endothelial cell behavior, especially in the immature brain. It rapidly crosses the BBB and it concentrates in the brain tissue [38]. Once inside the brain, lead enters into the neurons and the glial cells similar channels. Lead in the brain has a half-life of roughly 2 years [39, 40].

During neural development, lead can disrupt neuron and glial migration, differentiation and formation of synapses. Glutamate and dopamine systems appear to be preferentially vulnerable to the metal, leading to particular disruptions in the long-term potentiation regions of the hippocampus and in the cortical executive-functioning regions. In mice, lead preferentially accumulates in the hippocampus and cerebral cortex as already described in humans [40].

In neurons, lead mainly interferes with the neurotransmitter glutamate, which is critical for the learning process in the developing brain. The N-methyl-D-aspartate receptor (NMDA-r) is an ionotropic receptor for glutamate (Fig. 4.12). The non-competitive antagonism for the engagement of the NMDA-r receptor by lead in the hippocampus, is one of the most important mechanisms implicated in neurotoxicity. Prenatal exposure to lead affects the long-term potentiation machinery of the evolving brain by unsettling the NMDA-r expression in the hippocampal region [40]. Interaction of NMDA-r with lead occurs both synaptically and extra-synaptically [36].

Lead alters the expression of NMDA-r subunits (NR2A, NR2B) resulting in the dysregulation of calcium-sensitive signaling pathways of the hippocampus. Recent studies have revealed that lead could down-regulate calcium binding in the synaptic NR2A-NMDA-r and up-regulate the release of calcium in the NR2B-NMDA-r of the extra synaptic region. Once lead enters the hippocampal synaptic region, it blocks the NMDA-r function and interferes with the binding of the receptor with calcium. Therefore, all calcium-dependent processes like calmodulin-II (CAM-II), neuronal nitric oxide synthase (n-NOS), and c-AMP response element-binding protein (CREB) are inhibited. CREB mediates the expression of brain-derived neurotrophic factor (BDNF), a pathway that has been linked to the lead-induced cognitive deficits. Additionally, the metal blocks the release of calcium from mitochondria, leading to the formation of reactive oxygen species, mitochondrial "self-destruction," and apoptosis of the neuron. Lead also leads to neuronal death by enhanced lipid peroxidation and endotoxicity in glutamate-signaling neurons, a process resembling ischemic stroke (Fig. 4.12) [36].

Although, we know how lead interferes with neuronal function, the underlying mechanism responsible for the development of neurotoxicity after chronic lead exposure is not completely understood. Oxidative stress, altered cell signaling, and neurotransmission are a few important aspects. Lead toxicity leads to free radical damage by generation of reactive oxygen species (ROS) and the elimination of antioxidants. It binds to functional sulfhydryl groups of antioxidant enzymes making them nonfunctional, which contributes to an impaired oxidative balance (Fig. 4.13).

Oxidative stress kills neurons by stimulating the *FOXO3* (Forkhead box class O) transcription factor, a pivotal enzyme in the cell death and survival pathway of neurons [40]. Furthermore, it inhibits DNA transcription, vitamin D synthesis, and the function of enzymes that maintain cell membrane integrity [42].

White matter lesions and regional atrophy as well as gross abnormalities in brain morphometry are some of the progressive degenerative changes following lead exposure. In addition, adult lead exposure has been linked to abnormal ratios of brain metabolites in the hippocampus. In literature, a small sample of patients ($n = 31$) underwent magnetic resonance spectroscopy (MRS) after lead exposure. It was reported that brain metabolite ratios in the hippocampus and notably an increase in the ratio of myinsitol-to-creatinine. The same findings are detected in patients with Alzheimer's disease. These metabolic abnormalities are believed to signal the beginning of the hippocampal neuron loss and gliosis associated with Alzheimer's disease. High brain-lead concentrations and an increase of diffuse neurofibrillary tangles with calcification have been reported in autopsies of lead-exposed Japanese

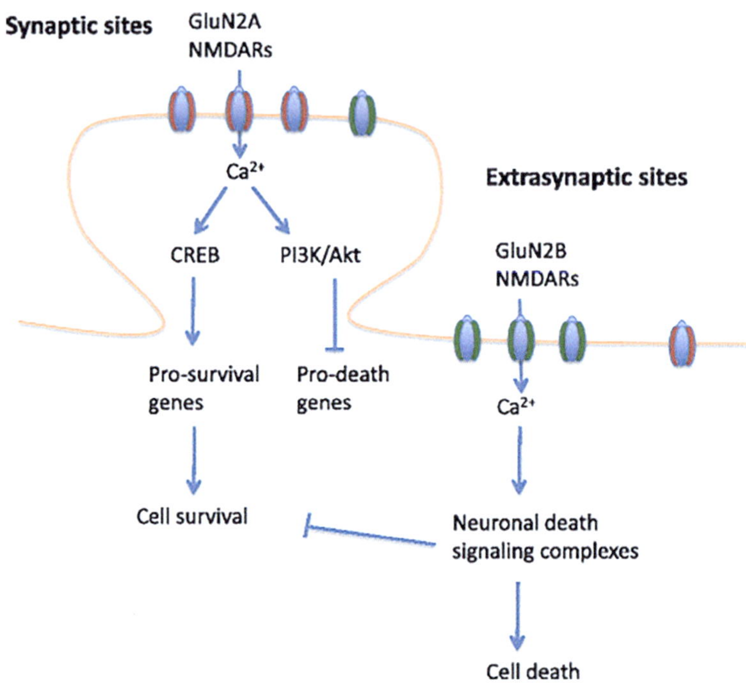

Fig. 4.12 Lead replaces calcium in the NMDA-receptor subunits, NR2A and NR2B, leading to dysregulation of the downstream pathway, a process resembling ischemic stroke

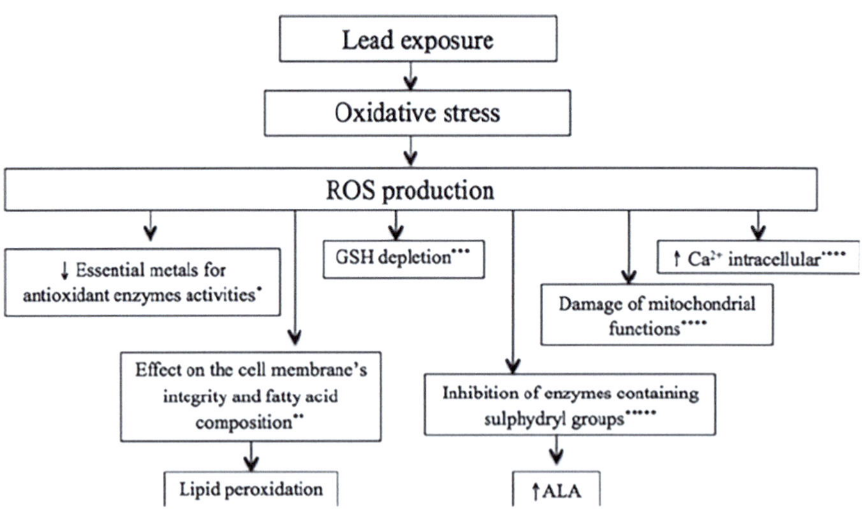

Fig. 4.13 Oxidative stress is the main mechanism of lead intoxication

workers with dementia [40]. Electrophysiological studies have shown decreased auditory sensitivity and visual motor performance in children exposed to lead as well as an increased latency in brain auditory-evoked potentials [41].

Autopsy of primate CNS tissue showed significantly greater AD-like pathology in the frontal cortex of lead-exposed monkeys ($n = 5$) than in the non-lead exposed control monkeys ($n = 4$), including the presence of diffuse amyloid-β plaques and neurofibrillary tangles. These pathological signs were accompanied by significantly elevated mRNA levels of amyloid-β-related genes (e.g., amyloid-β protein precursor (AβPP), transcription factor Sp1, and, marginally, elevated BACE1). DNA methylation patterns were indicative of enhanced brain aging and significant elevations in biomarkers of oxidative DNA damage, such as *8-OXO-dG*. Greater hyperphosphorylation of tau and pathological tau deposits were also observed [40].

As already mentioned, *bones* act as the primary reservoir of lead in the body, as normal human adults can deposit 80–85% and children 70% of available lead. The metal is stored in two compartments: the exchangeable pool, which corresponds to the bone's surface and the non-exchangeable pool, which is located deeper in the cortical bone. The surface lead can enter into plasma easily, but the non-exchangeable pool should move to the surface to enter the bloodstream, a process observed only when the bone is actively reabsorbed. Lead concentration in the cortical bone can be measured through X-ray fluoroscopy (XRF). The measurement can reflect the degree of long-term lead exposure and accumulation [36]. Patella and tibia lead concentrations are biomarkers of cumulative lead exposure and body burden, while blood lead is a biomarker of recent exposure [43].

Skeletal lead intoxication could lead to disruption of the development, formation, and resorption of bones and it is responsible for a variety of pathological changes. Skeletal malformations have been reported in animal models exposed to lead and they include the fusion of two or more vertebrae in the anterior part of the axial skeleton. Lead as low as <10 μg/dL is accompanied by low levels of osteocalcin (bone growth protein) and an impaired skeletal growth accompanied by an abnormal formation of the organic matrix components. In an experimental study, retarded new bone formation at epiphyseal ends and an increased osteoclastic activity was detected in animals intoxicated with lead [36]. Since lead could cross the placenta easily, human congenital lead toxicity has been reported to cause overt skeletal malformations such as a dense cranial vault, and a delayed skeletal and deciduous dental development at birth [36]. Post 7 months chelation therapy, the radiographs of these infants showed normal maturation, but delayed tooth eruption [36].

Vitamin D plays an important role in bone metabolism and there is considerable evidence that lead affects the synthesis of 1,25-dihydroxyvitamin D in the kidney. Mazumdar et al. observed low levels of dihydroxy Vitamin D (calcitriol) in a Bangladeshi population of jewelry factory workers. Similar findings have been observed in chronically lead-exposed children. Kemp et al. observed a seasonal variation in both blood-lead measurement and vitamin D levels [36].

Lead toxicity has been recognized as a risk factor for *cardiovascular disease* for more than a century. However, the mechanism of the effects to the cardiovascular system and blood vessels after lead exposure is still under investigation [39]. Several epidemiological and clinical studies have linked chronic lead exposure with elevated blood pressure. Adult occupational and residential exposure has been linked to increased rates of coronary heart disease, elevation in blood homocysteine levels and ischemic heart disease, peripheral arterial disease and atherosclerosis, and alterations in cardiac rhythm [40].

Evidence of the positive association was documented at the second NHANES II conducted from 1976 to 1980. The survey examined 543 white males between 40 and 59 years and it disclosed a significant relationship between blood lead and both systolic and diastolic blood pressure [36]. It has been shown that lead induces alpha adrenoceptor-mediated blood vessel changes [38]. Chronic lead exposure, even at low levels (<5 μg/dL), is detected to cause hypertension.

Long-term lead exposure promotes atherosclerosis in experimental animals; however, the mechanism remains unknown. Impaired antioxidant metabolism and oxidative stress may be implicated [38]. Additionally, cardiovascular disease due to high homocysteine levels, the thickening of cerebral arteries could lead to infarction and poor-perfusion-related hypoxia that, in turn, leads to cell death and the up-regulation of pathological protein pathways [40].

An epidemiological linkage study conducted from 1988 to 1994 used data from the NHANES-III. It performed a roughly 20-year follow-up on a representative cohort of lead-tested US adults (Mean age at baseline = 44.1). The study reported significantly elevated risk of cardiovascular disease-related and ischemic heart disease-related mortality for individuals at the 90th percentile of blood-lead levels at baseline compared to those at the 10th percentile, suggesting that cardiovascular problems from early lead exposure persist over time and they may result in premature death. It is suggested that children exposed to lead suffer greater rates of hypertension and cardiovascular disease in adulthood apart from a greater risk of cardiovascular-related degenerative brain pathology [40].

Experimental studies suggest that the pathology of cardiovascular disease due to lead exposure is induced by oxidative stress, inhibited endothelial repair, and impaired angiogenesis [40]. Cardiotoxicity is promoted by oxidative stress and an impaired antioxidant metabolism. Moreover, the modulation of the peripheral autonomic nervous system leads to chronic neuropathy affecting the cardiac tissue. It is detected that lead induces alpha adrenoceptor-mediated vasoconstriction and it modulates the renin and angiotensin production attributing to hypertension. This hypothesis was supported by Geraldes et al. who reported that chronic lead exposure in young adult rats promotes autonomic dysfunction with increased chemoreceptor sensitivity leading to a clear impairment of the autonomic outflow to the cardiovascular and respiratory systems even at levels below 10 mg/dL [36].

The adverse effects of lead poisoning on the *renal system* have been documented for over a century ranging from histologic changes to the clinical manifestations of renal dysfunction. Chronic, low-level lead exposure results in glomerular and tubulointerstitial changes that lead to glycosuria, proteinuria, chronic renal failure, and hypertension [36]. The findings are dose dependent [40].

Lead is mainly eliminated by the kidneys and its deposits initiate nephrotoxicity. The kidney excretes lead by means of tubular secretion and glomerular filtration. The metal has bidirectional transport across the tubular epithelium with a clearance rate from 1 to 3 mL/min that is relatively independent of kidney function. An algorithm for the pathogenesis of acute and chronic sequelae of the kidney due to lead toxification was proposed in 1997. It showed that acute lead poisoning leads to proximal tubular dysfunction and these changes usually disappear with chelation therapy or cessation of lead exposure. Chronic lead exposure (>60 mg/dL) can also affect glomerular function [38].

It has been described that exposure to low levels of lead early in life lead to glomerular hypertrophy and an increase in the volume of glomerular capillaries [40]. It may result in progressive tubulointerstitial nephritis, which is characterized by infiltration with leukocytes and by interstitial fibrosis and tubular atrophy [36]. Additionally, exposure of mice to lead acetate enhances the production of reactive oxygen species and reduces the mRNA expression of enzymes necessary to combat oxidative stress. Muntner et al. concluded that the low creatinine clearance (<4.2 mg/dL) combined with hypertension could be the cause of chronic kidney disease (CKD) in the US population [41].

In a longitudinal prospective Taiwanese study, Yu et al. observed that adults with CKD and high baseline levels of lead developed a rapid decline of kidney function. Chelation therapy slowed CKD progression compared to controls [40]. A longitudinal study of renal function and lead levels in middle-aged and elderly people showed that a tenfold increase in blood-lead level predicted a decline in renal function equivalent to 20 years of aging [41]. An appropriate urinary biomarker for lead-induced renal injury is the kidney injury molecule-1 (KIM1), and the alpha glutathione S-transferase (aGST) [38].

Additionally, the association of lead exposure to *hematopoietic system disease* is well known. A reduced hemoglobin synthesis and red cell life span leads to anemia. Regarding the production of erythrocytes, the enzymatic system involved in heme synthesis is adversely affected by both acute and chronic lead exposure. The three enzymes adversely affected by lead toxicity are d-aminolaevulinic acid dehydratase (ALAD), aminolaevulinic acid synthetase (ALAS), and ferro chelatase. However, the most significant effect is observed on the enzyme ALAD, which is clinically used to assess lead toxicity.

ALAD (d-Aminolaevulinic acid dehydratase) enzyme catalyzes the second step of heme synthesis in which two molecules of aminolaevulinic acid are asymmetrically added to form the monopyrrole porphobilinogen, the precursor of heme. Lead-exposed individuals carrying copies of the *ALAD2* allele have increased chances of retaining lead in their blood and tissues, which further increases the probability of adverse effects either due to lead-induced oxidative damage or through alterations in metabolites like aminolaevulinic acid (ALA) of heme synthetic pathway [36]. Heme biosynthesis is inhibited by the reduction of ALAD activity to 80–90% at a creatinine clearance of 55 mg/dL.

Another key enzyme of heme synthesis is ferro chelatase and its reduced activity leads to the accumulation of protoporphyrin in erythrocytes and the accumulation of coproporphyrin in urine. The accumulation is due to the replacement of iron by zinc giving rise to zinc protoporphyrin (ZPP), a useful parameter to assess lead toxicity [36].

However, the major mechanism of lead toxicity to the hematopoietic system manifests due to the increased generation of reactive oxygen species (ROS) and the interference with the antioxidant system, like glutathione superoxide dismutase. An increase in oxidative stress leads to damaged membranes and hemolytic anemia. Lead can also bind directly to phospholipids (particularly to phosphatidylcholine) of the red blood cell membranes reducing their levels and destabilizing them [38].

The lead toxicity of the *gastrointestinal (GI) tract* develops only when lead is ingested. Various biological functions of the system are impaired such as areas of lead absorption and the properties of transporters. Chronic or repetitive episodes of abdominal pain, nausea, vomiting, constipation, bloating, and anorexia are common symptoms observed, when lead levels range between 30 and 80 mg/dL. The interaction of lead with nutrients like calcium and iron affects lead bioavailability, which is mainly stored in the liver being, the largest soft tissue storehouse (33%). Chronic lead exposure results in significantly raised liver enzymes (ALP, ALT, AST, and GGT) as well as increased oxidative stress parameters. Elevated lead levels are also associated with increased LDL and decreased HDL. Occupationally exposed factory workers with blood-lead concentrations >70 mg/dL were found to have raised hepatic enzymes with concurrent suppression of cytochrome P450 activity [36].

The effects of lead exposure on the gut microbiome have been an area of research. The results demonstrated that lead exposure not only transformed the gut microbiome community and diversity, but also greatly affected metabolic functions. The oral administration of recombinant Lactococcus lactis, which constitutively expressed the fusion protein glutathione S-transferase (GST) showed promising results [36].

Lead toxicity has been extensively studied in relation to immunity in order to understand the role of heavy metals in the impairment of *immune function*. Lead has been reported to negatively affect the functions of neutrophils and macrophages of the innate immune system. Occupational exposure to lead affects chemotaxis and the generation of ROS in neutrophils. In addition, lead reduces phagocytosis, affects nitric oxide production and metabolism in mature macrophages. Experimental studies have further reported that bone marrow-derived macrophages have amplified production of TNF-a, IL-6, IL-12, and prostaglandin E2, and reduced production of the anti-inflammatory IL-10. However, others reported significantly higher plasma IL-10 levels in lead-exposed workers compared to non-exposed workers. In addition, lead modulates the classical complement pathway by increasing the expression of C4bp [36]. Thus, lead affects the first defense components of the immune system. It lowers host resistance to infections by various pathogens, including bacteria and viruses.

Moreover, it was reported that lead depressed total antibody levels. Low-level lead exposure induces skewing antibody isotype production. Lead results in the switching of B lymphocytes from producing IgM and IgG antibody isotypes to producing IgE isotopes. This switch is associated with allergic and hypersensitivity reactions, especially among children. IgA levels have also been found to be increased significantly in individuals with lead levels >10 lg/dL [36].

As far as specific immunity is concerned, T lymphocytes appear to be the most vulnerable to the deleterious effects of lead. Lead interferes with the critical TH1/TH2 lymphocyte balance. It is capable of moderating bone marrow dendritic cell development and consequently antigen-presenting cell functions leading to the predominance of TH2 phenotype [36]. CD4+ T-helper (TH) cells are particularly sensitive to lead-induced immunotoxicity. Mishra et al. reported the reduction of CD4+ T cells in lead-exposed individuals and he found a significant negative correlation between the CD4+ percentage and blood-lead levels.

The role of N-formyl kynurenine, a metabolite of tryptophan metabolism, is well documented in T lymphocytes viability. Lead exposure increases 2,3-dioxygenase expression which is required for N-formyl kynurenine formation. This metabolite encourages cell cycle arrest and apoptosis of T lymphocytes. The mechanism is also implicated in the differentiation of naive CD4 + T cells into regulatory T cells. However, these findings need further investigation [36].

Occupational exposure to lead has been associated with defective *lung function*. Pulmonary function test values were significantly reduced among lead-exposed workers compared to control subjects. Epidemiological studies have also revealed an increased incidence of asthma as well as increased bronchial responsiveness due to raised IgE levels. An association of asthma in childhood and lead exposure was further investigated in a recent study that showed that asthmatic children were five times more likely to have lead blood levels more than 10 µg/dL in comparison to controls, although results of other studies were controversial. In a large representative sample of the US population, a significant association between obstructive lung disease and lead concentrations was also observed [36].

Chronic environmental and occupational exposure to lead can induce defects on the *human reproductive system*. Epidemiological studies have observed increased lead levels among infertile men in comparison to fertile groups. Occupational lead exposure affects male fertility by decreasing sperm count, motility, morphology, and functional parameters. Toxic levels of lead in females can also lead to miscarriages, prematurity, low birth weight, and developmental problems during childhood [36].

Lead behaves like an *endocrine disruptor (ED)* molecule, as it can interfere with the endocrine system at certain doses. There are several reports detecting the interference to growth and puberty. Lead toxicity-related delayed puberty and menarche (adjusted for body size) were observed at levels as low as 3 mg/dL. The possible underlying mechanism of delayed puberty has been investigated in animal models. Studies in rats showed that prenatal, lactational, and prepubertal exposure to lead alters the synthesis of growth and sex hormones. Apart from delaying puberty, lead may also alter the onset of menopause [36].

The effect of lead compounds on the *hypothalamic–pituitary–adrenal axis function* blunting TSH, GH, and FSH/LH responses to their respective trophic hormone stimulation suggests that lead acts as a stressor and it targets brain mesocorticolimbic dopaminergic systems. It also affects the sympathetic nervous system directly and indirectly by alterations in neurotransmitter secretion. Furthermore, in experimental studies, lead acetate-treated rats showed regressive transformation and toxic effects on the adrenal cortex. Moreover, reduced basal plasma corticosterone concentrations and corticosterone responses to lead-induced stress were also reported [36].

Epigenetic Mechanism of Intoxification

It has already been discussed that lead expresses its intoxification by the production of ROS. Reactive oxygen species are implicated in the damage of neuronal, kidney cells, erythrocytes, etc. Moreover, ROS produce more permanent effects mediating epigenetic changes that could be transferred to the next generation.

In vivo and *in vitro* studies conclude that lead exposure is associated with changes in the DNA methylation pattern and the hypermethylation of specific genes. In general, all studies conclude that lead intoxification operates to global DNA hypomethylation. A Chinese population study ($n = 110$ participants) showed 86.3%, 78.6%, and 73.9% average LINE-1 methylation in the blood of lead exposed groups relative to lead levels <100, 100–200, and >200 μg/L, respectively. In 678 men from the US Normative Aging Study, the absolute difference in average LINE-1 methylation in circulating leukocytes was approximately −0.25% (95% CI −0.44, −0.05) in relation to patella lead concentrations [38]. Other studies use other genomic methylation parameters like Alu elements to evaluate the global methylation pattern. In one such study, inverse association of maternal tibia lead signal to the methylation of Alu and LINE-1 in newborns was observed, which was not observed in association to cord lead [38].

As far as specific gene methylation in the promoter regions is concerned, different studies detect a number of genes implicated. In the only epigenome-wide association study ($n = 24$), a CpG site in the *COL1A2* (*collagen a1*) gene showed decreased DNA methylation with elevated blood-lead exposure being under the established threshold [42].

The product of collagen type 1 alpha-2 (*COL1A2*) gene is a major component of connective tissue. Hanna et al. studied the CpG site (*COL1A2 P407*) within the *COL1A2* promoter region in women undergoing the first completed in vitro fertilization cycle. The methylation of this site was found to be increased in groups with low and decreased in groups with high levels of lead exposure, respectively. The alteration of DNA methylation at *COL1A2 P407* is a suitable candidate for detecting epigenetic changes associated with lead exposure and fetal development [38].

A Chinese study showed that lead exposure increased the *ALAD* gene methylation levels leading to the downregulation of ALAD transcription. Methylation frequencies were statistically significant among study groups and individuals with methylated *ALAD* gene showed an increased risk of lead poisoning [38].

In another study, high lead levels were associated with the methylation of the tumor suppressor *CDKN2A/TP16* gene, whereas only partial methylation was detected, when lead levels were low. In another experimental study, methylation of the *prostaglandin-endoperoxide synthase 2* gene was found to be reduced in lead-exposed carcinoma cell line A431 cells. Based on their findings, authors concluded that the duration of lead exposure is a decisive factor in determining changes in specific methylation parameters. The most vulnerable group remains the developing fetus, which may experience global methylation changes, because of maternal lead exposure [38].

To explain the long-term consequences of lead exposure and their implication to Alzheimer's disease, the Amyloid precursor protein (APP) was investigated. The protein is an integral membrane protein, undergoes proteolytic cleavage to produce beta-amyloid, which is a component of amyloid plaques, a hallmark finding of Alzheimer's disease. High levels of beta-amyloid can produce ROS, and epigenetic modulation in the methylation pattern of cytokines which interfere with the repair or oxidation potential of adjacent oxidized guanines. The expression of APP proteolytic enzyme b-site APP cleaving enzyme 1 (BACE1) and its transcriptional regulator Sp1A were found to be increased in a group of lead-exposed monkeys. Another experimental study on lead-exposed PC12 cells showed that among the 41 CpGs in APP promoter, 4 CpG dinucleotide sequences (located at positions 2284, 2280, 2261, and 165) were hypomethylated [38]. Thus, the APP promoter is a specific gene hypomethylation site in lead-exposed individuals.

In the US population, recent evidence suggests that these epigenetic changes of the APP promoter can accumulate and be expressed later in life. In a mouse study, *AβPP* gene expression was followed by months of normal gene expression levels. These levels abruptly increased toward the end of the study-mice lifespans and they were accompanied by accumulation of AβPP and amyloid-β in the affected cortex tissue [40].

Luo et al. investigated alterations in *histone modification* of the brain tissue in lead-exposed rats. They found chronic lead exposure possesses a significant increase of histone acetylation in the hippocampus. The transcriptional level of *p300*, a typical histone acetyltransferase, was also significantly increased upon higher dose of lead exposure. Authors concluded that increased levels of histone acetylation may be mediated by the enzymatic activity of *p300* and it is associated with the pathogenesis of the attention deficit hyperactivity disorder (ADHD) [40]. Lead is considered a risk factor in the development of ADHD disorder [43]. Other histone modifications observed in experimental models are transcription-related active histone modifications associated to DNA hypomethylation [43].

In a recent study, Sanders et al. identified distinct sets of miRNAs in cervical swabs from 60 pregnant women and established their association with bone and blood-lead level. While miR-297 and miR-188 were increased in association to lead level, six miRNAs (miR-320e, miR-22, miR-93, miR-769, miR-425, and miR-361) were decreased. miRNA-297 had increased expression associated with patellar lead level as well as bone and blood-lead level. In an earlier study, miR-222 expression had a positive association and miRNA-146a expression had a negative correlation with lead exposure in peripheral leukocytes in subjects exposed to lead via inhalation. Whereas miRNA 146a is predicted to play a role in innate immune response, it has also been found to have an association with inflammation in rodent models. The miRNA 222 is known to regulate cyclin-dependent kinase inhibitor 1B (p27Kip1), a tumor suppressor protein. Downregulation of miRNA 222 has been linked to altered cell proliferation associated to incidences of several cancers [38].

SNPs Related to Lead Sensitivity

Several single nucleotide polymorphisms (SNPs) and mutations in various genes have been reported in the literature to have a role in the susceptibility to lead toxicity [40].

Glutathione S-transferases (GSTs) are a family of phase II metabolic isozymes known for their ability to catalyze glutathione conjugation for the purpose of detoxification. There are five main classes of GST enzymes in humans. Sirivarasai et al. observed that Thai individuals with GSTM1 and GSTT1 null genotypes and the GSTP1-Val105 allele were more sensitive to lead exposure than controls. In the Normative Aging Study participants, a Boston-based prospective cohort of men, Eum et al. reported that the GSTP1-Val105 allele and GSTM1 null genotype modified lead-related cognitive function. Individuals with GSTP1-Val105 allele performed worse on cognitive assessments. The role of GST gene variations in susceptibility to lead toxicity has been linked to changes in blood pressure and inflammatory responses [38].

Metallothioneins are a group of low-molecular-weight, cysteine-rich proteins that are involved in metal transportation, distribution, and detoxification inside the body. These proteins are encoded by the *MT* gene and they are known to have four major isoforms (MT1–MT4). High numbers of thiol groups in metalloproteins enable them to bind to metals effectively and to modulate homeostatic functions like absorption, transportation, and elimination. Their role in the metal homeostatic regulation has made *MT* gene variants novel candidates for the assessment of susceptibility to lead toxicity [38].

A lot of studies detect an association of lead exposure and toxicity to SNPs. Yang et al. found that SNPs in the *MT1A* gene, rs11640851 (647 A < C) and rs8052394 (1245 A < G)], were positively associated with decreased urinary uric acid level and excretion of N-acetyl-beta-D-glycosaminidase in chronic lead-exposed workers of Taiwan. Fernandes et al. evaluated the association of chronically lead-exposed Brazilian workers and *MT2A* rs10636 (C < G). They found that individuals carrying at least one non polymorphic allele of rs10636 had higher lead levels in their blood than those homozygous for the allele. Their results were consistent with observations of Gundacker et al., who also found negative associations between rs10636 and lead blood levels in Austrian students [38].

Another polymorphism in the *MT2A* gene, rs28366003, characterized by A < G at 5 core promoter region, has also been studied in Polish and Turkish populations. In the Polish population, the presence of mutant G allele was associated with higher lead levels and an increased risk of prostate cancer incidence. These findings were consistent with the findings in the Turkish population, where it was observed that individuals with the GG genotype had higher lead levels than those carrying at least one polymorphic allele. Tekin et al. also found that pregnant women with the AG genotype had significantly high maternal lead levels than those with AA genotype [38].

No studies have reported the involvement of *MT3* gene variants in susceptibility to lead toxicity. MT4 rs396230 (216 A < G) analysis in lead-exposed workers revealed an association with susceptibility to lead toxicity, blood pressure, serum renal function parameters, and urinary uric acid excretion. The workers with G allele were more susceptible to the toxic effects of lead on systolic blood pressure and serum renal function parameters [38].

Lead absorption is inversely associated with body iron burden. This has led to genetic studies to evaluate the role of iron metabolism genes in the modification of body lead burden. The *HFE* gene, located on the short arm of chromosome 6 at location 6p21.3, encodes for the HFE protein, which regulates the uptake of circulating iron by modulating the interaction of transferrin receptor with transferrin. So far, two mutations namely C282Y and H63D in the *HFE* gene have been studied in relation to lead toxicity. C282Y relation is associated with the G < A mutation at nucleotide 845, leading to a cysteine to tyrosine substitution at amino acid position 282, whereas H63D relation with the G < C substitution at nucleotide 197, leading to a histidine to aspartic acid substitution at amino acid position 63. Each mutation affects HFE1 protein expression and function, thereby leading to an increase in intestinal absorption of iron and its transportation into cells [38].

Apart from *HFE*, transferrin (encoded by the *TF* gene), a monomeric glycoprotein that facilitates the transportation of ferric iron among the sites of absorption, storage, and utilization, has been identified as a modulator of lead toxicity. Roy et al. studied a SNP rs1049296 (P589S) in *TF* gene and found that children who carried the variant form were prone to lead-related IQ reduction, and they showed increased susceptibility to lead-induced toxicity. In another Mexican study, it was detected that when genetic effects of *HFE* and *TF* were combined, the rs1049296 variant carriers showed high lead levels in children's blood. The carriers of both *HFE* and *TF* variant alleles had 50% higher lead blood levels in comparison to wild alleles. In a recent Korean study, lead levels had an association with plasma homocysteine levels and SNPs of *TF* gene were found to modify that association [38].

Vitamin D receptor (VDR) protein in humans is encoded by the *VDR* gene. The gene is located in chromosome 12q13.1 and it contains six promoter regions (1a–1f) and eight protein coding exons (2–9). To date, over 470 polymorphisms have been reported in this gene. Most of the polymorphisms are named based on different restriction enzymes (BsmI, TaqI, and FokI) capable of genotyping them. The frequency and distribution of these SNPs show a lot of national variation. These SNPs can modify bone mineralization by affecting the amount or activity of a protein that is regulated by vitamin D3 which may further interact with lead. This probable mechanism makes these SNPs suitable candidates for further research of any association with susceptibility to lead toxicity [38].

The X-ray Repair Cross-Complementing group 3 (*XRCC3*), a DNA repair gene, codes for a protein involved in homologous recombination repair of damaged DNA and cross-link repair in mammalian cells. A SNP, rs861539, characterized by 241 T < C is associated with various diseases, especially cancer. Recently, association between *XRCC3* polymorphisms and susceptibility to lead has been studied in occupationally lead-exposed Chinese workers. Individuals with CT or TT genotype showed significantly higher lead levels than those with XRCC3–241CC [38].

Cadmium

The Metal, Its Forms, and Its Distribution in the Environment

Cadmium is a chemical element with the symbol Cd which is chemically similar to the two other stable metals zinc and mercury. Like zinc, it demonstrates oxidation state +2 in most of its compounds such as Cd-chloride (CdCl2), Cd-acetate, ($Cd(O_2CCH_3)$ 2(H_2O) 2) and Cd-carbonate ($CCdO_3$) and like mercury, and it has a lower melting point than other transition metals.

Cadmium, in its pure form, is a soft silver-white metal, which is present in the earth's crust in association with multiple different metals, and it is extracted as a secondary product during the processing of metals such as zinc (Zn), lead (Pb), or copper (Cu). Cadmium (Cd) is a high-profile example of a hazardous trace metal despite being found naturally in the Earth's crust at an abundance of 0.1–0.5 mg/kg dry weight. The presence of cadmium and its compounds in the environment is a consequence of both natural and anthropogenic processes. Natural sources include volcanic activity, weathering destruction of rocks, sea aerosols, forest fires, and mobilization in soils and landfills [44].

Due to the extended industrial activity during the last two centuries, cadmium (Cd) is now a ubiquitous environmental contaminant that ranks at the top of the ten chemicals of concern. It is ranked as a priority pollutant worldwide due to the very long biological half-life (10–30 years), substantive accumulation along the terrestrial food chains, and toxicity at concentrations one tenth that of lead (Pb), mercury (Hg), aluminum (Al), or nickel (Ni). It accumulates and affects multiple functions. It interferes with tissue functioning networks, gene expression and epigenotype [45].

As a derivative of anthropogenic activities, it is widely used in the manufacturing of batteries, pigments, plastic stabilizers, pesticides and fertilizers, photovoltaic devices as well as in rubber processing, galvanization process, fossil combustion, and waste incineration. It is widely used for the manufacturing of mobile phones and computer circuit boards. It can be released to soil, water, and air and it bioaccumulates in the organic matter by entering the food chain [44]. Cadmium was used for a long time as a corrosion-resistant plating on steel, and cadmium compounds were used as red, orange, and yellow pigments to color glass, and to stabilize plastic (Fig. 4.14).

A common way of intoxication with heavy metals is the contamination of the soil with organic and industrialized fertilizers which can be absorbed, transported, and accumulated into parts of plants ingested by humans. However, the content measurement of As, Cd, Cr, Cu, Ni, Pb, and Zn in soil were below the limits established by environmental regulatory agencies. Contents lower than 3 mg/kg for As, Cd, Cr, Ni, and Pb are normally observed in dry matter. A prolonged intake of food with high concentrations of these elements may promote accumulation in the human body and may cause disease. People are exposed by dietary intake of contaminated food (particularly cereals and grains, leafy vegetables, potatoes, and offal). It can be perfused in alcoholic beverages, although their consumption represents a significant source of metals only in heavy drinkers [46].

Cigarette smoking is the main source of non-occupational exposure to cadmium. Among environmentally exposed population, tobacco smokers are the most exposed subjects, since tobacco leaves accumulate large amounts of cadmium. Cadmium is heavily concentrated in tobacco leaves and 50% of inhaled cadmium by smoking can be absorbed into the body making it the highest toxic metal found in cigarette smoke [48]. Indeed, it has been estimated that smokers absorb about 1–3 μg of the metal from smoking one pack of cigarettes per day which is double the concentration of cadmium in the blood of non-smokers. Geometric mean blood plasma cadmium concentration for adults in smokers and non-smokers is 1.58 μg/L and 0.38 μg/L, respectively (Fig. 4.14) [44].

The occupational exposure occurs almost exclusively by inhalation of cadmium-polluted fumes or dust and by ingestion through dust-contaminated hands [44]. Owing to its coloring property, cadmium is used in different cosmetic products, such as lipsticks, skin, and hair creams (Godt et al. 2006). Gondal et al. (2010) found cadmium concentrations ranging from 4.9 to 10.6 μg/g in lip cosmetics. Low contents of the metal have also been identified in shampoos, soaps, hair gels, body shower, and skin creams [44].

Geographic distribution of cadmium has been detected and recent data based on residential addresses of pregnant women revealed geographic clustering of cadmium cases and other toxic metals in an inner-city neighborhood, suggesting contaminated soils of some urban landscapes as a source of pollution [47]. Environmental health agencies also point to contaminated dietary staples, water, and pollution-contaminated air from which cadmium can be ingested or inhaled [44].

Cadmium is a good representative of environmental exposure to *toxic metals* and their negative effects in health. Heavy metals represent one of the most common categories of pollutants in the environment and they are nearly ubiquitous, being released in the environment mainly by anthropogenic emissions. Their distribution into the biosphere due to their wide use in industry represents a threat to human

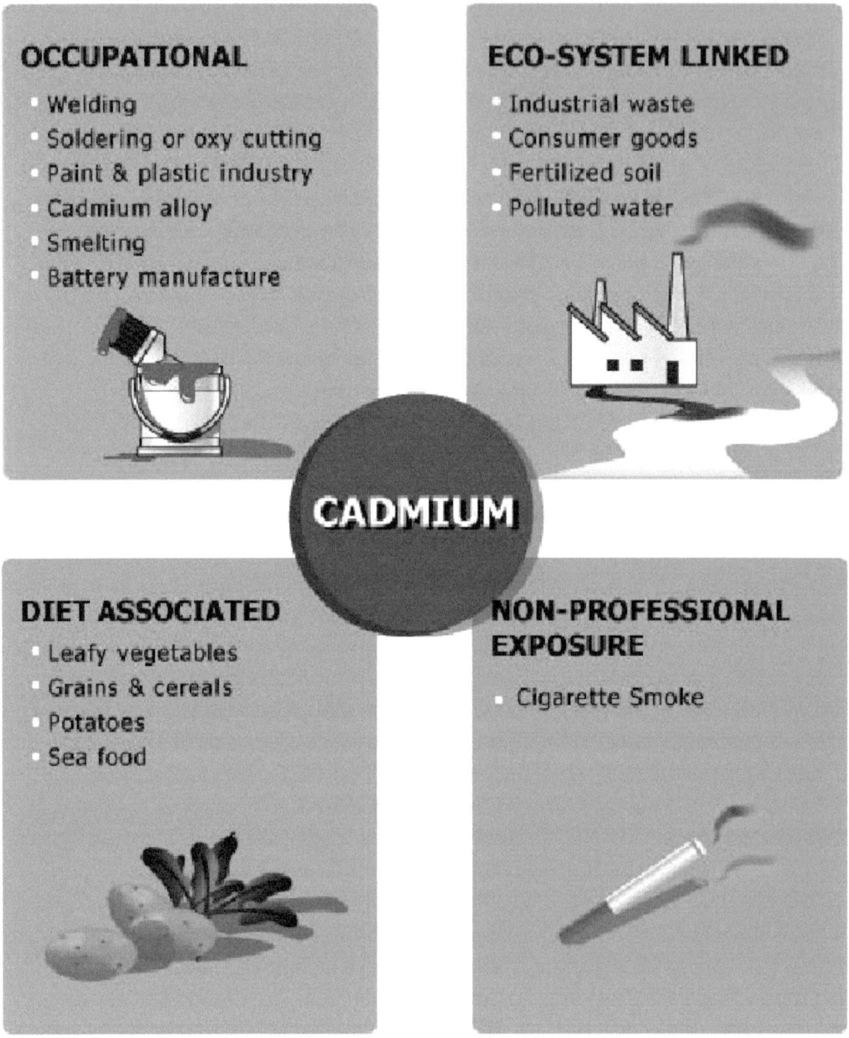

Fig. 4.14 The multiple sources of cadmium distribution

health, because of their long-term persistence in the environment and their accumulation in the organic matter (bioaccumulation) [49].

The route of intoxication from cadmium is both digestion and inhalation of the metal. The main characteristic of cadmium intoxification is bioaccumulation due to slow excretion rates. The accumulation affects mainly target tissues such as the liver and the pancreas, and long-term storage compartments, including the kidney, neural tissues, and bone. The half-life of cadmium in humans is impressively long and it is estimated to be between 10 and 30 years. Emerging data suggest that ethnic minorities may be at higher risk of non-occupational exposure due to contaminated residential soil and water. These people are frequently exposed to toxic levels of other metals such as arsenic (As) and lead (Pb) that are known to interact with Cd to exacerbate health effects [50].

In 2004, cadmium was officially categorized as a Class I human lung carcinogen based on epidemiological studies playing an important role in prostate, renal, liver, bladder, and stomach cancers [48]. Cadmium use generally starts to decrease, because of its toxic properties. It is specifically listed in the European Restriction of Hazardous Substances. For example, nickel–cadmium batteries have been replaced by nickel-metal hybrids and lithium-ion batteries. Its use in cadmium telluride solar panels is discussed.

The term "chronic exposure" is defined as continuous exposure to a particular compound over an extended period of time, whereas "post-chronic exposure" refers to the time after exposure when there is no longer exposure to the particular compound. Systematic research on post-chronic cadmium exposure can be extremely important, since it is relevant to ex-smokers and retired workers in high-risk occupations, but also to people from the general population who live in cadmium-polluted environments [47].

Low-level exposure to heavy metals often elicits no immediate symptoms and there is often a long latent period between exposure and the initiation of symptoms. Exposure may occur as early as the prenatal period and it is frequently associated with obesity that it may not become evident until middle childhood [51].

Cadmium levels in the blood above 5 μg/L and in the urine 10 μg/dL suggest cadmium intoxication. The metal bioaccumulates in the liver, placenta, kidney, lungs, brain, and the bones, where it substitutes calcium resulting in osteoporosis [47]. The higher concentration of absorbed cadmium (~99%) attaches to erythrocytes which have a lifespan of ~120 days [47].

Accidental Exposures in History

Cadmium is a well-known non-essential metal for the body recognized for its toxicity since nineteenth century. The disease Itai-Itai ("ouch ouch" in Japanese) is related to the toxic effects of high exposure to cadmium after the consumption of contaminated rice [48]. Itai-Itai disease causes unbearable pains of the bones and it is known as the cause of death of thousands of people in the area of Jintsu River Valley of Japan due to consumption of contaminated water close to zinc mines [44].

Effects on Health

The ingestion of the metal causes *acute toxicity*. The fatal dose is approximately 1 g. The symptoms of acute poisoning with cadmium are nausea, vomiting, cramps, and difficulty in breathing. *Chronic exposure* is related to alopecia, anemia, arthritis, learning difficulties, headaches, emphysema, osteoporosis, lack of taste and smell, anorexia, and cardiovascular problems as well as developmental failure in children. The chronic exposure may cause skin tumors and kidney damage. It can also cause bone brittleness and therefore increase in the risk of bone fractures (Pereira and Pereira 2018) [44]. Cadmium levels above 5.00 μg/dL cause muscle dysfunction and twitching, lung damage, and elevated blood pressure. Long-term exposure is carcinogenic and mutagenic, since cadmium acts as an endocrine disruptor [50].

A significant trend towards a worldwide decline in *human fertility* has been reported by international literature over the last years. The progressive deterioration of couple fertility cannot be justified exclusively by changes in lifestyle (increase in women occupational rate, increase in women age at pregnancy), prevention of undesired parenthood (use of contraceptives, intentional abortions), or economic issues (decrease of income, increase of expense). Environmental factors are implicated in the emerging problem. A significant overall decline in male fertility is the consequence of a progressive impairment of semen quality, reported since early '90s. Meta Analysis of studies published between 1938 and 1991 reports show that the fertility of couples declines. Nevertheless, a recent re-analysis of the same studies revealed heterogeneous results, mainly attributable to marked geographical variations in semen quality [44].

Cadmium is related to problems of reproduction in young couples in both males and females. The important finding of research is that it can be toxic even at low levels of exposure. Observational studies reported conflicting data; however, research categorize the metal as an endocrine disruptor [44].

In males, cadmium significantly increased the weight of the prostate gland and of the seminal vesicles. Increasing evidence confirm that the hypertrophy is related to endocrine abnormalities. Cadmium activates the androgen receptor, it suppresses the expression of the LH receptor and it binds to estrogen receptor. Consequently, it alters the effective concentrations of gonadotropins. More specifically it interferes to steroidogenesis enzymes such as StAR, cholesterol C20-22 desmolase, 17α-hydroxylase, 17β-hydroxysteroid resulting in hormonal imbalance. Cadmium exerts strong estrogenic- and androgenic-like actions, both *in vivo* and *in vitro* [44]. The name of the metals presenting with properties similar to the function of steroid hormones is "metallohormones."

Testis is particularly susceptible to cadmium, and testis injury occurs at high, but even low levels of exposure. Cadmium reproductive toxicity is mediated by multiple mechanisms, including structural damage of testis vasculature and of blood–testis barrier, inflammation, cytotoxicity on Sertoli and Leydig cells, oxidative stress, apoptosis, epigenetic actions, and disturbance of the hypothalamus–pituitary gonadal axis [44].

The estrogenic effects of cadmium are mediated by the high affinity binding to the ligand-binding domain of estrogen receptor α and the receptor isoform that drives the mitogenic actions of E2 in target organs. It induces proliferation and it promotes the expression of estrogen-induced genes *in vitro*. Moreover, cadmium is able to activate the non-genomic estrogen receptor α pathway, by extracellular regulated kinase 1/2 and protein kinase B, in breast cancer cells [44].

Independently from the dietary intake, *females* seem to be more prone to cadmium-related health effects apart from infertility, suggesting that a gender difference might exist in the susceptibility. The probable explanation of gender differences is different absorption. Indeed, the gastrointestinal absorption of dietary cadmium (about 5% in men and equal or more than 10% in women) varies among individuals and it is influenced by dietary intake of essential nutrients, including iron, zinc, and selenium [44].

Besides the damage on the endothelium, mainly expressed in the sensitive testis tissue, cadmium acts on the cells of other tissues. Reported health effects of high occupational levels are consistent with target tissue injury (e.g., nephrotoxicity, neurotoxicity, etc) [47].

Mechanism of Intoxification

The main intoxification mechanism of cadmium is the generation of ROS and the increase of oxidative stress within the cell. It interferes with the glutathione antioxidant system and it causes depletion of intracellular glutathione reserves and total thiol pools. It also interferes with the antioxidant function of different vitamins. The imbalance of redox equilibrium causes insufficiency to the energy production by inhibition of the synthesis of mitochondrial adenosine triphosphate (ATP). The effects of cadmium are non-specific. However, an increased lipid peroxidation and functional inhibition of antioxidative enzymes have been observed in numerous *in vivo* and *in vitro* studies [49].

The participation in free radical processes is evidenced by the fact that the supply of exogenous antioxidants (e.g., tannic acid, a-tocopherol, ascorbic acid, zinc, selenium) can inhibit the metal's toxic effects and it can increase the organism's overall antioxidative potential. After short-term exposure of rats to cadmium, increased activity of antioxidative enzymes in the animals' tissues was observed, which may indicate the cells' attempt to adapt to conditions of strong oxidative stress [49].

Direct toxicity is also described in the testis. Cadmium severely affects testis structure by damaging vascular endothelium and blood–testis barrier integrity and by inducing inflammation and apoptosis. The effect on the vascular endothelium is due to structural, metabolic, and functional damage of endothelial cells at high cadmium doses resulting in cell death [44].

The main structural change of intoxification is a direct change in the adherent junctions between the endothelial cells of the testis vessels. In particular, research has shown cadmium-induced alterations in the expression and function of the calcium-dependent cell adhesion molecule vascular endothelial cadherin (VE-cadherin) at the cell–cell contacts and a reorganization of the actin cytoskeleton. *In vitro* studies on human endothelial cells show that the metal at low concentrations demonstrates a direct inhibitory effect on cell metabolism, cell migration, and tube formation in testis, through a redistribution of VE-cadherin and a decrease in nitric oxide production by the endothelial cells. At high concentrations, cadmium produces acidification and permeabilization of lysosomes, followed by the release of active DNAse II and DNA degradation, leading to programmed cell necrosis of the endothelial cells [44].

Specific signaling mediators, such as the p38 mitogen-activated protein kinase and the transforming growth factor β2 and β3 molecular pathways, that are involved in occludin production and in the vessel-assembly, have been found to be implicated in cadmium-induced destabilization of membranes of the testis. The re-localization of occludin and of the protein involved in cell-cell adhesion, named focal adhesion kinase (FAK), at the Sertoli–Sertoli interface leads to endocytosis and, perhaps, recycling. A study demonstrated that CdCl2 treatment induced endocytosis of occludin and FAK, which were removed from the Sertoli-Sertoli interface and transported towards the cytoplasm [44].

In addition to direct harm of the cell membrane, experimental studies *in vivo* showed that cadmium was capable to disturb the balance between cell proliferation and apoptosis of the cells of Sertoli glands. It enhances the expression of the pro-apoptotic genes *p53* and Bcl2-Associated X (*Bax*) and it suppresses the expression of the antiapoptotic function of the Bcl2 protein. Similar effects were seen in mice germ cells, since cadmium treatment strongly induced germ cells apoptosis, driven by the downregulation of the antiapoptotic protein Bcl2-like 1 isoform 1 (*Bcl-XL*), and the upregulation of the proapoptotic proteins Bax and caspase 3. Further studies in mice demonstrated that both mitochondrial pathway and endoplasmic reticulum pathway are involved in cadmium-induced germ cells apoptosis [44].

As already mentioned, the toxicity of the metal is influenced by concentration of other metals in the body, primarily Cu, Fe, and Se [44]. *Ionic mimicry* is called the molecular similarity of the binding sites of cell receptors or enzymes to different metals. Two different mechanisms of mimicry have been identified: the ionic and the molecular. The ionic mimicry is the ability of "free" cadmium ions to imitate other chemical compounds and to substitute their function. For instance, cadmium has the ability to enter the cell by means of membrane protein carriers or channels used by other ions. The molecular mimicry is the ability of cadmium to substitute ions in their position in biological molecules interfering with the molecule's function. Ionic mimicry could modulate the concentrations of important ions within the cell and molecular mimicry could modify the function of biological molecules. In conclusion, cadmium could change other ions absorption, bioability, tissue distribution and excretion. Mimicry phenomena have been extensively described between Cd and Zn, whereas interference has been described between Cd, Zn and Se [44].

For example, high concentrations of cadmium in the blood are related with molecular mimicry phenomena in Zn receptors of the cells. Cadmium could replace Zn in receptors and enzymes, although the ZN ion is bigger that Cd. Cd and Zn share similar physical and chemical properties and mainly bind to biological macromolecules containing sulphydryl, hydroxyl, and nitroxyl groups. Cd in excess has a higher affinity for sulphydryl-containing proteins and nucleic acids and it quite easily substitutes Zn by molecular mimicry. On the basis of Cd/Zn similarities, Cd can potentially interfere to several Zn-mediated biological processes, including the reproductive function [44].

A putative route of cadmium entry in the cells is suggested to be transporters belonging to the ZIP family of Zn transporters. Cadmium intracellular concentration may rise by the activation of ZIP family Zn transporters by active transportation from the extracellular and intracellular milieu in a zinc competitive fashion. The active transportation is used in multiple levels: intestinal absorption, bioavailability, tissue distribution, and excretion [44]. On the other hand, Zn was detected to exert protective effects from Cd poisoning, by reducing its intestinal absorption and retention, as confirmed by experimental *in vivo* studies in birds exposed to Cd and fed adequate-Zn or low-Zn diets. *In vivo* studies on rats demonstrated that treatment of cadmium exposed rats with Zn protected the animals from the Cd-induced vascular injury and attenuated the severe pathological changes and the oxidative stress induced by cadmium [44].

In fact, the presence of Zinc finger motifs in steroid hormone receptors and other DNA binding molecules offers targets for cadmium binding with subsequent alterations in their function. In some cases, such as with the estrogen receptor, cadmium interaction will actually activate the receptor [48].

Cadmium shows molecular mimicry to selenium (Se). Se is an essential trace element with pivotal functions in the maintenance of male reproduction. It influences histological structure of the testis, spermatogenesis, semen quality, and, ultimately, male fertility [44]. The adequate dietary intake of Se, in both organic or inorganic form, is required for a proper protection of the male reproductive system, since Se and selenoproteins deficiency or excess results in oxidative stress, abnormal sperm maturation, and morphology and pathological alterations of the testis [44].

Several experimental studies in animals showed that cadmium might interfere with Se function at multiple levels. A diet rich in Se could ameliorate the toxicity of cadmium on the testis endothelium by a number of enzymes, including the acute regulatory protein (StAR), 3β-hydroxysteroid dehydrogenase, and 17β-hydroxysteroid dehydrogenase activity suggesting that the proper concentration of Se should be maintained in order to prevent Cd-induced damage [44].

It is already mentioned that cadmium interferes with the antioxidant mechanisms of the cells and it reduces the energy supplies of the cell. It is also mentioned that it reduces the concentration of glutathione and vitamins. As far as glutathione is concerned, cadmium affects the GSH peroxidase (GSH-Px) concentration. Se is a constituent of GSH-Px, a selenoprotein, which is protective against hydrogen peroxide, organic peroxides and specifically phospholipid peroxides. Optimal Se concentrations promote GSH-Px activity and it functions to reduce intracellular stress. Cadmium by substituting Se of the enzyme leads to dysfunction of its antioxidant activity.

Cadmium molecular mimicry to Calcium (Ca) interferes to Ca homeostasis, signaling and regulation of transcription of genes. Calcium is a crucial mediator involved in the regulation of sperm motility and acrosome reaction. Calcium and cAMP pathways are strictly and reciprocally linked, thus the effects of cadmium on Ca signaling also indirectly contribute to molecular changes downstream to cAMP.

The Cd-induced modulation of the expression of different calcium channel variants may be considered as additional cause of its reproductive toxicity [44].

Cadmium-induced alterations in the ubiquitin proteasome system are accompanied by the activation of stress signaling pathways, in particular by increasing the phosphorylated form of the stress-activated protein kinase/c-Jun N Terminal kinase (SAPK/JNK) and p38 [44].

Overall, cadmium is suggested to affect both testis spermatogenesis and endocrine function by several mechanisms icluding (1) a direct functional impairment of Sertoli and Leydig cells, (2) oxidative stress in both somatic and germ cells, and interference with essential ions involved in ROS scavengers' activity, and (3) apoptosis in germ cells, and (4) ionic and molecular mimicry and (5) Epigenetic changes were also demonstrated in Cd-treated Leydig cells and testis extracts from animals, although these changes were not directly linked to reproductive dysfunction. Lastly, (6) a direct disturbance of the hypothalamic-pituitary-gonadal axis after cadmium treatment was also reported [44].

The accumulative low exposure to cadmium is related to *carcinogenicity*. Metals with carcinogenic potential include As, Be, Cd, Cr, and Ni, while other candidates are Cu, Fe, and Pt. How these metals induce cell transformation of normal cells, it is proposed that a range of molecular mechanisms are implicated depending on the diverse chemical properties of these metals [52]. The oxidative concept in metal carcinogenesis proposes that complexes of the metals (Co, Cr, Cu, Fe, Ni) formed *in vivo* undergo redox cycling forming high levels of metal ions which generate reactive oxygen species and they oxidize DNA sequence. Some of the products of oxidative DNA damage, including 8-oxoguanine and strand breaks, induce mutations, which may lead to neoplastic transformation. The establishment of metal-binding modes in the cell nucleus and their reactivity is crucial for the understanding of molecular events in metal carcinogenesis [52].

There are four major Cd-induced carcinogenesis mechanisms: (1) oxidative stress, (2) attenuation of apoptosis, (3) inhibition of DNA damage repair, and (4) alterations in gene expression. In addition, cadmium may exert carcinogenic effects by (5) endocrine disruption, and (6) aberrant DNA methylation.

Epigenetic Mechanism of Intoxification

Cadmium is a genotoxic and mutagenic agent. Although it has a weak DNA binding activity, it can cause chromosomal aberrations and DNA damage. In addition, cadmium interferes with the apoptosis mechanisms protecting from oncogenesis [48]. Moreover, cadmium exposure is associated to the activation of oncogenes which promote cell proliferation (i.e., *c-myc*, *c-jun*, and *c-fos*) as well as to the inhibition of tumor suppressor genes such as *p53* and *p27* [48]. As a prominent tumor suppressor gene, Cd-induced inactivation of *p53* has been shown to disrupt cell cycle arrest and apoptosis. Specifically, cadmium is able to alter *p53* structure and function of by displacing zinc. The acquisition of apoptotic resistance by *p53* inactivation may be critical for Cd-induced carcinogenesis [48].

Cadmium interferes with at least 60 key signaling pathways, including important cancer-related pathways such as HIF-1α, NF-κB, RAS, and PI3K-Akt. Signaling pathways involving TGF-β, a membrane receptor-binding protein implicated in various types of malignant cell transformation, also exhibited increased activity following Cd exposure. Cadmium exposure can also induce aneuploidy and loss of expression of tumor suppressor genes such as *RASSF1A* and *P16* by promoter hypermethylation [48].

In this context, the Cd-selective metallothionein (*Cd-MT*) gene deserves particular consideration since it can be activated by dietary cadmium levels as low as 2 mg/L. Its function is generally inducible, unlike constitutively expressed genes with housekeeping functions. The *Cd-MT* gene contains CG sites, but there is no information available about its methylation status [51].

The similarities in physical and chemical properties of cadmium and Zn enable them to antagonize each other. The displacement of Zinc finger binding domain of TET protein leads to conformational and activity changes of the protein with the potential of alterations of the methylation status of DNA (Fig. 4.15).

Cadmium was positively correlated to global DNA hypermethylation, as measured by the LINE-1 methylation status in men, and when adjusting results to smoking habits, this correlation was maintained. Similar results were also described in male newborns exposed prenatally to cadmium. LINE-1 hypermethylation has been associated with an impairment of semen quality, in particular an impairment of total

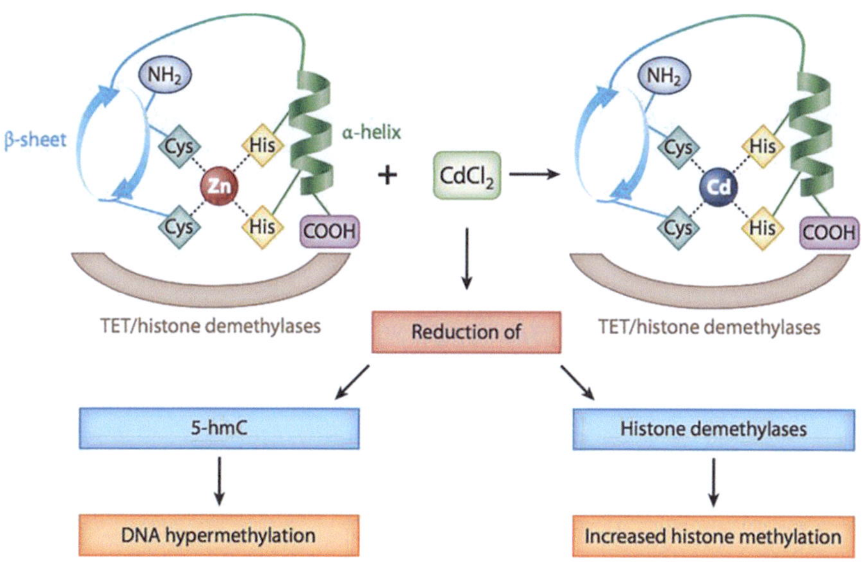

Fig. 4.15 Zn substitution by Cd in the DNA binding domain of TET protein leads to different epigenetic alterations [48]

motility and a progressive loss of function in humans. Nevertheless, to the best of our knowledge, no studies directly evaluated the epigenetic signature in Cd-exposed versus non-exposed male subjects, nor directly correlated epigenetic changes to semen quality and fertility potential [44].

Other studies suggest that cadmium can inhibit DNA methyltransferases, thereby inducing DNA hypomethylation. It has been reported that short-term cadmium exposure reduced DNA methylation, by inhibition of DNMT1. Without the activity of this enzyme, methylated cytosines in DNA will be lost due to passive DNA demethylation, since this enzyme preserves parental cytosine methylation on the daughter strand [48]. The discrepancy between the opposite effects of long- or short-term cadmium exposure on global DNA methylation is not understood, although a changing landscape of chromatin methylation in either direction may initiate signaling pathways responsible for many downstream processes [48].

The oxidative stress is not directly related to cadmium toxicity. However, brain is a target of cadmium-mediated oxidative stress effects [50]. The association between prenatal programmed obesity and cadmium exposure has been described. The mechanism by which heavy metal exposure might lead to obesity may involve the effects of metal neurotoxicity on brain function and signaling related to appetite and satiety. Since brain development is affected by both lead and cadmium, a disruption in energy balance could result to a dysregulated appetite and satiety response, with consequent increased caloric intake [51]. The increased oxidative stress and inflammation are associated with childhood obesity due to early-life exposure to cadmium [50].

In humans, early-life cadmium and lead exposure is also associated with higher risk of attention-deficit/hyperactivity disorder, a neurodevelopmental condition that is linked to obesity. The *PLAGL1* gene was found to be involved in neocortical development. It has been detected a positive association between its methylation and lead exposure, while its reduced expression is associated with obesity [50].

Studies for the explanation of infertility have focused on methylation modifications during spermatogenesis in cadmium-exposed cells. A complex and precise epigenetic reprogramming takes place in germ cells and it is essential for spermatogenesis. An improperly elevated global hypomethylation, detected in LINE repetitive elements, a proxy marker of global DNA methylation status, and a locus specific hypermethylation of imprinted and non-imprinted genes, was demonstated in poor human semen quality. Moreover, a locus specific (i.e., *H19*, *MEST*, *BRDT*, *MTHFR*) alteration of methylation patterns was associated with the impairment of semen quality or azoospermia in human subjects [44].

It was suggested that cadmium exposure could cause heritable changes in chromatin structure which are related to rapid transcription activation and to the establishment and maintenance of a bivalent chromatin domain at the *MT-3* promoter (Martinez Zamudio and Ha 2011). In Cd-transformed urothelial cells, levels of H3K4me3, H3K27me3, and H3K9me3 occupancy at the *MT-3* promoter were increased compared to untransformed cells, indicating chronic cadmium exposure may alter transcriptional responses by histone modifications [53].

In addition, cadmium has the ability to alter the expression of various DNA repair genes, and it is known to be able to impair the DNA repair system by the disruption of proteins containing zinc finger motifs. Evidence support the interference to the repair pathways of single-stranded DNA (ssDNA) damage, e.g., base excision repair (BER), nucleotide excision repair (NER), and mismatch repair (MMR), but there are very limited for interference to double-stranded DNA (dsDNA) damage such as homologous recombinational repair (HRR) and non-homologous end-joining (NHEJ). In order to better understand the mechanisms of cadmium carcinogenesis, it is important to analyze the impact of cadmium exposure to the DNA repair machinery, allowing the accumulation of cells with DNA damage which in turn will produce mutations during cell division [47].

Research has analyzed the impacts of cadmium exposure to a cluster of imprinted genes, the DMRs (Demodulation Reference Signal). The cluster includes the paternally expressed *DLK1* (Delta-like 1) and the maternally expressed *MEG3* and *MEG8* genes at chromosome 14q32.2. *MEG3* encodes a long noncoding RNA (lncRNA) that is normally expressed in multiple tissues including the placenta, the brain, the adrenal gland, the pancreas, the spleen, and the liver. The expression of these genes is controlled by CpG methylation at the germline. More specifically, the *MEG3*- IG methylation of the paternal allele is established in sperm and the *MEG3* is established post-fertilization [47]. Hypermethylation of the maternally imprinted *MEG3* DMR has been associated with multiple forms of cancer, many in organs targeted by Cd [49].

MEG3 functions as a tumor suppressor gene and it is related to the activation of *p53*. *p53* is a crucial tumor suppressor checkpoint involved in cancer. It responses to cellular stressors, acting as a transcription factor and inducing the expression of genes involved in senescence, apoptosis, and cell cycle arrest. *MEG3* has be shown to interact with the *p53* DNA-binding domain at various *p53* target genes. Ectopic expression of *MEG3* in human cancer cell lines restores tumor suppressive activities and it results in accumulation of *p53* protein, activation of *p53* downstream targets, and inhibition of cell proliferation by inducing cell cycle arrest and apoptosis. It is therefore plausible that Cd-induced decreases in *MEG3* transcripts that could result in loss of *MEG3* tumor suppressive function and to increase susceptibility to cancer [47].

In mice, deletion of the *MEG3-IG* on the maternal chromosome results in an epigenotype switch, where the domain starts exhibiting the transcriptional profile of the paternal chromosome. In a similar way, aberrant hyper-methylation of the *MEG3-IG* or *MEG3 DMRs* corellates with greatly decreased *MEG3* transcripts, which in turn is associated with increased expression of the reciprocally imprinted *DLK1*, cell proliferation *in vitro*, more aggressive and higher-grade human tumors including meningioma, colon, nasopharyngeal, liver and leukemia [47].

Current studies have not explored other types of epigenetic mechanisms including histone posttranslational modifications. The role of miRNA deregulation in Cd-induced carcinogenesis has not been well studied [48]. Only a study among electric-furnace steel plant workers, experiencing metal-rich (lead, cadmium, chromium) particulate matter exposure has resulted in significant alterations of miRNA expression in peripheral blood leukocytes [52].

Aluminum

The Metal, Its Forms, and Its Distribution in the Environment

Aluminum is a chemical element with the symbol Al and it is a silvery-white, soft, non-magnetic, and ductile metal in the boron group. Aluminum makes up about 8% of the Earth's crust and it is the most abundant metal. It occurs naturally in the ore bauxite, various clays, and alumino-silicate minerals, and it has a preferred oxidation state +3. Aluminum metal is highly reactive and the native specimens are rare and limited to extreme reducing environments. Instead, it is found combined in over 270 different minerals.

Until recently people believed that aluminum could be easily washed out from the biosphere, because it is inert and insoluble and thus there are no health dangers from its use. Therefore, aluminum compounds are widely used. Aluminum is already ubiquitous in daily uses and in occupational environments due to its special physical and chemical properties. It forms deposits in acid environment and a lot of aluminum compounds are widely used in a soluble form [55].

Aluminum can enter the human body by ingestion and inhalation after environmental exposure, which is mainly by the diet, drinking water, beverages, or medicines. The most common form of human exposure is the gastrointestinal tract and and the metal is absorbed across all parts of the intestine including the colon [59]. Aluminum is a nonessential element for the human body and it does not have any physiological function. However, more detailed analysis of its properties showed that the body is difficult to eliminate the metal, which bioaccumulates. The presence in high levels has recently been related with a number of diseases [55].

It has been suggested that genetic polymorphisms predispose to a greater susceptibility to its adverse effects. A group, that shows susceptibility to exposure, is children, pregnant mothers, and women of child-bearing age. Therefore, it is important these groups to avoid unnecessary exposure to the metal.

Aluminum cookware may be replaced by safer alternatives, while aluminum-containing antiperspirants, potentially implicated in the rise of cases of breast cancer, may be replaced by non-aluminum products. The use of aluminum salts in medical products is a more controversial issue, but antacids should be avoided [56]. Aluminum levels may also be elevated in patients on hemodialysis [57].

A known, but still unrecognized for its significance, source of exposure to high levels of aluminum salts are cosmetic preparations, especially in western societies. Aluminum chloride, aluminum chlorhydrate, and aluminum zirconium chlorhydrate glycine complexes are licensed to be used at levels up to 15%, 20% and 25% by weight, respectively. As antiperspirants, aluminum salts are left on the skin around the underarm and breast area, not washed off products as shampoos and soaps, which allows the continuous exposure [58].

However, the main source of exposure to aluminum is drinking contaminated water. A lot of research programs have found a strong positive correlation between the level of aluminum in drinking water and the incidence of Alzheimer's disease in countries such as the United Kingdom, Canada, Norway, and France. The most recently reported association was published by Rondeau et al. in 2009 who demonstrated that high daily consumption of aluminum in drinking water was associated with a significantly increased risk of developing mild cognitive impairment. The French cohort study involved 1925 recruits and follow-up extended for up to 15 years [59].

In 2011, the World Health Organization and the Food and Agriculture Organization of the United Nations revised the acceptable limits of aluminum to 2mg/kg based on research in humans and rats exposed to the metal. It noticed that 900mg/L concentration in drinking water "could be derived" after treatment with aluminum-based coagulants. Drinking water with high aluminum concentrations above 900 μg/L has been reported in the United States, Argentina, Bangladesh, Brazil, Ethiopia, Ghana, India, Iran, Kosovo, Lesotho, Sri Lanka, Sweden, Tanzania, Turkey, and Uganda [59].

Effects on Health

Acute poisoning with aluminum is not reported in the literature, even though products with soluble aluminum cover a lot of products in the market. Chronic as well as prenatal exposure is implicated in the development of neurogenerative disorders and this topic will be discussed in detail [55].

The accumulation of metals and mainly aluminum and iron in the brain has been identified with microprobe studies on characteristic neurofibrillary tangles in cases of Alzheimer's disease. Similarly, increased concentrations of iron and aluminum have been noted in the neurons of the substantia nigra in cases of Parkinson's disease patients (Good et al. 1992). It is not considered likely that these elemental accumulations play an etiologic role for these diseases, although they may contribute to their pathogenesi [57].

Alzheimer's Disease (AD) is a common, progressive, irreversible, neurodegenerative disease in the elderly. The main clinical feature of AD is the progressive impairment of memory, judgment, decision-making, orientation to physical surroundings, and language. The typical pathological hallmarks of AD are the presence of extracellular senile plaques (SPs) containing the amyloid protein (Aβ) and neurofibrillary tangles (NFTs) composed by hyperphosphorylated tau protein. The deposition of the plaques leads to irreversible loss of neurons, particularly in the cortex region and hippocampus [55].

The extension of average lifespan of the human population in the industrialized world it is accompanied by rapidly increasing numbers of people suffering from AD that it has become a major public health problem of the elderly age. Nowadays, AD has been the most common form of dementia and it accounts for around 70% of dementia cases. It is estimated that its prevalence will increase significantly in the coming decades. Despite improvements in the knowledge and understanding of the disease, there are currently no effective approaches to prevent, cure, or even slow down its progression, because its exact etiology and pathological mechanisms remain to be determined. The late-onset form of Alzheimer's disease (onset ≥65 years) is the more common form of AD, accounting for 90–95% of cases, and occurs sporadically [55].

Research on the genetic etiology of AD revealed candidate genes for the development of the familiar form of AD. Many of these genes are associated with both the expression and metabolism of APP protein. The mutations in the presenilin 1 (*PSEN1*), presenilin 2 (*PSEN2*), and *APP* (amyloid precursor protein) gene are the well-known genetic cause of familial AD, while the mutations in the apolipoprotein E (*APOE*) ε4 allele show high susceptibility to the development of sporadic AD. However, the same genes, *PSEN1* and *PSEN2*, *APP* genes, which all encode proteins involved in APP breakdown and Aβ generation, have been firmly implicated in the pathophysiology of other neurogenerative diseases also. AD-linked mutations within these three genes exhibit high penetrance (>85%). Mutations in 21 other genes and an 18q deletion syndrome have also been reported to be associated with tau pathology. The gene susceptibility and mechanisms leading to (or accompaning) the impairment of the central APP processing and tau networks are widely accepted as the major contributors to a diseased state [55].

A small amount of aluminum exposure on rats has the potential to redirect APP cleavage from its non-amyloidogenic pathway, forming sAPPα, to its amyloidogenic pathway, forming Aβ. The susceptible rats chronically exposed to the metal by drinking water or food show a slowly progressing cognitive function impairment [59]. APP/tau triple-transgenic (3xTg-AD) mice are reported to have higher aluminum levels in their brains than controls, even without supplementation. These findings indicate that there is a close relationship between overexpression of APP and PSEN1 genes and aluminum overload [59]. Chronic or cumulative exposure reflected by increased levels in the serum and in the cerebrospinal fluid (CSF) could also be implicated in the pathogenesis and pathophysiology of other diseases also, such as Multiple Sclerosis and Parkinson's disease [56].

Metal neurotoxicants, and aluminum as a central one, are suggested as main environmental risk factors of AD. A lot of studies demonstrate that the levels of the metal are significantly elevated in serum and CSF of patients with AD [55]. In 1973, it was first shown that gray matter from brains of AD patients contains more aluminum than the brains of non-demented age-matched controls. The analysis of brain autopsy samples from AD patients has also indicated that excessive exposure may contribute to the beginning and/or progression of AD. Moreover, chelation therapy to reduce the aluminum burden has been reported as beneficial. However, these first results were inconsistent between studies.

There is considerable *in vitro* and *in vivo* evidence demonstrating that aluminum ions inhibit the dephosphorylation of tau, potentiate the development of neurofibrillary tangles (NFTs), cause the accumulation of amyloid beta protein and accelerate the formation of amyloid plaques [56]. A specific study proposes that environmental factors in combination with genetic factors, may play a very important and controllable role in the development of AD, especially in neurobehavioral changes and neural cells loss [55].

Other studies have not replicated the above data and the enthusiasm, relating aluminum exposure to the etiologic implication of aluminum in the pathogenesis of AD, waned. Extensive analysis data suggested that aluminum levels were no higher in the brains and in the cores of amyloid or neuritic (senile) plaques of AD patients than in healthy controls. Furthermore, there was a failure to detect aluminum in neurofibrillary tangles (NFTs) and amyloid plaques in post-mortem tissue at higher levels in AD than in age- and sex-matched controls [56].

It is interesting to note that similar factors seem to be implicated for the autism spectrum disorders in childhood. *Autism spectrum disorder* (ASD) refers to an increasingly common group of heterogeneous disorders identified by the presence of impairment in social interactions and communication together with a restrictive range of repetitive and stereotypical behaviors [56].

Most neurodevelopmental and neurodegenerative diseases are now thought to arise from a combination of genetic and environmental factors, so they are categorized under the term multifactorial diseases. In general, none of these factors has complete penetrance and only the combination leads to the onset of the disease. It is speculated that individual genetic characteristics predispose for the harmful effects of environmental exposures [55].

A lot of interest has focused on the perspective that aluminum immunization adjuvants could potentially contribute in the development of ASD. A recent analysis in the USA, applying Hill's criteria for establishing causality, reported that children, living in countries with the highest prevalence of ASD, appear to have the greatest exposure to vaccine-based aluminum. It recorded after a 20-year analysis a significant correlation. A wider analysis revealed the presence of significant positive correlation between the levels of aluminum in adjuvants in vaccines administered to preschool children at around 3–4 months old and the existing ASD prevalence in seven major Western countries [56]. Another analysis detected a significant association only between the first measles, mumps, and rubella (MMR) immunization before the age of 36 months in African-American males and a diagnosis of ASD. However, these analyses were challenged by the opposite results. This last study has since been retracted [56].

There was a considerable research and media concern, regarding an increased number of cases diagnosed with autoimmune or other immune-mediated illnesses, previously immunized for the hepatitis B virus. The largest such study evaluated the medical records of 93 patients and reported prevalence rates of different manifestations as follows: neurological 67%; general symptoms 60%; musculoskeletal 60%; gastrointestinal 51%; fatigue 42%; ophthalmological 32%; muco-cutaneous 30%; sleep disturbance 19%; psychiatric 16%; and local reactions 11% [56].

However, in spite of the publicity for the immunization threats related to autism, some questions could not be answered. For example, that if aluminum or other adjuvants per se provoke chronic central nervous system and peripheral pathology, why autism is not equally distributed in different countries in accordance with the level of immunization and why the prevalence is rapidly increasing although the state of immunization remains stable. Given the ubiquity of vaccination, one would expect that almost every child would be affected in western societies. It rather seems that adjuvants are highly unlikely to be related to the cause of ASD [56].

Aluminum salts have been used as adjuvants due to their ability to activate innate immunity. They induce the production of the interleukins IL-1β, IL-8, and IL-18 in a TLR-stimulated manner of dendritic cells and macrophages. The inclusion of adjuvants in vaccines increase the immunization results by the activation of inflammasome and specifically the activation of the nucleotide-binding oligomerization domain-like receptor pyrin domain-3 (NLRP3). Aluminum salts can activate the NLRP3 inflammasome with a number of different routes. These include destabilization of phagosomes, acidification of lysosomes, and increases in reactive oxygen species (ROS) levels [56].

Aluminum adjuvants provoke an immune response by activating DAMPs (Damage-associated molecular patterns), mainly uric acid and host DNA (outside cell nuclei and mitochondria). There is a body of evidence demonstrating that uric acid and DNA are released *in vivo* following aluminum hydroxide injection. Uric acid is synthesized during purine nucleotide catabolism and its concentration increases during cellular stress, such as at the site of injection. The functional relevance of uric acid levels in increasing T-cell priming and the instigation of humoral immune responses has been repeatedly demonstrated. Both uric acid and aluminum hydroxide can independently activate the NLRP3 inflammasome, inducing the secretion of IL-1β. Uric acid crystals administered in the form of an adjuvant can also induce complement responses and a T helper type 2 (Th2) cell differentiation pattern. Host DNA released into the intracellular space following cellular necrosis also acts as a DAMP. Evidence to the highly immunogenic nature of double-stranded DNA was reported. Double-stranded cytosolic DNA is sensed by TLR-9 and it causes the activation of nuclear factor (NF-κB) and interferon-beta (IFN-β) pathways and the activation of irf3 [56]. *In vivo* studies also show that aluminum hydroxide appears to induce dendritic cell and T-cell activation by NLRP3 activation, although alternative routes such as immunoreceptor tyrosine-based activation motif (ITAM) and interferon response factor 3 (irf3) activation could be involved [56].

Autism is a disease with an important health burden in society. The etiology of the disease is not known. Genetic predisposition is only applied for few cases of patients with the disease. Chronic inflammation and dysregulated immune response to environmental insults during critical periods of fetal development are currently implicated in the development of disease. Autism is a disease with an important health burden in society. The etiology of the disease is not known. Genetic predisposition is only applied for few cases of patients with the disease. Chronic inflammation and dysregulated immune response to environmental insults during critical periods of fetal development are currently implicated in the development of disease [56]. Evidence show dysregulated immune responses in combination with specific polymorphisms genetically inherited contribute to the chronic activation of immune and inflammatory pathways ultimately destroys microglia and astrocytes in the brain [56].

Another issue of concern is the fact that aluminum can act as an endocrine disruptor. Estrogens are a well-established influence in breast cancer. Results reported demonstrate that aluminum in the form of aluminum chloride or aluminum chlorhydrate can interfere with the function of estrogen receptors of *MCF 7* human breast cancer cells both in terms of ligand binding and of estrogen-regulated gene expression. Aluminum acts on intracellular receptors which function as ligand-activated zinc finger transcription factors. Overall, aluminum is added to the increasing list of metals capable of interfering with estrogen action [58].

Mechanisms of Intoxification

Aluminum exposure is related to *neurotoxicity* by the combination of several mechanisms. It is the result of oxidative stress and inflammatory processes which final result to apoptosis and neuronal damage. Oxidative stress in the mitochondria of the cells and especially of neurons is responsible for many of the pathogenetic findings of neurodegenerative and neurodevelopmental diseases [55].

Oxidative damage associated with mitochondrial dysfunction constitutes the major vehicle underpinning aluminum-induced neurotoxicity. It is maintained by many enzyme pathways that impair mitochondrial function and toxicity indirectly by mechanisms inducing endoplasmic reticulum (ER) stress. The induced by prolonged aluminum exposure is evidenced by increased lipid peroxidation and by depleted antioxidant defenses and it is morphologically obvious in the prefrontal cortex, the cerebellum, the hippocampus, and the brainstem. The increased lipid peroxidation subsequent to aluminum-induced mitochondrial dysfunction is enabled by the increase of lipogenic enzymes such as acetyl CoA carboxylase. The increase in activity of these enzymes is accompanied by decreases in the activity of key enzymes of the electron transport chain and the tricarboxylic acid (Kreb's) cycle, such as succinate dehydrogenase and AKG (alpha ketogluaric acid). The reduced activity of these energy-associated enzymes leads to a significant decrease in the levels of ATP produced by oxidative phosphorylation [56].

The loss of AKG activity in mitochondria and cytosol results in significant decrease in energy generation within the glial cells [56].AKG acts to stabilize hypoxia-inducible factor-1-alpha (HIF-1α) in the cytoplasm and it prevents its translocation to the nucleus. However, in a cellular environment of increased aluminum cations and lipid peroxidation, AKG levels reduce and HIF-1α translocates to the nucleus provoking increases in the transcription of hexokinase, pyruvate kinase, lactate dehydrogenase, and glucose-6-phosphate dehydrogenase. The ultimate effect is the switching from energy production by oxidative phosphorylation to the phylogenetically more ancient pathway of energy production by glycolysis [56].

The exposure to aluminum ions leads to a significant decrease in energy production within the cell. It specifically inhibits the activity of cytochrome C oxidase, NADH, and succinate dehydrogenase with a subsequent decrease in state 3 (ADP stimulated) and state 4 mitochondrial respiration reactions. The conformational changes of these enzymes are a direct result of aluminum ion binding. Aluminum impairs mitochondrial biogenesis by decreasing levels of peroxisome proliferator activated receptor gamma coactivator 1α (PGC-1α) activity, either directly or indirectly as a result of inducing elevated levels of oxidative stress. Aluminum ions also display the capacity to bind to the phosphate groups of ATP and ADP and the dephosphorylation of the former and the inhibition of the phosphorylation of the latter molecule and this together with the capacity to inhibit a wide range of kinase and phosphatase enzymes can grossly impair energy homeostasis [56].

Furthermore, aluminum exposure induces mitochondrial dysfunction by provoking release of calcium ions from intracellular stores. The functional and physical relationship between the endoplasmic reticulum (ER) and mitochondria is well documented in apoptosis, but perhaps under-discussed in the case of ER stress, which is sub-lethal to the cell. In the latter environment, the unfolded protein response in general and protein kinase RNA-like ER kinase activity in particular, leads to a state of chronic mitochondrial underperformance rather than cellular death. It is noteworthy that aluminum-induced oxidative damage and disruption of calcium ion homeostasis is similar in pattern to that seen in AD [56].

Calcium dyshomeostasis is equally detrimental to mitochondrial and to neuronal function since calcium is of vital importance in maintaining the matching of mitochondrial energy production to demand. In particular, modest elevations of calcium ions in the cytosol, following increases in neural activity act as the "gas pedal" (or "accelerator") to increase energy production and they maintain ATP homeostasis, hence impaired calcium homeostasis can have profound adverse effects on neural function, even in the absence of frank apoptosis [56].

Aluminum is a highly charged element that binds strongly to proteins and acts as a cross-linking stabilizer. Many of the proteins comprising the intraneuronal inclusions that characterize the neurodegenerative diseases, the neurofibrillary tangle being the most prominent example, are highly cross-linked and thus resistant to degradation. Whether aluminum has a role in the natural history of neurogenerative disorders remains unclear, although recent evidence appears to support this concept [57].

Apart from oxidative stress and dysfunction of calcium homeostasis, aluminum also impairs the antioxidant defense mechanisms in the brain. It induces significant alterations to glutamate/glutamine recycling within astrocytes leading to increased glutamine to glutamate conversion coupled with increased uptake of glutamate and increased intracellular levels of glutamine. The depletion of glutathione appears to be affected by reduced activity of glutathione synthase, glutathione reductase and changes of the plasma membranes resulting in a reduced influx of glutathione from the lumen to the mucosa. This aluminum-induced depletion of glutathione impairs the activity of calbindin-D9k resulting in decreased of the transcellular absorption of calcium ions. Given the positive role played by calcium ions in maintaining epithelial barrier integrity, depletion of glutathione could well underpin the increases in intestinal inflammation and intestinal barrier permeability caused by prolonged consumption of aluminum [56].

Aluminium decreases levels of this thiol by inhibiting the activity of NADPH dependent isocitrate dehydrogenase in mitochondria and malic enzyme and NADPH isocitrate dehydrogenase in the cytosol. Thus, it modulates glutamatergic and GABA-ergic neurotransmission, but may also have significant bioenergetic consequences given that increased levels of glutamate within astrocytes act as a further stimulus for increased glycolysis. These unclear observations are important from the perspective of potentially impaired neurodevelopment as astrocytes play an important role in the development of the brain by regulating processes involved in synaptic transmission, neuronal migration, synaptogenesis, and maybe even myelination. The weight of evidence also indicates that the activity of these glial cells is of paramount importance in the development and maintenance of neural networks and circuits [56].

Chronic or prolonged exposure to aluminum can induce astrocyte apoptosis mediated by *p53* leading in DNA and chromatin damage. Apoptosis could also be induced by inhibited mitochondrial function and ATP production, ultimately causing necrosis, which can have profound and prolonged neuro-inflammatory consequences. Prolonged aluminum exposure can also induce significant metabolic changes in astrocytes compromising their function even in the absence of degeneration. Metabolic abnormalities include the decreased activity of γ-butyrobetaine aldehyde dehydrogenase and γ-butyrobetaine aldehyde dioxygenase and the reduced levels of α-ketoglutarate (AKG) leading to low levels of L-carnitine and subsequently impaired fatty acid beta-oxidation, mitochondrial dysfunction, reduced ATP production, and increased lipogenesis [56].

Moreover, a consequence of oxidative stress due to chronic aluminum exposure stress is the *activation of microglia immunity* towards inflammation by the excretion of TNF-α, IL-6, and cytokine-inducible nitric oxide synthase (iNOS or NOS-2). There is overwhelming evidence that the inflammation of microglia compromises microglia function, which has an indispensable role in the development of the brain by regulating processes such as synaptic pruning, synaptic plasticity, synaptogenesis, neuronal development, and other vital processes of neurogenesis [56].

The stepwise development of microglia is regulated by the activity of several genes and the disruption in the expression of these genes can occur as a result of prenatal immune activation due to environmental factors. However, an important pathway impaired after metal intoxification is multifunctional molecular complex of the adapter protein, DNAX activating protein of 12 kD (DAP12) with TREM2, which is expressed on the surface of glial cells [56].

The *DAP12/TREM2 protein complex* plays a major role in the regulation of central nervous system homeostasis and detoxification. TREM2 binds a range of liposaccharides, phospholipids, and polyanions with the subsequent activation of DAP12 and an array of downstream kinases cumulating in cellular activation. Activation of the *DAP12/TREM2* axis has an important role in limiting neurotoxins following TLR activation by PAMPs and DAMPs, and it promotes the survival and proliferation of microglia and other cells of the myeloid lineage. However, in recent years, the bulk of research on the *DAP12/TREM2* complex has focused on its role in enabling and regulating microglial phagocytosis. TREM2 acts as a molecular sensor of macromolecular debris and it plays an indispensable role in its immunogenically silent clearance, thereby resolving potentially inflammatory reactions and acting as an impediment to the development of neuro-inflammation [56] (Fig. 4.16).

Microglial phagocytic processes play an essential role in removing damaged or stressed neurons and synaptic structures, apoptotic cells, and other cellular debris from the brain, which have the potential of being immunogenic, and thus eliminate the development of neuro-inflammation. Unsurprisingly, several research teams have reported that the failure of these unceasing microglial phagocytic processes have serious and potentially catastrophic innate-immune, pro-inflammatory, and neuropathogenic consequences. The impairment of the activity and/or the expression of *TREM2* causes gross impairment of microglial phagocytic activity leading to a broad spectrum of central nervous system pathology including neuro-inflammation, synaptic, and neuronal loss [56].

Apart from neurotoxicity, chronic exposure to aluminum has also detrimental effects to other tissues. Chronic exposure to inhaled aluminum contributes to the development of a *chronic lung disease named "pneumonoconiosis"* [57]. The

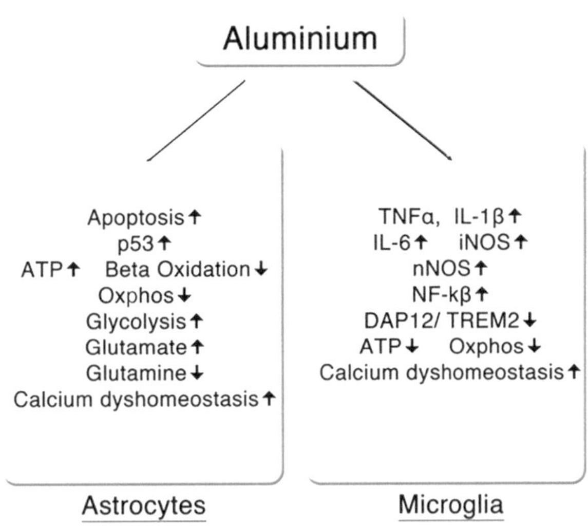

Fig. 4.16 The results of oxidative stress in astrocytes and microglia. Altered function of mitochondria leads to the dysfunction of astrocytes, while inflammatory processes turn microglia towards the inflammatory phenotype [57]

respiratory system acts like a filter by removing particles and bacteria from inhaled air and leaving exhaled air essentially free of contaminants. The diseased and blackened lungs of miners who die of coal worker's pneumoconiosis contain less than 2% of the dust originally deposited in the lungs. This illustrates the efficiency of the respiratory tract as a filter. However, the normal defense mechanisms of the lung can be overwhelmed by high exposure levels to inhaled foreign material [57].

Inhaled aluminum particles can produce a variety of injury patterns in the lung, which include diffuse alveolar damage due to fume exposure, dust macules with or without small airway fibrosis, diffuse interstitial fibrosis, alveolar proteinosis, granulomatous interstitial pneumonitis, giant-cell and desquamative interstitial pneumonitis, and lung cancer [57].

The diffuse alveolar damage (DAD) is histologically similar to adult respiratory distress syndrome (ARDS), which is the most common serious reaction to inhaled gases and fumes. Microscopically, it is characterized by hyaline membranes lining edematous alveolar septa. Practically, any noxious gas or fume inhaled in sufficient concentration can potentially cause diffuse alveolar damage. The dust macules are nonpalpable, peribronchiolar interstitial aggregates of pigmented dust and dust-laden macrophages. Initially, macules may have little associated fibrosis; however, with sufficient exposure, peribronchiolar fibrosis may occur. This pattern of injury generally has little functional impact, but due to the radiodensity of the dust, it is usually associated with an abnormal chest radiograph [57]. The diffuse interstitial fibrosis pattern of injury, an uncommon complication of metal exposure, can be seen with iron (mild fibrosis), aluminum, hard metals (containing cobalt), copper, rare earths, and silicon carbide exposure. Asbestos and mixed dusts containing silicates can also produce interstitial fibrosis [57].

Patients under hemodialysis can develop *cardiotoxicity* due to exposure to aluminum-covered filters. Although multifactorial, the mechanism of cardiac hypertrophy in patients with end-stage renal disease may involve direct aluminum cardiotoxicity. Aluminum may be demonstrated by special stains in the myocardium of such patients. There is little evidence reported that aluminum toxicity plays a role in idiopathic dilated cardiomyopathy [56].

Evidence demonstrating the development of *chronic autoimmune or autoinflammatory* conditions following environmental exposure to aluminum salts and other adjuvants is increasingly becoming a matter of concern. Much of this evidence has been presented in the context of the "autoimmune (auto-inflammatory) syndrome induced by adjuvants" (ASIA), which encompasses a broad spectrum of immune-mediated illnesses triggered by exposure to medical, cosmetic, or environmental adjuvants such as aluminum salts, silicon compounds, or indoor mold. ASIA is characterized by specific and non-specific manifestations such as chronic fatigue, myalgia, arthralgias, neurocognitive impairment, respiratory symptoms, gastrointestinal symptoms, dermatological signs, and the development of autoantibodies [56].

Epigenetic Mechanism of Intoxification

It is increasingly acknowledged that epigenetic phenomena may be a crucial component in the development of complex brain disorders. Moreover, data regarding genetic and environmental risk factors which contribute to the establishment of epigenetic marks suggest that AD risk is established during early life [55].

Wang et al. described a genome-wide decrease in DNA methylation in AD. Recent evidence shows that AD patients have an elevated DNA methylation state of repetitive elements [60].

Learning and memory can be broadly defined as lasting alterations of a behavioral output produced in response to a transient environmental input. In order for a transient stimulus to induce a lasting change in behavior, cells must undergo a complex set of stimulus-specific cellular and molecular changes that will consolidate a memory into an everlasting trace. This epigenetic process of gene expression is critical to learning and memory. One of the most important findings that support the importance of epigenetics in the functioning of the brain has been the discovery that neuronal activity per se modifies DNA methylation and histone modification patterns. These modifications are important regulators of the memory process. A research study in the blood of a total of 366 aluminum-exposed workers showed that mild cognitive impairment was significantly associated with the global DNA methylation. Another study detected that exposure of rats to aluminum maltolate may induce learning and memory impairment and decrease genome-wide methylation rate [60].

Additionally, aluminum decreases the methylation level of pectin and consequently results in higher aluminum binding in the cell wall. Many pathogenic genes (*APP*, *PSEN1*, *APOE*, *BACE*) in AD and other AD-related susceptibility genes contain methylated CpG sites in their promoter region suggesting gene silencing [60].

The promoter region of the *APP* gene is hypomethylated contributing to a potential enhancement of Aβ production. An epidemiological study in aluminum workers showed that reduced methylation of the promoter region of the *APP* gene may be associated with increased serum aluminum level and down-regulated methylation of the promoter region of *APP* gene may accelerate *APP* gene transcription. Animal experiments also found that aluminum chloride might cause *APP* promoter methylation decline, which affects the APP mRNA production, increases the APP expression, and results in Aβ deposition in the hippocampus in rats [60].

Furthermore, *in vitro* hypomethylation of the *PSEN1* gene increased the cleavage of APP and the production of Aβ in a neuroblastoma cell line. However, a sign of potential deregulation of the epigenomic machinery in the brains of affected individuals may be the observation that aberrant DNA methylation patterns are not uniform and they occur either as demethylation or as de novo methylation [60].

Besides, the trivalent aluminum modulates methylation reactions, including DNA methylation in SH-SY5Y human neuroblastoma cells, by inhibiting insulin-like growth factor 1(IGF-1) and the dopamine-stimulated methionine synthase activity, as well as folate-dependent phospholipid methylation, by a PI3-kinase- and MAP-kinase dependent mechanism. However, the methylation pattern of DNA is very complex and the alterations do not seem convenient when the target gene was Al-induced type or Al-repressed type [60].

The phenomenon of global DNA hypomethylation in the brain, as a driver of altered gene expression in ASD children, has also been reported. It is reported that hypomethylation of the above genes, and the consequent upregulation of complement and tumor necrosis factor-alpha (TNF-α) genes, are involved in the regulation of the neurodevelopmental role of microglia and synaptic scaling, respectively [56].

In the nervous system, histone acetylation has been unequivocally associated with facilitating learning and memory. It is becoming more evident that histone acetylation plays a key role in the etiology of AD. Histone acetylases and histone deacetylases (HDACs) are the well-known covalent enzymes that modify the reversible acetylation of lysine residues in histone amino-terminal domains. Acetylation of histones is generally characterized by an elevation in gene expression; conversely, deacetylation is associated with a decrease in gene expression, which is a commonality in AD.

The most compelling evidence on the role of epigenetics on AD comes from the results of treatment of AD patients with inhibitors of HDAC. Some studies have found that a decrease in histone acetylation has been observed in animal models of AD due to increased histone deacetylase (HDAC)-2 function. Some recent studies indicate that HDAC inhibitors are neuroprotective by regulating memory and synaptic dysfunction in cellular and animal models [60].

Histone acetyltransferase *p300* plays a critical role in controlling the expression of AD-related genes by regulating the acetylation of their promoter regions, suggesting that *p300* may represent a novel potential therapeutic target for AD. Other studies have found that histone phosphorylation exists in brain tissue from AD patients. However, the role of these enzymes remains controversial [60].

Some findings emphasize the potential significance of histone lysine methylation for brain development and they suggest lysine methylation could be used as a molecular toolbox to study chromatin function *in vivo* and *in vitro*. The H3K4-specific methyltransferase MLL is essential for hippocampal synaptic plasticity and it might be involved in cognitive dysfunction. A recent study demonstrated *in vivo* and *in vitro* that chronic aluminum exposure reduces expression of histone H3K4 tri-methylation (H3K4me3) and the activity and expression of MLL. In these cases, chronic aluminum exposure resulted in cognitive dysfunction, causing deficits in exploratory behavior and learning and memory in a dose- and time-dependent manner. Taken together, these results indicate that chronic aluminum exposure may reduce H3K4me3 levels by suppressing activation of MLL, which in turn affects cognitive ability [61].

Considerable number of data suggests miRNA interference in the development of AD. *TREM2* expression is regulated by a number of miRNAs, most notably miR-34a. Microglial miR-34a expression is upregulated by activation of the pro-inflammatory transcription factor NF-κB. Aluminum in adjuvant or environmental form upregulates NF-κB and thereby induces a number of NF-κB-sensitive pro-inflammatory miRNAs apart from miR-34a which in turn downregulate *TREM2* expression in the membranes of glial cells leading to a profound deficit in their phagocytic capability [56].

Upregulation of NF-κB induces the synthesis of other inflammatory miRNAs such as miR-9, miR-125b, miR-146a, and miR-155, which are recognized drivers of the pathology in several neurodegenerative diseases including AD. While aluminum clearly could provoke chronic pathology in people with a high or prolonged exposure, it should be further emphasized that the likelihood of adjuvant use could be the main cause of ASD in children, or indeed harmful to the vast majority of children with normal immune responses. However, these suggestions should be further examined [56].

Abnormalities in the performance of PPRs (Pattern Recognition Receptors) are associated with the development of neurodegenerative and autoimmune conditions. In particular, functional polymorphisms in the genes, encoding proteins involved in the immune response, they increase the susceptibility and/or severity of a range of neurological, auto-immune, inflammatory, and infective illnesses, including SLE, rheumatoid arthritis, asthma, sepsis, and hepatitis. Their action is established after ligation to TLR-4, TLR-7, and TLR-9. Unsurprisingly, functional polymorphisms in cytokine genes, such as those encoding for TNF-α, IL-β, IL-6, IL-10, and IL-4, also increase the susceptibility to develop a range of these diseases [56].

The presence of these polymorphisms (1) influence susceptibility of the individual to disease and (2) influence the severity and ultimate trajectory of the disease in any given individual. (3) There is a accumulating evidence that functional polymorphisms in genes encoding cytokines and/or their receptors increase the risk for neuro-inflammatory and neurodegenerative pathologies such as stroke, AD, and multiple sclerosis (MS). (4) Functional polymorphisms can also lead to altered structure and function of the receptor and the effector proteins involved in a broad range of inflammasome responses. (5) In general, "pathological" polymorphisms with nuclear localization in leucine-rich-repeat protein-1 (NLRP1) pathways lead to an exaggerated and/or prolonged inflammatory response [56].

It is of interest that recent research has revealed a significant association between the cumulative presence of immune gene polymorphisms and the increased risk of developing ASD. Single nucleotide polymorphisms (SNPs) in immune genes in many children with ASD also display a signature pattern of PPRS, which is characteristic of autoimmune diseases. Unsurprisingly, polymorphisms in the HLA system, notably HLA-DR4 and HLA-A2, are associated with a significantly increased likelihood of developing the syndrome. It is also noteworthy that the concomitant presence of abnormally expressed genes governing the regulation and delivery of immune and inflammatory responses can predict the development of ASD with some 83% accuracy [56].

The pathogenesis of AD, ASD, and other neurodegenerative and neurodevelopmental disorders remains incompletely understood, and many environmental and occupational exposures have shown varying associations with the disease. Recent epidemiological, neuropathological, and biochemical studies have investigated aluminum as a contributor to the development and progression of these diseases [61].

Manganese

The Metal, Its Forms, and Its Distribution in the Environment

Manganese is a chemical element with the symbol Mn. It is found as a free element in nature; it is also found in minerals in combination with iron. Manganese is a transition metal with a multifaceted array of industrial alloy uses, particularly in stainless steel.

Manganese is the twelfth most abundant element in the Earth's crust and the fourth most widely used metal in the world. Eight million tons of manganese metal are extracted annually, of which 94% are employed in the manufacturing of steel. Manganese is also used in the manufacturing of batteries. Potassium permanganate is a widely employed bactericidal and fungicidal agent in water purification processes. Additionally, methylcyclopentadienyl manganese tricarbonyl (MMT) is an organic manganese compound that has been used as an anti-knock additive to gasoline. Mining and industrial processes expose workers at manganese toxic levels [62].

Manganese occurs naturally in both surface and groundwater, as a result of weathered and solubilized manganese from soil and bedrock. Manganese is also deposited into waters from human activities. According to the Geological Survey of Sweden (SGU, unpublished data), manganese concentration in Swedish groundwater used for household purposes is on average 150 ± 510 μg/L (median 60 μg/L) with maximum values as high as 30,000 μg/L. The U.S. Environmental Protection Agency (2003) reported median manganese concentrations in groundwater at 5 μg/L, with higher levels at 2900 μg/L [63]. "A review of typical Western and vegetarian diets found average adult manganese intakes ranging from 0.7 to 10.9 mg/day which represents an overexposure to the element" (WHO 2004) [63].

People could be exposed to high levels of manganese by the consumption of well drinking water or inhalation of atmospheric air with manganese released as a combustion product of methylcyclopentadienyl manganese tricarbonyl (MMT). Manganese exposure by contaminated water rather than diet has been associated with significant deposition of manganese in hair samples of children [64].

Manganese is an essential element for the organism. The adequate daily intake is reported to be about 2 mg/day (mean values range from 1.3 to 3.6 mg/day) in children, whereas adult diets resulted in a slightly higher intake of about 2.7 mg/day (mean range 1.5–3.9 mg/day). The World Health Organization (WHO) guideline value for manganese in drinking water is 400 μg/L (2006) [63]. The current health-based guideline value of 400 μg/L for manganese in water is based on an estimated no observed adverse effect level (NOAEL) for manganese in food [63].

Dietary intake depends on the contamination of the soil of the area with manganese, dietary preferences, and cofactors. Both children and adults with a vegetarian diet have reportedly higher values of daily manganese intake than those with an omnivorous diet, which can be explained by the higher manganese concentration in plants than in meat and fish. The daily intake also seems to differ with country of residence, likely because of different eating habits [63]. Cofactors that interfere with the absorption are total caloric and fat intakes as well as the parallel intake of calcium, magnesium, iron, lycopene, and methyl-donating dietary B vitamins (including folate, vitamin B-6, and vitamin B-12). Furthermore, other factors that may disturb absorption are smoking packs/years, antihypertensive drug use, anti-inflammatory drug use, statin use, and elevated fasting blood glucose, which are potential confounders [65] (Table 4.2).

Accidental Exposures in History

The manuscripts of Mena and coworkers documented the development of psychiatric and parkinsonian features in a group of manganese miners in Chile, and more recent papers from Taiwan have documented cases of intoxification associated with smelting [62].

Table 4.2 Sources of high content in manganese

Route of exposure	Basis for exposure	Relative potential for exposure
Ingestion, diet	Manganese in breast milk, infant formula, and leafy vegetables, fruit, and fruit juices	Relatively lower for general population, but higher prenatal and infant exposure
Ground water, drinking water, surface water	As a natural constituent of water, but elevated level at some water sources because of pollution or bedrock formation	Lower
Soil	Manganese may present in soil but is not a major contributor to exposure	Lower
Inhalation, ambient air	Higher air concentration nearby industries (mining operations, metal processing plants, etc.) (MMT as additive in gasoline) is cause of concern for residential population	Higher
Indoor air	Significantly lower but associated with Mn in outdoor dust and the distance from industrial point sources	Relatively lower, but implications in childhood exposure [65]

Effects on Health

Acute poisoning or *chronic exposure* to manganese mainly affect the central nervous system. Neurological disorders similar to schizophrenia and Parkinson's disease (PD) have been observed in exposed individuals exposed to high concentrations [64].

Epidemiological observation and research studies show that "manganism" and "parkinsonism" are the two manifestations of high manganese exposure. These neurodegenerative manifestations stand as two extreme conditions of exposure depending on the intensity, duration, and route of exposure. Manganism is the result of high acute exposure whereas parkinsonism happens due to low chronic exposure. According to the WHO, manganism can occur when the airborne concentration of manganese in inhaled particles is above 1 mg/m^3, while parkinsonism may result from lifetime exposure to much lower concentration of around 100 ng/m^3 of manganese in respirable dust particles. The delayed and long-term toxicity may result, because of cumulative accumulation of the metal along with other known neurotoxicants, such as pesticides and lead. Brain may be unable to compensate to the exposure to multiple neurotoxicants leading to persistent and cumulative damage. Thus, lifetime chronic exposure to manganese from prenatal to an older age may be considered as a risk factor for development of parkinsonism [64].

The clinical features of *manganese neurotoxicity* closely resemble acute psychosis and Parkinson's disease, but have more prominent dystonic features. Initial stages typically include psychiatric disturbances, such as behavioral abnormalities, hallucinations, and, at times, frank psychosis. This syndrome is referred as "manganese madness" or "locura manganica." Not all cases pass the psychiatric phase. Some patients present with mainly extrapyramidal manifestations. The extrapyramidal features consist of bradykinesia, gait disturbance, postural instability, rigidity, micrography, masked facies, and speech disturbances. Although tremors may be seen, they usually are not the typical "pill-rolling" tremors seen at rest in Parkinson's disease, but they are described as more as intention type of tremors. Dystonic features can be rather prominent with facial grimacing and plantar flexion of the foot, which results in a very characteristic gait disturbance. The gait of manganese poisoning consists of a patient keeping the foot dorsiflexed and walking with elbows flexed and spine erect. This rather unique and characteristic gait is commonly referred as "coq au pied" or "cock walk," as it resembles the strutting of a rooster. In most cases, exposure by inhalation is the main root of contamination [62].

Parkinson's disease (PD) is among the most common and well-known neurodegenerative diseases. Cumulative studies have correlated the occurrence of PD with inherited susceptibility. However, the increase of the disease in industrialized areas entail that environmental factors could contribute to the appearance of PD in susceptible patients. Long-term chronic exposure to certain metals like copper and manganese individually or in combination has been significantly found to be linked to an increased risk. At the cellular level, the disease is often characterized by eosinophilic Lewy bodies in dopaminergic neurons of substantia nigra of the brain which are aggregates of normal and misfolded proteins [64].

Familial cases of PD are associated with genetic mutations in several genes including the *SNCA*, *parkin*, *UCHL-1*, *PINK1*, *DJ1*, and *LRKK2* genes and they contribute only for 5–10% of PD cases, while the majority of PD cases are idiopathic. Recessive mutations to other PD linked genes such as *parkin/PARK2* and *DJ1* leads to loss of their function and it is reported to be responsible for the early onset of parkinsonism [64].

α-Synuclein aggregation in neuronal cells in the form of Lewy bodies is the pathological hallmark of Parkinson's disease. In more detail, Parkinson's disease is characterized by the presence of α-synuclein proteinaceous inclusions, known as Lewy bodies (LBs) and Lewy neurites (LNs), in the dopaminergic neurons of brain. Manganese has been found to accelerate the α-synuclein aggregation in neuronal cell lines by inducing the overexpression of the gene. The significance of elevated α-synuclein expression can be understood by the fact that a 50% increase in α-synuclein expression is suggested to cause idiopathic Parkinson's disease [64].

The Unified Parkinson's Disease Rating Motor Scale 3 (UPDRS 3) is used to access the risk of parkinsonism in exposed workers to welding fumes of manganese. The magnetic resonance imaging (MRI) of patients with a history of exposure to manganese indicated increased intensities in globus pallidum and substantia nigra regions of the brain. Positron emission tomography and radiotracer FDOPA-based investigation of welding workers with mild parkinsonism signs have shown the evidence of dopaminergic dysfunction in the caudate nucleus. PET neuroimaging studies, carried out in animal models chronically exposed to manganese, reported movement abnormalities in the absence of nigrostriatal neuronal degeneration by inhibition of striatal dopamine release. These studies revealed that dysfunction of the dopaminergic system is the cause of the motor deficit. However, the detailed mechanism regarding the impact of the metal on motor abnormalities in the absence of dopaminergic neurodegeneration needs to be further addressed using advanced molecular biology tools [64].

Additionally, due to similar mechanisms as in Parkinson's disease, exposed workers to manganese could develop other neurological disorders like Huntington's disease and mood disorders (depression and anxiety) [64].

Differences in behavior (hyperactivity, oppositional behavior) were also reported in children with different manganese concentrations in the tap water. A study included 46 children, of whom 28 lived in houses with elevated manganese levels (average 600 μg/L) in the tap water. The water in the houses of the control children had average manganese concentration of 160 μg/L. The children in the first group had elevated hair manganese concentrations, which correlated with the water manganese concentration, suggesting that manganese in tap water was a significant exposure source [63].

Although no single study is entirely convincing regarding the level of toxicity in infants and young children, the number of indicative studies makes it reasonable to assume that children are at considerably higher risk than older individuals (6). The performance of 92 matched-paired 11- to 13-year-olds was described. The manganese concentration in the drinking water consumed by the exposed group of children was 240–350 μg/L compared to 30–40 μg/L of the control group. The exposed children had elevated hair and blood manganese concentrations. They had lower scores on neurobehavioral tests and also lower scores in mathematics and language. Subsequent studies confirm these results [63].

Mechanism of Intoxification

The generation of oxidative stress, the mitochondrial dysfunction leading to energy imbalance and the defective ubiquitin-mediated proteasomal cellular pathways contribute to manganese intoxification. ROS have the potential to influence epigenetic processes by oxidative stress mechanisms and to be responsible for more permanent changes [64].

Whether the increased deposition of iron participates in the neuronal damage by oxidative damage or it is secondary to the parenchymal destruction remains unclear. The substantia nigra pars compacta is the location of the dopaminergic cells that project to the striatum and it is the main targets of damage in Parkinson's disease. However, these regions remain intact in manganese poisoning. Lewy bodies, the characteristic microscopic feature of Parkinson's disease, they do not encounter in cases of manganese-induced parkinsonism. Presumably, dysfunction of the extrapyramidal system occurs due to secondary damage of the globus pallidus and striatum, the major targets of the striatonigral projections [62].

Morphological, neurochemical, and behavioral alterations were reported in rats exposed to $MnCl_2$/Mn $(OAc)_3$ by the inhalation route. Entering the brain, manganese accumulates in several regions particularly in globus pallidus, substantia nigra pars compacta, and other ganglia structures. Manganism appears due to accumulation of manganese specifically in globus pallidus region of the brain, whereas parkinsonism results from accumulation in other brain regions, including globus pallidus and substantia nigra. Although globus pallidus is a major site of manganese deposition in manganism, recent evidence suggests that it may also occur in substantia nigra. As already mentioned, substantia nigra is a specific brain region implicated in dopamine production. The progressive loss of dopaminergic neurons, due to manganese accumulation in this region, is responsible for the pathophysiology of parkinsonism [64] (Fig. 4.17).

Additionally, manganese has been reported to alter the catabolism of dopamine in dopaminergic cells resulting in oxidation of dopamine and the generation of reactive oxygen/nitrogen species ROS/RNS. Free radicals cause structural and functional modifications of proteins, DNA, and RNA along with lipid peroxidation. Additionally, free radicals cause release of inflammatory mediators like prostaglandin contributing to apoptotic cell death. The pathways that result in oxidative stress, inflammation, and apoptosis upon manganese exposure are linked with each other and they are responsible for the pathophysiology of similar neurodegenerative diseases. The cascade of activation of functional dopaminergic neurons includes a number of genes [64].

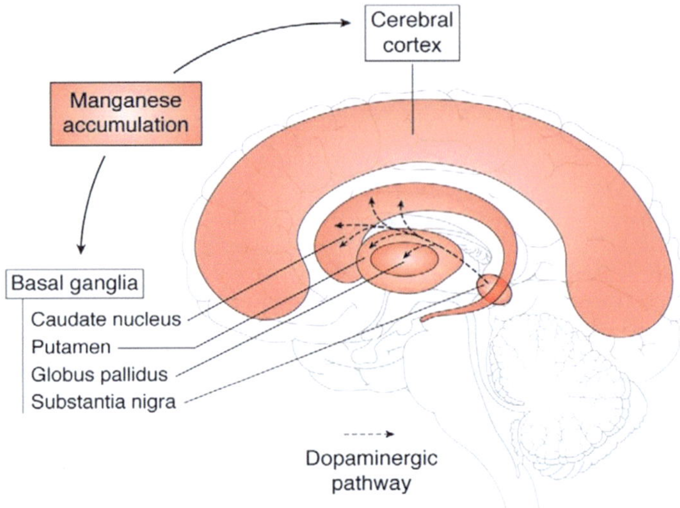

Fig. 4.17 Areas in the brain where manganese accumulates [64]

The toxic effects of manganese after entering the cells, target three different cellular functions. First, manganese may induce endoplasmic reticulum stress, especially in mitochondria. The accumulation of manganese in mitochondria results in the inhibition of oxidative phosphorylation and the generation of reactive oxygen species (ROS). Second, manganese was found to interact with complex I and complex II and ATPase and third, it may also interfere with Ca^{2+} activated ATP production. The intramitochondrial Ca^{2+} regulates the rate of ATP production by binding a series of Ca^{2+}-sensitive sites. Ca^{2+}-activated dehydrogenases which are associated with the Krebs cycle such as pyruvate dehydrogenase (PDH), isocitrate dehydrogenase (ICDH), α-ketoglutarate dehydrogenase (αKGDH), and ATP synthase. ATP is an important paracrine signaling molecule mediating the intracellular communication between the astrocytes by transmission of Ca^{2+} waves. Ca^{2+} signaling is a critical process for regulating the synaptic function, metabolism, and cerebral blood flow in the CNS [64].

The neurodegenerative diseases, such as Parkinson's and Alzheimer's diseases, have been associated with dysfunction in the Ca^{2+} signaling. The transient receptor potential channel TRPC3 was linked with pathophysiological activation of astrocytes. Recently, the ability of manganese to affect Ca^{2+} signaling was studied in striatal astrocytes and manganese was found to inhibit the Ca^{2+} influx in astrocytes by the receptor potential channel TRPC3, most likely by competition for Ca^{2+} influx in cation channels. However, more detailed electrophysiological studies are needed to precisely detect the mechanism by which manganese induces disruption of the Ca^{2+} signaling in astrocytes [64].

Manganese in low concentrations in the blood interferes with the production of metabolites of oxidative stress in tissues. The redox-active metal manganese plays a key role in cellular adaptation to oxidative stress. As a cofactor for manganese superoxide dismutase or by formation of non-proteinaceous manganese antioxidants, this metal can combat oxidative damage. In either case, the antioxidant properties of manganese are vulnerable when iron is present. Cellular pools of iron can outcompete manganese for binding to manganese superoxide dismutase, and by Fenton chemistry, iron may counteract the benefits of non-proteinaceous manganese antioxidants [66].

In biology, iron and manganese play important roles in oxygen chemistry. Both metals are used in oxygen production by photosynthesis and they can also serve as cofactors for enzymes that remove harmful byproducts of O_2 metabolism such as superoxide and hydrogen peroxide. However, the properties of these metals are quite different regarding their ability to react with oxygen. Iron is well known for its reactivity with peroxide, generating the highly reactive hydroxyl radical by the so-called Fenton reaction. Manganese is less prone to such chemistry due to a higher reduction potential. Iron is often considered a pro-oxidant in biology under situations in which manganese is an antioxidant. Without the deleterious side effects of Fenton reaction, manganese can safely operate as a cofactor for superoxide dismutase (SOD) enzymes and also provide oxidative stress resistance by formation of non-proteinaceous manganese-based antioxidants [66].

On the contrary, orally ingested manganese in high doses induces inflammation processes. *In vivo* and *in vitro* studies generally support the inflammatory potential of manganese with observed increases in the production of circulating biomarkers of inflammation including interleukin 6 (IL-6) and tumor necrosis factor alpha (TNF-α). The nuclear factor kappa-light-chain-enhancer of the active B cell (NF-κβ) pathway is a major regulator of cytokine production and it is affected by manganese [65] (Fig. 4.18).

In detail, the oxidative stress-mediated apoptotic pathways by which manganese causes the dopaminergic cell death is described. Manganese enters the neuronal cells by the transferrin/DMT1 receptor. Within the cell, manganese primary accumulates in mitochondria leading to the inhibition of complex I/II and ATPase and interferes in Ca^{2+}-activated ATP production. The result is the loss of mitochondrial potential and the release of Cyt C. Cyt C activates caspase 3, a key regulator of apoptosis. Caspase 3 activates the protein kinase C δ (PKCδ) which activates the ERK/MARK pathway in the cytoplasm and the expression of proapoptotic genes after translocation to the nucleus. PKCδ generates protein phosphatase 2A which is a critical enzyme in dopamine production. In addition, mitochondrial dysfunction causes the generation of free radicals including ROS. Free radicals and the dopamine autoxidation process result in the damage lipid membranes and it causes the release of inflammatory mediators that activate the ERK/MARK pathway. The activated kinases phosphorylate the enzyme FoxO preventing its translocation to the nucleus and the activation of antioxidant enzymes. As described, manganese also induces endoplasmic reticulum (ER) stress. All these parameters result in the activation of caspase 12 which in turn activates other downstream caspases culminating in apoptotic cell death (Fig. 4.18).

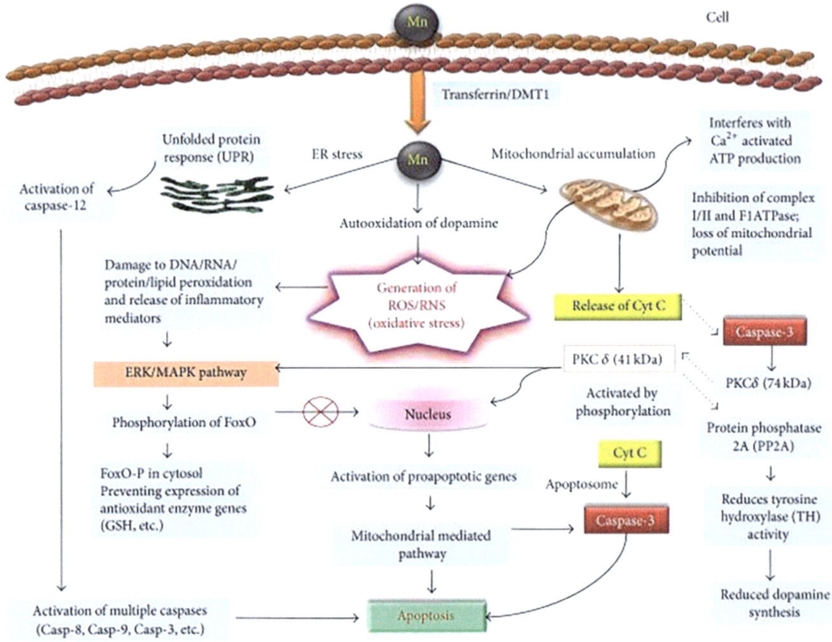

Fig. 4.18 The pathways of toxicity induced by magnesium

Protein phosphatase 2A (PP2A) is known to regulate the activity of tyrosine hydroxylase enzyme which is responsible for dopamine synthesis. The dopamine depletion is dependent on the concentration and duration of manganese exposure [65]. Dopamine-producing cells exposed to manganese have been reported to reduce glutathione and glutathione peroxidase synthesis, an important antioxidant in cellular defense against oxidative stress. These cells were found to be rescued when supplemented with the antioxidants N-acetyl cysteine and glutathione. This suggests the involvement of ROS and the depletion of antioxidants in apoptotic cell death [64].

Neuronal cell rescue from manganese intoxification is evident by a complex of antioxidant genes. Expression of antioxidant genes is tightly regulated inside the cell. A forkhead box transcription factor class O (FoxO) is a critical transcriptional factor that regulates antioxidant genes. Phosphorylation and dephosphorylation events of FoxO decide the fate of its activity. The phosphorylated form is located in the nucleus, where it mediates the transcription of antioxidant/protective genes on manganese exposure. On the other hand, phosphorylation of FoxO prevents its translocation to the cell nucleus, thus inhibiting the transcription of antioxidant genes contributing to neuronal cell death [64].

MAPK kinase (ERK, JNK, and p38 MAPK) regulates a variety of cellular processes such as differentiation, proliferation, and inflammatory response by acting as a connecting link between the extracellular signals and intracellular signaling pathway through intermediate signaling molecules. ROS, generated as a result of mitochondrial respiratory dysfunction on manganese exposure, has been reported to trigger the activation of ERK and p38-MAPK analogous, while ultraviolet radiation-induced ROS generation leads to MAPK activation [64].

Epigenetic Mechanism of Intoxification

The epigenetic changes, linked to Parkinson's disease due to manganese intoxication, are similar to other degenerative processes. Although, the epigenetic effects of manganese are not well understood, epidemiologic studies from occupational settings suggest manganese can affect the regulation of DNA methylation [65].

Two studies, until today, have examined the DNA methylation pattern among workers who were occupationally exposed to airborne manganese and the main changes were related to the Nf-kb pathway, the initiation of inflammation, and the production of specific interleukins. In a study of welders ($n = 201$), gene methylation at *NOS2*, which encodes inducible nitric oxide synthase, a regulator of oxidative stress, was significantly lower in individuals exhibiting symptoms of parkinsonism compared to unaffected controls. Furthermore, in a study of 63 healthy steel workers, airborne manganese exposure was positively associated with peripheral blood leukocyte methylation of promoter regions in *APC* and *RASSF1A* tumor suppressor genes. While occupational studies have not specifically assessed the role of manganese in DNA methylation of inflammatory or NF-κβ pathways, they do suggest that manganese has the potential to disrupt these processes. Individuals with higher estimated dietary manganese intake had higher gene methylation of *NKAP* and *NKAPP1*, which encode proteins involved in the activation of the transcription factor NF-κβ [65].

Previous studies have shown that overexpression of NKAP-induced activation of NF-κβ P65 and the activation of this transcription factor were associated with increased interleukin production. In mouse models, inhibition of the expression of IL-1β was detected by reducing transcription factor binding to the *IL1B* promoter region. *In vitro* studies have shown that cells transfected with *NF-IL6* had strong synergistic activation of IL-6 and IL-8. Additionally, a binding site is specific to the activation of IL-8 expression. While there are recognize multiple factors that control cytokine and interleukin production, the higher methylation of the gene bodies of *NKAP* and *NKAPP1* offer a potential mechanism to explain the observed increase in the production of inflammatory cytokines, although additional research on this potential mechanism is warranted [65].

In addition, higher estimated dietary manganese intake was associated with higher methylation of *VCAM1*, but there was not observed a relationship with circulating VCAM-1 concentrations. *VCAM1* encodes the vascular cell adhesion 1 protein which mediates the adhesion of white blood cells to the vascular endothelium. These seemingly paradoxical findings could be explained by the unmeasured influence of other epigenetic mechanisms such as histone modifications or silencing

RNAs, which could also regulate expression of *VCAM 1* and they potentially have a stronger influence on circulating VCAM-1 concentrations compared to DNA methylation [65].

In summary, there was found that estimated dietary manganese intake was positively associated with circulating concentrations of three interleukins (IL-1β, IL-6, and IL-8), and with methylation of two gene bodies (*NKAP* and *NKAPP1*) that regulate the NF-κβ pathway, a major contributor to interleukin production. Chronic and excessive production of inflammatory cytokines is increasingly being recognized as a risk factor for a multitude of chronic diseases. Our results highlight the inflammatory potential of ingested manganese [65].

Divalent metal transporter-1 (DMT1) is a major transporter protein for manganese and it is reported to be present in the inner ear cells, although the exact cellular location of DMT1 is still not determined. Manganese was reported to be toxic for neurons and sensory cells in cochlear cultures from postnatal rats, suggesting that hearing deficits may occur with excess manganese exposure. Divalent metal transporter DMT1 has broad specificity for transport of metal ions but plays a significant role in the transport of manganese. DMT1 is a substrate for parkin, and hence, it is involved in maintaining a steady-state level of this transporter [64].

Parkin is a E3 ubiquitin ligase that regulates protein functions by targeting them for proteasomal degradation. Loss of parkin function makes the cell susceptible to elevated uptake of manganese. DJ1 is an oxidative stress sensor in the neuronal cells of the brain. The inactive oxidized form of DJ1 was found to be elevated in sporadic PD brain patients. The inactivation of DJ1 can make the cell more vulnerable to the oxidative stress [65]. Recently, mutations in the parkin gene were reported in PD patients of south India, without the development of Lewy bodies [64].

Alpha-synuclein (α-syn) is a small soluble protein (14 kDa) expressed primarily at presynaptic terminals and it is natively unfolded, although it gets folded in cellular membranes and when it binds to lipid molecules. α-syn has gathered considerable interest recently due to the fact that the alteration in its gene expression is reported to be involved in the etiology of several neurodegenerative diseases including Parkinson's disease. In vitro investigations have revealed that α-syn is a major component of Lewy bodies. α-syn aggregates as a result of excessive oxidative stress generated by autooxidation of the neurotransmitter dopamine. Alpha-synuclein overexpression can significantly contribute to the increased risk of sporadic PD. PD patient's brain samples have shown the hypomethylation of *SNCA* intron 1 correlates with an increased expression of α-syn [64].

α-Synuclein has been recognized to play dual role in neuronal cells. When expressed at normal levels, α-synuclein is neuroprotective, while overexpression beyond a certain threshold level results in toxicity. α-Synuclein has been known to play a significant role in the dopaminergic cell death as observed in Lewy body diseases, such as Parkinson's disease. Manganese exposure generates oxidative stress and it causes dopamine oxidation. A recent finding has demonstrated the involvement of ERK1/ERK2 MAPK pathway is responsible for α-synuclein overexpression. The nuclear localization of α-synuclein was reported to cause the down-regulation of proapoptotic PKC*δ*. Additionally, hypomethylation of α-synuclein

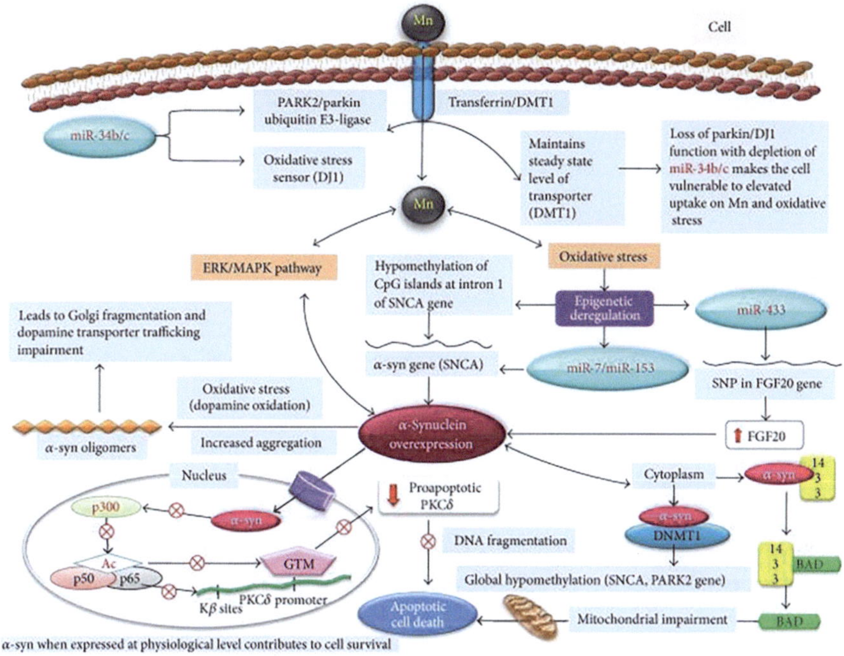

Fig. 4.19 Schematic shows the potential role of epigenetic mechanism that may be involved in manganese-induced chronic neurotoxic effects and in the pathology of Parkinson's disease

gene promoter has been demonstrated to cause overexpression of α-synuclein in PD and related disorders. Thus, it may be hypothesized that alterations in methylation pattern may be responsible for manganese-induced α-synuclein overexpression [64] (Fig. 4.19).

Recent *in vivo* and *in vitro* studies provide evidence that α-synuclein interacts with histone proteins (*p300*) and reduces the histone acetyltransferase activity. Subsequent activation of *p65* suppresses the binding of general transcription machinery (GTM) to PKCδ promoter and represses its gene expression. PKCδ is an oxidative stress kinase and a critical regulator of apoptosis in the central nervous system, recognized as downstream substrate for caspase-3. Proteolytic activation of PKCδ was reported to be involved in the dopaminergic cell death in cellular models of Parkinson's disease. The mechanism by which α-synuclein interacts with and disrupts p300 acetyltransferase activity is not yet clear. However, administration of HDAC inhibitors was reported to restore the acetylation state and protect neuronal cells against α-synuclein toxicity [64].

Simultaneous *miRNA analysis* in differentiated cell lines and PD brain samples revealed depletion or reduction in miR-34b/c. Further, this reduction was found to be associated with reduced parkin and DJ1 protein expression. Dysregulation in parkin and DJ1 function makes the cell prone to oxidative stress and mitochondrial dysfunction that compromises cell viability [64] (Fig. 4.20). Other miRNAs were also reported to regulate the α-syn expression at posttranscriptional level. Specifically, miR-7 and miR-153 were reported to regulate α-syn expression. Additionally, impairment in miR-64, miR-55, let-7, miR-10a, miR-10b, miR-132, miR-7, miR-153, and miR-435 expressions in symptomatic transgenic PD mice and PD disease models was found to be linked with increased α-syn level [64].

Epigenetic changes in the promoters of genes after exposure to mangenese have also been related to neoplasias. Hypermethylation of glutathione-S-transferase P1 (*GSTP1*) gene was reported in intraepithelial neoplasia that makes the cells more susceptible to oxidative stress-mediated DNA damage. Gene-specific promoter methylation in response to oxidative stress has been reported in the tumor suppressor gene *RUNX3*. Methylation of *E-cadherin* promoter was reported in human colorectal and hepatocellular carcinoma cells [64].

The single nucleotide polymorphism in genes determines the susceptibility level of an individual in response to toxicant exposure. How the epigenetics is linked with genetic susceptibility might be the area of research to be investigated in future. The study of epigenetic mechanism may further elaborate the present knowledge about manganese neurotoxicity. The definitive pathways may be identified which can be used as target for disease progression and therapeutics in relevance to manganese exposure [64].

Epidemiological studies report the positive correlation between single nucleotide polymorphisms (SNPS) and susceptibility to manganese. It is proposed that differences in genotype might be responsible for a variability of the blood levels of manganese in people similarly exposed to the metal. Cytochrome P450 *CYP2D6* is a susceptibility biomarker for identifying miners exposed to manganese who are at higher risk for developing manganism. The gene has a protective role by perturbation of manganese metabolism and the increase or decrease of enzymatic activity.

Magnesium

The Metal, Its Forms, and Its Distribution in the Environment

Magnesium is a chemical element with the symbol Mg. The metal is quite abundant and it is known merely for its deficient status in diet rather than its toxic properties.

Effects on Health

The role of magnesium to normal heart function and the relation of deficiency of magnesium to cardiovascular disease and acute myocardial infarction is a controversial issue. Low levels of magnesium and fluoride in drinking water were found to be associated with increased risk for acute myocardial infarction in reports from Finland and Sweden. Opposite findings were reported in England, where magnesium in drinking water showed a protective effect for ischemic heart disease, but not for acute myocardial infarction [67].

Over the past 100 years, evidence from multiple sources confirm that a number of diverse control mechanisms, including physiological, nutritional, and biochemical factors, are responsible for a functional cardiovascular system. Homeostatic factors maintain the patency of blood vessels, the cardiac output, and the fluidity of the blood. Pathophysiological changes in peripheral blood vessels and the chambers of the heart can result in alterations of vascular wall geometry, disturbances in oxygenation, and the nutritional status of the blood vessels and the myocardium as well as the tissues they perfuse. Although multiple theories and hypotheses have generated to account for the hypertrophy and the development of resistance of blood vessels, which contribute to the aging process, there is no agreement as to the precise mechanisms related to the process [68].

Diet habits are known to promote lipid deposition and to accelerate the growth and transformation of smooth muscle cells in the vascular walls. It is impressive that autopsies of children, who have died as a consequence of accidents, they demonstrate early signs of atherogenesis (i.e., fatty streaks) on aortic and carotid arterial blood vessel walls as early as 5–6 years of age [68].

Several epidemiological studies in North America and Europe have shown that people consuming Western-type diets which are low in magnesium (Mg) content have a predisposition to cardiovascular events. Eating habits of the 60–80% of Americans contain only 185–235 mg/day of Mg. Low Mg content in drinking water, found in areas of soft water and Mg-poor soil, is associated with high incidences of ischemic heart disease, coronary vasospasm, hypertension, and sudden cardiac death [68].

Mechanism of Action

Manganese plays an essential role in more than 350 enzymatic reactions in the body and it is required for oxidative phosphorylation and all energy-generating reactions. It is a natural calcium (Ca^{2+}) channel blocker on myocardial and vascular smooth muscles and it is a natural statin that lowers blood cholesterol, LDL, and triglycerides acting as a vasodilator agent. Hypermagnesemic diets were detected to ameliorate hypertension and atherogenesis [68].

Research showed, even several years ago, that a variation in free manganese ions cause sustained changes in membrane phospholipids and several second messenger molecules, as well as the activation of apoptotic pathways. Lack of magnesium causes membrane oxidation and truncation of membrane fatty acids. Decreases in extracellular Mg^{2+} produces a fall in membrane sphingomyelin (SM), whereas increases in Mg^{2+} results in increases in membrane SM and phosphatidylcholine, intracellular ceramide formation, and activation of neutral-sphingomyelinase

(N-SMAse). Ceramides, released as a consequence of sphingomyelinases (SMAses) acting on SM, are now thought to play an important role in fundamental processes such as angiogenesis, cell proliferation, immuno-inflammatory responses, atherogenesis, programmed cell death, and the aging process [68].

Epigenetic Changes

Telomerase length known to be related with aging is influenced by manganese deficiency. Short-term magnesium deficiency is associated with downregulated telomerase activity in all chambers of the heart as well as in aortic smooth muscle in animal models. These results indicate a strong correlation and a cross-talk of telomerase downregulation in cardiovascular tissues with serum-ionized Mg levels, N-SMAse, p53, and 8-OHdG. The increased formation of 8-OHdG in all cardiovascular tissues and cells studied indicate an increased oxidation of DNA [68].

Chronic stress has been long-associated with manganese deficiency and recently with decreases in telomere length and decreased telomerase activity. Oxidative stress, associated with increased levels of 8-OhdG, is known to promote telomeric erosion. Short-term Mg deficiency causes a marked reduction in the cellular levels of the free radical scavenger glutathione in cardiovascular tissues and cells. The reduction of glutathione together with the activation of several nitric oxide synthetases could cause fragmentation of DNA and subsequently mutations in the DNA of many cell types [68].

Interestingly, recent findings in Mg-deficient animals, tissues, and different cell types have shown elevated levels of many of the proinflammatory cytokines such as IL-1, IL-6, TNF-alpha, among others. TNF-alpha is known to be associated negatively on telomerase activity in some cell types. Recent experiments in short-term Mg-deficient animals and vascular smooth muscle cells have clearly demonstrated cross-talk between cytokines and chemokines, ceramide generation, proto-oncogenes, *p53*, and activation of nuclear factor-kB pathways. The present study supports the hypothesis that even short-term manganese deficiency probably leads to the decline and activation of multiple molecular pathways which would contribute (or be responsible for cardiovascular dysfunction) in the normal aging process [68].

Moreover, oxidative stress generates the activation of *p53* which is a "driver" of ceramide synthesis by the activation of N-SMAses and ceramide synthase in diverse cardiac and vascular smooth muscle cells. Interestingly, atherosclerotic plaques in vascular walls of hypertensive subjects demonstrate considerable DNA damage, activation of DNA repair pathways, increased expression of *p53*, oxidation, and apoptosis as well as increased levels of ceramide [68].

Supplementation of 40–100 mg/Mg added to the drinking water can completely prevent all of the vascular and molecular anomalies produced by Mg-deficient diets reported. We could suggest that, the results imply that water intake (e.g., from tap waters, well waters, bottled waters, beverages using tap/well/spring waters, or desalinated waters) in humans should contain at least 25–40 mg Mg^{2+}/L [68].

Nickel

The Metal, Its Forms, and Its Distribution in the Environment

Nickel (Ni) belongs to the group 10 of the periodic table along with iron, cobalt, palladium, platinum, and five other elements; elemental nickel is a silver-white solid metal with high thermal and electrical conductivity.

In nature, nickel is found in combination with antimony, arsenic, and sulfur. Nickel exists in +2 (Ni^{2+}) oxidation state; other states (+3 and +4) may be found, even if they are less frequent. It is resistant to the air, water, and alkali corrosion, but it is readily soluble at $pH < 6.5$ and it dilutes in oxidizing acids. Nickel generates salts with strong acids (chloride, nitrate, and sulphate) and organic salts, which are easily water-soluble, whereas, nickel salts with weak inorganic acids, nickel sulfides, and nickel oxides, are poorly water-soluble. Compounds of great commercial importance are nickel chloride, carbonate, nitrate, sulphate, acetate, hydroxide, and oxide. Its physicochemical properties include resistance to very high temperatures, corrosion, and oxidation; in addition, it is very ductile, it alloys readily and it is fully recyclable [69].

Natural sources of contamination with nickel include volcanic eruptions, soils, ocean floors, and ocean water [70]. Most natural nickel is not accessible, because it is placed in the iron–nickel alloy molten outer core of the earth. Natural and anthropogenic distribution of the metal is abundant. Natural sources of atmospheric nickel include wind-blown dust derived from weathering of rocks and soils and forest fires. Direct leaching from rocks and sediments produces high concentrations of nickel in water, where it is present in the divalent nickel form, as well as in suspended insoluble particles [69].

Anthropogenic distribution of the metal in the air derives from the combustion of coal, diesel oil, and fuel oil and the incineration of waste and sewage. Nickel released is emitted as oxides, sulfides, soluble compounds, and to a lesser content, as metallic nickel. The most important source of nickel compounds in air derives from the combustion of fossil fuels [69]. Due to its chemical properties, gloss and low price, nickel and nickel compounds have many industrial and commercial uses. Nickel Raney (in powder or in granular form) is a catalyst employed in reduction reactions, such as the hydrogenation of unsaturated compounds. Nickel alloys are present in solders, surgical steel instruments (5–20% Ni), white gold (10–15% Ni), German silver (10–15% Ni), pigments and catalysts and sterling silver. Nickel compounds are used in electroplating, electroforming, and production of nickel-cadmium batteries and electronic equipment. Nickel alloys, such as stainless steel, are largely used in the production of tools, machinery, armaments, and aerospace equipment. Besides industrial use, nickel is used in the production of coins, keys, paper clips, clothing fasteners (such as zippers, snap buttons, and belt buckles), medical prostheses, and orthodontic materials [69].

Compounds with other metals, such as gold, silver, titanium, chromium, and copper, are ubiquitous in our environment and they are widely used in costume jewelry, coins, and mobile phones. Additionally, nickel is used in inexpensive jewelry, stainless steel household utensils, and in food processing industries. It is also noted that kitchen kettles may release nickel into drinking water [69].

Of great concern is the exposure to orthodontic materials, since patients are exposed to a considerable amount of nickel, cobalt, titanium, and other metals. The acid oral environment enriched by a spectrum of bacteria combined with saliva, intake of drinks, food, and mouthwashes facilitate corrosion and result in the release of ions from orthodontic appliances into oral tissues. They could topically cause contact dermatitis, hypersensitivity, cytotoxicity, and DNA damage [69].

Moreover, nickel-containing cobalt and titanium alloys have become ubiquitous in the manufacturing of neurovascular medical devices. In fact, aneurysm clips, endovascular devices, and stents containing varying proportions of nickel (14–35%) are used in treatment of cerebral aneurysms. Allergic reactions to nickel have been implicated in complications related to coronary stents, including migraine headaches, fever, dyspnea, dermatitis, and pericarditis [69].

Tobacco smoking is a source of exposure to nickel. It has been demonstrated that each cigarette contains an amount of 1.1–3.1 μg of the metal and nickel in tobacco smoke may be present as nickel carbonyl, which is extremely hazardous to human health. Another source of nickel exposure in the general population is by dietary exposure. In fact, some vegetables (spinach, asparagus, carrots, broccoli and green beans, tomato) as well as cocoa, chocolate, and nuts contain high amounts of the metal. Nickel is also involuntarily added to the diet by food processing with stainless steel equipment or by hand to mouth contact [69].

Occupational exposure of several million workers worldwide, predominantly in mining, refining, alloy production, electroplating, and welding [71], has given rise to high levels of nickel in blood, urine, and body tissues, especially in the lung. Nickel is considered as an occupational hazard. Finally, nickel nanoparticles (NiNPs) are among the most widely used nanomaterials, which are employed in many fields, such as catalysts, magnetic materials, biological medicines, conductive pastes, and additives to lubricants [72] (Table 4.3).

Humans can be exposed to nickel by inhalation, ingestion, and skin contact [71]. Exposure to nickel is associated with a multitude of human health risks, including allergic contact dermatitis, asthma, inflammation, bronchitis, pulmonary fibrosis, pulmonary edema, as well as diseases of the kidney and cardiovascular system. Moreover, epidemiological studies indicate development of cancer as the most serious outcome of nickel exposure [72].

Accidental Exposures in History

An example of chronic exposure to nickel is the Jinchuan Group Co, Ltd (JNMC) workers in Gansu Province, China. The area is the largest producer of nickel in China and its nickel output accounts for more than 90% of the total production.

Table 4.3 Nickel-containing foods and items

Nickel-containing foods	Hazelnuts; peanuts; almonds; cocoa and dark chocolate; fruits (dates, figs, pineapple, plums, raspberries); grains (bran, buckwheat, millet, whole grain bread, oats, brown rice, sesame seeds, sunflower seeds); seafood (shrimps, mussels, oysters, crab, salmon); vegetables (beans, savoy cabbage, leeks, lettuce, lentils, peas, spinach, cabbage), tea from drink dispensers; soya and soya products; licorice; baking powder.
Nickel-containing items	Inexpensive jewelry; cosmetics; keys; cell phones; eyeglass frames; paper clips; orthodontic braces; stainless steel articles; nickel-plated articles; clothing fasteners (zippers, snap buttons, belt buckles); electrical equipment; armaments; alloy; metallurgical and food processing industries; pigments; catalysts.
Nickel toxic effects	Contact dermatitis; headaches; gastrointestinal manifestations; respiratory manifestations; lung fibrosis; cardiovascular diseases; lung cancer; nasal cancer; epigenetic effects [72]

Workers in the JNMC are routinely exposed to nickel and several other contaminants, including industrial dust, particulate matter and gaseous pollutants, cobalt, covellite, and chromium. Data of examination of workers exposed were collected between 2011 and 2012. In parallel, lifestyle parameters were examined (smokers, alcohol drinkers, and tea drinkers referred 39.1%, 19.7%, and 55.2%, respectively) as well as a prevalence of hypertension, allergic rhinitis, and cholecystitis, the top three prevalent diseases (11.7%, 11.0%, and 8.9%, respectively) [71]. The causes of death and mortality of the workers in JNMC were analyzed, indicating that cancer was the most important disease affecting their health accounting for 35.9% of the total cancer-related deaths with the highest standard mortality of 163.62/105. Lung cancer deaths accounted for 33.4% (309/925) of the total cancer-related deaths with a standard mortality of 54.22/105. It has been shown that exposure to nickel was closely related to the increase of lung cancer risk [72].

Oral cancer incidence occurs at high frequencies in Taiwan. Changhua County, located in central Taiwan, has one of the highest rates of oral cancer in the country and this region also contains many metal-related industries and metal-contaminated soil. A study found that blood levels of nickel were significantly higher in oral cancer patients compared to control patients indicating that exposures to nickel in this region of Taiwan may play a role in the development of oral cancer [70].

Effects on Health

Acute intoxification to nickel even in a soluble form has not been described. However, accumulation of nickel and nickel compounds in the body by chronic exposure may be responsible for health problems such as lung fibrosis, kidney, and cardiovascular diseases and cancer of the respiratory tract. High incidence of nasal and lung cancer in workers exposed to nickel and nickel compounds has been observed [69].

Among the most common side effects of exposure to nickel are *skin disorders*. The list of nickel compounds exhibiting dermal toxicity has been well catalogued. Such metals include compounds used in medicinal products, industrial processes, pesticides, cosmetics, dyes, and jewelry. Dermal toxicity is the result of local tissue responses due to direct contact of the metal with the skin or, alternatively, it may represent manifestations of systemic toxicity following ingestion or inhalation [73]. In 1930–1935, observations of nickel dermatitis produced by objects in everyday use were reported by Rothman and Preininger about coins, by McAlester AW and McAlester about spectacle frames, and by Du Bois about wrist watch [69]. A particular problem is due to the use of nickel for coinage, such as the European one- and two-euro coins [69].

Clinically, nickel allergy sometimes occurs when nickel-containing items are in direct and prolonged contact with the skin leading to corrosion of nickel by sweat and release of ions to the skin. The symptoms include contact dermatitis, lichen planus, dyshidrotic eczema, angular cheilitis as well as periodontitis, stomatitis with mild to severe erythema, loss of taste, and numbness, when the source is orthodontic processes. Nickel is a component of many orthodontic materials, as a nickel-titanium alloy with a concentration of nickel of about 50%. Nickel-containing devices used in medicine like coronary stents and endoprosthesis are a source of continuous exposure. Surgical stainless-steel instruments contain nickel at a concentration of 10–20% and a release rate of more than 0.2 $\mu g/cm^2$/week. It is present in stainless steel in arch-wires and brackets with a concentration of approximately 8% [71].

Approximately 10–15% of the population suffers from nickel allergy. Literature reports show that women are more prone to dermatitis than men. In fact, approximately 13–18% of females and 3–6% of males are allergic to nickel as a consequence of contact with nickel-containing items, such as jewelry, buttons, certain shampoos, detergents, and pigments. According to dermatologists, the frequency of nickel allergy is still increasing and it can be explained by the fashionable piercing. Nickel allergy apart from skin manifestations, it can present as recurrent headaches, gastrointestinal, and respiratory problems [72].

Prevention strategies and reduced exposure could ameliorate symptoms of people suffering from contact dermatitis to nickel. It is hypothesized that prevention of exposure could diminish the number of those who are sensitive to the metal by one-quarter to one-third. Alternative ways to reduce contact should be applied, for example objects with sentimental value (heirlooms, wedding rings) could be treated with an enamel or rhodium plating, and stainless-steel devices could preferably be replaced by titanium-molybdenum alloys. In 2011, Vemula and co-authors showed that the penetration of nickel ions into the skin could be prevented, by applying a thin layer of glycerin containing nanoparticles from calcium carbonate or calcium phosphate. The nanoparticles may capture nickel ions by cation exchange and remain on the surface of the skin, which could further be removed by washing with water. Therefore, the use of nanoparticles with diameters smaller than 500 nm in topical creams may be effective in limiting the exposure that can cause skin irritation [69].

However, the route of exposure with the higher risk to nickel is *inhalation*. The absorption of nickel particles deposited in the alveolar, tracheobronchial, and nasopharyngeal regions of the respiratory system depends on various factors, first of all on the diameter of the inhaled particles, and moreover, on the solubility, the quantity deposited, the ventilation rate, and the retention rates. Only particles with a diameter less than 100 μm can be inhaled to settle along the respiratory tract. Particles with a diameter of less than 4 μm are deposited in the lower alveolar region of the respiratory system; particles deposited in the tracheobronchial region are between 4 and 10 μm in size, and finally, particles with a diameter between 10 and 100 μm are deposited in the nasopharyngeal region [69].

Water-soluble nickel compounds are absorbed in the respiratory track and removed by the kidneys. They can cause nose and sinuses irritation and they may also lead to loss of the sense of smell and even to nasal septum perforation. The insoluble nickel compounds deposit and remain in the lungs for a long time and they are responsible for many lung diseases. The ingestion of food containing nickel or nickel compounds may cause adverse systemic reactions [69].

Mechanism of Intoxification

The underlying mechanism of developing *contact dermatitis* is hypersensitivity reactions type IV. Nickel released from various products is a potent allergen or hapten that can trigger skin inflammation. Nickel penetrates the skin and activates epithelial cells to produce cytokines or chemokines. The reaction involves the activation of antigen-presenting cells and T cells in complex immune responses. Activated antigen-presenting cells migrate to the lymph nodes where nickel presents as an allergen or hapten to naive CD4+ T cells. Following the re-exposure, these T cells become stimulated leading to visible signs of hypersensitivity 48–72 h after allergen exposure. After repeated exposures to nickel, the clones of T-cells reach a "threshold" value and the skin develops a rash that can manifest as acute, subacute, or chronic eczema-like skin patches. The skin reaction can take place at the site of contact, or sometimes diffuse to the rest of the body. Cutaneous exposure can cause localized erythematous, pruritic, vesicular, and scaly patches [69].

The mechanisms underlying nickel relation to *other diseases*, including carcinogenicity and cardiorespiratory disease, are still largely unknown. Compounds containing nickel have been found to cause oxidative stress, hematotoxicity, and immunotoxicity [72]. It has been proposed that nickel may replace other metals within catalytic enzymes, for example, to replace magnesium in DNA interactions, enhance chromatin condensation, and trigger de novo DNA methylation [73].

The nervous system is one of the main target organs for nickel toxicity. The metal can accumulate in the brain. The molecular mechanisms of Ni-induced *neurotoxicity* are still not clear, but researchers think that oxidative stress and mitochondrial dysfunction have a primary and important role. Mitochondrial damage induced by nickel may occur first as mitochondrial membrane potential damage, then as mitochondrial ATP concentration reduction, and finally as mitochondrial DNA destruction. Damage of mitochondrial function interferes with the mitochondrial transport chain, amplifies ROS, and exacerbates oxidative stress [69].

According to Song et al., the use of antioxidant molecules, such as L-carnitine, taurine, and melatonin, which are molecules that stimulate and amplify the antioxidant enzyme activity, could prevent nickel-induced neurotoxicity and carcinogenicity. These antioxidants play a significant role in neurotransmission, detoxification, and mitochondrial energy homeostasis, reducing oxidative stress and ROS production [69].

Nickel nanoparticles are associated with *reproductive toxicity*. Among other processes, high temperature of manufacturing, such as welding, is used in the production of nanoparticles. Nickel nanoparticles in rats significantly decrease the activity of superoxide dismutase (SOD) and catalase (CAT) and they increase the levels of ROS of malondialdehyde (MDA, lipid peroxidation marker) and of nitric oxide (NO) in comparison with control groups. In a study, mitochondrial swelling and disappearance of cristae in the ovaries of the exposed groups was detected. The mRNA expression of caspases, a group of cysteine proteases, and the expressions of Cyt C, Bax, and Bid proteins were significantly increased, while the expression of B-cell lymphoma-2 (Bcl-2) protein was drastically decreased [69].

IARC (The International Agency for Research on Cancer) classified soluble and insoluble nickel compounds as Group 1 carcinogen to humans and nickel alloys as Group 2B (possibly carcinogenic to humans). The toxic and carcinogenic effects of nickel are related to the route of intoxification. Potential toxicity of nickel and nickel compounds is dependent on their physico-chemical characteristics, as well as the amount, duration of contact, and route of exposure [69].

Nickel, as an environmental and occupational toxicant, is associated with *nasal and lung cancers*. Earlier studies on human lung epithelial cells showed that nickel exposure causes alterations in the expression of hundreds of genes, which do not revert to pre-exposure status, even after the termination of exposure. The mechanism of transformation to malignancy is the property of the cells to undergo epithelial–mesenchymal transition (EMT) and to gain invasive and migratory characteristics. EMT is the process during which the epithelial cells lose their polarity and cell–cell adhesion and acquire a migratory and invasive phenotype. The above-mentioned phenotype is associated not only with cancer and metastasis, but also milder diseases, such as asthma, bronchitis, and fibrosis. Thus, the persistent transcriptional changes, caused by chronic nickel exposure, could be the predisposing factor for the impact of nickel exposure on human health [74].

Epidemiological studies have demonstrated increased mortality from cancer of nasal cavities and lung cancer in nickel refinery workers who are chronically exposed to nickel-containing dusts and fumes. Insoluble nickel sulfide (Ni_2S_3) is a carcinogen agent for the respiratory tract. When it is inhaled, particles of nickel sulfide accommodate themselves in the lungs, where they remain in contact with epithelial cells. These nickel particles are removed by macrophages to the digestive tract. Under high exposure to nickel, the macrophage activity and the potential of removal could be interrupted. Ni_2S_3 particles could be engulfed by epithelial cells by endocytosis. Thereafter, nickel particles derive to the nucleus of lung epithelial cells, causing a heritable change in chromosomes and inducing double- and single-stranded DNA changes [69].

The EU Nickel Directive in agreement with the European Chemical Agency (ECHA) poses limits on the amount of nickel that may be released from jewelry and other products. These limits are known as migration limits and they include 0.2 μg/cm^2/week for post assemblies which are inserted into pierced ears and other pierced parts of the human body and 0.5 μg/cm^2/week for other products intended to come into direct and prolonged contact with the skin [69].

Moreover, in the cell, nickel can lead to protein cleavage or oxidize side chains to produce Tyr-Tyr cross-likings. Thus, it can destabilize cell membranes. In addition, the chemistry of the metal is prone to produce metal-associated ligands. Cellular nickel does not exist in a fully aquatic state and the variety of adventitious ligands which bind to nickel can have a profound impact on the ability of the metal to produce reactive oxygen species. Indeed, nickel binds to cysteine, histidine, glutathione, small peptides, and amine-containing compounds. The ability of a ligand to act as a prooxidant depends on the low ratio to nickel which is more likely to provide an open coordination site for the binding of the oxidant. In addition, the metal can stimulate lipid oxidation in the presence of thiols or when combined with iron. For such reactions, nickel uses a variety of oxidants including oxygen, superoxide, hydrogen peroxide, and organic hydroperoxides. Finally, nickel can participate in DNA cleavage, interstrand cross-links, and base modification. However, nickel is a rather poor generator of oxidative damage, when compared to iron or copper; nevertheless, oxidative stress is not the prominent reason for carcinogenicity [75].

Another mechanism related to nickel toxic effects is the ability to substitute other metals in enzymatic reactions. The group of metals with the property of substitution is called *metalloenzymes*. As a metalloenzyme, nickel modifies or inactivates a number of intracellular enzymes. For example, nickel substitution of iron in a particular iron oxygenase could hinder growth on the cognate carbon source. Nickel is known to bind to catalytic residues of several enzymes, cysteine, histidine, glutamate, and lysine, blocking the catalytic activity. In addition, the substitution of other metals could lead to binding to secondary or allostatic sites affecting activity. Although scarce reports exist, nickel interaction with enzymes could be related with the inhibitory activity of the metal [75] (Fig. 4.20).

Chen et al. propose that the enzymes iron- and 2-oxoglutarate-dependent dioxygenases are an important target that mediates the toxicity and carcinogenicity of nickel. The structural motif of this dioxygenase family is a triad of 2-histidine-1-carboxylate that coordinates the Fe^{2+} ion in the catalytic site. According to their hypothesis, two different classes of enzymes in this iron- and 2-oxoglutarate-dependent dioxygenase family, including JMJD1A histone demethylase and DNA repair enzyme ABH2, are all sensitive to inhibition by nickel ions due to the fact that nickel replaces iron in their catalytic sites [69] (Fig. 4.21).

In literature, metallopeptidases with Zn^{2+}, Cu^{2+}, and Co^{2+} at the active site and sensitivity to nickel substitution have been described in many species. One example

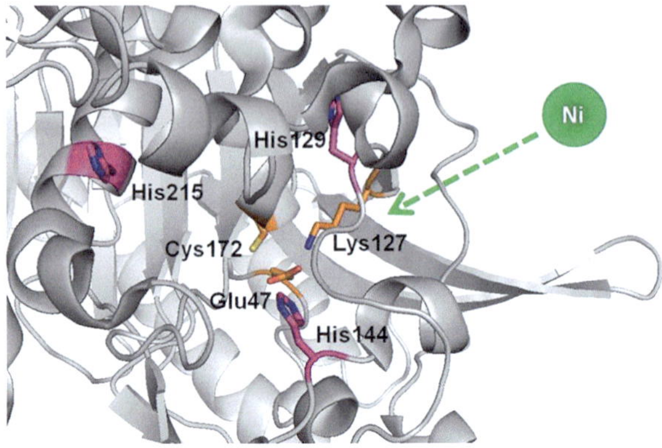

Fig. 4.20 Specific substitution sites of other metals usually contain 2 histidine and 1 glutamate molecules as binding proteins for nickel

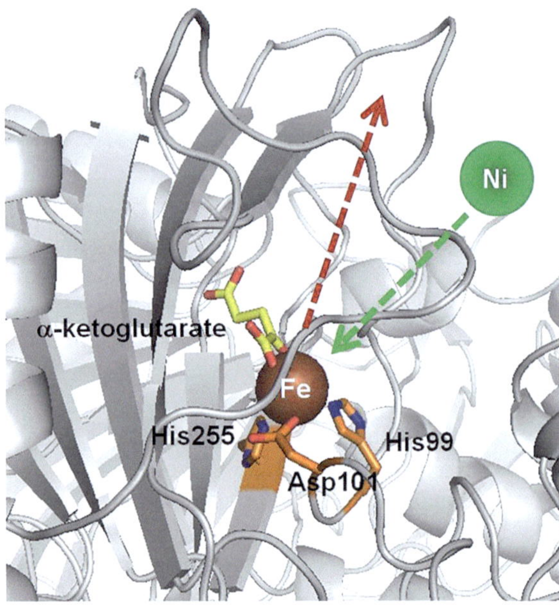

Fig. 4.21 Substitution of iron in 2-oxoglutarate-dependent dioxygenases

is *Trypanosoma cruzi*, the causative agent of the tropical parasitic American trypanosomiasis Chagas disease. Its Zn-carboxypeptidase cleaves the C-terminal aminoacid residue from peptides and proteins, but it can lose 54% of its activity upon treatment with Ni^{2+}. In the active center of this enzyme, there are two histidine molecules and one glutamate residues to bind Zn^{2+} ion [76].

Evidence related to the hypothesis that nickel directly causes oxidative stress, is much less compelling. If nickel toxicity does indeed lead to oxidative stress, then it will be necessary to determine whether this is a direct or indirect effect. A likely mechanism by which nickel could indirectly cause oxidative stress is by the increase of cellular iron. Iron has long been shown to be the primary metal involved to oxidative stress. The increased demand for iron import during nickel overload coupled to nickel's ability to damage iron proteins suggest that nickel-toxicity leads to an increase in the intracellular free iron pool. This indirect enhancement of oxidative stress was also hypothesized for copper-generated oxidative stress *in vivo* [76]. Oxidative stress caused by nickel exposure is significantly lower compared to other toxic metals. In contrast to genotoxic metals, nickel does not form DNA adducts or DNA breaks and therefore is considered non-mutagenic [74].

Epigenetic Mechanism of Intoxification

A number of studies suggest that epigenetic alterations underlie nickel pathogenesis. *In vitro* studies showed that treatment with nickel resulted in both increased global methylation and promoter hypermethylation, but extensive studies are not available [76]. A large number of genes are persistently altered. The majority of gene expression changes remain after the termination of nickel exposure [74] due to the irreversible effects of the *epithelial-mesenchymal transition (EMT)* [71].

During EMT, cells lose their junctions and polarity and become migratory and invasive. EMT is an essential cellular process involved in developmental events such as heart and neural crest development and wound healing. EMT, as already mentioned, is a major pathogenic process associated with airway remodeling in asthma, fibrosis, development of premalignancy, cancer progression, and metastasis. The persistent transcriptional alterations and EMT induction that are identified in Ni-exposed cells could be important in the development of Ni exposure-associated diseases [72].

A gene altered by nickel exposure is the EMT master *regulator gene, ZEB1 (Zinc Finger E-Box Binding Homeobox 1)* which is one of the most highly upregulated genes in all the exposed cells. It is suggested that persistent upregulation of *ZEB1* could play an important role in the acquisition of the state of EMT in nickel-exposed cells [73]. *ZEB1* gene possesses four hypoxia response elements (HREs) at its proximal promoter. HIF1α directly binds to *ZEB1* HRE-3 and increases its expression. Ni compounds are hypoxia mimetics, which could cause HIF1α stabilization even under normoxia. The hypoxia effect of nickel exposure could potentially play a role in *ZEB1* upregulation and subsequent EMT transformation [73]. In addition to *ZEB1*, nickel exposure causes persistent upregulation of other EMT master regulators, such as the *SNAI1* and *TWIST1*, although the upregulation is not as dramatic as with *ZEB* [71] (Fig. 4.22).

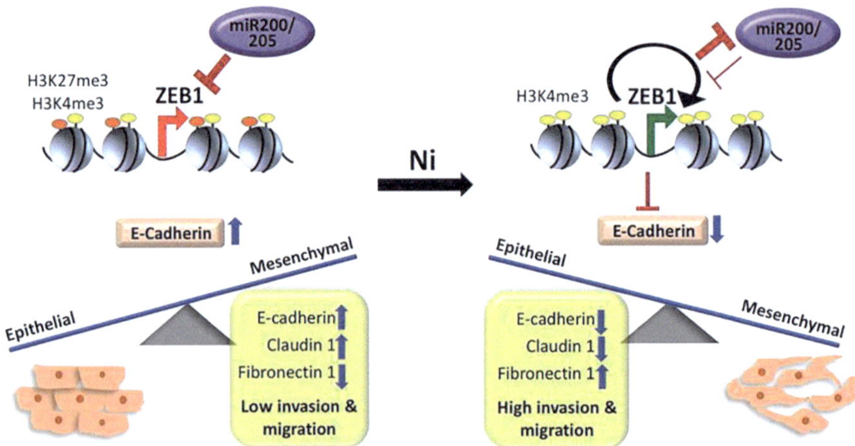

Fig. 4.22 Transition to epithelial-mesenchymal state

ZEB1 reveals Ni-induced bivalency resolution at its promoter by the loss of H3K27me3. The monovalent chromatin structure is irreversible even after the termination of exposure, resulting in the persistent activation of *ZEB1*. Thus, it could be speculated that PU-A genes are predominantly bivalent and they are activated by the resolution of repressive bivalency by loss of H3K27me3, while the H3K4me3 levels remain constant. On the other hand, PU-B genes are largely devoid of H3K4me3 before nickel exposure, while being enriched with H3K27me3, suggesting that these genes exist in a repressive chromatin environment prior to nickel exposure. Significant increase in the levels of H3K4me3 occurs after the termination of exposure, which coincides with the increase in gene expression [74]. Members of microRNA (miR) are miR-200 family genes and miR-205 are *ZEB1* repressors. On the other hand, *ZEB1* can downregulate miR-200 and 205 expression by binding to their promoters and recruiting transcriptional repressors [71].

To summarize, the above data show that the genes that are upregulated in the presence of nickel (PU-A genes) display high levels of H3K4me3 levels prior to nickel exposure. This suggests open chromatin environment. On the other hand, upregulation of gene expression that occur only after the termination of nickel exposure (PU-B genes) is clearly associated with increase in H3K4me3. This suggests that these genes may exist in a relatively inaccessible chromatin conformation and the increase of H3K4me3 could be important for creating accessible chromatin and gene activation [74, 76] (Fig. 4.23).

Nickel ions are able to induce *heterochromatinization* by binding to DNA-histone complexes and initiating chromatin condensation. Nickel carcinogenic potential is thought to be due to the ability to precipitate to these epigenetic changes. Both water-insoluble nickel sulfide (NiS) and water-soluble nickel sulphate ($NiSO_4$) and nickel chloride ($NiCl_2$) are human carcinogens, although insoluble nickel

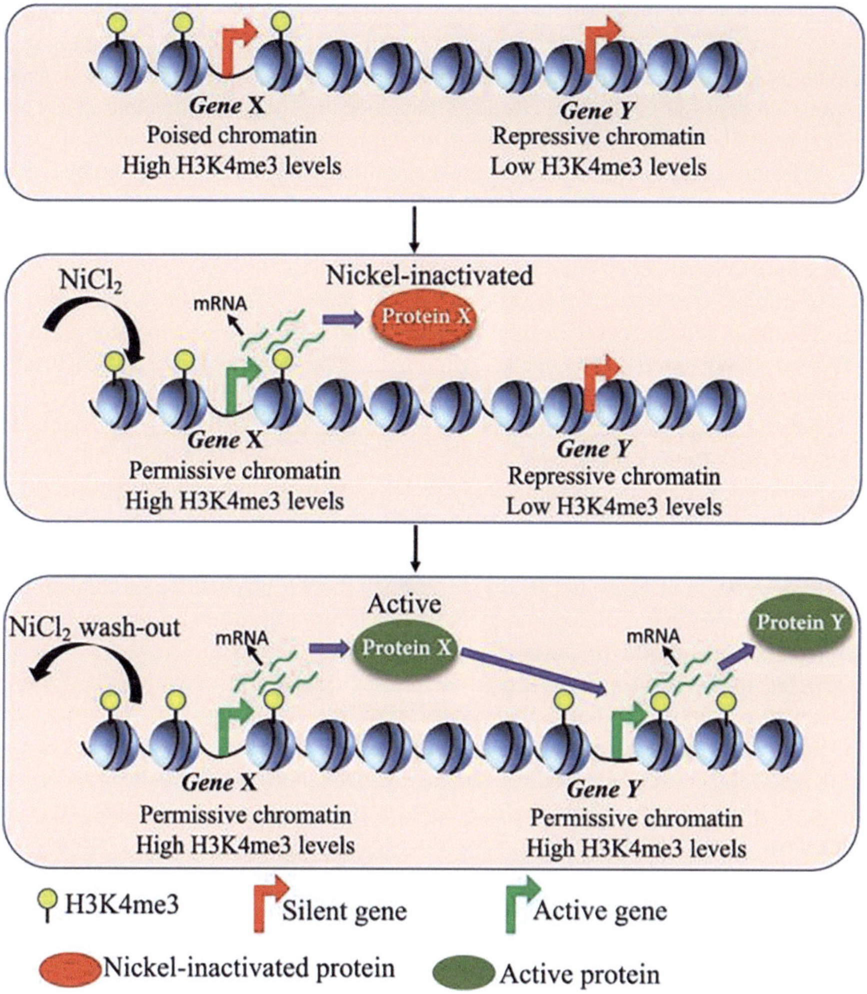

Fig. 4.23 In the untreated with nickel cells, Gene X, a potential chromatin regulator, is transcriptionally silent. The Gene X promoter is enriched with H3K4me3 in a poised chromatin environment. Gene Y is also transcriptionally silent, expresses low H3K4me3 levels and exists in a repressive chromatin state. Nickel exposure induces the first wave of transcription and it persistently activates Gene X by induction of upstream transcriptional regulations and loss of bivalency. The result is loss of repressive chromatin modifications. The protein X (red) remains inactive in the presence of Ni^{2+} ions both by inhibition of its activity and alteration of its DNA binding affinity. Upon Ni wash-out protein X renders its activity, but it influences the chromatin environment of gene Y. Gene Y gains high levels of H3K4me3 and protein Y is transcriptionally activated which leads to a second wave of transcription activated by Ni exposure

compounds are more potent carcinogens than the soluble ones. Nickel compounds can produce histone hyperphosphorylation (H3S10), hypermethylation (H3K4), and hyper-ubiquitination (H2A and H2B), inducing epigenetic effects that affect gene expression [76].

In a study, it was observed that nickel in the presence of Mg^{2+} condenses chromatin to a greater extent than magnesium. Given that chromatin condensation is a necessary step towards heterochromatinization, Ni^{+2} is associated with various gene expression-repressive changes. Reports conclude that the carcinogenic effect of nickel might be mediated by modulation of chromatin, such as heterochromatinization. The authors speculated that heterochromatinization could be followed by DNA methylation and the outcome might be cell transformation, tumor progression, and oncogenesis, if these silenced chromatin regions contained tumor suppressor or senescence-related genes [76].

Additionally, *in vitro*, nickel exposure has been associated with various histone modifications and decreased histone H4 acetylation [72]. Exposure to soluble $NiCl_2$ has been shown to reduce histone acetylation, increase demethylation of H3K9, and increase mono-ubiquitination of H2A and H2B [73].

Nickel ion exposure has been shown to increase global H3K9 mono- and demethylation, both of which have been associated with increased DNA methylation and long-term gene silencing. In human lung cells exposed to soluble nickel compounds, three major changes in histone modifications have been observed: (i) loss of acetylation of H2A, H2B, H3, and H4; (ii) increased H3K9 demethylation; (iii) increased ubiquitinoylation of H2A and H2B [76]. Among the various histone modifications, H3K4me3, a histone modification associated with transcriptionally active genes is strongly predictive of expression changes [71] (10).

Nickel ions inhibit histone acetyltransferases causing pervasive deacetylation. Nickel also inhibits H3K9 demethylases resulting in increase in the levels of H3K9me2. In addition, nickel disrupts repressive chromatin domains marked by H3K9me2 and causes its spreading. Increase in DNA methylation has been observed in nickel-exposed cells. These findings suggest epigenome as an important target of nickel exposure [74] (10).

Nickel can cross the placenta and has *embryotoxic and teratogenic properties* [71]. Apart from an increase of prenatal and perinatal mortality, nickel may cause a defined type of malformation in the embryos. It was hypothesized by Leonard et al. that the prenatal nickel effects could partially be due to the changes in mitosis, leading to cellular death. Sanderman and co-authors studied the effects of nickel carbonyl, Ni $(CO)_4$, and nickel sulfide (Ni_2S_3) administered to pregnant hamsters and female rats. Progeny of dams treated with Ni $(CO)_4$ resulted in fetuses with cystic lungs, exencephaly, fused ribs, and anophthalmia as well as cleft palate. In addition, in the progeny of dams treated with nickel carbonyl, there were two fetuses with hydronephrosis and one with fused ribs. However, administration of organic nickel compounds decreased the number of live pups due to chromosomal damage during the meiosis of spermatogenesis [69].

Another distinguished property of nickel ions is the ability to induce *apoptosis* that is a protective mechanism for the cell. Nickel ions allow the release of Cyt-C from mitochondria to the cytosol, where Cyt C cleaves procaspase-9 and activates caspase-9, which in turn activates caspase-3, -6, and -7. These caspases act on PARP (poly ADP ribose polymerase) and they induce apoptosis. On the cell surface, the nickel ions promote the interaction between Fas and FasL with the formation of the death-inducing signaling complex, which contains FADD and procaspase-8 and -10 that are activated to caspase-8 and -10. In the cell, caspase-8 and -10 cleave and activate the effector proteases, such as caspase-3, -6, and -7 followed by apoptosis. In addition to its importance as a biological phenomenon, defective apoptotic processes have been involved in a wide variety of diseases. Excessive cell death is responsible for many neurodegenerative diseases, whereas failure to undergo apoptosis results in autoimmune diseases and uncontrolled cell proliferation, such as cancer [69].

Calcium

Calcium is an essential nutritional agent. It is a basic metal for the cell function, especially for the function of muscle cells and heart muscle as it participates in muscle contraction. Actually, there are no side effects from high exposure to calcium. However, it is described with other metals due to the complex relation to atherosclerosis [77].

Calcium homeostasis is highly dependent on the metabolism of vitamin D. The precursor, vitamin D_3, could be synthetized on exposed skin by the UV-B (290–315 nm) radiation of sunlight. The energy of the UV radiation is used in a non-enzymatic reaction to open the B-ring of the cholesterol precursor 7-dehydrocholesterol, creating the thermodynamically unstable molecule pre-vitamin D_3, which then further isomerizes into vitamin D_3 (Fig. 4.24) [78]. The greater the exposure to sunlight the greater the production of vitamin D_3. The ability of endogenous synthesis of vitamin D_3 implies that the term "vitamin" is not the appropriate name for vitamin D_3.

However, nowadays humans tend to stay preferentially indoors and they cover their skin by textile outdoors; thus, insufficient UV-B exposure results in low endogenous vitamin D_3 production [78] and the vitamin should be acquired by diet [78]. Since worldwide, the average human diet does not contain much vitamin D_3, an estimated number of more than a billion individuals are vitamin D deficient. Fatty fish, such as tuna, sardines, salmon, and mackerel, are the best dietary sources of vitamin D_3 followed by liver, cheese, egg yolk, and mushrooms. Moreover, in some countries, foods, such as milk and other dairy products, orange juice, cereals, and margarines, are fortified with vitamin D_3. In addition, direct vitamin D supplementation is recommended in many countries [78].

The biologically inert molecule vitamin D_3 provided by food is converted in the liver to 25-hydroxyvitamin (25(OH)-D_3), which is the most stable and abundant vitamin D metabolite in human serum, and it is traditionally used as a biomarker for the vitamin D status of individuals. Further hydroxylation at carbon 1, creates the 1α, 25-dihydroxy-vitamin (1,25 $(OH)_2$-D_3) at the proximal tubule cells of the kidneys. This molecule is the active molecule, which functions as an endocrine hormone, serving as a high-affinity ligand agent to the transcription factor vitamin D receptor (VDR). On a molecular level, the effects of vitamin D are mediated by the

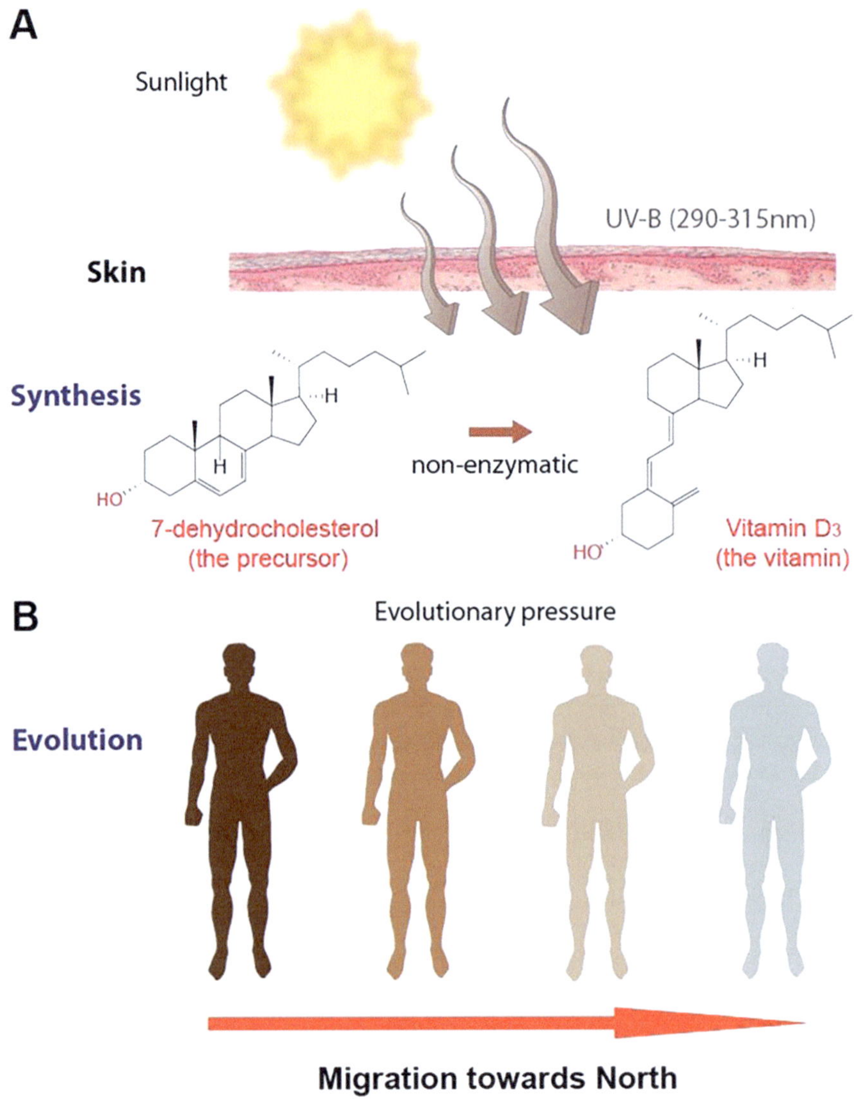

Fig. 4.24 Conversion of cholesterol to pre-vitamin D3

vitamin D receptor (VDR). The VDR is a transcription factor and a member of the nuclear receptor superfamily [78]. The main source of the endocrine production of 1,25 $(OH)_2D_3$ is the kidneys, but in a paracrine or autocrine fashion, monocytes, macrophages, and dendritic cells of the innate immune system, osteoblasts within bones, and keratinocytes of the skin are also able to produce the hormone [78] (Fig. 4.25).

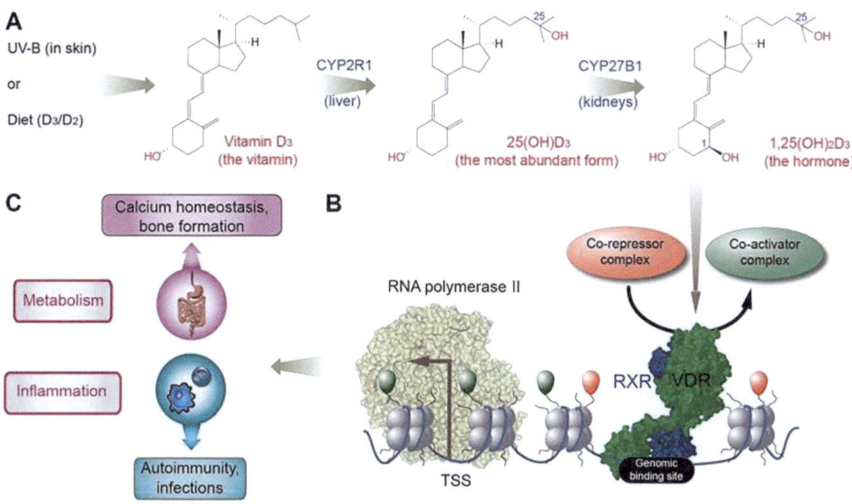

Fig. 4.25 The process for the production of 1α, 25-dihydroxyvitamin D_3 [79]

Serum levels of 25-hydroxyvitamin D [25(OH)D] induced by exposure to sunlight never exceed 60 ng/mL (149.76 nmol/L). According to the Mayo medical laboratories reference ranges, the total serum level of 25(OH)D normally ranges from 25 to 80 ng/mL. A level of 25(OH)D level less than 30 ng/mL is defined as vitamin D insufficiency and a level of less than 20 ng/mL defined as vitamin D deficiency, while 80 ng/mL or greater is defined as possible toxicity [79].

Effects on Health

Vitamin D is an effective regulator of several physiologic processes, such as calcium (Ca) homeostasis and innate and adaptive immunity.

Calcium is frequently deposited in the extracellular matrix of advanced atherosclerotic plaques. However, there is no clear association between dietary calcium or serum levels of calcium and the risk for development of coronary atherosclerosis [77]. On the contrary, high levels of serum calcium may have a mildly protective effect in atherosclerosis. It is detected that there is a borderline inverse relationship between serum calcium and the extent of coronary atherosclerosis. However, there is no clear association between calcium in drinking water and the incidence of deaths due to acute myocardial infarction deaths, as demonstrated in studies in England and Scandinavia [77]. An exception to these results is a study from Finland that demonstrated a protective effect of high levels of calcium in drinking water to decrease the incidence acute myocardial infarction in postmenopausal women [77].

Epidemiological and experimental studies reveal that the VDR and its ligands are promising targets for the prevention and possible treatment of cancer, autoimmune diseases, and infections as well as bone and mineral disorders. Ca levels are inversely correlated with the level of the cancer antigen 19-9 (CA 19-9) tumor marker, but not with the carcinoembryonic antigen (CEA) or alpha fetoprotein (AFP) tumor markers or the CaSR genotypes in colorectal cancer. It seems that there is positive relation with the initiation of this type of cancer with Ca and vitamin D levels [78].

Asbestos

The Metal, Its Forms, and Its Distribution in the Environment

Asbestos refers to a group of naturally occurring hydrated mineral silicate fibers including two major forms: serpentine, represented by chrysotile (white asbestos) and amphibole, represented by crocidolite (blue asbestos). Other forms include amosite (brown asbestos), anthophyllite, actinolite, and tremolite [80]. Thus, "asbestos" is a generic term for a group of natural crystalline silicates [81].

This element is actually a stone which is heat-, acid-, and friction-resistant. Besides, asbestos production is of low cost and its mining is worth-efficient. Therefore, it was abundantly used all over the world during the last century. However, it was recognized, even until the 1920s, that a rare type of cancer appeared in the workers in the mining for asbestos [80].

The widespread use of asbestos starts from ancient times. It has been recognized that cloths covering Egyptian mummies contained asbestos. In addition, a samurai, pharmacologist, writer and inventor, Gen-nai Hiraga, discovered this fibrous stone in the Edo Period in the mountainous area of Chichibu on the suburbs of Tokyo and he produced what is called burn-free textile (Kakanpu in Japanese) [82].

Many developed countries have early banned the use of asbestos, but not the USA. However, there have been multiple efforts in the USA to establish strict exposure standards, to limit asbestos use and to seek compensation by the courts for asbestos-injured workers' [83]. The National Institute for Occupational Safety and Health (NIOSH) has led these efforts and in 1976 was the first US federal agency to recommend a ban on asbestos in the workplace. As a consequence of these initiatives, asbestos use has declined dramatically in the USA, despite the absence of a legally mandated ban [84].

The carcinogenicity of asbestos was evaluated by the second IARC (International Agency for Research on Cancer) Working Group in 1976. The conclusions from experimental studies with animals remained largely the same and human research evidence explicit, that exposure to chrysotile, amosite, crocidolite and mixed fibers, increased the risk of lung cancer and mesothelioma. Increased risks of cancers of the larynx and the gastrointestinal tract were also noted for the first time [84]. To date, total or partial asbestos ban legislation has been adopted by 54 countries [81].

However, it was not until 1987 that the IARC designated asbestos as a Group 1 carcinogen and as a definite carcinogen to humans. In 2009, IARC Working Group concluded that there was sufficient evidence that "all forms of asbestos," especially chrysotile, crocidolite, amosite, tremolite, actinolite, and anthophyllite, are carcinogenic. The U.S. National Toxicology Program also evaluated the carcinogenicity of asbestos around the same time [84]. In the fall of 2012, Canada stopped mining chrysotile [82]. However, many developing countries produce and use asbestos [80].

Currently, about 90% of world asbestos comes from four countries: Russia, China, Brazil and Kazakhstan. Approximately, 50% of world asbestos is used by two countries, China and India, followed by Brazil, Indonesia, and Russia [81].

Asbestos minerals are used in a variety of small- and large-scale industrial applications all over the world [83]. Although current regulations mention only six commercial mineral fibers (actinolite, amosite, anthophyllite, chrysotile, crocidolite, and tremolite), several other mineral fibers found in the environment are related to disease. For example, erionite is more carcinogenic than chrysotile and crocidolite asbestos [83].

Products of asbestos have many applications. Some examples of asbestos containing products that were allowed to remain in US markets include cement corrugated sheet, cement flat sheet, cement shingle, cement pipe, roofing felts, non-roofing coatings, roof coatings, vinyl floor tile, clutch facings, millboard, automatic transmission components, friction materials, disk brake pads, drum brake linings, brake blocks, gaskets, and clothing pipeline wrap [83]. The asbestos-containing products banned by EPA under the Toxic Substances Control Act (TSCA) were limited to a mere handful of product types, including corrugated paper, roll board, commercial paper, specialty paper, and flooring felt [84]. Banning of asbestos production and use and the adoption of asbestos abatement programs over recent decades has not yet eliminated its use [83].

Occupational exposure has long been considered the form of exposure involved in the highest risk of developing asbestos-related diseases (ARDs). Asbestos workers themselves can contaminate their homes and cars, exposing their family members and housemates to asbestos. Additionally, living near plants of production is a high-risk exposure [83].

Repair, renovation, and demolition of asbestos-containing buildings also pose a threat through soil contamination, environmental pollution, and bystander exposure [83]. Some schools and public buildings still use asbestos as insulation material. Resuspension of materials from asbestos-containing ceilings has been shown to be the main source of asbestos pollution in old, poorly maintained buildings [83].

It is worth mentioning that there is no safe level of exposure. Asbestos fibers are fibrous mineral silicates that have been associated with the development of pulmonary fibrosis, bronchogenic carcinoma, and malignant mesotheliomas in both humans and experimental animals. The fact that asbestos, a well-established carcinogen, has been used extensively in industry and households makes it an important health concern. It has been suggested that the danger of developing asbestos-related diseases extends beyond that of a simple occupational hazard [82].

Accidental Exposures in History

"*Asbestos dust is, perhaps, the best-documented industrial toxin in history…*". The earliest recorded description of asbestos related diseases was in the USA on Belvidere Mountain near Lowell, VT before 1823. The Vermont deposits run northwest to Quebec in Canada. Not much interest was shown in these asbestos deposits in the early 1800s, because commercial mining of asbestos did not begin until the 1890s, when chrysotile was mined from open pits at three locations on Belvidere Mountain near the towns of Eden and Lowell [84].

During the 1920s, *asbestosis* was coined as the name for the form of pneumoconiosis, the dust disease affecting the lungs. The disease affects asbestos-exposed persons and the curious bodies, later known as asbestos bodies, were first demonstrated. Today, it is known that asbestos bodies can form in extra-pulmonary sites, such as the liver and spleen [84]. A major breakthrough in understanding, the rate of diseases caused by asbestos occurred in 1960, when a large series of case reports was published involving the rare pleural tumor mesothelioma (MM) [84].

All of the reported cases had in common the exposure to crocidolite asbestos in a mining region of the NW Cape of South Africa. By 1963, Wagner had collected 120 mesothelioma cases, which he reported to the 14th International Congress on Occupational Health [84]. Besides asbestos, some other cofactors have been linked to MM development, e.g., Simian virus 40 (SV40), ionizing radiation, genetic predisposition, and certain mineral compositions, such as erionite. Such factors may also contribute to the significant variations in MM incidence in different geographic areas [80].

Another clustering region with the disease is three small villages in Cappadocia, Turkey, an unprecedented mesothelioma epidemic causes 50% of all deaths. The incidence of MM in those villages is about 1:100/year compared to 1–20 cases/million in the western world. Recently, Luo and colleagues reported that in a province called Da-yao in southwestern China, the annual mortality rate of mesothelioma is up to 365 cases/million, compared to 2–3 cases/million in the general population of China [80].

The incidence of MM has increased significantly, and currently there are about 2000–3000 cases/year in the USA. However, about 70,000 new MM cases are expected in the USA over the course of the next 20 years [83]. Asbestos kills 12,000–15,000 Americans each year and results in innumerable cases of disease and disability. It is reported that in the USA in 2015, there were 2597 deaths due to mesothelioma representing only about 15–20% of overall asbestos-related deaths. The majority of asbestos-related deaths other than those caused by mesothelioma are due to lung cancer ($\approx$60%) and asbestosis ($\approx$10%), with additional deaths due to cancers of the larynx, ovary, and complications of asbestos-related non-malignant respiratory diseases [84].

In Europe, the anticipated peak year of MM incidence was in the period of 2015–2020, with a predicted incidence of 250,000 cases over the next 40 years [80]. In Japan, they are detected approximately 1500 new MM cases each year, and incidence rates are increasing. This figure may be small in comparison to approximately 80,000 cases each year worldwide and MM carries a social impact, because 80% of cases are thought to be associated with former asbestos exposure [80].

Effects on Health

Asbestos fibers are inhaled from the polluted air and as they are extremely thin, they frequently pass from the nasal cavity and the trachea to the lung (Fig. 4.26).

A number of diseases of the lungs develop due to asbestos exposure, which includes bronchitis and CVD (cardiovascular disease) due to continuous deposit of fibers. Pneumoconiosis is the most serious consequence of asbestos exposure and it is considered to be a prodromal stage for the development of mesothelioma [84].

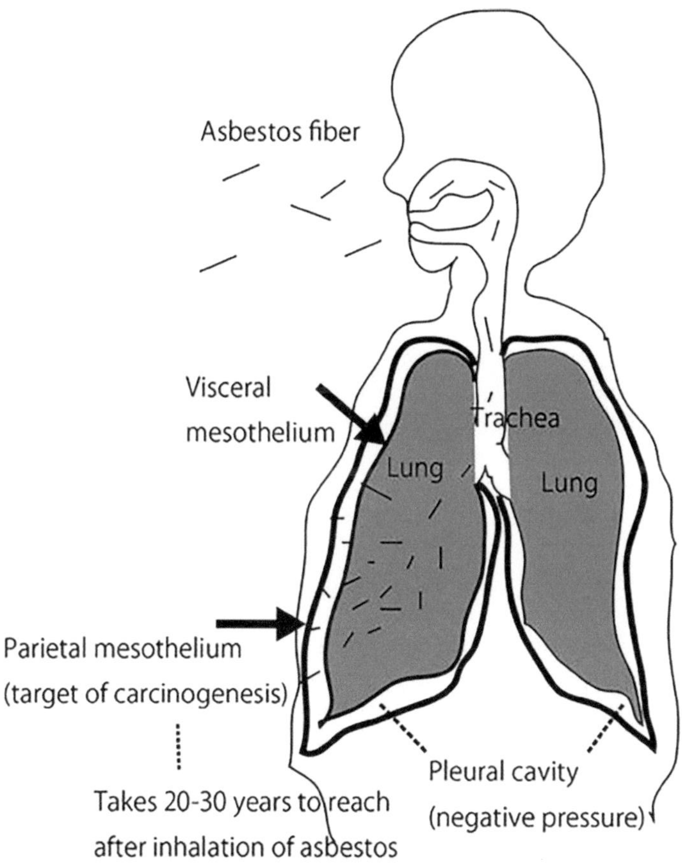

Fig. 4.26 The deposits of asbestos fibers in the respiratory tract

The term *pneumoconiosis* is referred to the parenchymal lung diseases that arise from inhalation of usually inorganic dusts, especially at work. Initially, these dusts are biologically inert, but visible on a chest X-ray or CT scan. Thus, they are radiologically alarming, before the appearance of clinical symptoms and the deficits of the pulmonary function. The asbestos inorganic dusts, for example crystalline silica, are fibrogenic and the damage they cause is through fibrosis induced by the inhaled dust. Classically, these give rise to characteristic radiological patterns and restrictive deficits in lung function, mainly reduction in diffusion capacity. Importantly, they may progress long after the exposure to the causative mineral has stopped [85].

In the UK and similar countries, asbestosis is the commonest form of pneumoconiosis, but in less developed parts of the world asbestosis is less frequent than silicosis. Other, rare types of pneumoconiosis include stannosis (from tin fume), siderosis (from iron), berylliosis (from beryllium), hard metal disease (from cobalt), and coal workers pneumoconiosis. The chest X-ray in all these cases reveals several pleural plaques, probably calcified, but no other abnormalities [85].

Asbestosis is a form of pneumoconiosis that arises due to exposure to asbestos in the workplace. The diagnosis is applied when there is radiological evidence of pulmonary fibrosis on a background of heavy occupational exposure to any type of asbestos. The diagnosis is not necessarily straightforward and it is often a matter of judgement [85]. In addition, patients with asbestosis suffer not only from the respiratory complications, such as chronic bronchitis, emphysema, and lung fibrosis, but also from autoimmune disorders [85].

The amount of asbestos required to induce fibrosis is relatively high and it is usually acquired after 20–30 years of high exposure. Common occupations associated with asbestosis include boiler lagging with asbestos, work with asbestos textiles, and the manufacture of asbestos-containing building materials [85].

The "Helsinki criteria" suggest that a dose of at least 25 fiber/mL/years is necessary to cause asbestosis. This cumulative exposure metric is similar to the "pack years" used to measure cigarette smoking; thus, 10 years work in an average airborne concentration of 2.5 fiber/mL is equivalent to one year at 25 fiber/mL. This latter concentration is very high and it was readily detected in the asbestos textile industry during the 1950s and 1960s [85].

The *pathognomonic feature* of asbestosis is the formation of pleural plaques detected in X-rays. Plaques could develop even after quite low asbestos exposure, they are more common in men from industrialized countries, they are thought to be benign and they are not responsible for any symptoms and loss of lung function, although they cause a lot of anxiety. They do not need follow-up and they are considered the first manifestation of exposure [85]. The identification of plaques is not the plaques per se, but the fact that they reflect asbestos exposure with its attendant risks. While plaques do not themselves increase the risk of cancer, the exposure that gave rise to them does; the risk depends in part on the dose and time of exposure and it is usually around 5% [85].

The *radiological features* of asbestosis are non-specific and they closely resemble "interstitial pneumonia." They also characterize the disease as "idiopathic pulmonary fibrosis" (IPF). Changes are far more readily seen on the CT scan than on a

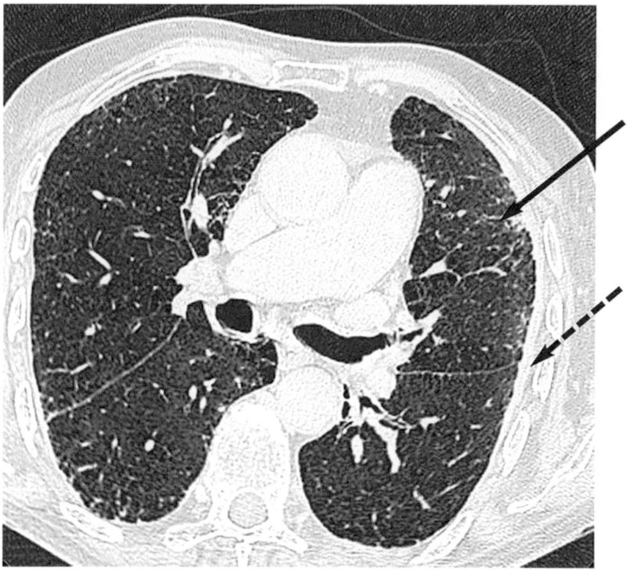

Fig. 4.27 Interstitial fibrosis in CT scan due to asbestosis (the arrows show the plaque formation)

conventional chest X-ray. In fact, the increasing use of thoracic CT in both respiratory and cardiac care is probably responsible for the detection of early signs of asbestosis [85]. Other radiological features of asbestos exposure such as pleural thickening are often seen and they can be used to help making the distinction between asbestosis and idiopathic lung disease [85] (Fig. 4.27).

Most patients with asbestosis should be followed-up for years and clinical assessment and spirometry could be helpful to detect the early signs of the disease. There is no effective drug treatment for asbestosis and in most cases no pharmacological treatment is offered, such as bronchodilators or corticosteroids. N-acetyl cysteine was proved to be of some benefit in idiopathic pulmonary fibrosis and it is widely used. Advanced cases may benefit from oxygen supplementation at home [85].

The course of the disease depends largely on its extent and on the quantity of asbestos retained in the lungs. In mild cases, there is little evidence of progression for many years. In advanced disease, there is a steady decline towards respiratory failure. A precipitous deterioration, marked by increasing breathlessness and sometimes cyanosis on exercise, is usually indicative of secondary pulmonary hypertension [85].

Many patients with asbestosis die from lung cancer, a process accelerated by concurrent smoking. It is probably a good practice to advise those with asbestosis, who smoke to quit smoking, although the benefit is uncertain. Around one in ten patients will develop and die from mesothelioma and a similar number will die from the complications of asbestosis [85].

Malignant mesothelioma (MM) is a rare, but very aggressive tumor, which arises from mesothelial cells lining the pleural, peritoneal, and pericardial cavities. This aggressive, lethal cancer arises is located on pleural (80–90%), peritoneal (10–15%), and pericardial cavities (<5%). Its long latency (≥30–60 years) and non-specific symptoms often lead to a late diagnosis and poor survival [83]. MM is sub-typed into three forms according to the histological morphology: epithelial, sarcomatoid, and biphasic. The prognosis of MM is poor and the median survival time for these three types is 18, 8, and 11 months after diagnosis, respectively [80].

The latency period between the time of initial exposure and diagnosis is about 30 years and ranges from 20 to 50 years. Two-thirds of MM patients are between 50 and 70 years of age. Males are at a much higher risk for MM than females, likely due to occupational exposure [80].

Diagnosis is based on X ray and CT investigation that are indicative of the disease. However, definite diagnosis relies on pleural biopsy, pathology, and immunohistochemistry. It is characterized by the histological presentation into epithelioid (50–60% of cases), sarcomatoid (10–20%), biphasic (25–35%), and other, less common subtypes. The differential diagnosis from benign proliferations and other malignancies is highly complex [83].

Apart from lung cancers, asbestos is implicated in many other cancers, especially of the gastrointestinal tract [85]. There is evidence that asbestos exposure is related to the cancers of stomach, pharynx, and colorectal cancer. The overall scientific consensus supports that asbestos is a proven human carcinogen and that there is no safe level of asbestos exposure [84].

Mechanisms of Intoxicification

The mechanism of the transformation of epithelial lung cells to malignant cells, due to asbestos exposure, is not entirely clear. Various *in vitro* and *in vivo* studies suggest that the fiber dimensions, surface properties, and physical durability are important factors that promote the carcinogenicity of asbestos fibers [83].

Mesothelial cells cover the three somatic cavities (pleural, peritoneal, and pericardial cavity) forming a villous surface with flat single-layered morphology. Mesothelial cells do not face the outer surface of the cavity and they are in continuation with lymphatic vessels. The main function of mesothelial cells is to reduce the friction during functional movements. They secrete hyaluronic acid which helps with pulsating and peristaltic movement. However, not much is known regarding the damage, repair, and proliferation of mesothelial cells. Recent *ex vivo* cultures of the cells provide evidence that they move extensively and they develop into a cuboidal shape after sensing damage nearby [80].

Mesothelial cells have the unique characteristic of engulfing anything, whether a solid, liquid, or gas, thus they are implicated in phagocytosis of asbestos fibers. Interestingly, they actively engulf asbestos fibers which are not harmful for the cells. Phagocytosed asbestos fibers likely attach to chromosomes when they are dividing, leading to massive genomic alterations such as deletion and amplification after some repair processes. However, compounds such as amosite asbestos cause direct mesothelial cell destruction. Mesothelial cells are 10 to 100 times more sensitive to the cytotoxic effects of asbestos than normal human bronchial epithelial or fibroblastic cells [80].

Alveolar macrophages are responsible for the disposal of asbestos fibers, but fail to dispose them, if the fibers are too long (length > 20 μm) and/or too thin (diameter < 250 nm) [83]. Macrophages that phagocytose fibers are destroyed by a mechanism called apoptosis [80]. After death of macrophages, residual proteins and other molecules are adsorbed and they accumulate in neighboring cells. Thus, they produce "asbestos" bodies, which additionally contain other metals such as abundant iron. It is characteristic that asbestos including bodies move toward the pleural cavity due to negative pressure. Thereafter, they reach the visceral pleura, penetrate it, and finally reach the parietal pleura. Fibers engulfed in the parenchyma can cause pleural plaques, i.e., asymptomatic focal thickenings that are the hallmark of asbestos exposure and abnormal fluid collections, while fibers trapped between the pleural layers and the wall of the chest cavity induce oxidative stress and chronic inflammation, thus promoting carcinogenesis [83]. It is well recognized that most cases of MM occur at the parietal pleura [82].

Asbestos bodies are responsible for the local production of reactive oxygen species (ROS). Although ROS have been indicated as one of the key determinants of asbestos-induced mutagenesis and carcinogenesis, the types and the origin of these free radicals remains elusive. It has been suggested that ROS, such as superoxide anions (O_{2-}) and hydrogen peroxide, originate not only from redox reactions on the fiber surface, but also the incomplete phagocytosis of fibers in phagocytic, mesothelial, and rat lung epithelium cells. The protective involvement of antioxidant enzymes, such as catalase and superoxide dismutase, as well as other radical scavengers has been studied in various cell lines [83].

These reactive ROS, mainly hydroxyl radicals (OH-), act directly or indirectly to the damage of neighboring biomolecules such as DNA and membrane lipids. There is evidence to suggest that the array of DNA and chromosomal damages may lead to a broad spectrum of mutations. Most of them are substitutions, deletions, rearrangements, insertions, sister chromatid exchanges, disturbance of mitosis, and other chromosomal aberrations [83]. Cytogenetic and loss of heterozygosity analyses of mesothelioma cells detect deletions in specific regions within chromosomes 1p, 3p, 4p, 4q, 6q, 14q, 15q, and 22q [81].

Furthermore, a significant contribution to the transformation of lung cells to malignant has the molecular changes in gene level. Mesothelial cells acquire gene mutations, alteration of growth signaling, deregulation of methylation status, changes in the heterochromatinization of histones resulting to the production of high affinity and aberrant miRNA expression. Thus, epigenetic modifications contribute to carcinogenesis [80]. Another parameter that contributes to the transformation is inflammatory processes due to asbestosis and accumulation of asbestos bodies [83] (Fig. 4.28).

Certain *tumor suppressor genes* are implicated in asbestos carcinogenesis, especially the ones located in the above-mentioned chromosomal regions, for example, *CDKN2A/ARF* at 9p21 and *NF2* at 22q12 regions. Mutations of the *p53* were

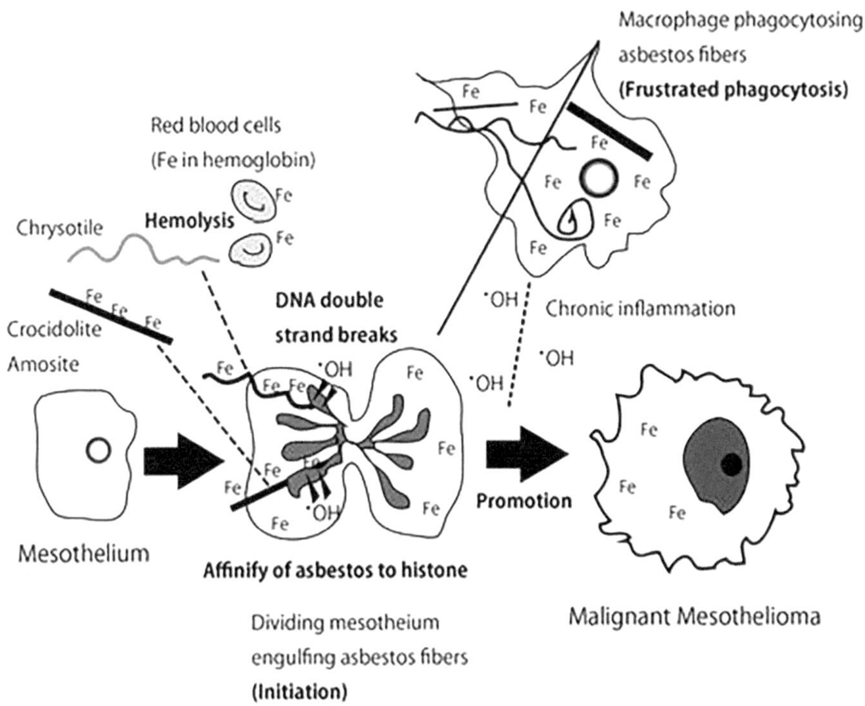

Fig. 4.28 Mechanisms of asbestos-induced mesothelial carcinogenesis

occasionally observed. Loss of function or inactivation of these tumor suppressor genes may play a role in the development and progression of mesothelioma. The *CDK2A/ARF* gene encodes the tumor suppressor *p16* (*INK4a*), a cyclin-dependent kinase inhibitor and *p14* (*ARF*), a component of the *p53* cell cycle checkpoint, while homologous deletions of the *CDKN2A/ARF* locus in the mesothelioma cells impair both the retinoblastoma (*Rb*) and *p53* pathways. The *NF2* gene product is called merlin and it represses cyclin D1 expression, while loss of function or inactivation of *NF2* in mesothelioma leads to cell cycle progression in connection with upregulation of cyclin D1. Merlin also inhibits Rac/Pak and focal (FAK) signaling, which play a role in cell migration and metastasis, while inactivation of *NF2* in mesothelioma cells promotes invasiveness and metastasis [81]. Moreover, very recent evidence indicates that asbestos causes the release of High Mobility Group Box Protein-1 (HMGB1), a key molecule in the inflammatory process that leads to fibrosis and carcinogenesis [83].

A role of *impaired immunity* due to asbestos compounds has been suggested. Recent work demonstrated a critical role for tumor necrosis factor-alpha (TNF-α) and NF-κB signaling in mediating responses of human mesothelial cells to asbestos. *In vivo* studies have revealed that, following asbestos exposure, there is an inflammatory reaction with a large component of mononuclear phagocytes. Upon differentiation into macrophages, these cells phagocytize asbestos and, in response, release TNF-α. At the same time, asbestos induces human mesothelial cells to express the TNF-α receptor (TNF-R1). Thus, mesothelial cells also stimulate the secretion of TNF-α (both paracrine and autocrine effects). TNF-α binds to its receptor and activates the NF-κB pathway, which increases the percentage of human mesothelial cells that survive asbestos exposure. The activation of the NF-κB pathway allows mesothelial cells with asbestos-induced DNA damage to divide rather than die, and if sufficient specific genetic damage accumulates to eventually develop into a MM [81].

In addition to TNF-α, other growth factors and cytokines have been implicated in asbestos carcinogenesis and their role in MM pathogenesis is being investigated. These include: transforming growth factor beta (TGF-β), which might have a role in stimulating tumor growth; platelet-derived growth factor (PDGF), which may act as a regulatory factor in MM cell proliferation; insulin-like growth factor (IGF), which promotes tumor proliferation and cell migration; interleukins such as IL-6 and IL-8, which may promote tumor growth and the development of new capillaries; vascular endothelial growth factor (VEGF), which also promotes tumor angiogenesis; and hepatocyte growth factor (HGF), which stimulates mesothelioma cell migration and tumor invasiveness [81].

An example of this process is figured in Fig. 4.30. Crocidolite is a very toxic compound and it accumulates macrophages in the pleura and lung. These macrophages release TNF-a when they encounter asbestos. At the same time, induces mesothelium secretion of TNF-a with paracrine and autocrine effects. TNF-a leads to translocation of the NF-kb subunit, which in turn increases mesothelium cell survival and the risk for cancer (Fig. 4.29).

It is mentioned also that asbestos could trigger other cell-signaling pathways apart from NF-kb also. For example, crocidolite fibers can induce autophosphorylation of the epidermal growth factor receptor, etc. [81].

We have mentioned that carcinogenicity is different depending on the components of asbestos. The association between amphibole asbestos exposure and MM development is well accepted. Crocidolite is generally considered to be the most mutagenic type of asbestos. The long and thin fibers (especially ≥8 μm in length ≤0.25 μm in width) are thought to be more dangerous, because they have longer bio persistence in the pleura. These fibers are able to penetrate the lung and they cause repeated damage, tissue repair, and local inflammation [81].

Chrysotile is the most common type of asbestos and accounts for about 90% of the world's asbestos production. Whether chrysotile causes MM is still controversial. Some authors suggested that chrysotile may cause MM, but at a lower rate compared to amphibole asbestos. Hodgson and Darnton suggested that the exposure specific risk of MM is broadly dependent on the ratio of 1:100:500:500 for chrysotile, amosite, and crocidolite, respectively [81].

Among different types of mineral fibers, erionite is the most potent compound to induce MM. Erionite has been detected in the lungs of villagers in several towns in Cappadocia, Turkey, where 50% or more of deaths are caused by MM [81].

Approximately 60% of iron in humans exists as heme in hemoglobin within erythrocytes. It has been reported that crocidolite (blue asbestos) and amosite

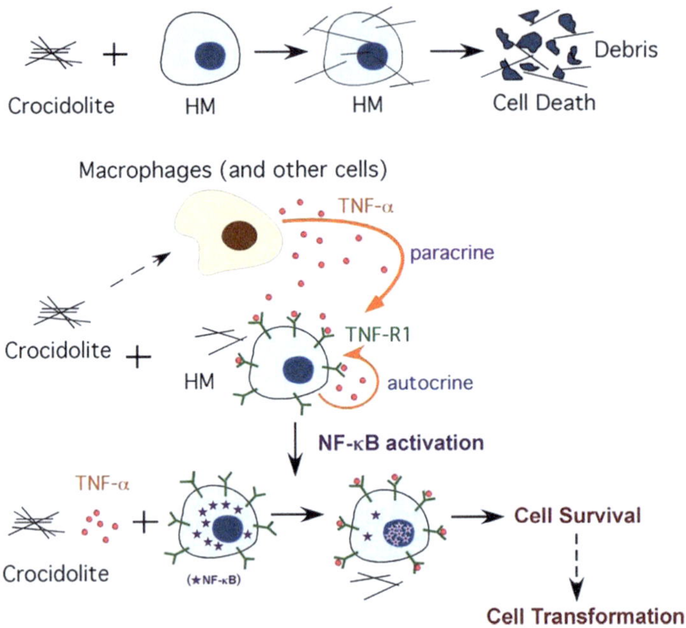

Fig. 4.29 Crocidolite induction of mutagenesis in mesothelial cells (HM)

(brown asbestos) have an affinity to hemoglobin, which appears to be the reason that asbestos accumulates iron [82]. Chrysotile (white asbestos) itself contains little iron. Notably, chrysotile is potent in causing hemolysis after attaching to hemoglobin. The catalytic activity of chrysotile is highly increased in the presence of hemoglobin and, furthermore, chrysotile interacts with DNA molecules. The toxicity of chrysolite is probably related to iron absorption. Therefore, chrysotile, due the ability to activate iron-related processes, is more dangerous for carcinogenesis than formerly thought [82].

Direct damage of mesothelial cells by asbestos fiber is an important initiation process for asbestos-induced mesothelial carcinogenesis. Here, we could add the additional damage starting because of local iron overload. Iron-coated asbestos fibers function as a 'sharp knife' to cause DNA-double-strand breaks in the mesothelial cells in addition to local inflammation that combined promote carcinogenesis. Considering that we cannot remove thin asbestos fibers after inhalation, local iron deposits removal could be a target treatment for the prevention of asbestos-induced mesothelial carcinogenesis [82].

Another factor that should be mentioned is that lymphatic vessels are in continuity with mesothelial cells. When the length of asbestos fibers is sufficiently large, they stuck at the orifice of lymphatic vessels, which may be the starting point of mesothelial carcinogenesis. Currently, there is no way to clean the human lung after inhalation of asbestos; thus, it is extremely difficult to remove asbestos fibers *in vivo* after inhalation [82].

Epigenetic Mechanism of Intoxification

The epigenetic changes contributing to the development of MM and other cancers due to asbestos exposure are mainly in the miRNA level. Overexpression of the *EZH2* gene, which encodes core components of polycomb repressor complex-2 (PRC-2), involved in the pathogenesis of different cancers, is responsible for the low expression of its mRNA regulators, miR-101 and miR-26, in mesothelial cells [81].

Analysis of the combined expression of miR-15 and miR-16 documented their significant downregulation and tumor-suppressing function as well as inhibition of proliferation in MM. It has also been reported that miR-23a and miR-27a modulation in MM induce silencing of *ZIC1*, a potential tumor suppressor gene involved in apoptosis. The *PVT1* locus is another oncogene acting as a non-coding RNA by different, alternatively spliced isoforms. Its frequent copy number gain in MM cell lines, combined with miR-1204 depletion at the same locus, promotes overexpression of anti-apoptotic genes and the malignant phenotype. Significant miR-31 downregulation has been reported in macro-dissected tissue vs. reactive mesothelial proliferations, whereas miR-31 upregulation is significantly associated with a worse prognosis in patients with sarcomatoid MM. Down-regulated miR-223 expression has been reported to modulate *STMN1*, which has an important role in microtubule dynamics; both miR-31 and miR-223 are involved in the JNK signaling pathway. Comparison of miR-192, miR-193a-3p, and miR-200 family expression in normal pleura and MM specimens has highlighted a significant reduction in miR-192 and miR-193a-3p in MM tissue. Finally, restoration of miR-193a-3p levels has been reported to inhibit MM cell growth and xenograft tumors *in vivo* [83]. An aberrant methylation status and silencing of miR-34b and miR-34c has also been described in MM specimens. Extensive functional investigation of miR-34b/c activity has suggested that downregulation of miR-34 family members induces proliferation and invasion of human mesothelial cells, thus playing an important role in carcinogenesis [81].

Preclinical evaluation of adenovirus-mediated miR-34b/c gene therapy has shown promise in the treatment of malignant pleural mesothelioma (MPM). The therapeutic approach based on miR-34 family members promotes radiation-induced apoptosis [83].

Silica

The Metal, Its Forms, and Its Distribution in the Environment

Silicon dioxide, also known as silica, is an oxide of silicon with the chemical sign SiO_2, most commonly found in nature as quartz and in various living organisms. In many parts of the world, silica is the major constituent of sand. Silica is one of the most complex and most abundant family of materials, existing as a compound of several minerals and as a synthetic of many products. Notable examples include fused quartz, fumed silica, silica gel, and aerogels. It is used in structural materials, microelectronics (as an electrical insulator), and as components in the food and pharmaceutical industries.

Crystalline silica is found in many types of stone, but sandstone constitution is approximately 70% from silica. Stonemasons, particularly those using angle grinders, which generate large quantities of respirable dust, are at a high risk of developing silicosis. Other groups at risk are quarry workers and tunnellers, foundry and pottery workers, as well as construction workers (such as pavers, who frequently cut or break stone, concrete, or brick) [86].

Effects on Health

Inhaling finely crystalline silica is toxic and it can lead to severe inflammation of the lung tissue, causing bronchitis, silicosis, lung cancer, and systemic autoimmune diseases, such as lupus and rheumatoid arthritis. Inhalation of amorphous silicon dioxide in high doses leads to short-term inflammation followed by healing.

Silica particles are known as typical environmental and occupational causatives of pneumoconiosis [87]. Exposed persons suffer from respiratory complications such as chronic bronchitis, emphysema, and lung fibrosis. Advanced disease may be complicated by pulmonary hypertension, cor pulmonale, and respiratory failure, and any evidence of these should trigger a pulmonary transplant assessment [86].

Silicosis results from the accumulation of respirable particles of crystalline silica in the lung [87]. Classic silicosis is the most common presentation and typically follows 10–20 years of work exposure, during which time the individual often remains asymptomatic. The typical plain chest X-ray appearances are similar to those of coal worker's pneumoconiosis with a profusion of small nodules in the upper and mid zones. Hilar and mediastinal lymph node enlargement may be present (Fig. 4.30).

High resolution CT (HRCT) scan provides greater diagnostic confidence showing bilateral well-defined 3–5 mm nodules in a centrilobular and subpleural distribution with a posterior bias. Additionally, progressive massive fibrosis and coalescence of nodules into irregular masses, plus lymph node involvement is frequently detected (Fig. 4.31) [87].

However, substantial lung damage should be interpreted with caution. Pitfalls include the absence of symptoms and normal lung function, especially before the age of 30s. In addition, FEV1 based on predicted average values for a given age, height, and sex could remain within the normal range [87].

Non-specific findings include peripheral blood lymphopenia and raised serum angiotensin converting enzyme (ACE). Thus, always consider differential diagnosis from sarcoidosis [87]. Lung biopsy may be considered, if the diagnosis remains uncertain on clinical and radiological grounds. Findings include acellular whorls of

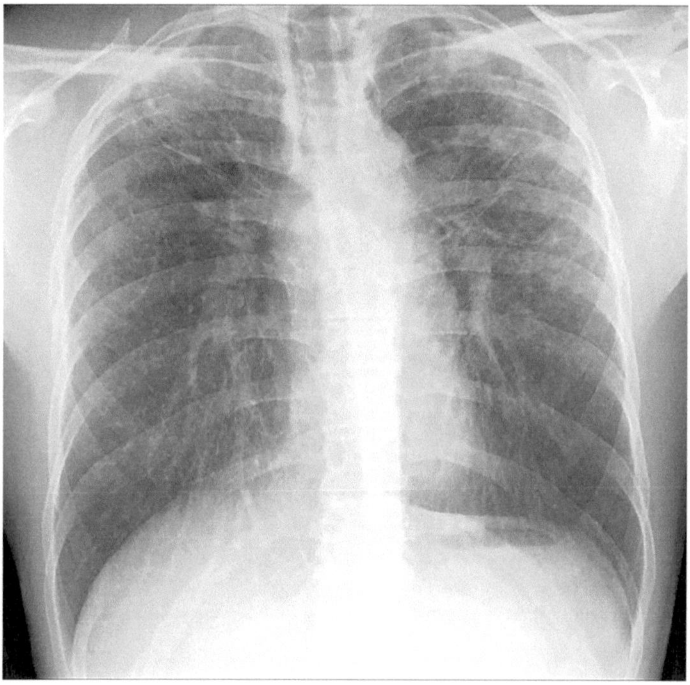

Fig. 4.30 Silicosis: the chest X-ray shows multiple small nodules in the upper zones with evidence of upper lobe fibrosis causing loss of volume in the upper lobes and elevation of the hila [87]

hyaline collagen and, when viewed under polarized light, the presence of bi-refringent crystals confirming the presence of silica and other silicates [87].

Interestingly, patients with silicosis suffer from multiple autoimmune diseases. The typical autoimmune disorders include rheumatoid arthritis (well known as Caplan syndrome), systemic lupus erythematosus (SLE), systemic sclerosis (SSc), and anti-neutrophil cytoplasmic antibody (ANCA)-related vasculitis. These patients, even asymptomatic, often exhibit higher titers of autoantibodies and they quite often show higher titers of anti-nuclear antibodies (ANA) compared with that of healthy controls [86].

Other autoantibodies detected in the blood of these patients include anti-topoisomerase I autoantibody (also known as anti-Scl-70 autoantibody) which is a marker for lung fibrosis, diffuse sclerotic skin, and peripheral vasculitis in systemic sclerosis patients. This specific antibody is associated with particular HLA types, such as DQB1-0301 in the Caucasian population and DQB1:0601 in the Japanese population as well as a higher frequency of DQB1:0402. There was also reported positive anti-desmoglein autoantibody in patients with silicosis patients and these patients usually suffered from pemphigus vulgaris, but without any symptoms of bullous disease [86].

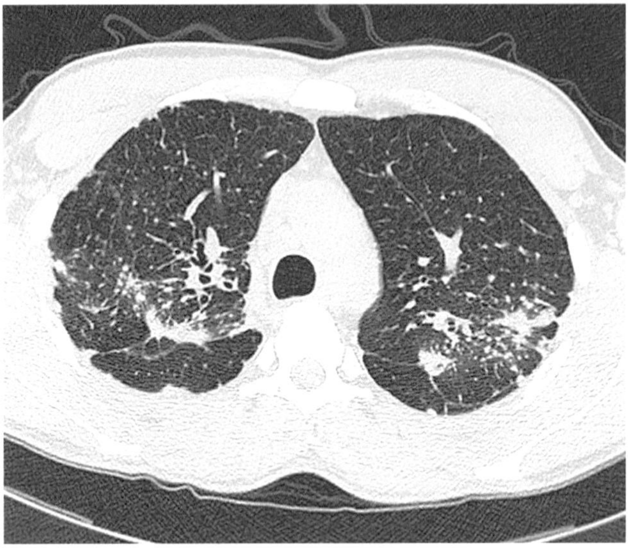

Fig. 4.31 Silicosis. The HRCT shows multiple small pulmonary nodules with a posterior bias and it confirms the presence of irregular conglomerate masses, consistent with progressive massive fibrosis

Mechanisms of Intoxification

The mechanism of the appearance of autoimmunity in patients with silicosis is not completely understood. The *anti-CD95/Fas ligand pathway* is implicated. The Fas/anti-CD95 autoantibody is detected in one quarter of silicosis patients. The CD95/Fas molecule is well known as a cell death receptor that causes apoptosis. The genetic alteration of Fas and its ligand (Fas ligand) is known to cause the autoimmune lymphoproliferative (ALPS) syndrome. The non-functional Fas may cause longer survival of responder T lymphocytes, which may react with self-antigens and the long survival may induce clinical symptoms of a chronic inflammatory process [86].

The Fas/Fas ligand pathway mediates apoptosis by activation of caspase-8, which is the key molecule that causes DNA fragmentation. Once Fas is activated after binding to the Fas ligand, initiator caspases (such as caspase 8, 10, and 2) are also activated in a specific cascade and they activate the effector caspases (such as 3, 6, and 7) and processing molecules which are responsible for DNA fragmentation. Anti-caspase 8-autoantibody is found in increased titers also among silicosis patients. The positivity of this autoantibody is also related to specific HLA types, such as DRB1, DQB1, and DPB1 alleles [86].

As mentioned above, the Fas molecule is very important in regulating the fate of lymphocytes and associated apoptosis. Normally, the activated T lymphocytes during inflammation terminate their function by apoptosis (activation-induced cell death; AICD), when the insult is eradicated. However, upon dysfunction of the pathway Fas/Fas ligand, T lymphocytes continue to be activated leading to the production of self and non-self-antibodies. Thus, if the Fas-mediated apoptosis is inhibited by certain factors, some T cell clones will continue to react and they could survive for a longer period, thereby creating the basic conditions for the occurrence of clinical manifestations of autoimmune disorders [86]. All of these findings indicate the altered regulation of autoimmunity in silicosis patients, which is the cause of development of autoimmune diseases, especially when Treg T cells are observed [86].

Furthermore, mRNA expression of soluble Fas in peripheral blood mononuclear cells derived from silicosis patients was significantly higher compared to that of healthy donors. In addition to soluble Fas, alternatively spliced variants that contain the Fas ligand-binding domain, but lack the transmembrane domain were also detected more frequently. Additionally, expression of the Decoy Receptor 3 (DcR3) molecule, whose function is known to be similar to that of soluble Fas, that inhibits Fas-mediated apoptosis, was also higher. All of these alternative molecules indicate that Fas-mediated apoptosis in lymphocytes from silicosis patients was inhibited to a greater extent than that recorded for healthy donors [86].

In addition, the serum levels of soluble interleukin (IL)-2 receptor (sIL-2R), which was initially considered a marker for T cell lineage malignancies and autoimmunity, were also demonstrated to be higher in serum of patients with silicosis [86] (Fig. 4.32).

Following the rapid development of nanotechnology, the potential harmful *health effects of nanomaterial* have aroused the public's widespread attention during the past decades. Due to the unique physicochemical properties of nanoparticles (NPs), they can reach various tissues and organs in the whole body and they frequently penetrate the body's protective barriers, such as the blood–brain barrier [88].

Silica nanoparticles (SiNPs) are one of the most widely used engineered nanomaterials. SiNPs due to their favorable properties (biocompatibility and biodegradability) have been widely exploited in the cosmetics, food, pharmaceutical industries, and in the biomedical fields, such as in disease diagnosis, drug delivery, and cancer therapy [88]. With the widespread prevalence of SiNPs, potential dermal, oral, intranasal exposure, or intravenous chances of exposure have increased [89]. SiNPs are currently on the priority lists for toxicological evaluation by the Organization for Economic Cooperation and Development (OECD) [88]. Therefore, it is necessary to study the detailed toxic mechanisms of SiNPs, which helps to evaluate their risk assessment and safety evaluation [89].

Silica nanoparticles (SiNPs) are generally of two types: rigid (nonporous) and mesoporous nanostructures. Rigid SiNPs have attracted increasing attention as an efficient host material for cellular cargo, typically enzymes, and they are usually immobilized by absorption or covalent cross-linking methods. Mesoporous silica nanoparticles have numerous pores that are suitable to load cargo. In addition, lipid bilayer coatings or organic modifications are applied at nanoparticle surfaces to

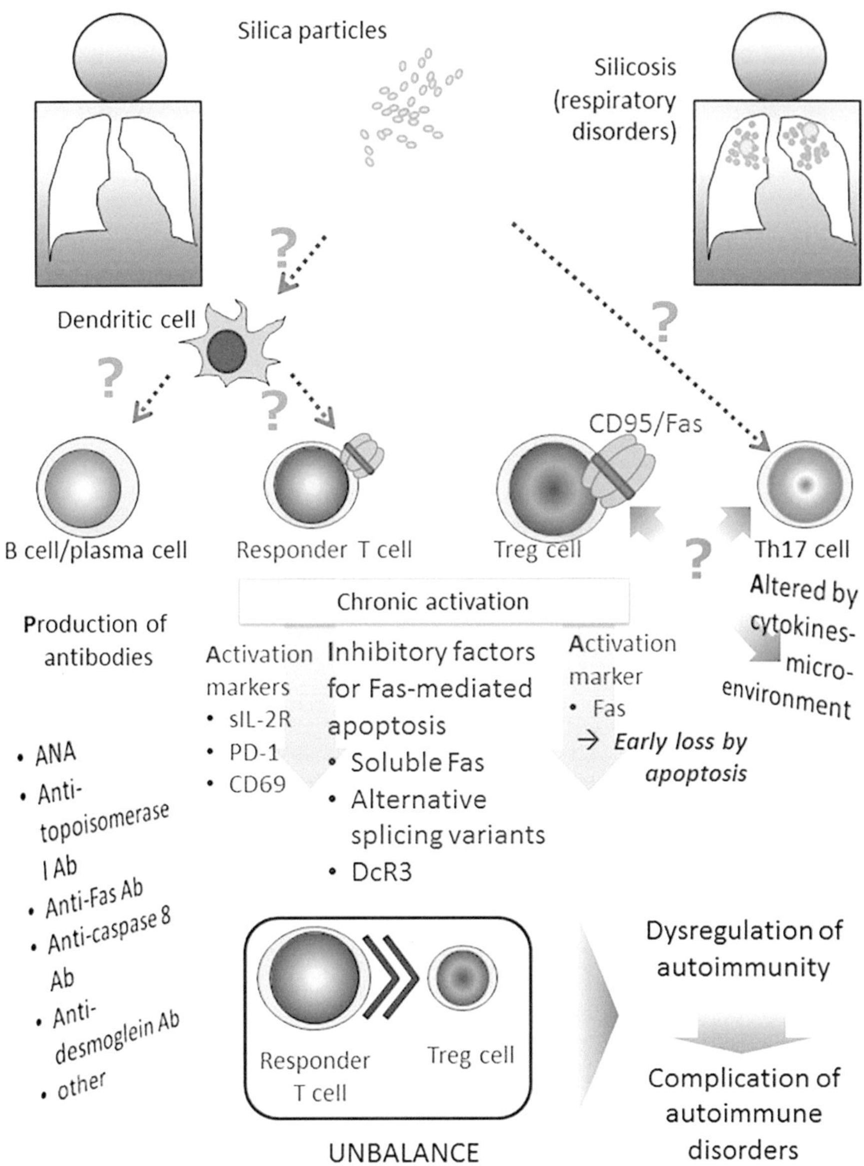

Fig. 4.32 The pathways of initiation of autoimmunity in silicosis

protect or to release control of cargo. Recently, various hybrid nanocomposites containing SiNPs have been synthesized and applied for controlled drug delivery and targeted imaging agents [90].

Like all other metals, silica nanoparticles induce toxicity, because they can produce excessive reactive oxygen species (ROS) in the body by glutathione depletion and ROS generation. In the case of silica apart from oxidative stress, excessive ROS affects endoplasmic reticulum (ER) homeostasis leading to ER stress, which results to the unfolding of proteins in the cytoplasm and initiates the autophagy process in the nucleus [88].

The *endoplasmic reticulum (ER)* is an important intracellular organelle involved in the secretory pathway. The ER regulates the biosynthesis of proteins or lipids and maintains calcium homeostasis. ER stress is induced by various pathophysiological conditions such as oxidative stress, glucose deprivation, DNA damage, and viral infection [90].

Endoplasmic reticulum (ER) stress is a type of sensitive cellular stress that can be induced by external or internal factors, such as metabolic disturbances, inflammation, and oxidative stress [89]. It is induced by various pathophysiological conditions, such as the production of excessive ROS, the accumulation of misfolded or unfolded proteins and the imbalance of intracellular Ca^{2+} [88].

SiNPs are reported to stimulate ER stress and consequently inflammation and to damage proteins and subcellular organelles. Oxidative stress and ER stress are interconnected. When ER stress occurs, a well-orchestrated signaling network called the unfolded protein response (UPR) is activated. The UPR is regulated by three ER unique sensors: *EIF2AK3* (eukaryotic translation initiation factor 2 alpha kinase 3), *ATF6* (activating transcription factor 6), and *ERN1/IRE1* (endoplasmic reticulum to nucleus signaling). All these pathways activate distinct transcription factors: *EIF2AK3* pathway activates *ATF4* (activating transcription factor 4) and the *ATF6* pathway activates *DDIT3* (DNA damage inducible transcript 3), which induce the autophagy process, while *ERN1/IRE1 pathway* activates spliced *XBP1* (X-box binding protein 1). Other nanoparticles have been reported to induce ER stress, such as TiO_2NPs, AgNPs, and iron oxide nanoparticles [89].

Another function of the cell that is important to understand due to the fact that it plays a significant role in the toxicity of nanoparticles is *autophagy*. Autophagy refers to an intracellular degradation process of macromolecules and organelles, and it is critical to maintain cellular homeostasis. It is a widespread phenomenon in nanoparticle-induced biological responses. Plenty of nanoparticles have been reported to induce autophagy, such as quantum dots, silver nanoparticles (AgNPs), gold nanoparticles, silica nanoparticles, rare earth oxides, and polystyrene particles [89].

Nanoparticle-induced autophagy includes the induction of autophagy initiation, the increase of autophagosome synthesis, the inhibition of autophagic flux, and blockage of autophagosome degradation which altogether consist a complex system of regulation of the excretosome of the cell. Autophagy dysfunction is caused by extensive autophagosome accumulation (Fig. 4.34, autophagy) [89].

To restore ER function, ER stress activates molecular chaperones in the ER, such as *GRP78* (glucose-regulated protein 78), *CHOP* (CCAAT/enhancer binding protein homologous protein), and *ERO1α* (ER oxidoreduclin 1α). It is important to know that severe or prolonged ER stress can trigger cell apoptosis. CHOP, as an apoptotic transcriptional factor induced by ER stress, is an essential marker for the assessment of ER stress-induced apoptosis [88] (Fig. 4.33).

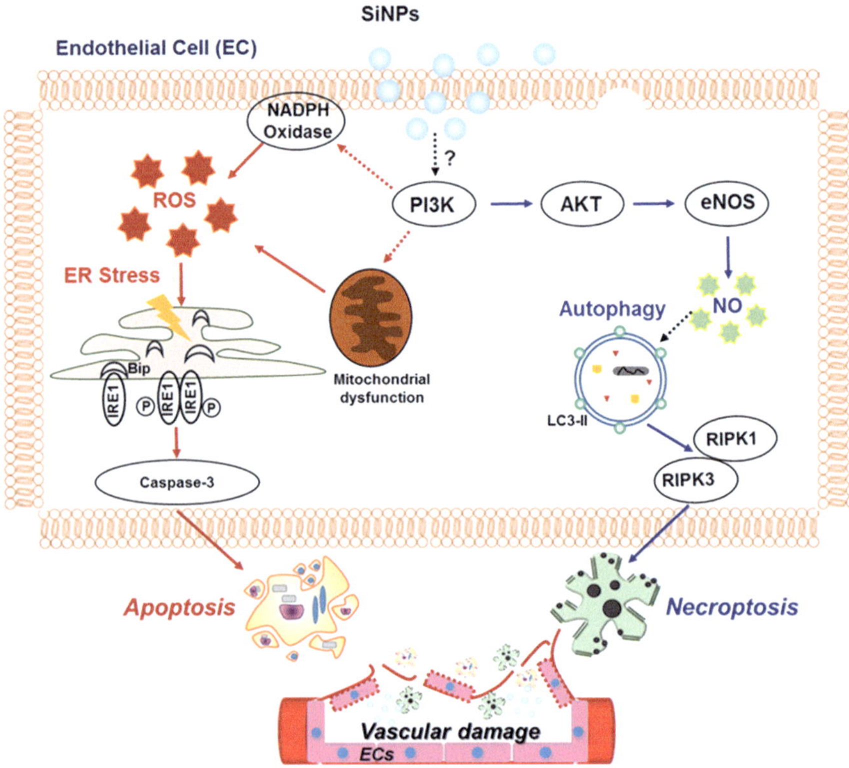

Fig. 4.33 Endoreticulum stress and activated pathways for destruction of the cell

Most evidence about autophagy dysfunction has focused on the termination stage of autophagic flux. For example, several nanoparticles such as AgNPs, graphene oxide quantum dots, and polymeric nanoparticles, could suppress the degradation of autophagosomes by lysosomal impairment. In addition, it is reported that nanoparticles could disrupt the cytoskeleton and prevent lysosomal trafficking, leading to flux blockage. Moreover, a recent study indicated that conversion nanoparticles induce auto lysosomal membrane turnover anomalies leading to abnormal autophagy [89].

Autophagy is a catabolic process related to the death process of the cell, as a process of pathogen or tumor clearance. Antigen presentation, is a mechanism of the cell to keep homeostasis after exposure to nanoparticle medication. Autophagosomes markedly accumulate in various cell lines treated with nanoparticles and the exposure of blood vessels to them induce endovascular damage [90].

Copper

The Metal, Its Forms, and Its Distribution in the Environment

Copper is a chemical with the symbol Cu (from the Latin cuprum). It is a soft, malleable, and ductile metal with very high thermal and electrical conductivity. Copper is used as a conductor of heat and electricity, as building material, and as a constituent in metal alloys, such as sterling silver used in jewelry, cupronickel used to make marine hardware, and coins and constantan used in strain gauges and thermometers for temperature measurements.

Copper is one of the few metals that can occur in nature in a directly usable metallic form (native metals). Commonly encountered compounds are copper II salts, which often acquire blue or green colors to minerals such as azurite, malachite, and turquoise, and they have been used widely historically as pigments. Copper compounds are used as bacteriostatic agents, fungicides, and wood preservatives.

In the first place, copper is an essential nutrient for life, since it is used in a diverse array of cellular functions. Copper deficiency can occur when dietary sources of copper (yeast bread, potatoes, tomatoes, ready-to-eat cereal, and beef) are scarce, or when intestinal absorption is impaired [91].

Copper is the third most abundant essential transition metal in humans, where the highest concentrations are found in the liver and brain. Copper participates in several physiological processes, among others, skin pigmentation, preservation of the integrity of blood vessels, myelination, iron homeostasis, antioxidative defense, and neurotransmitter synthesis. The essentiality of copper in a variety of biological processes is due to its role as a cofactor or structural component in numerous cuproproteins [92].

People get copper from their diet or through inhalation of copper particles. High concentrations of copper are ingested by the consumption of acidic contaminated water and inhaled during mining or grinding copper containing metal [91]. When copper enters blood, it is delivered by organs, cells, and proteins that require its presence to exert their physiological functions [92].

Ceruloplasmin is the main protein involved in copper transportation in blood, being able to bind six copper ions per protein. Other proteins in the plasma, such as albumin, also participate in copper transport. In the cell, copper is transferred by copper transporter Ctr1. Within the cell, the free copper is mostly in the cuprous CuI form, while the compounds of copper in the cell are called metallothioneins including low-molecular-weight ligands, such as glutathione. These compounds keep the concentration of free copper at extremely low levels (10^{-18} M). Cytosolic copper can be then transferred to its final destinations by the action of numerous and specific chaperones [92].

At least six proteins, the soluble proteins Cox17, Cox 19, and Cox 23 and the mitochondrial inner membrane proteins Sco1, Sco2, and Cox11 have been classified as copper chaperones and they transfer copper to the mitochondrial cytochrome

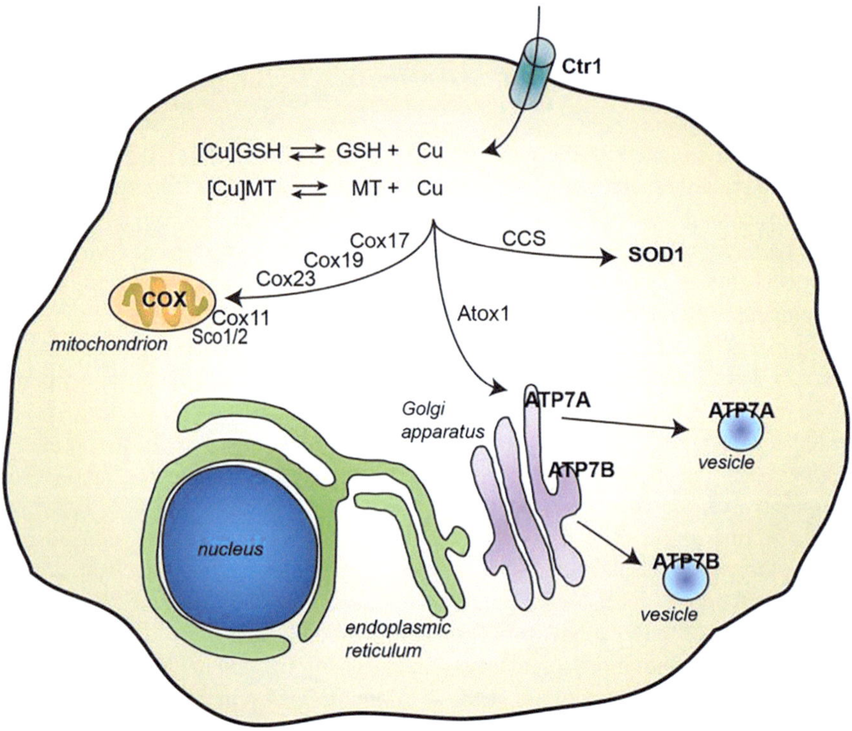

Fig. 4.34 Cytoplasmic transportation of copper by the use of chaperones

C oxidase complex from cytosol. Finally, the metal is incorporated as ion into two different subunits of complex IV. The copper chaperone CCS for superoxide dismutase 1 (SOD1) is responsible for the maturation of SOD1, which eventually leads to SOD1 activation, while the copper chaperone Atox1 delivers copper to the P-type copper-transporting ATPases ATP7A and ATP7B. These proteins are typically located in the trans-Golgi network, where they are responsible for loading copper into cuproenzymes, including dopamine-β-hydroxylase, lysyl oxidase, tyrosinase, and ceruloplasmin [92] (Fig. 4.34).

The liver acts as the critical organ controlling *Cu homeostasis*. The correct handling of Cu is regulated by the activity of two classes of proteins: (i) cuproenzymes that use Cu as a cofactor and (ii) Cu trafficking proteins that transport the metal [93].

It is important to mention that Cu shows functional duality. When not used for vital catalytic functions, it takes part in which Cu (II)/Cu(I)-driven redox cycles and it generates ROS. It is well established that Cu which is not bound to ceruloplasmin (nCpCu) is associated with this redox-mediated toxicity. Within cytoplasm department, Cu is binding to metallothioneins, where it remains in an inactivated state, whereas the free Cu is used in radical reactions. Copper in metallothioneins, as in nCp-Cu, comprises 5–10% of the plasma copper and it is in dynamic equilibrium with other transfer proteins such as albumin, peptides, and amino acids. nCp-Cu permeates tissues and organs, including the brain, where it constitutes the "labile" Cu pool [93].

The significance of copper in a variety of biological processes is due to its role as a cofactor or structural component in numerous cuproproteins. More specifically in cuproproteins, copper changes its oxidative state from Cu(I) to Cu (II) and the potential of this process can be finely tuned by the protein scaffold that binds the metal ion [92].

Copper toxicity is extremely rare, but can result when copper exceeds and when is lower than normal levels [91].

Effects on Health

Acute copper toxicity may result from ingestion (either suicidal or accidental) of copper sulfate which is used as an emetic or for the treatment of burns. Acute toxicity due to ingestion of copper ions from equipment made of copper and acid pH conditions for the body (e.g., malfunctioning hemodialysis equipment or vending machines) is rare. Acute poisoning could be related to high levels in drinking water [93]. The largest series of cases of acute poisoning with copper sulfate, ingested for suicidal purposes, was reported in New Delhi, India [93].

Acute toxicity is a quite rare event as the amount of copper required to induce adverse effects is rather high. The symptoms are mainly from the gastrointestinal, hepatic, renal, hematological, and cardiovascular systems and they include abdominal pain, nausea, vomiting, hypotension, tachycardia, intravascular hemolysis, liver, and renal failure [92]. Copper poisoning leads to hemolytic anemia due to inhibition of glucose-6-phosphate dehydrogenase activity, the inhibition of glycolysis, the denaturation of hemoglobin (with Heinz body formation), and the oxidation of glutathione [94].

Chronic copper toxicity may result from occupational environmental or domestic exposure. An example of occupational exposure is the chronic exposure of vineyard workers to fungicide sprays containing copper sulfate. The main symptom is chronic hepatic injury and the formation of noncaseating granulomas, intraacinar and periportal fibrosis, cirrhosis, and, rarely, angiosarcoma [94]. Chronic exposure to copper is reported to be the setting for Indian childhood cirrhosis, and a similar disease reported from outside the Indian subcontinent in children who had ingested large quantities of copper in their drinking water [94].

Moreover, chronic copper dyshomeostasis is reported as a result of genetically determined disorders, such as Menkes disease and Wilson's disease. *Menkes disease* is an X-linked recessive disorder, caused by mutations in the *ATP7A* gene coding for the copper-transporting ATPase, ATP7A. It is a fatal metabolic disease characterized by progressive neurological degeneration, connective tissue abnormalities, muscular hypotonia, hypothermia, and abnormalities of the skin and hair. A copper deficiency is observed in blood, kidney, liver, and brain, while copper deposits are present in the intestinal enterocytes, due to the requirement of ATP7A for the transport of copper through the intestinal epithelial membrane [92].

Wilson's disease (WD) is an autosomal recessive disease caused by copper accumulation, mainly in the liver and in the brain, where the copper content has been detected approximately eight times higher than controls [92]. The genetic defect is mutations affecting the copper transporter gene, *ATPase copper transporting beta* (*ATP7B*). The clinical presentation includes hepatic, neurologic, and psychiatric manifestations [95].

The severity of these symptoms varies widely. Liver manifestations range from mild elevation of liver enzymes to acute liver failure and cirrhosis. The neurological effects can be similar to those of Parkinson's disease and they include tremors, dysarthria, and ataxia. The psychiatric symptoms include anxiety, depression, disinhibition, and personality changes [95].

The age of onset of WD also widely varies. Liver manifestations that are related to WD have been reported to first appear in patients aged as young as 9 months and in those aged over 70 years. The course of the disease tends to be benign, if treated early by chelation therapy with penicillamine. The response to copper chelation or zinc treatment varies and it is accompanied by serious side effects [95].

In healthy individuals, ATP7B contributes to copper trafficking within the hepatocyte and it is required for copper excretion into the biliary tract. However, while the genetic basis is now better understood, WD still represents a clinically challenge and it is often an unrecognized condition due to incomplete understanding of in-depth pathogenic mechanisms, lack of a gold standard diagnostic test, and limited treatment options. The lack of a clear correlation between DNA sequence mutations and clinical presentation can be attributed to the presence of more than 500 mutations related to ATP7B copper transporter activity and the potential presence of concomitant modifier genes. The most striking evidence, that factors aside from genetic mutations affect the phenotype, is derived from multiple case reports describing monozygotic twins affected by WD, but presenting different phenotypes [95].

Thus, genetic diseases are associated with copper transportation and mutations in the ATP7A and ATP7B genes. When copper levels increase within the cell, ATP7A translocates to the plasma membrane to facilitate copper export. ATP7A protein is also responsible for the transport of copper through the intestinal epithelium and thus its absorption. Additionally, ATP7A seems to play a crucial role also in copper release at synapses. In contrast, ATP7B contributes to the elimination of excessive levels of copper in the liver by delivering copper from the liver into the bile [92].

Recent genetic studies reported a prevalence Wilson's disease up to 1:7.026, meaning that the disease is more frequent than initially described (1:30,000). Over 500 mutations in the *ATP7B* gene have been identified, 380 of which have been deemed to be disease-causing mutations. Most of them are missense mutations, small insertions and deletions in the coding region, or splice junction mutations. Whole-exon deletions, mutations in the promoter region, and multiple mutations within the same gene also occur, but are less common. The result is a decrease of ATP7B protein levels that cause improper protein localization and reduce substrate binding and activity [95].

The clinical variability of Wilson's disease and the different response to treatment is regulated by modifier genes. In other words, other genes affect the expression of ATP7B protein by altering expressivity or the degree to which a particular trait is expressed. *ATP7B* modifier genes are believed to influence tolerance to copper accumulation and copper storage capacity [95]. Although liver and brain are the most affected tissues in WD, copper also increases in blood plasma leading to alterations of humoral and cell-mediated immunity [95].

Alterations of copper blood levels has been related to neurodegenerative diseases and it is described for Parkinson's disease [92]. *Parkinson's disease* is characterized by the preferential degeneration of dopaminergic neurons and growing evidence points out a possible role of dopamine itself in promoting cell death. In contrast to neurons containing other neurotransmitters, dopamine can make dopaminergic neurons particularly susceptible to oxidative damage. After its synthesis, dopamine is almost completely sequestered inside synaptic vesicles, where it can reach concentrations up to 1 Mm. However, the cytosolic fraction of dopamine can undergo a spontaneous autoxidation process, which leads to the formation of both reactive oxygen species and dopamine quinones (DAQs). This latter pathway is confirmed by the presence of neuromelanin, a dark polymer, which is formed by the polymerization of DAQs and DAQ-modified proteins. Dopamine quinones could incorporate metal ions and copper. Neuromelanin is considered to provide a protective mechanism that prevents dopaminergic neurons from metal toxicity [92].

Copper has been demonstrated to increase the oxidation process of dopamine leading to a variety of potentially toxic species. Interestingly, the locus coeruleus and substantia nigra, where neuromelanin is mostly found, are the brain regions with the highest levels of copper (8). Moreover, a promoting role in oxygen production could implicate the protein α-synuclein. The production of reactive oxygen species, mediated by α-synuclein oligomers, which could be inhibited by the presence of copper chelating agents, supports the suggestion that copper is related to Parkinson's disease, although *in vivo* evidence is lacking [92]. In addition to the direct reactivity of copper and dopamine, oxidation processes have been described for ligands and peptides involved in other neurodegenerative processes (Fig. 4.35).

Since copper is involved in the stimulation of the immune system and the protection against infections as well as in the repair of injuries and healing, it is widely used in nanotechnology. Therefore, it is used in face masks, wound dressings, and socks. It is applied in conductive films, lubrication, nanofluids, gas sensors, solar cells, lithium batteries, and catalysts. Cu-based nanostructures are also used in various electronic devices, e.g., inkjet printers. The global production of these particles is estimated to increase to 1600 tons in 2025 [96].

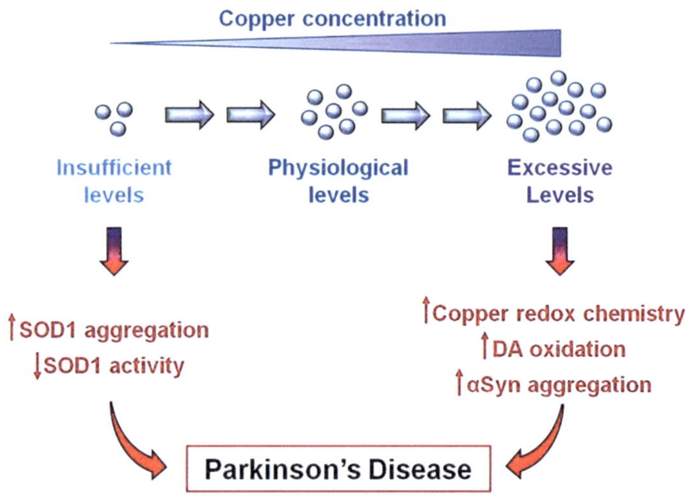

Fig. 4.35 The mechanism of implication of copper to the development of Parkinson's disease (8)

Copper nanoparticles are a health concern from the use of copper in modern technologies. The biological effects of these products have already been described. The majority of studies reported inflammatory responses in the lungs that included neutrophil and eosinophil infiltration after acute exposure and neutrophilic, lymphocytic, eosinophilic, fibrotic, granulomatous, and fibrotic granulomatous inflammation after chronic exposure. Inhaler administration may even cause systemic effects affecting the liver, kidney, or spleen. Inflammatory processes may induce DNA damage, which may cause mutations and tumor formation in the lungs [96].

Mechanism of Intoxification

The mechanism that copper induces ROS and cell damage is by depletion of SAM. Methylation reactions rely on the availability of the universal methyl donor S-adenosylmethionine (SAM), most of which is produced in the liver. Methionine and ATP produce SAM by methionine adenosyl transferase (MAT). SAM is then converted to S-adenosylhomocysteine (SAH) by the bi-directional enzyme S-adenosylhomocysteine hydrolase (AHCY). ACHY is a key enzyme that regulates the amount of SAM that is available for methylation reactions. Copper inhibits ACHY leading to the accumulation of SAH (S-adenosylhomocysteine), a reduction in SAM, the inhibition of MAT, and ultimately impaired methylation [97].

Mitochondria are integral part of metabolism, constituting the site at which the tricarboxylic acid (TCA) cycle, electron transport chain (ETC), and β-oxidation of fatty acids take place. The morphology of mitochondria from patients with WD is altered, manifesting as separated outer and inner mitochondrial membranes, cristae dilations, and finally giant mitochondria. Although alterations in neuronal mitochondria have not been investigated in WD patients, animal models show that copper intoxification promote neuroinflammation and development of behavioral abnormalities [97]. The structural changes in hepatocyte mitochondria might be caused by copper-induced disulfide bond (thiol)formation [97].

Furthermore, copper inhibits key mitochondrial enzymes disrupting the production of metabolites that are involved in epigenetic regulation [97]. The process of intoxification from copper and the environmental influence are described for patients with Wilson's disease (Fig. 4.36).

Mitochondrial cytochrome C oxidase, also known as Complex IV, is an extremely important copper-containing enzymatic complex. Additionally, copper participates in iron homeostasis by its interaction with ceruloplasmin. In fact, ceruloplasmin

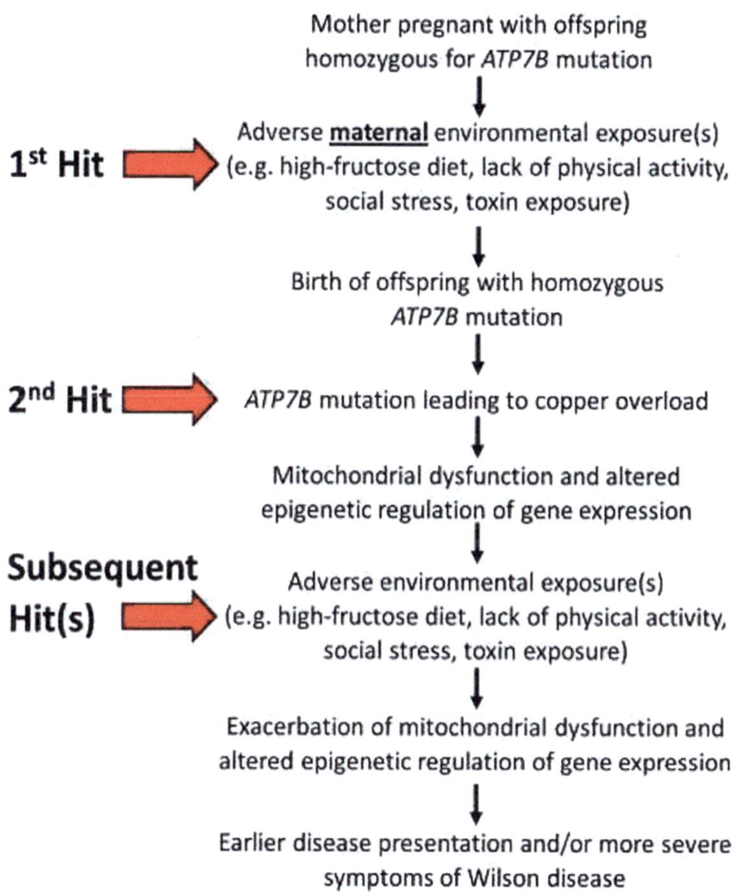

Fig. 4.36 Apart from genetic predisposition, influences of the environment participate into the severity of copper intoxification in WD disease

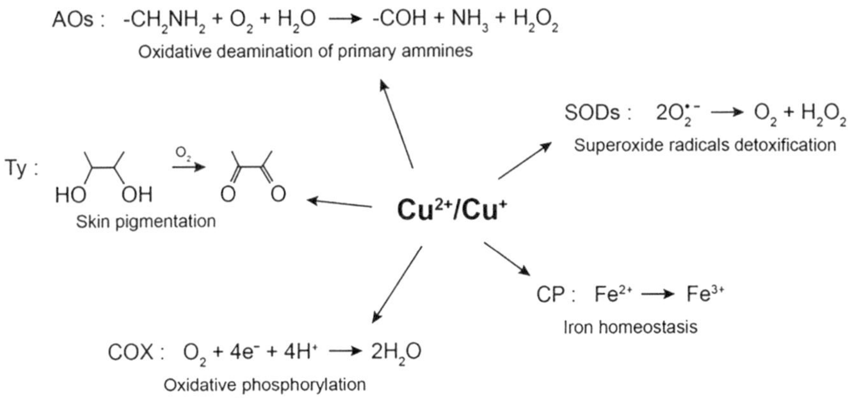

Fig. 4.37 Copper is implicated in many metabolic pathways in the body

exhibits a copper-dependent oxidase activity, which transforms Fe (II) to Fe (III), thus assisting the transferrin-mediated transport of iron in the plasma. In addition, copper has been demonstrated to play a key role at synapses in the brain by functioning as a cofactor in numerous redox enzymes and by its interaction with synaptic proteins and neurotransmitter receptors. Coherently, by modulating synaptic activity, excitotoxic cell death, and signaling cascades induced by neurotrophic factors, copper participates in several neural cell functions. Considering the numerous cellular pathways that require the presence of copper, it is not surprising that failures in the homeostatic processes leading to either copper excess or deficit can lead to severe disorders [92] (Fig. 4.37).

Epigenetic Mechanism of Intoxification

Due to the lack of robust genotype–phenotype correlations in WD, epigenetic regulation of gene expression might be involved in disease penetration. Case studies have reported twins and siblings who have the same WD genotype presenting with disparate phenotypes, implicating factors other than genetics in the presentation of WD [97].

Genome-wide investigation shows methylation changes occurring over liver-specific enhancers and transcription factor binding sites, demonstrating that WD is a genetic disease whose progression and, potentially, pathogenesis are associated with epigenetic changes. Furthermore, DNA methylation differences distinguish patients who will present with mild or severe hepatic and/or neurological involvement [95].

Cobalt

The Metal, Its Forms, and Its Distribution in the Environment

Cobalt is a chemical element with the symbol Co. Like nickel, cobalt is found in the earth's crust only in chemical compounds found in alloys of natural meteoric iron. The free element, produced by reductive smelting, is a hard, glossy, silver-gray metal.

Cobalt occurs naturally as only one stable isotope, cobalt-59. Cobalt-60 is a commercially important radioisotope, used as a radioactive tracer and for the production of high-energy gamma X-rays. Radioactive isotopes of cobalt are used in industry, medicine, and nuclear research [98]. Cobalt is primarily used in lithium-ion batteries and in the manufacturing of magnetic, wear-resistant, and high-strength alloys. The compounds cobalt II silicate (CoSi) and cobalt II aluminate ($CoAl_2O_4$) give a distinctive deep blue color to glass, ceramics, inks, paints, and varnishes. Cobalt is massively used in the steel industry, being a major constituent of hard metal alloys. Other industrial uses include diamond polishing with Co-containing disks and the production of drying agents and catalysts [98].

Cobalt is the active element of a group of coenzymes called cobalamins. Vitamin B_{12}, the best-known example of the group, is an essential vitamin for all animals. Cobalt in inorganic form is also a micronutrient for bacteria, algae, and fungi.

Exposure by inhalation, either occupational or accidental, to metallic cobalt or cobalt oxide particles has increased recently due to the increase in their industrial use. Following absorption by inhalation, cobalt is eliminated in the urine [100]. In the case of the poorly soluble Co_3O_4P, the increased risk of intoxification following inhalation is enhanced by the fact that it can be retained for periods of time ranging from months to years, thus prolonging its toxic potential [99].

In the USA, more than a million workers are potentially exposed to cobalt and its compounds. In nuclear power plants, ^{59}Co-containing alloys can be activated into radioactive ^{60}Co oxides. Thus, it can be dispersed in the cooling water and to contaminate workers. A study measuring the ambient air in cobalt powder production reported concentrations of cobalt ranging from 0.675 to 10 mg/m^3 [98].

The industrial application of cobalt nanoparticles ranges from supercapacitors and pseudo capacitors to sensors, while in biomedicine they are mainly employed in magnetic resonance imaging and as nonviral DNA carriers in gene therapy. Although the use of cobalt particles has improved many industrial and biomedical applications, their biocompatibility and permanence in tissues and cells remain an open issue [99]. When dispersed in aqueous solutions, cobalt nanoparticles undergo leaching and release cobalt ions (Co^{2+}) [99].

Effects on Health

Occupational exposure to cobalt occurs mainly by inhalation leading to various lung diseases, such as pneumonitis, fibrosis, and asthma. Similar to human exposure, animal exposure to cobalt-containing aerosols causes pronounced respiratory effects [98].

Cobalt toxicity is related to dilated cardiomyopathy. There is a documented epidemic cardiopathy in 1967 in Canada associated with increased consumption of beer with cobalt sulfate or cobalt chloride as a stabilizer of beer foam. Beer toxicity caused increased cardiac muscle vulnerability to free radicals that was also associated with variability in the dietary habits of the people. Lethal cardiomyopathy was reported in people who consumed large quantities of beer containing cobalt as a foam stabilizer (0.04–0.14 mg cobalt/kg/day) and emergency transfer due to cardiac failure accounted for 18% of deaths [98]. The histologic features of cobalt cardiomyopathy differ from the findings of idiopathic cardiomyopathy. There is a considerable myofibrillar loss and atrophy, less fibrosis, and myocyte hypertrophy [100].

Recently, cobalt is used in hip arthroplasty to treat joint failure caused by osteoarthritis. Metal-on-metal bearings were originally introduced over the last 20 years, because of their lower volumetric wear rates in comparison to conventional metal-on-polyethylene bearings. However, hip prosthesis due to continuous motion could cause the generation of metal particles and ions, mainly cobalt (Co) and chromium (Cr). These metals lead to systemic or local toxicity such as neuro-ocular toxicity, cardiotoxicity, bone loss, tissue damage, metal hypersensitivity, and chromosomal changes [101].

Mechanisms of Intoxification

The mechanism of Co^{2+} toxicity is not completely understood. Starting from Co_3O_4P, it is still unclear whether the toxic potential of Co_3O_4P is intrinsic or it is due to ionic release in solution. Co^{2+} derived from soluble cobalt chloride ($CoCl_2$) induced a less severe cytotoxicity, but it is was harmful for bladder, hepatic, and lung derived cells. $CoCl_2$ was significantly more cytotoxic than Co_3O_4P and CoOP. Co nanoparticles displayed a higher cytotoxicity at short exposure times and they induced genotoxicity and neoplastic transformation, whereas $CoCl_2$ was more efficient in the induction of primary DNA damage [99].

It has been suggested that Co_3O_4P and $CoCl_2$ exert genotoxicity by different mechanisms. Co_3O_4P seems to induce primary single DNA strand breaks (SSB) and free Co-ions exert double DNA strand breakages (DSB) [99].

Regarding nanoparticles, their genotoxic potential has been linked to their ability to induce oxidative stress. For example, soluble CoOP exerted significant induction of ROS in human lymphocytes. The depletion of glutathione and the induction of membrane lipid peroxidation was higher in the presence of poorly soluble Co_3O_4P than $CoCl_2$ [99].

Newly identified genes linked to cobalt acute toxicity include potential cobalt carriers (*FBXL2*, *ZNT1*, *SLC12A5*) and tumor suppressors or transcription factors (*MAZ*, *DLG1*, *MYC*, AXL). Some of these genes provide new hypotheses for elucidating the mechanisms of cobalt intracellular chemical toxicity. Targeted biological approaches might confirm their biochemical role in the cobalt response [98].

A property that distinguishes cobalt from the toxic effects of other metals is that cobalt can function as an "hypoxia-simulating agent" binding with a number of genes to induce apoptosis, glycolysis, angiogenesis, and erythropoiesis [98]. Hypoxia-inducible factor (*HIF-1*) is a transcription factor which acts as the main regulator of oxygen homeostasis [98].

Epigenetic Mechanism of Intoxification

Cobalt ions were shown to induce epigenetic changes and histone modifications in the lung. The involvement of epigenetics due to cobalt chloride (CoCl2) exposure showed that it reduces the acetylated levels of histone H3 and H4 by intracellular ROS-p38-mitogen-activated protein kinase (MAPK) signaling and histone deacetylation [100]. The same changes appear from poorly soluble Co_3O_4P in lung, liver, kidney, intestine, and the immune cells [99]. MAPK including p38-MAPK, extracellular regulated kinase (ERK), and c-jun N-terminal kinase (JNK) play pivotal roles in a large number of physiological processes such as apoptosis, proliferation, and differentiation [98].

Selenium

The Metal, Its Forms, and Its Distribution in the Environment

Selenium is a chemical element with the symbol Se. It is a metalloid with intermediate properties as sulfur and tellurium. It rarely occurs in its elemental state or as pure ore compound in the earth's crust.

Selenium is found in metal sulfide ores, where it partially replaces the sulfur. Commercially, selenium is produced as a byproduct in the refining of these ores. Minerals that are pure selenide or selenate compounds are rare. The main commercial uses for selenium today are glassmaking and pigments. Selenium is a semiconductor and it is also used in photocells. Applications in electronics, once important, have been mostly replaced by silicon semiconductor devices. Selenium is still used in a few types of DC power surge protectors and one type of fluorescent quantum dot.

Selenium (Se) is a naturally occurring element essential to some organisms, including bacteria, some green algae, and mammals. Although higher plants do not require selenium, but only in low quantities, they take it up from the soil due to its chemical and physical similarities to sulfur [102]. In plants, high doses of selenium can give rise to symptoms like chlorosis, necrosis, leaf and root withering, and drying [102]. Beef and bread products are recognized as good sources of selenium and bioavailability of selenium from these sources is relatively high. Consumption of white rice is negatively correlated with selenium levels [103].

Selenium is an essential element for humans. The World Health Organization (WHO) recommends a dose of selenium 30–40 μg/day and it is states that daily intake up to 400 μg selenium shall be considered safe. Recommended dose of selenium varies in different countries and it should take into consideration geographical differences, race, and life style of the specific population [104]. The serum level of the metal is influenced by non-dietary factors, such as gastrointestinal absorption, tissue uptake, protein incorporation, and tissue metabolism. Other factors that influence serum levels are older age, alcohol intake, and cigarette smoking [103].

Selenium has been proposed to be a beneficial supplement for many metabolic disorders. Supplementation of selenium may improve atherosclerosis, hypercholesterolemia, type 1 diabetes mellitus, and phenylketonuria, but its function remains controversial for type 2 diabetes mellitus. While regulation of hyperphenylalaninemia may implicate an independent mechanism, hyperglycemia and hyperlipidemia may have shared mechanisms with a series of common genes involved [104].

Although high doses of selenium can definitely cause toxicity, the rational intake is safe and useful. Optimal dose of selenium intake in patients with metabolic diseases is difficult to estimate, but it seems that daily supplementation of 31.5–200 μg of Se is beneficial [104]. Toenail selenium levels provide an excellent biomarker of long-term selenium exposure due to correlation with long-term changes in dietary intake and supplement use [103].

Accidental Exposures in History

Selenium deficiency is related to disease. Rare endemic diseases occur, where the soil is extremely low in selenium, for example in parts of China. These diseases include Keshan cardiomyopathy and Kashin-Beck disease. Keshan cardiomyopathy is a form of dilated cardiomyopathy which is characterized by multifocal necrosis and by replacement of the cells of the myocardium with fibrous tissue, which results in acute or chronic heart failure. Kashin-Beck disease is a deforming arthritis [104].

Effects on Health

Acute poisoning from overdose of selenium can be lethal and it is called selenosis. It could cause amyotrophic lateral sclerosis. *Chronic exposure* to environmental selenium is also a high-risk factor for the development of disease [105]. In humans, daily intake of selenium at 4.9 mg/person/day is considered toxic as measured in Chinese and Indian populations [105].

Selenium deficiency causes, as mentioned, dilated cardiomyopathy and chronic heart failure. It is also implicated in coronary artery disease due to oxidative stress and in an increase in platelet aggregation [104]. Moreover, epidemiological studies have provided some evidence for the role of selenium deficiency in the etiology of atherosclerosis. Angiographic studies suggest that low levels of selenium are associated with coronary stenosis and patients with acute myocardial infarction tend to have lower serum selenium compared to controls [102].

In Europe, a mild selenium deficiency has been related to the high-prevalence of diseases, such as cancer and heart disease. The role of mildly low selenium intake in the development of reproductive disease, mood disorders, and thyroid function inflammatory disease has been recently reviewed [102].

Mechanism of Intoxification

In first place, selenium is an antioxidant. It combines with cysteine as a component of selenoproteins, many of which have antioxidant properties [102]. *Selenoproteins* are important components of several critical antioxidant systems, such as glutathione peroxidase (GPx), that actively protects against damage from free radicals and reactive oxygen species [103]. Selenium-related proteins may interact with xanthine metabolism and superoxide-producing enzymes in metabolic diseases [105].

The "soft" nature of the selenium in selenocysteine and the many specific structural aspects of the active site of GPx facilitate the rapid reduction of hydroperoxides. Hydroperoxides react with the ionized selenol (−Se−1) on the active site of GPx, forming a selenenic acid (−SeOH). This removes the hydroperoxide, forming water and the corresponding alcohol. The oxidized active site is recycled by a two-step mechanism replacing the two electrons lost in the oxidation by gathering electrons from two glutathione molecules [106].

The major function of GPx1 is to protect cells against fluxes of H_2O_2 as well as organic hydroperoxides. These oxidants can be generated by biochemical processes, such as xenobiotic metabolism, immune responses, and lipid peroxidation. Superoxide is also a contributor to the flux of H_2O_2, where superoxide dismutases (SOD) facilitate its dismutation forming H_2O_2. Superoxide is also formed during mitochondrial respiration, immune responses, and many other biochemical mechanisms [106].

However, in the same time, cellular exposure to high-dose selenium can cause elevation of intracellular ROS, which is considered as the main mediator of selenium-induced cell toxicity. Although, selenium in suggested doses is overall regarded as an essential factor of antioxidant enzyme production, in high doses indirectly generates ROS. The antioxidant function is mainly due to its capability to react and form intramolecular disulfide bonds (S-Se) with essential thiol groups or cysteine resides in the substrates [107].

Thus, selenium chemistry presents an interesting paradox between both pro- and antioxidant activities, depending on the dosage and cell types exposed. Selenium compounds may cause oxidative stress, DNA damage, and apoptosis or rescue cells from oxidative damage, heat shock, cadmium toxicity, and neural toxins [107] (Fig. 4.38).

It is suggested that consuming foods that induce oxidative stress could deplete selenium levels due to high demand. It is also possible that foods such as white rice, which are typically refined and depleted of phytochemicals, could replace other higher selenium foods in the diet. Interestingly, intakes of coffee, dairy products, and eggs were each inversely associated with selenium levels [103]. However, the mechanism underlying the action of selenium, which may be dependent or independent to internal antioxidant defense system, shall be studied by more experimental investigations [105].

The mechanism of cardiomyopathy and arteriosclerosis in selenium deficiency is unclear but it is likely related to protection of low-density lipoproteins against oxidative modification, regulation of glutathione peroxidase, modulation of prostaglandin synthesis and platelet aggregation, and protection against toxic heavy metals [102].

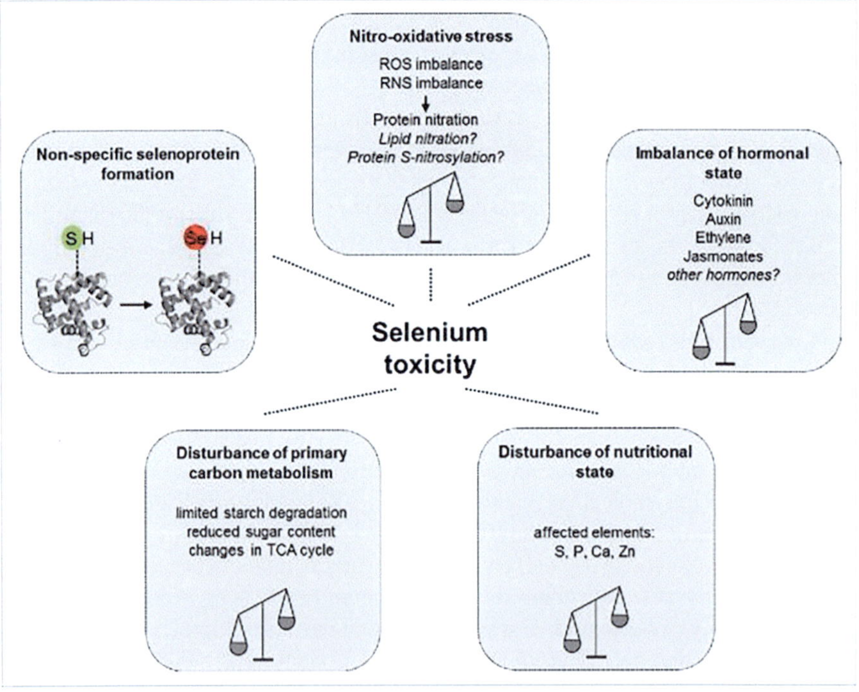

Fig. 4.38 Selenium toxicity [104]

Iron

The Metal, Its Forms, and Distribution in the Environment

Iron is a chemical element with symbol Fe (from Latin: ferrum). It is common in nature forming much of Earth's outer and inner core. It is the fourth most abundant element in the earth's crust. In its metallic state, iron is rare, limiting its deposition to meteorites. On the contrary, iron ores are abundant, but the extraction of the useful metal requires kilns or furnaces capable of reaching 1500 °C (2,730 F).

Iron is an essential element in the diet of humans and it participates in numerous fundamental cellular processes. The body of an adult human contains about 4 g (0.005% body weight) of iron, mostly in the form of hemoglobin and myoglobin [108].

The biological functions of iron include oxygen transport, mitochondrial respiration, nucleic acid replication, xenobiotic metabolism, host defense, and cell signaling. These biological functions depend on iron's property as a transition metal to readily donate or accept electrons in oxidation–reduction reactions. Iron performs these essential functions as a constituent of proteins in the form of heme, iron sulfur clusters, or other functional groups, thereby targeting and limiting its reactivity [108].

The cycle of iron is the process of gaining and loosing iron from the body. Iron is used to generate 200 billion erythrocytes per day (~20 mg iron), accounting for 80% of daily iron needs. Erythrocytes and other cells obtain iron from the plasma, which contains only 2–4 mg/L iron. Plasma supplies with iron come from reticuloendothelial macrophages that salvage iron from aged erythrocytes (20–25 mg daily) and from duodenal enterocytes that absorb dietary iron (1–2 mg daily), with the remainder provided by body stores. Iron losses (~1–2 mg/day) occur mainly by desquamation and blood loss in irregular processes. When iron losses are increased, iron homeostasis is mainly controlled by regulating iron supplies [108].

Dietary elemental iron is transported across the enterocyte membrane by divalent metal transporter 1 (DMT1) after reduction by ferriductases, such as duodenal feroportin B. Additionally, mechanisms exist for heme iron uptake, but the specific transport mechanisms are unknown. Intrinsically, reticuloendothelial macrophages phagocytose aged erythrocytes by degradation in phagolysosomes, followed by iron release by heme oxygenase 1. In both enterocytes and macrophages, produced iron can be either used for the cell's own metabolic processes or stored in an inert form in ferritin or exported into the plasma for use by other tissues. The excess iron for the cells as well as transfer of iron is maintained by transferrin [108].

Within cells, iron in excess of cellular needs is oxidized to the ferric (Fe^{+3}) form. The storage proteins when iron is in excess are *ferritin* and *hemosiderin*. When it is stored in ferritin, it is kept in a relatively inert state. When cellular iron stores exceed ferritin capacity for synthesis and storage, iron enters a secondary storage state in hemosiderin, which is a membrane-enclosed structure that contains ferric iron which resembles a residual body of a lysosome. When systemic demand for iron increases (e.g., due to enhanced erythropoiesis), the storage of iron is mobilized and exported from hepatocytes by ferroportin, entering the circulation as diferric (Fe^{+2}) transferrin. Because there is no regulatory mechanism for the excreting process of iron from the cell, the excess of iron is regulated by the uptake of the metal from the cell [109]. The dysfunction of the mechanisms that regulate iron uptake can cause excess iron deposition in the liver and other organs, such as the pancreas and the heart [109].

Hepatocytes are rich in heme-containing enzymes. The major route of entry of iron into hepatocytes is by the classical system of receptor-mediated endocytosis of diferric transferrin by transferrin receptor-1 (TFR1), which is a common enzyme in all cells in the body. After endocytosis of transferrin binded to its receptor complex, the iron is released from transferrin in the acidic lysosomal compartment, which is an important intracellular storage area for iron due to its acidic pH (iron is soluble at low pH). The iron in the lysosomal compartment can be delivered to other organelles, such as the mitochondrion for incorporation into macromolecules containing heme or iron-sulfur centers [109].

The liver plays a major role in whole-body iron metabolism. It is a major storage site for iron, it takes up transferrin- and non-transferrin-bound iron and it releases iron back into the circulation. The uptake of iron from the hepatocyte is regulated by the hormone hepcidin and its receptor ferroportin. The liver is a sensor of whole-body iron stores, regulating intestinal absorption of dietary iron and the release of recycled iron from macrophages in response to multiple demand factors [109].

A key regulator of these processes is the hormone *hepcidin*, which is produced by hepatocytes. Hepcidin levels increase in response to elevated transferrin saturation or inflammatory cytokines, which results in decreased uptake of iron across the intestinal epithelium and in diminished release of iron from macrophages [109]. The receptor for hepcidin, *ferroportin*, is the only known mammalian iron "exporter" and its degradation after binding hepcidin prevents iron reflux. Inflammatory stimuli have been shown to downregulate expression of ferroportin, indicating that ferroportin could be used as an acute phase index for inflammation. Interestingly, ferroportin could be localized to the nucleus, where it may have an important role in intracellular iron trafficking, when hepatocytes have increased metabolic activity. Decreased levels of iron decrease the expression of hepcidin enhancing erythropoietic drive. In the latter case, erythropoietin production suppresses hepatic expression of hepcidin. Dysregulation of hepcidin is presumed to be a common denominator in conditions characterized by iron overload [109] (Fig. 4.39).

After discovery of the hepcidin–ferroportin axis, it quickly became clear that excessive or insufficient hepcidin production plays a central role in most major disorders of systemic iron homeostasis. Whereas hepcidin suppression by anemia or hypoxia has a physiologic role to increase iron availability, excessive hepcidin suppression by ineffective erythropoiesis contributes to pathologic iron overload in thalassemia and other iron-loading anemias. Whereas hepcidin induction may be protective to sequester iron during infection, excessive hepcidin induction

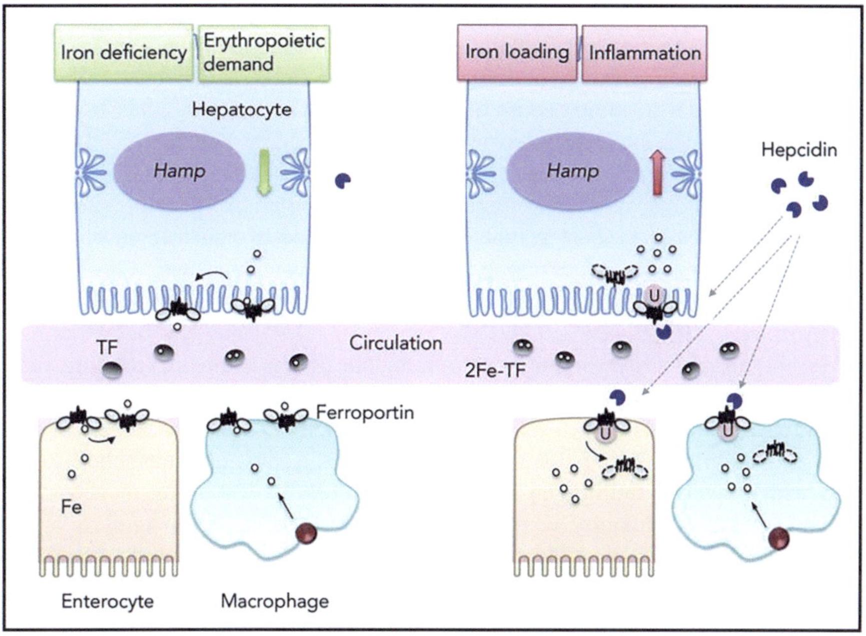

Fig. 4.39 The cycle of iron in the body [108]

contributes to iron-restricted erythropoiesis and anemia in chronic inflammatory diseases [109].

Thus, the sensitive balance of intake, storage, use, and excretion is mainly regulated by hepcidin of the hepatocyte. This balance called iron homeostasis is maintained by very sensitive hepatic cell sensors [108].

When excess "free" iron is present, this redox cycling can catalyze the production of free radicals, which can damage DNA, proteins, and lipids. To meet the dual challenge of providing adequate iron to perform essential functions, while limiting the toxicity of excess iron, organisms have evolved specialized proteins and homeostatic mechanisms to tightly control iron availability [108].

Effects on Health

The genetic condition related to the accumulation of iron in the liver due to defective transport is *hereditary hemochromatosis (HH)*. Liver related pathology (cirrhosis, hepatocellular carcinoma) has been reported to be an important cause of mortality in patients with HH. Our understanding of this condition has evolved since 1996 when homozygosity of mutations of the *HFE* C282Y gene were associated with the disease [109].

Iron is not within the list of toxic metals. However, it could be a toxic metal and dangerous poison, when accidentally or on purpose ingested. *Acute toxicity* after accidental overdose of iron containing drugs and of dietary supplements (in tablet or capsule form) is a leading cause of fatal poisoning of children under 6 years of age. However, occasional cases have been reported in adults [110].

Since 1986, US poison control centers have received reports of more than 110,000 incidents of children younger than 6 years, who accidentally swallowed iron tablets. Almost 17% of pediatric deaths reported to poison control centers between 1988 and 1992 were due to iron poisoning, compared with 12% between 1984 and 1987. Death has occurred from ingesting as little as 200 mg to as much as 5.85 g of iron [110].

Symptoms of acute poisoning with iron include abdominal pain, vomiting, diarrhea, shock, and coma. Additional findings are coagulopathy, metabolic acidosis, hyperglycemia, and leukocytosis [111]. Hepatotoxicity is a known sequela of iron poisoning and it is associated with 50% mortality. The cause of death is hepatotoxicity which leads to multiorgan failure. Serum levels of 300 μg/dL or above can present as moderate toxicity, while serum iron concentration of 500 μg/dL or above represents severe poisoning. Hepatotoxicity is found to be dose dependent. Autopsy reveals enlarged liver and histopathology shows periportal necrosis and bile cholestasis. On Perl's stain, iron deposits were noted in the liver, lungs, and spleen, suggestive of iron overload [111].

Mechanism of Intoxification

Iron is an intracellular toxic metal, and it exerts its toxic effect on mitochondria by shunting electrons to the electron transport chain, uncoupling the oxidative phosphorylation, thus leading to anaerobic metabolism [111]. The underlying mechanism for the toxicity of iron derives from the ability of iron to interact with H_2O_2 to generate highly reactive hydroxyl radicals. Iron that is not tightly bound to ferritin or hemosiderin (sometimes called "free" or labile iron) may participate in reactions that generate highly reactive and damaging products in biological systems. As a consequence, under physiological circumstances, tight regulation of intracellular iron plays an important role in preventing oxidative damage to macromolecules such as proteins, lipids, and nucleic acids [109].

Furthermore, iron has been implicated in promoting hepatic inflammation. A well-studied mechanism linking iron to inflammation involves the effects of iron on nuclear factor-κB (NF-κB) in Kupffer cells. NF-κB is a redox-sensitive transcription factor that stimulates the transcription of pro-inflammatory cytokines such as tumor necrosis factor-alpha (TNFα) and interleukin-6 (IL-6). Evidence *in vitro* and *in vivo* supports a role for iron in activating NF-κB [109].

Hepatic stellate cells are the mesenchymal cells of the liver. These cells have an important role in the development of hepatic fibrosis. Since fibrosis is the final common pathway of many forms of chronic liver disease, it is important to consider the effects of iron on these pro-fibrogenic cells. Under normal conditions, these cells are quiescent and store vitamin A, but when activated, they differentiate into myofibroblasts and produce collagen, thus leading to matrix deposition [109].

Iron is the most abundant trace metal in the human brain, present in neuronal and glial cells. It acts as a catalytic center for many enzymes and it supports the synthesis of DNA, neurotransmitters, and myelin. Furthermore, it participates in oxygen transport, neurotransmitter metabolism, and mitochondrial respiration [112].

It is recognized that a number of metallic elements, such as iron, copper, and manganese, are essential to life and they play an important role in the function of the central nervous system (CNS). Nevertheless, these metals have *neurotoxic* properties acknowledged since ancient times. The accumulation of metals in the brain has been associated with several age-related neurodegenerative disorders. The two metals most often cited are aluminum and iron [110].

Iron plays a role in age related diseases such as Parkinson's disease (PD), Alzheimer's disease (AD), amyotrophic lateral sclerosis (ALS), or prion diseases. As protein aggregation is a hallmark among many of these diseases, it is suggestive that a mutual interplay of iron and protein aggregation amplifies the detrimental effects of aggregation. It has been discussed how toxification with aluminum, copper and manganese contribute to the pathogenesis of neurogenerative diseases and iron participates to the toxicity of these metals [113]. For example, iron induces α-synuclein aggregation by direct binding or by oxidation. Indirectly, iron also influences α-synuclein on its transcriptional and translational level. α-synuclein acts as a ferrireductase and it can induce iron accumulation by overexpression. Iron fosters aggregation of both Aβ und Tau by binding. Whereas Aβ reduces levels of ferritin bound iron, an overexpression of mitochondrial ferritin reduces Aβ toxicity. That is to say, there is evidence for both Aβ induced iron accumulation and Aβ induced iron depletion [112].

Iron induced oxidative stress is involved in various pathological conditions, including heart failure, cardiomyopathy, coronary artery disease (CAD), and atherosclerosis. In the early 1980s, this led to Sullivan formulating the "iron hypothesis" [113].

Ferroptosis is an example of an iron-mediated mechanism with pathological effects on the heart. Ferroptosis is an iron dependent form of regulated cell death that is distinct from apoptosis and other types of programmed cell death. Ferroptosis is a key factor in the pathogenesis of several diseases such as hepatocellular carcinoma, renal injury, and cell death of cardiomyocytes. However, the role of ferroptosis in cardiovascular diseases is not well characterized [113].

Additionally, iron metabolism is implicated in the development of *atherosclerosis*. Monocytes and macrophages are the inflammatory cells with a key role in the formation of atherosclerotic plaques. Circulating monocytes enter the subendothelial space of the arterial intima and they differentiate into macrophages. Macrophages are known to have two types: pro-inflammatory M1 macrophages and anti-inflammatory M2 macrophages. Both of them have been identified in atherosclerotic plaques [113].

M1 macrophages express scavenger receptors on their surface and they deposit oxidized lipoproteins to atheromatous plaques, thus worsening atherosclerosis. On the other hand, the role of M2 macrophages in atherogenesis is complicated as they have a variety of cellular and molecular targets and effects. M2 macrophages produce anti-inflammatory cytokines such as IL-4, IL-10, and IL-13, which inhibit inflammation of endothelial cells and the function of TH1 and macrophage populations of endothelial cells, Th1 lymphocytes, and M1 macrophage populations. Animal models showed that iron overloading conditions with persistent red cell extravasation increase proinflammatory M1 macrophage populations and perpetuate inflammation. It is likely that iron accumulation from red cells may cause chronic inflammation in vessels, resulting in exacerbation of atherosclerosis [113].

Various risk factors such as dyslipidemia, diabetes, smoking, and obesity have been recognized as predictors of atherosclerosis and coronary artery disease (CAD). Iron may possibly be another additional risk factor involved in cardiovascular diseases; however, the evidence for this issue is still controversial [113].

Experimental data support the role of iron in the process of lipid peroxidation, which is hypothesized to be involved in early stages of atherosclerosis. The exact mechanism by which endothelial cells and macrophages interact with iron and low-density lipoprotein are still unknown, as already mentioned. There is speculation that catalytically active iron, in the form of non transferrin bound plasma iron and hemoglobin, it oxidizes low density lipoproteins which further interact with the macrophage oxidized low density lipoprotein receptor. At a clinical level, the role of iron in lipid oxidation has been questioned. Some studies have failed to show an association between serum ferritin or dietary iron intake with low density lipoprotein oxidation. Several studies correlating body iron stores with clinically measured atherosclerosis have been positive, which demonstrates a synergistic effect with low density lipoprotein levels [113].

In addition, many studies suggest a role of iron in the development of the microbiome. Iron is an essential, growth-limiting nutrient for many gut bacteria, which along with the host, compete for dietary iron. For gram-negative, *Salmonella* and *Shigella* or pathogenic forms *Escherichia coli*, iron acquisition plays an essential role in virulence and colonization. In contrast, *Lactobacilli*, a significant group of beneficial bacteria do not require iron. Dysbiosis, a shift in the ratio of protective to pathogenic bacteria, is described to contribute to the development of colorectal cancer [114].

The involvement of iron in the development of *cancer* is discussed. Due to higher rates of proliferation and DNA synthesis, neoplastic cells have higher requirements of iron than normal cells. Functioning in various electron transfer systems, iron is indispensable in maintaining the activity of iron- or heme-containing enzymes. Thus, it possesses active role as a site-dependent molecule or labile iron in DNA replication and repair as well as in the epigenetic modification and mitochondrial oxidases involved in respiratory complexes. Furthermore, iron plays a crucial role in regulating cell cycle by affecting both the formation and activity of the cyclin proteins (cyclin A, B, D, and E) and cyclin-dependent kinase (CDKs) complexes. Interestingly, $p53R_2$ is vulnerable to iron depletion a fact that has been used in cancer therapies. ROS generated by oxidative stress produced by iron can attack DNA, causing mutations and damages, inactivating tumor suppressor genes or activating oncogenes [115].

Moreover, iron plays an important role in matrix degradation and cancer metastasis by stimulating or stabilizing some metalloprotease activities. Iron overload increases metalloprotease-2 (MMP-2) and metalloproteinase-9 (MMP-9) activity in a dose-dependent manner partly by activation of AP-1 and ERK/Akt pathway. On the other hand, some studies indicate that iron may inhibit vascular endothelial growth factor (VEGF) which induces endothelial cell proliferation, migration, tube formation, and sprouting. The suppressor N-myc downstream regulated gene 1 (NDRG1) is a well-known metastasis suppressor that decreases metastases and it improves patient prognosis in breast, prostate, pancreas, and colon cancer. Therefore, iron participates in oxidative stress related to carcinogenesis and it participates in the micoenvironment of the tumor which related to its metastatic potential [115].

The *tumor microenvironment* refers to the extracellular matrix (ECM) containing other non-malignant cells, immune cells, surrounding blood vessels, signaling molecules, and cytokines around the tumor cells. Cancer initiation and progression largely depend on extrinsic signaling from its cell environment. In the past years, a more detailed understanding of the interaction between iron metabolism and tumor microenvironment has been addressed. Inflammatory stressors in the tumor microenvironment play a critical role in controlling iron metabolism and homeostatic pathway. Elevated iron in cancer cells and peri-carcinomatous compartments protect cancer cells from natural killer cell phagocytosis by upregulating ferritin expression and by antagonizing tumor necrosis factor (TNFα)- and NO (Nitric Oxide) cytotoxicity. Immunologic factors like interleukin-6 cause significant upregulation of the hepcidin by activation of the IL-6 pathway and they result in intestinal iron uptake suppression and serum ferritin levels elevation [115].

Fe^{3+} is reduced to Fe^{2+} by STEAP1–4 (six transmembrane epithelial antigen of prostate) (*STEAP1–4*) member of reductase family. *STEAP1* and *STEAP2* are highly expressed in various human cancer types, such as colon, breast, cervix, prostate, pancreas, bladder, ovary, testis, and Ewing sarcoma. *STEAP3* is overexpressed in malignant gliomas. *STEAP4* is activated under hypoxia conditions and it leads to mitochondrial iron imbalance, it enhances ROS production, and it increases the incidence of colitis-associated colon cancer. Monoclonal antibodies, antibody-drug conjugates, DNA and small noncoding RNAs vaccines targeting STEAP enzymes are a promising treatment for the above cancers [115].

Epigenetic Mechanism of Intoxification

Iron depletion has been shown to induce global histone and DNA methylation changes in cancer cells without affecting expression levels of histone and DNA methyltransferases or demethylases, implying the influence of iron-related enzymes. Iron and 2 oxoglutarate histone demethylation mediated by JmjC family are enzymes, whose dynamic processes have been linked to the epigenetic regulation of tumor formation and development [115].

Beryllium

The Metal, Its Forms, and Its Distribution in the Environment

Beryllium is a silver-gray group II metallic element. It has a crystal structure and it is well known for its unique chemical properties. It is the lightest of all solid and chemically stable substances and it has a very high melting point as well as an increased strength to weight ratio. Beryllium is lighter than aluminum, but it is over 40% more rigid and approximately one-third more elastic than steel. It is insoluble in water, but soluble in acids and alkali. It has an excellent electrical and thermal conductivity and it is not magnetic. At ordinary temperatures, beryllium resists oxidation in the air; however, a thin film of beryllium oxide forms on the surface, making it highly resistant to corrosion. In alloys, beryllium contributes to hardness, strength, and high electrical and thermal conductivity and it enhances resistance to corrosion (IPCS 1990; IARC 1993; HSDB 2009) [116].

Many compounds of beryllium are known such as beryllium chloride, fluoride, hydroxide oxide, metaphosphate, orthophosphate, sulfate, sulfate tetrahydrate with soluble differences [116]. Approximately 50 beryllium minerals occur in nature and over half of these minerals are silicates. The most common beryllium silicates are beryl ($Be_3Al_2Si_6O$), chrysoberyl ($BeAl_2O4$), and phenacite (Be_2SiO_4). These minerals are most commonly found in deposits formed by pegmatites (coarse-grained igneous rocks) and by hydrothermal alteration [117].

Beryllium's unique properties, mainly the fact that it is a light metal with a very high melting point, make it very useful in industry. In alloys, it increases thermal and electrical conductivity and strength; addition of just 2% beryllium to copper forms alloys increase strength up to six times than copper alone (IARC 1993). A 2010 U.S. Geological Survey Mineral Commodities Survey reported that based on sale revenues, nearly half of beryllium use was in computer and telecommunications products and the remainder was in aerospace and defense applications, appliances, automotive electronics, industrial components, and other applications [116].

Beryllium oxide is the most important high purity commercial beryllium chemical produced. It is used in high technology ceramics, etc., and as an additive to glass, ceramics, and plastics (IARC 1993; ATSDR 2002). Beryllium oxide was used in the past in the manufacture of phosphors for fluorescent lamps [116].

The medical community relies on beryllium for production of pacemakers, lasers, high-resolution X-ray images, and dental alloys for the manufacturing of crowns, bridges, and dental plates. Most recently, the sports industry has used the metal for the construction of golf clubs and bicycle frames [117].

Beryllium has unique applications in the nuclear defense industry, because of its neutron adsorption and reflection properties [117]. The US government began using beryllium in weapons components during the 1940s to support efforts in World War II. Pure beryllium metal is used in aircraft disc brakes, space vehicle optics and instruments, aircraft and satellite structures, missile parts, nuclear reactor neutron reflectors, nuclear weapons, fuel containers, precision instruments, rocket propellants, and navigational systems [116].

The USA is the world's largest producer of beryllium, accounting for roughly 86% of world production in 2009; total US production was 120 metric tons. Other countries producing beryllium are China, Mozambique, Portugal, Madagascar, and Brazil [116]. Bertrandite is mined primarily in the state of Utah, United States, while beryl is imported to the USA primarily from Argentina and Brazil. In the USA, 65% of the beryllium resources occur in deposits of Utah and the Seward Peninsula of Alaska. The USA is the world's leading producer, processor, and consumer of beryllium products [117].

Environment is contaminated by the metal mainly by disposal of waste materials. Red mud, a residue of beryllium processing, is the most significant beryllium solid waste form. Another viable pathway to produce beryllium is by organic complexation in surface and subsurface waters, which can result in beryllium enrichment in coal and organic surface soil horizons [117].

The primary route of human exposure to beryllium is by inhalation of dusts and fumes (ATSDR 2002). Beryllium may also be ingested in drinking water or in food. Beryllium is measured in high levels in some fruits like 0.1 μg/L in a pineapple to 74.9 μg/L in papaya. The highest levels of human exposure to beryllium are by occupational exposure, which may occur by inhalation of beryllium dust or dermal contact with products containing beryllium [116].

As a means of protecting the environment, beryllium concentrations are regulated in groundwater and waste materials sent for land disposal. The groundwater standard for beryllium (0.004 mg/L) is lower than for most other hazardous metals including arsenic (0.05 mg/L), barium (2 mg/L), cadmium (0.005 mg/L), chromium (0.1 mg/L), and lead (0.015 mg/L). Only mercury, with a standard of 0.002 mg/L, is lower than that of beryllium. Disposal of some wastes containing beryllium are regulated under the Resource Conservation and Recovery Act (RCRA) [117].

Effects on Health

Beryllium is a toxic element. It is responsible for a pulmonary disease named chronic beryllium disease (CBD) or berylliosis. and is listed as a Class I IARC carcinogen [117]. The first identified cluster of CBD cases was termed "Salem Sarcoidosis" when former workers from a fluorescent lamp manufacturing plant were found to have a chronic lung disease with features similar to sarcoidosis [118].

Acute beryllium disease (ABD), the cause of chemical pneumonitis is not common today [118]. The *chronic exposure* to beryllium, due mainly to occupational reasons, causes berylliosis or chronic beryllium disease (CBD). The manifestations of the disease are pulmonary and extrapulmonary and this is the reason that the disease is frequently indistinguishable from sarcoidosis. The classic symptoms of CBD are dry cough, progressive dyspnea on exertion, fatigue, and night sweats which are symptoms that overlap with other respiratory diseases. Profound anorexia and weight loss, one of the prominent symptoms of sarcoidosis, are not common. Progression is insidious in most cases with the onset of symptoms from months to 40 years after exposure. However, beryllium workers who are exposed to high levels of the metal can express more fulminant fatal disease due to the development of cor pulmonale [118].

CBD is a granulomatous lung disease that develops from a cell-mediated immune response to inhaled beryllium. It affects 1–17% of exposed individuals. The discrepancy between people exposed and those who develop the disease is due to genetic predisposition [117]. Research regarding genetic susceptibility has shown a high degree of association between individuals with CBD and a variation in the HLA-DPB1 locus, which is believed to code the beryllium-binding protein. Furthermore, a more recent study shows a correlation between the presence of the HLA-DPB1 marker and the type of beryllium exposure. Additionally, 30% of CBD victims possess homozygous glutamic acid in polymorphism Glu69, whereas only 1.3% of beryllium-exposed workers that were not sensitized or affected by CBD possess a variation of chain alleles [117].

As mentioned before, berylliosis is a granulomatous lung disease that may be pathologically and clinically indistinguishable from pulmonary sarcoidosis. Extrapulmonary manifestations are similar to sarcoidosis, although with a narrower spectrum. However, these are only noted in historic cohorts, since they are very rare today [118].

Diagnosis of CBD is established on a clinical basis in beryllium exposed patients with consistent radiographic imaging and clinical course. Definitive diagnosis of CBD is established by evidence of immune sensitization to beryllium and diagnostic bronchoscopy with bronchoalveolar lavage and transbronchial biopsy. The beryllium lymphocyte proliferation test based on T cell sensitization is a relatively reliable testing for berylliosis [118].

Our understanding of the health effects of beryllium became apparent after its use in industry in the early part of the twentieth century. Initially, workers involved in the primary extraction of beryllium were exposed to very high levels of beryllium salts as well as to the metal, ceramic, and alloy products. Subsequently, cases of CBD were noted in industries that used beryllium products, including the fluorescent lamp and light bulb industries, ceramic producers, manufacturers of X-ray equipment, radio, and electronic tubes, the atomic industry, and other applications in the electronics industry [118].

Corticosteroids have always comprised the mainstay of treatment of CBD. The first trials of adrenocorticotrophic hormone were undertaken in 1949 to treat CBD, followed by use of cortisone, triamcinolone, and prednisone. The response was variable. Although a good number of patients demonstrate notable improvement with therapy, complete resolution is rare [118]. Beryllium disease is treatable but incurable [117].

Beryllium is known to impact enzyme function, DNA synthesis, protein phosphorylation, and cell division. It is well established as a carcinogen in several animal models. Beryllium metal, beryllium aluminum alloy, beryl ore, and beryllium as a chloride, fluoride, hydroxide, sulfate, and oxide have all been related to lung tumors in rats, while beryllium oxide and beryllium sulfate cause lung carcinomas in monkeys. Moreover, beryllium metal, beryllium carbonate, beryllium oxide, beryllium phosphate, beryllium silicate, and zinc-beryllium silicate all produced osteosarcomas in rabbits. Only nickel and chromium were ranked more carcinogenic than beryllium [117].

Tin Poisoning: An Example of Selective Toxicity of Organic Metal Complexes

Although tin canning was thought to be an innocent practice, some data have proved the contrary. Metallic tin and its organic salts have been used in medical preparations in high doses since the sixteenth century. However, organic tin compounds are highly lipid soluble and toxic to the nervous system. The symptoms are similar to neurotoxicity enhanced by methylmercury. The first cases described in history were during a serious outbreak of organic tin poisoning in 1954 in France due to the introduction of the drug Stalinon for the treatment of staphylococcal infections. Each Stalinon capsule contained 15 mg of diiodide–diethyltin. Of those exposed to this drug, approximately half died, one-third recovered, and the remainder suffered from a wide variety of chronic neurologic deficits. Autopsies on the acute fatal cases showed evidence of an unusual form of diffuse edema that was confined to the brain white matter. Subsequent experimental work has shown that triethyltin produces a characteristic and selective white matter edema, presumably by its toxic effects on the functioning of oligodendroglial membranes. Interestingly, animals exposed to trimethyltin fail to show evidence of white matter edema but instead develop a selective necrosis of neurons of the hippocampus [119].

Executive Summary

Mercury

- Mercury exists in the elemental and oxidative form as well as organic methylmercury in nature and it is produced by anthropogenic sources
- It accumulates in the food chain and people are exposed in high levels when they consume contaminated fish
- Accidental exposures in history are the Minamata Japan, the Elmedon miners and people working in the tanning industry in Victorian times

- It is mainly a neurotoxic chemical related to impaired vision, tremor, irritability, and behavioral disturbances in adults and impairment of cognitive, learning, language development, attention deficit problems, sensoric/motor dysfunction, low IQ and motivation behavioral disorders in children, symptoms resembling autistic spectrum disorder
- It causes dilated cardiomyopathy as well as renal, visual, auditory problems in addition to endocrine, reproductive, and immune problems
- It alters oxidative stress in the cell by inhibiting metabolism of glutathione pathway
- It causes significant hypomethylation of *Alu (Yb8)* in peripheral blood, hypermethylation of *8-OHdG* (one of the major products of DNA oxidation), of *Oxoguanine glycosylase (OGG1)* and *of X-ray repair cross complementing 1 (XRCC1)* genes as well as hypomethylation of the *seleno-protein P plasma 1 (SEPP1)* promoter, and changes of methylation of *SEPP1* and *SOD2 genes* of the antioxidant pathway of glutathione
- It is related with modifications of histone H3K4 trimethylation (H3K4me3)
- WHO health-based drinking water guideline for mercury is 6 μg/L of inorganic mercury

Chromium

- Chromium has multiple oxidation states, ranging from Cr (0) (elemental chromium) to Cr (VI) (hexavalent chromium)
- The Cr (VI) is the toxic form which is actively transported within the cell
- High accidental exposures to chromium were in New Jersey, USA, Liaoning province, China, India, and Oinofita region of Greece
- The International Agency for Research on Cancer (IARC) has classified Cr (VI) as a human carcinogen related to gastrointestinal cancer
- It has also genotoxic effects
- Metabolic intermediates Cr (V), Cr (IV), and the final product Cr (III) are highly reactive and readily form Cr-DNA adducts
- Intoxification causes global
- Global hypomethylation, *p16* hypermethylation, hypermethylation of three genes: *WIF1* (WNT protein), *APC* (Adenomatous polyposis coli), and *MLH1* (mismatch repair genes) as well as hypermethylation of the *APC* and *MGMT* gene promoters
- Various histone marks including, decreased H3K27Me3, and increased levels of H3K4Me3 as well as H3K9Me2 and H3K9Me3 are related to disease

Arsenic

- Inorganic trivalent arsenic compounds are considered to demonstrate the greatest toxicity
- Class I human Carcinogen for the development of cancer of the skin, bladder, lung, angiosarcoma of the liver, prostate, and colon
- Genotoxicity

- Acute poisoning: severe abdominal pain, nausea and vomiting, diarrhea, muscle cramps, metallic taste and extreme thirst, followed by stupor, coma, cardiovascular collapse, and death in children. Death can occur within 24 h of exposure, but with sublethal doses, survival is possible and liver injury may arise 24–48 h after the acute ingestion
- Chronic poisoning: arsenical keratosis, arsenosis, ischemic heart disease, nephrotoxicity, diabetes mellitus, autoimmune, and autoinflammatory diseases
- DMPS (2,3 Dimercaptopropanesulfonate) may be the chelating agent of choice for arsenic
- Intoxification due to overproduction of ROS
- Intoxification is related to Global hypermethylation in lower doses and hypomethylation in higher, hypermethylation in the promoter regions of tumor suppressor *p16* and *DAPK* genes, hypermethylation of *CDKN2A* (cyclin-dependent kinase inhibitor 2A), *RASSF1A* (suppress the proliferation in tumor cells), and *PRSS3* (in the locus of T cell receptor beta variable orphans) in bladder cancer, activation of histone kinases such as the nuclear mitogen and stress-activated protein kinase 1 (*MSK1*), aberrant alternative splicing (miR-210, miR-22, miR-34a, miR-221, and miR-222).

Lead

- Relatively unreactive post-transition metal
- Lead in gasoline and lead paint are the main source of intoxification
- The World Health Organization and the U.S. Centers for Disease Control and Prevention announced that blood-lead levels >10 μg/dL may affect health
- Acute poisoning: lead encephalopathy, a life-threatening condition characterized ataxia, seizures, stupor, coma, and often death. In adults, high doses due to acute poisoning cause motor peripheral neuropathy
- Chronic poisoning: *In children* distractibility, attention deficit problems, and memory problems. *In adults*: neuropathy consisting of weakness of distal muscles as well as marginal sensory involvement due to axonal degeneration. Relation to neurodegenerative diseases, Alzheimer's and Parkinson's diseases
- Affects the cardiovascular, renal, neurological, skeletal, hematological, immune, respiratory, gastrointestinal, reproductive, and endocrine systems
- Bones act as storage, skeletal abnormalities
- Exposure related to oxidative stress
- Hypermethylation of LINE-1
- Hypermethylation of *COL1A2* P407, *ALAD*, tumor suppressor *CDKN2A/TP16*, *BACE1* (proteolytic enzyme b-site APP cleaving enzyme 1) alterations
- *P300* alterations and histone modifications
- (miR-320e, miR-22, miR-93, miR-769, miR-425, and miR361) decreased expression and miRNA-297 increased expression associated with patellar lead level
- Different susceptibility of individuals depending on polymorphisms of Glutathione S-transferases, metalloproteins, *HFE* and transferrin genes, Vitamin

D receptor (*VDR*) protein, X-ray Repair Cross-Complementing group 3 (*XRCC3*) genes

Cadmium

- It is widely used in the manufacturing of batteries, pigments, plastic stabilizers, pesticides and fertilizers, photovoltaic devices as well as from rubber processing, galvanization process, fossil combustion, waste incineration, manufacturing of mobile phones and computer circuit boards
- Cigarette smoke is the main source of non-occupational exposure
- In 2004, cadmium was officially categorized as a Class I human lung carcinogen
- *Acute toxicity*: Fatal dose is approximately 1 g. The symptoms of acute poisoning with cadmium are nausea, vomiting, cramps, and difficulty in breathing
- *Chronic toxicity*: alopecia, anemia, arthritis, learning difficulties, headaches, developmental failure, emphysema, osteoporosis and increased risk for bone fractures, lack of taste and smell, anorexia, and cardiovascular problems. Skin tumors and kidney damage
- Cadmium is currently related to problems of reproduction in young couples in both males and females
- Mechanisms of action: (1) a direct functional impairment of Sertoli and Leydig cells, (2) oxidative stress in both somatic and germ cells, and interference with essential ions involved in ROS scavengers' activity, and (3) apoptosis in germ cells, (4) ionic mimicry. (5) Epigenetic changes, although not directly linked to reproductive dysfunction. Lastly, (6) a direct dysfunction of hypothalamic-pituitary-gonadal axis
- Carcinogenesis
- Mechanisms of action: (1) oxidative stress, (2) attenuation of apoptosis, (3) inhibition of DNA damage repair, and (4) alterations in gene expression, (5) endocrine disruption, cell proliferation, and (6) aberrant DNA methylation
- Genotoxic and mutagenic
- Weak DNA binding activity and DNA damage, altered methylation pattern leading to global DNA hypermethylation in LINEs and global hypomethylation, activation of oncogenes (*c-myc*, *c-jun*, and *c-fos*) and inhibition of tumor suppressor genes (*p53 and p-27*, *RASSF1A* and *P16*), interference with signaling pathways, hypermethylation of *PLAGL1* gene and histone increased methylation, bivalent chromatin domain at the *MT-3* promoter, dysfunction of many DNA repair genes, hypermethylation of the maternally imprinted *MEG3*

Aluminum

- Aluminum is a nonessential element for the human body and its toxicity is related to bioaccumulation.
- In 2011, WHO permissible intake of aluminum to 2 mg/kg body weight/weekly
- Accumulated within the characteristic neurofibrillary tangles in Alzheimer's disease and neurons of the substantia nigra in Parkinson's
- Autism spectrum disorder gained publicity in relation to the use of aluminum as adjuvant in vaccinations, but results of studies are controversial
- Aluminum functions as an endocrine disruptor

- It is related to pneumoconiosis, cardiotoxicity, and chronic autoimmune and autoinflammatory conditions
- Neurotoxicity of aluminum is likely to be the result of oxidative stress, inflammatory processes, and induction of apoptosis and neuronal damage
- Genome-wide hypo or hypermethylation in AD, hypomethylation of pectin, and hypermethylation of *APP*, *PSEN1*, *APOE*, *BACE*, histone acetylation, histone lysine methylation, activation of miR-34a, miR-9, miR-125b, miR-146a, and miR-155 upregulation
- The role of recognition pattern receptors (PPRs)

Manganese

- High exposure in the manufacturing of steel, batteries, methylcyclopentadienyl manganese tricarbonyl (MMT, an anti-knock additive to gasoline), and potassium permanganate employed as bactericidal and fungicidal agents in water purification processes
- People exposed by inhalation, ingestion of contaminated water and vegetarian type of diet
- Acute high intoxification: "manganese madness" or locura manganica: psychiatric disturbances, such as behavioral abnormalities, hallucinations, and, at times, frank psychosis. Chronic exposure: neurotoxicity closely resembling Parkinson's disease, but with more prominent dystonic features. Extrapyramidal manifestations: bradykinesia, gait disturbance, postural instability, rigidity, micrography, masked facies, and speech disturbances, intension tremor, facial grimacing, plantar flexion of foot. Characteristic gait. In children, behavioral problems (hyperactivity, oppositional behavior) and low performance.
- Oxidative stress, mitochondrial dysfunction leading to energy imbalance, oxidation in dopaminergic neurons and inflammation leading to neuronal apoptosis are key elements
- The epigenetic changes include: Hypermethylation of NF-κβ pathways and *NKAP* and *NKAPP1* genes, epigenetic upregulation of α-Synuclein, disrupted *p300* acetyltransferase activity and acetylation of histones, depletion of miR-34b/c , impairment in miR-64, miR-55, let-7, miR-10a, miR-10b, miR-132, miR-7, miR-153, and miR-435
- Individual susceptibility detected by SNPs

Magnesium

- Significant role of magnesium to normal heart function, relation of deficiency to cardiovascular disease and acute myocardial infarction
- Essential for 350 enzymatic reactions
- Epigenetics: maintenance of telomere length
- Deficiency related to proinflammatory state of the cell

Nickel

- Natural and anthropogenic distribution of the metal is abundant
- IARC classified soluble and insoluble nickel compounds as Group 1 (carcinogen to humans) and nickel and alloys as Group 2B (possibly carcinogenic to humans)

- Dermal toxicity is a result of local tissue exposure, allergy, cancer
- Chronic intoxification: lung fibrosis, kidney and cardiovascular diseases, and cancer of the respiratory tract as well as cancer of oral cavities
- Neurotoxicity
- Teratogen in rats
- Type IV hypersensitivity reactions on the skin, production of compounds with proteins as well as substitution of other metals in the structure of important enzymes
- Promoter hypermethylation and global hypermethylation, epithelial mesenchymal transition that remains after cessation of exposure and upregulation of the *EMT* master regulator gene, *ZEB1* (*Zinc Finger E-Box Binding Homeobox 1*) leads to loss of H3K27me3 histone, miR-*200* family genes, and *miR-205*

Calcium

- Essential nutritional agent, especially for muscle contraction of muscle cells and heart muscle
- Extracellular matrix of advanced atherosclerotic plaques
- Metabolism of vitamin D_3 and calcium homeostasis

Asbestos

- "Asbestos" is a generic term for a group of natural crystalline silicates
- Heat-, acid-, and friction-resistant properties, in addition to economical mining, in addition to economical mining helped its abundant use
- The International Agency for Research on Cancer (IARC) designated asbestos as a Group 1 carcinogen
- Asbestosis is a pneumoconiosis related to asbestos exposure in the workplace: chronic bronchitis, emphysema and lung fibrosis, autoimmune disorders and predisposition to lung cancer
- Malignant mesothelioma (MM) is a rare, but very aggressive tumor that arises from mesothelial cells lining the pleural, peritoneal, and pericardial cavities
- ROS and direct DNA damage leading to multiple aberrations of mesothelial cells, activation of oncogenes such as *CDKN2A/ARF*, *NF2* and *p53* as well as concurrent inflammation activated by High Mobility Group Box Protein-1 (HMGB1) and release of tumor necrosis factor-alpha (TNF-α) and NF-κB signaling
- mi RNA interference

Silica

- Silica is the oxide of silicon SiO_2
- In nature: sandstones, fused quartz, Anthropogenic: fumed silica, silica gel, aerogels, microelectronics (as an electrical insulator), and as components in the food and pharmaceutical industries
- Silicosis is silica pneumoconiosis: chronic bronchitis, emphysema, and lung fibrosis. Advanced disease may be complicated by pulmonary hypertension, cor pulmonale, and respiratory failure
- Increased incidence of autoimmune diseases due to inhibition of the Fas/Fas ligand pathway-apoptosis

- Concerns about silica nanoparticles (SiNPs), one of the most widely used engineered nanomaterials
- Nanoparticles, ROS in the body by glutathione depletion and ROS generation

Copper

- Copper is used as a conductor of heat and electricity, as building material and as a constituent in metal alloys
- Copper is a nutrient essential for life used in a diverse range of cellular functions: skin pigmentation, preservation of the integrity of blood vessels, myelination, iron homeostasis, antioxidative defense, and neurotransmitter synthesis
- Acute copper toxicity: gastrointestinal, hepatic, renal, hematological, and cardiovascular systems and the symptoms include abdominal pain, nausea, vomiting, hypotension, tachycardia, intravascular hemolysis, liver, and renal failure
- Chronic copper toxicity: chronic hepatic injury and the formation of noncaseating granulomas, intraacinar and periportal fibrosis, cirrhosis, and, rarely, angiosarcoma
- Menkes disease: (X-linked recessive disorder, *ATP7A* gene coding for the copper-transporting ATPase in the liver) leading to deficiency of copper. It is a fatal metabolic disease, characterized by progressive neurological degeneration, connective tissue abnormalities, muscular hypotonia, hypothermia and abnormalities of the skin and hair
- Wilson's disease (WD): (autosomal recessive disease, transporter gene *ATP7A* causes copper accumulation mainly in the liver and in the brain). Liver: a range from mild elevation of liver enzymes to acute liver failure and cirrhosis, symptoms of Parkinson's disease: tremors, dysarthria, and ataxia, psychiatric symptoms: anxiety, depression, disinhibition, personality changes
- Oxidation of dopamine is the cause of symptoms of Parkinson's
- Epigenetic mechanisms may be implicated in different phenotypes of Wilson's disease
- Inhibits methyltransferase and affects DNA methylation

Cobalt

- Used in lithium-ion batteries, magnetic, wear-resistant and high-strength alloys, Cobalt aluminate ($CoAl_2O_4$) give a distinctive deep blue color to glass, ceramics, inks, paints, and varnishes, massively used in the steel industry, production of drying agents and catalysts, nuclear power plants, hip arthroplasty
- Occupational exposure: lung diseases, such as pneumonitis, fibrosis, and asthma
- Lethal cardiomyopathy in drinkers of beer
- Co^{2+} toxicity related to genotoxicity and neoplastic transformation
- Histone modifications, reduction of the acetylated levels of histone H3 and H4 in the lung liver, kidney, intestine, and the immune cells
- Production of ROS, hypoxia-simulating agent leading to increased apoptosis, glycolysis, angiogenesis, and erythropoiesis

Selenium

- Selenium is found in metal sulfide ores and as a product in photocells
- Essential to some organisms, including mammals
- Supplementation of selenium may improve atherosclerosis, hypercholesterolemia, type 1 diabetes mellitus, and phenylketonuria
- Selenium deficiency: China, Keshan cardiomyopathy (a form of dilated cardiomyopathy which is characterized by multifocal necrosis and replacement fibrosis of the myocardium) and Kashin-Beck disease, a deforming arthritis. Implicated in coronary artery disease
- Acute poisoning: selenosis, can be lethal, possible causes amyotrophic lateral sclerosis
- Bidirectional function, in low doses it has antioxidant properties, while in high doses generates the production of ROS

Iron

- Iron is an essential element required for numerous fundamental cellular processes
- The cycle of iron includes absorption by enterocytes by metal transporter 1 (*DMT1*) or release with heme oxygenase 1 by destroyed erythrocytes in macrophages. It is transferred to different tissues by transferrin
- Liver is the major storage organ of the body and the axon of the hormone hepcidin with its receptor ferroportin regulate the cycle of iron
- Some tissues such as liver, spleen, pancreas, and spleen accumulate and store iron in the form of ferritin and hemosiderin. Excess of storage is due to high hepcidin like in impropriate hemopoiesis
- Accidental overdose: abdominal pain, vomiting, diarrhea, shock, and coma. Additional findings are coagulopathy, metabolic acidosis, hyperglycemia, and leukocytosis
- Hepatotoxicity is known sequela of iron poisoning and is associated with 50% of mortality. Cause of death severe hepatotoxicity and multiorgan failure dose-dependent
- Serum levels of 300 μg/dL or above can present as moderate toxicity, while serum iron concentration of 500 μg/dL or above represents severe toxicity
- "Free" iron when is low has antioxidant properties due to the function of antioxidant enzymes, while in excess produces a redox cycling and ROS, which can damage DNA, proteins, and lipids and inflammation
- Iron-induced oxidative stress is involved in various pathological conditions, including neurodegenerative diseases, heart failure, cardiomyopathy, coronary artery disease (CAD), and atherosclerosis
- Consists part of the cancer microenvironment
- Iron depletion influences enzymatic activity of DNA methyltransferases or demethylases inducing histone and DNA methylation changes

Beryllium

- It has excellent electrical and thermal conductivity and it is not magnetic
- Abundant metal in industry: alloys, it increases thermal and electrical conductivity and strength; nearly half of beryllium use was in computer and telecommuni-

cations products; aerospace and defense applications, appliances, automotive electronics, and industrial components

- Medical use for production of pacemakers, lasers, high-resolution X-ray images, and dental alloys for manufacturing of crowns, bridges, and dental plates. Most recently the sports industry is interested.
- Occupational exposure by inhalation is the main source of intoxification
- Acute berylliosis: fulminant fatal disease due to the development of cor pulmonale (rare). Chronic exposure to beryllium: dry cough, progressive dyspnea on exertion, fatigue, and night sweats. Profound anorexia and weight loss are not common. Extrapulmonary manifestations similar to sarcoidosis

References

Mercury

1. Mossman BT. Environmental pathology. Toxicol Pathol. 1999;27(2):180–6.
2. Sanders AP, Burris HH, Just AC, et al. Altered miRNA expression in the cervix during pregnancy associated with lead and mercury exposure. Epigenomics. 2015;7(6):885–96.
3. Frisbie SH, Mitchell EJ, Sarkar B. Urgent need to reevaluate the latest World Health Organization guidelines for toxic inorganic substances in drinking water. Environ Health. 2015;14:63.
4. Abelsohn A, Gibson BL, Sanborn MD, et al. Identifying and managing adverse environmental health effects. Canadian Med Assoc J. 2002;166:1549.
5. Abelsohn A, Gibson BL, Sanborn MD, et al. Persistent organic pollutants. CMAJ. 2002;166(12):1549–54.
6. Buch AC, Sisinno CLS, Correia MEF, et al. Food preference and ecotoxicological tests with millipedes in litter contaminated with mercury. Sci Total Environ. 2018;633:1173–82.
7. Blossom SJ, Gilbert KM. Epigenetic underpinnings of developmental immunotoxicity and autoimmune disease. Curr Opin Toxicol. 2018;10:23–30.
8. Ruiz-Hernandez A, Kuo C-C, Rentero-Garrido P, et al. Environmental chemicals and DNA methylation in adults: a systematic review of the epidemiologic evidence. Clin Epigenetics. 2015;7(1):55.
9. Morris G, Puri BK, Frye RE, et al. The putative role of environmental mercury in the pathogenesis and pathophysiology of autism spectrum disorders and subtype. Mol Neurobiol. 2018;55(6):4834–56.
10. Narváez DM, Groot H, Diaz SM, et al. Oxidative stress and repetitive element methylation changes in artisanal gold miners occupationally exposed to mercury. Heliyon. 2017;3(9):e00400.
11. Basu N, Goodrich JM, Head J. Ecogenetics of mercury: from genetic polymorphisms and epigenetics to risk assessment and decision-making. Environ Toxicol Chem. 2014;33(6):1248–58.
12. Rudgalvyte M, Peltonen J, Lakso M, et al. Chronic MeHg exposure modifies the histone H3K4me3 epigenetic landscape in Caenorhabditis elegans. Comp Biochem Physiol C Toxicol Pharmacol. 2017;191:109–16.

Chromium

13. Sun H, Brocato J, Costa M. Oral chromium exposure and toxicity. Curr Environ Health Rep. 2015;2(3):295–303.
14. Des Marais TL, Costa M. Mechanisms of chromium-induced toxicity. Curr Opin Toxicol. 2019 Apr;14:1–7.
15. Zhitkovich A. Chromium in drinking water: sources, metabolism, and cancer risks. Chem Res Toxicol. 2011;24(10):1617–29.
16. Chen QY, Marais TD, Costa M. Metals and mechanisms of carcinogenesis. Annu Rev Pharmacol Toxicol. 2019;59:537–54.
17. Wang X, Mandal AK, Saito H, et al. Arsenic and chromium in drinking water promote tumorigenesis in a mouse colitis-associated colorectal cancer model and the potential mechanism is ROS-mediated Wnt/β-catenin signaling pathway. Toxicol Appl Pharmacol. 2012;262(1):11–21.
18. Wang Z, Jianjun W, Humphries B, et al. Upregulation of histone-lysine methyltransferases plays a causal role in hexavalent chromium-induced cancer stem cell-like property and cell transformation. Toxicol Appl Pharmacol. 2018;342:22–30.
19. da Couto RR, Faversani J, Ceretta CA. Health risk assessment and soil and plant heavy metal and bromine contents in field plots after ten years of organic and mineral fertilization. Ecotoxicol Environ Saf. 2018;153:142–50.

Arsenic

20. Baker BA, Cassano VA, Murray C. Arsenic exposure, assessment, toxicity, diagnosis, and management. J Occup Environ Med. 2018;60(12):e634–9.
21. da Rosa R, Couto JF, Ceretta CA, et al. Health risk assessment and soil and plant heavy metal and bromine contents. Ecotoxicol Environ Saf. 2018;153:142–50.
22. Di X, Beesley L, Zhang Z, et al. Microbial arsenic methylation in soil and uptake and metabolism of methylated arsenic in plants: a review. Int J Environ Res Public Health. 2019;16(24):5012.
23. Hu Y, Li J, Lou B, Wu R, et al. The role of reactive oxygen species in arsenic toxicity. Biomol Ther. 2020;10(2):240.
24. Bjørklund G, Aaseth J, Chirumbolo S, et al. Effects of arsenic toxicity beyond epigenetic modifications. Environ Geochem Health. 2018;40(3):955–65.
25. Bilal M, Iqbal HMN. An insight into toxicity and human-health-related adverse consequences of cosmeceuticals a review. Sci Total Environ. 2019;670:555–68.
26. Xu L, Mondal D, Polya DA. Positive Association of Cardiovascular Disease (CVD) with chronic exposure to drinking water arsenic (As) at concentrations below the WHO provisional guideline value: a systematic review and meta-analysis. Int J Environ Res Public Health. 2020;17(7):2536.
27. Coryell M, Roggenbeck BA, Walk ST. The human gut microbiome's influence on arsenic toxicity. Curr Pharmacol Rep. 2019;5(6):491–504.
28. Wang X, Mandal AK, Saito H, et al. Arsenic and chromium in drinking water promote tumorigenesis in a mouse colitis-associated colorectal cancer model and the potential mechanism is ROS-mediated Wnt/β-catenin signaling pathway. Toxicol Appl Pharmacol. 2012;262(1):11–21.
29. Martin EM, Stýblo M, Fry RC, et al. Genetic and epigenetic mechanisms underlying arsenic-associated diabetes mellitus: a perspective of the current evidence. Epigenomics. 2017;9(5):701–10.
30. Christensen BC, Marsit CJ. Epigenomics in environmental health. Front Genet. 2011;2:84.

31. Tam LM, Price NE, Wang Y. Molecular mechanisms of arsenic-induced disruption of DNA repair. Chem Res Toxicol. 2020;33(3):709–26.
32. Sun H, Brocato J, Costa M. Oral chromium exposure and toxicity. Curr Environ Health Rep. 2015;2(3):295–303.
33. Arslan B, Djamgoz MBA, Akün E, et al. ARSENIC: a review on exposure pathways, accumulation, mobility and transmission into the human food chain. Rev Environ Contam Toxicol. 2017;243:27–51.
34. Cardenas A, Koestler DC, Andres Houseman E, et al. Differential DNA methylation in umbilical cord blood of infants exposed to mercury and arsenic *in utero*. Epigenetics. 2015;10(6):508–15.
35. Ye Y, Gaugler B, Mohty M. Old dog, new trick: trivalent arsenic as an immunomodulatory drug. Br J Pharmacol. 2020;177(10):2199–214.

Lead

36. Juhász ML, Marmur ES. A review of selected chemical additives in cosmetic products. Dermatol Ther. 2014;27(6):317–22.
37. Sanders AP, Burris HH, Just AC, et al. Altered miRNA expression in the cervix during pregnancy associated with lead and mercury exposure. Epigenomics. 2015 Sep;7(6):885–96.
38. Mitra P, Sharma S, Purohit P, et al. Clinical and molecular aspects of lead toxicity: an update. Crit Rev Clin Lab Sci. 2017;54(7–8):506–28.
39. Mossman BT. Environmental pathology. Toxicol Pathol. 1999;27(2):180–6.
40. Reuben A. Childhood lead exposure and adult neurodegenerative disease. J Alzheimers Dis. 2018;64(1):17–42.
41. Chen QY, Des Marais T, Costa M. Metals and mechanisms of carcinogenesis. Annu Rev Pharmacol Toxicol. 2019;59:537–54.
42. Rhee SY, Hwang Y-C, Woo J-T, et al. Blood lead is significantly associated with metabolic syndrome in Korean adults: an analysis based on the Korea National Health and Nutrition Examination Survey (KNHANES), 2008. Cardiovasc Diabetol. 2013;9(12):9.
43. Ruiz-Hernandez A, Kuo C-C, Rentero-Garrido P, et al. Environmental chemicals and DNA methylation in adults: a systematic review of the epidemiologic evidence. Clin Epigenetics. 2015;7(1):55.

Cadmium

44. de Angelis C, Galdiero M, Pivonello C, et al. The environment and male reproduction: the effect of cadmium exposure on reproductive system and semen quality and its implication in fertility. Reprod Toxicol. 2017;73:105–27.
45. Nica DV, Popescu C, Draghici GA, et al. High-level dietary cadmium exposure is associated with global DNA hypermethylation in the gastropod hepatopancreas. PLoS One. 2017;12(9):e0184221.
46. da Rosa R, Couto JF, Ceretta CA, et al. Health risk assessment and soil and plant heavy metal and bromine contents in field plots after ten years of organic and mineral fertilization. Ecotoxicol Environ Saf. 2018;153:142–50.
47. House JS, Hall J, Park SS, et al. Cadmium exposure and *MEG3* methylation differences between Whites and African Americans in the NEST Cohort. Environ Epigenet. 2019;5(3):dvz014.
48. Chen QY, Des Marais T, Costa M. Metals and mechanisms of carcinogenesis. Annu Rev Pharmacol Toxicol. 2019;59:537–54.

49. 49. de Angelis C, Galdiero M, Environment CPT, Reproduction M. The effect of cadmium exposure on reproductive system and semen quality and its implication in fertility. Reprod Toxicol. 2017;73:105–27.
50. Tan HW, Liang Z-L, Yao Y, et al. Lasting DNA damage and aberrant DNA repair gene expression profile are associated with post-chronic cadmium exposure in human bronchial epithelial cells. Cells. 2019;8(8):842.
51. Kulkarni AC, Kuppusamy P, Parinandi N. Oxygen, the lead actor in the pathophysiologic drama: enactment of the trinity of normoxia, hypoxia, and hyperoxia in disease and therapy. Antioxid Redox Signal. 2007;9(10):1717–30.
52. House JS, Hall J, Park SS, et al. Cadmium exposure and *MEG3* methylation differences between Whites and African Americans in the NEST. Cohort Environ Epigenet. 2019;5(3):dvz014.
53. Park SS, Skaar DA, Jirtle RL, Hoyo C. Epigenetics, obesity and early-life cadmium or lead exposure. Epigenomics. 2017;9(1):57–75.
54. Bal W, Kasprzak KS. Induction of oxidative DNA damage by carcinogenic metals. Toxicol Lett. 2002;127(1–3):55–62.
55. Nakayama SMM, Nakata H, Ikenaka Y, et al. One year exposure to Cd- and Pb-contaminated soil causes metal accumulation and alteration of global DNA methylation in rats. Environ Pollut. 2019;252(Pt B):1267–76.
56. Jirtle RL, Skinner MK. Environmental epigenomics and disease susceptibility. Nat Rev Genet. 2007;8:253.

Aluminum

57. Wang F, Kang P, Li Z, et al. Role of MLL in the modification of H3K4me3 in aluminum-induced cognitive dysfunction. Chemosphere. 2019;232:121–9.
58. Morris G, Puri BK, Frye RE, et al. The putative role of environmental aluminum in the development of chronic neuropathology in adults and children. How strong is the evidence and what could be the mechanisms involved? Metab Brain Dis. 2017;32(5):1335–55.
59. Cicero CE, Mostile G, Vasta R, et al. Metals and age-related neurodegenerative diseases. Environ Res. 2017;159:82–94.
60. Pandey A, Jatana GK, Sonthalia S. Cosmeceuticals. Treasure Island: StatPearls Publishing; 2021.
61. Frisbie SH, Mitchell EJ, Sarkar B, et al. Urgent need to reevaluate the latest World Health Organization guidelines for toxic inorganic substances in drinking water. Environ Health. 2015;14:63.
62. Liang R. Cross talk between aluminum and genetic susceptibility and epigenetic modification in Alzheimer's disease. Adv Exp Med Biol. 2018;1091:173–91.
63. Darbre PD. Aluminium, antiperspirants and breast cancer. J Inorg Biochem. 2005;99(9):1912–9.

Manganese

64. Mossman BT. Environmental pathology. Toxicol Pathol. 1999;27(2):180–62.
65. Ljung K, Vahter M. Time to re-evaluate the guideline value for manganese in drinking water? Environ Health Perspect. 2007;115(11):1533–8.

66. Tarale P, Chakrabarti T, Sivanesan S, et al. Potential role of epigenetic mechanism in manganese induced neurotoxicity. Biomed Res Int. 2016;2016:2548792.
67. Kresovich JK, Bulka CM, Joyce BT, et al. The inflammatory potential of dietary manganese in a cohort of elderly men. Biol Trace Elem Res. 2018;183(1):49–57.
68. Dafhne Aguirre J, Culotta VC. Battles with iron: manganese in oxidative stress protection. J Biol Chem. 2012;287(17):13541–8.

Magnesium

69. Mossman BT. Environmental pathology. Toxicol Pathol. 1999;27(2):180–6.
70. Shah NC, et al. Short-term magnesium deficiency downregulates telomerase, upregulates neutral sphingomyelinase and induces oxidative DNA damage in cardiovascular tissues: relevance to atherogenesis, cardiovascular diseases and aging. Int J Clin Exp Med. 2014;7(3):497–514.

Nickel

71. Genchi G, Carocci A, Lauria G, et al. Nickel: human health and environmental toxicology. Int J Environ Res Public Health. 2020;17(3):6792.
72. Brocato J, Costa M. Basic mechanics of DNA methylation and the unique landscape of the DNA Methylome in metal-induced carcinogenesis. Crit Rev Toxicol. 2013;43(6):493–514.
73. Jose CC, Jagannathan L, Tanwar VS, et al. Nickel exposure induces persistent mesenchymal phenotype in human lung epithelial cells through epigenetic activation of ZEB1. Mol Carcinog. 2018;57(6):794–806.
74. Bai YN, Yang AM, Hong Quan P, et al. Nickel-exposed workers in China: a cohort study. Biomed Environ Sci. 2014;27(3):208–11.
75. Sun H, Brocato J, Costa M. Oral chromium exposure and toxicity. Curr Environ Health Rep. 2015;2(3):295–303.
76. Jose CC, Wang Z, Tanwar VS, et al. Nickel-induced transcriptional changes persist post exposure through epigenetic reprogramming. Epigenetics Chromatin. 2019;12:75.
77. Macomber L, Hausinger RP. Mechanisms of nickel toxicity in microorganisms. Metallomics. 2011;3(11):1153–62.
78. Ellen TP, Kluz T, Harder ME, et al. Heterochromatinization as a potential mechanism of nickel-induced carcinogenesis. Biochemistry. 2009;48(21):4626–32.

Calcium

79. Jomova K, Vondrakova D, Lawson M, Valko M. Metals and age-related neurodegenerative diseases. Mol Cell Biochem. 2010;345(1–2):91–104.
80. Carlberg C. Nutrigenomics of vitamin D. Nutrients. 2019;11(3):676.
81. Al-Ghafari AB, Balamash KS, Al HA, Doghaither. Relationship between serum vitamin D and calcium levels and vitamin D receptor gene polymorphisms in colorectal cancer. Biomed Res Int. 2019;2019:8571541.

Asbestos

82. Yang H, Testa JR, Carbone M. Mesothelioma epidemiology, carcinogenesis and pathogenesis. Curr Treat Options in Oncol. 2008;9(2–3):147–57.
83. Lemen RA, Landrigan PJ. Toward an Asbestos ban in the United States. Int J Environ Res Public Health. 2017;14(11):1302.
84. Toyokuni S. Iron overload as a major targetable pathogenesis of asbestos-induced mesothelial carcinogenesis. Redox Rep. 2014;19(1):1–7.
85. Micolucci L, Akhtar MM, Olivieri F. Diagnostic value of microRNAs in asbestos exposure and malignant mesothelioma: systematic review and qualitative meta-analysis. Oncotarget. 2016;7(36):58606–37.
86. Lee S, Matsuzaki H, Kumagai-Takei N, et al. Silica exposure and altered regulation of autoimmunity. Environ Health Prev Med. 2014;19(5):322–9.
87. Cullinan P, Reid P. Pneumoconiosis. Prim Care Respir J. 2013;22(2):249–52.

Silica

88. Lee S, Matsuzaki H, Kumagai-Takei N, et al. Silica exposure and altered regulation of autoimmunity. Environ Health Prev Med. 2014;19(5):322–9.
89. Cullinan P, Reid P. Pneumoconiosis. Prim Care Respir J. 2013;22(2):249–52.
90. Chen F, Jin J, Jiahui H, et al. Endoplasmic reticulum stress cooperates in silica nanoparticles-induced macrophage apoptosis via activation of CHOP-mediated apoptotic signaling pathway. Int J Mol Sci. 2019;20(23):5846.
91. Wang J, Li Y, Duan J, et al. Silica nanoparticles induce autophagosome accumulation via activation of the EIF2AK3 and ATF6 UPR pathways in hepatocytes. Autophagy. 2018;14(7):1185–200.
92. Lee K, Lee J, Kwak M, et al. Two distinct cellular pathways leading to endothelial cell cytotoxicity by silica nanoparticle size. J Nanobiotechnol. 2019;17:24.

Copper

93. Kennedy E, Everson TM, Punshon T, et al. Copper associates with differential methylation in placentae from two US birth cohorts. Epigenetics. 2020;15(3):215–30.
94. Bisaglia M, Bubacco L. Copper ions and Parkinson's disease: why is homeostasis so relevant? Biomol Ther. 2020;10(2):195.
95. Sensi SL, Granzotto A, Siotto M, et al. Copper and zinc. Dysregulation in Alzheimer's disease. Trends Pharmacol Sci. 2018;39(12):1049–63.
96. Cicero CE, Mostile G, Vasta R, et al. Metals and age-related neurodegenerative diseases. Environ Res. 2017;159:82–94.
97. Kieffer DA, Medici V. Wilson disease: at the crossroads between genetics and epigenetics—a review of the evidence. Liver Res. 2017;1(2):121–30.
98. Rossner Jr P, Vrbova K, Rossnerova A, et al. Gene expression and epigenetic changes in mice following inhalation of copper (II) oxide nanoparticles. Nanomaterials (Basel). 2020;10(3):550.
99. Mordaunt CE, Kieffer DA, Shibata NM, et al. Epigenomic signatures in liver and blood of Wilson disease patients include hypermethylation of liver-specific enhancers. Epigenetics Chromatin. 2019;12:10.

Cobaltium

100. Malard V, Berenguer F, Prat O, et al. Global gene expression profiling in human lung cells exposed to cobalt. BMC Genomics. 2007;8:147.
101. Uboldi C, Orsière T, Darolles C, et al. Poorly soluble cobalt oxide particles trigger genotoxicity via multiple pathways. Part Fibre Toxicol. 2016;13:5.
102. Hattori S, Kamiya T, Hara H, et al. CoCl2 decreases EC-SOD expression through histone deacetylation in COS7 cells. Biol Pharm Bull. 2016;39(12):2036–41.
103. Li D, Tong W, Liu D, et al. Astaxanthin mitigates cobalt cytotoxicity in the MG-63 cells by modulating the oxidative stress. BMC Pharmacol Toxicol. 2017;18:58.

Selenium

104. Cicero CE, Mostile G, Vasta R, et al. Metals and age-related neurodegenerative diseases. Environ Res. 2017;159:82–94.
105. Park K, Rimm E, Siscovick D, et al. Demographic and lifestyle factors and selenium levels in men and women in the U.S. Nutr Res Pract. 2011;5(4):357–64.
106. Hadrup N, Ravn-Haren G. Acute human toxicity and mortality after selenium ingestion: a review. J Trace Elem Med Biol. 2020;58:126435.
107. Kolbert Z, Lehotai N, Molnár Á. "The roots" of selenium toxicity: a new concept. Plant Signal Behav. 2016;11(10):e1241935.
108. Stolwijk JM, Falls-Hubert KC, Searby CC, et al. Simultaneous detection of the enzyme activities of GPx1 and GPx4 guide optimization of selenium in cell biological experiments. Redox Biol. 2020;32:101518.
109. Chung S, Roy AK, Webster TJ. Selenium nanoparticle protection of fibroblast stress: activation of ATF4 and Bcl-xL expression. Int J Nanomedicine. 2019;14:9995–10007.

Iron

110. Wang C-Y, Babitt JL. Liver iron sensing and body iron homeostasis. Blood. 2019;133(1):18–29.
111. Cicero CE, Mostile G, Vasta R, et al. Metals and age-related neurodegenerative diseases. Environ Res. 2017;159:82–94.
112. Kobayashi M, Suhara T, Baba Y, et al. Pathological roles of iron in cardiovascular disease. Curr Drug Targets. 2018;19(9):1068–76.
113. Bloomer SA, Brown KE. Iron-induced liver injury: a critical reappraisal. Int J Mol Sci. 2019;20(9):2132.
114. Sane MR, Malukani K, Kulkarni R, et al. Fatal iron toxicity in an adult: clinical profile and review. Indian. J Crit Care Med. 2018;22(11):801–3.
115. Joppe K, Roser A-E, Maass F, Lingor P, et al. The contribution of iron to protein aggregation disorders in the central nervous system. Front Neurosci. 2019;22(13):15.
116. Omaye AT, Omaye ST. Caveats for the good and bad of dietary red meat. Antioxidants (Basel). 2019;12;8(11):544.
117. Wang Y, Yu L, Ding J, et al. Iron metabolism in cancer. Int J Mol Sci. 2019;20(1):95.

Beryllium

118. Bal W, Kasprzak KS. Induction of oxidative DNA damage by carcinogenic metals. Toxicol Lett. 2002;28;127(1–3):55–622.
119. Taylor TP, Ding M, Ehler DS, Foreman TM, et al. Beryllium in the environment: a review. J Environ Sci Health A Tox Hazard Subst Environ Eng. 2003;38(2):439–69.
120. Mayer AS, Hamzeh N, Maier LA, et al. Sarcoidosis and chronic beryllium disease: similarities and differences. Semin Respir Crit Care Med. 2014;35(3):316–29.

Tin Poisoning

121. Wilson AS, Power BE, Molloy PL. DNA hypomethylation and human diseases. Biochim Biophys Acta. 2007;1775(1):138–62. (1)

Chapter 5
Organic Pollutants and Disease

Persistent Organic Pollutants (POPs)

The Chemicals, Their Forms, and Their Distribution in the Environment

Persistent organic pollutants (*POPs*), also known as "forever chemicals," is a general term for carbon-containing compounds that share similar properties. The characteristics of these compounds include resistance to photolytic, biological, and chemical degradation, high toxicity, and especially persistence in the environment meaning that they present with a half life of decades to years before they degrade into less toxic forms. They are lipophilic and they accumulate in the fat of living organisms and thus they bio-accumulate in the food chain in ecosystems. The highest concentrations of POPs are thus found in organisms at the top of the food chain. As they are stored in fatty tissue, they possess a threat to human health and wildlife [1]. Bioaccumulation leads to biomagnification, as POPs are absorbed by lower trophic organisms, such as phytoplankton, which is consumed by zooplankton and they accumulate in the fatty tissues of higher organisms. This concentration effect reaches its maximum level in top predator species, such as humans and other mammals [1].

Another reason for bioaccumulation, apart from storage in the lipoid tissue, is the so-called enterohepatic cycle, which is one of main reasons for long half-lives of POPs in the human body. Lipophilic chemicals, such as POPs, are secreted into the bile from the liver conjugated with molecules such as glutathione and they are reabsorbed from the intestinal lumen to enter blood circulation returning back to the liver. Thus, the continuous recycling of POPs through enterohepatic circulation could lead to persistence in circulation and to chronic depletion of liver glutathione. There is a strong correlation between serum concentrations of POPs and age in

A. Salavoura, *Chemical Environmental Pollutants and their Effect on Health*,
https://doi.org/10.1007/978-3-031-80271-3_5

the general population, POPs provide examples of chemicals that can synergistically increase the risk of global hypomethylation with aging [2].

Most POPs are semi-volatile, which means that they can travel in the air thousands of miles from their source, before they settle. They may travel long distances and they may be found everywhere, including the Arctic, Antarctica, and even remote Pacific Islands [2]. POPs are chemically and environmentally stable chemicals [3].

POPs are generally divided intentionally and unintentionally produced byproducts of industrial processes. Some of the most popular POPs, used from multiple sources in everyday life include polyhalogenated industrial chemicals, such as polychlorinated biphenyl compounds (PCBs), polybrominated biphenyls (PBBs), organophosphorus compounds (OCPs) such as dichloro-diphenyl-trichloroethane (DDT), aldrin, and heptachlor and industrial by-products such as PAHs [2], polychlorinated dibenzo-p-dioxins (PCDD), and dibenzofurans (PCDF), commonly known as "dioxins" (Fig. 5.1).

A common classification of these broadly used chemicals is to the ones produced intentionally for specific use and the ones produced unintentionally as byproducts of combustion (Fig. 5.2).

Thus, according to WHO, POPs are chemicals of global concern due to their potential for long-range transport, persistence in the environment, ability to biomagnify and bio-accumulate in ecosystems, as well as their significant negative effects in human health and the environment. Many products used in our daily lives may contain POPs, which have been added to improve product characteristics, such as flame retardants or surfactants. As a result, POPs can be found virtually everywhere in measurable concentrations.

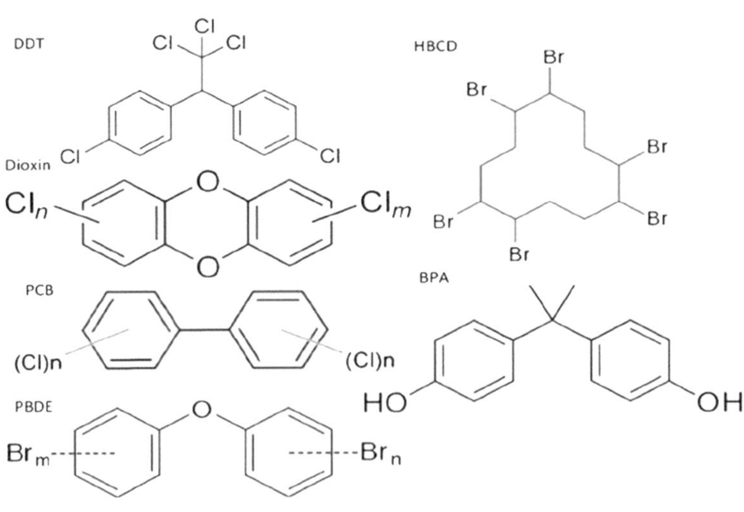

Fig. 5.1 The chemical structure of known POPs

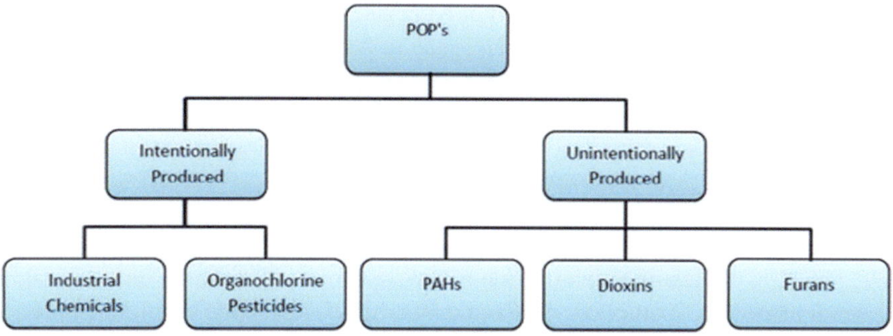

Fig. 5.2 POPs classification

In the 70s, many countries banned or severely restricted the use of the eight pesticides and four industrial chemicals and implemented pollution control strategies to reduce the amount of dioxin and furan released in the environment. However, it is reported that these products are still used in many countries today [1]. On 22 May 2001, in what can be considered a major achievement, governments met in Sweden to sign and adopt an international treaty concerning POPs. This is *the Stockholm Convention* on POPs [4].

The Stockholm Convention started with 12 substances, "the dirty dozen," which included eight chlorinated pesticides (aldrin, dieldrin, endrin, mirex, chlordane, heptachlor, DDT, and toxaphene), two industrial chemicals (PCBs and HCB) and two by-products (PCDFs and PCDDs). The convention entered into force and it became an international law on 17 May 2004. In May 2009, nine more substances were added [4]. In addition, there were granted as exemptions for use, some chemicals such as DDT in certain countries, where malaria poses a health threat, or some PCBs used in old electric transformers and capacitors in developing countries, where the alternatives are too expensive or too complicated to produce [2].

Other important international agreements include the *Basel International Convention* and the *Rotterdam Convention* that regulate chemicals, pesticides, and hazardous wastes at the global level. The PPPR (No. 1107/2009) and the Biocidal Products Regulation (No. 528/2012-BPR) both banned substances with endocrine disrupting properties based on scientific data [4].

People are exposed to POPs by inhalation (dust), ingestion (dairy and animal products), and skin contact [5]. Human exposure could start prenatally due to the fact that many POPs can cross the placenta. After birth, exposure also occurs by breast feeding [6].

Although everyone is exposed to a background level of POPs, certain people may have higher levels of POPs exposure because of their eating habits. Eating more fish than the general population as Southeast Asians, Canadians, Native Americans, sport anglers, and hunters who regularly eat large amounts of fatty fish or wildlife from the top of the food chain, such as waterfowl and waterfowl eggs, turtles and turtle eggs, muskrat, otter, moose, and deer, may be at risk of high exposure. Northern Aborginians, Inuit of Nunavik, who consume the fat of seals and beluga and narwhal whales, have been found to have higher body burdens of POPs [6] (Fig. 5.3).

Effects on Health

The potential effects of POPs include a wide spectrum of diseases including skin rashes to endocrine disruption, developmental delays, metabolic syndrome and diabetes, and cancer, depending on the type of compound and type of exposure [2]. Generally speaking, the main mechanism of these products to induce disease is epigenetic changes mainly by DNA methylation or the ability to function as endocrine disruptors (EDs).

Concerns about the health effects of exposure to POPs arose initially from studies of wildlife that showed reproductive, developmental, endocrine, immunologic, and carcinogenic effects. The wildlife tested had high rates of malformed genitalia,

Type	Environmental sources	Examples of dietary sources
Dioxins, furans	Byproducts of petrochemical industry and chlorine bleaching in pulp and paper mills; hospital and municipal incinerators	Meat, poultry and dairy products, sport fish (e.g., lake trout, salmon, walleye), wildlife (e.g., waterfowl and waterfowl eggs, muskrat, otter, moose, deer)
PCBs	"Fire resistant" synthetic products made before 1977, old electrical equipment, leaky containers in PCB disposal sites	Great Lakes fish (e.g., lake trout, salmon), Arctic marine mammals, breast milk
Aldrin	Pesticide used against insects in the soil, to protect crops such as corn and potatoes	Dairy products, meat, fish, oils and fats, potatoes, root vegetables
Chlordane	Broad-spectrum contact insecticide used on vegetables, grains, maize, oilseed, potatoes, sugar cane, beets, fruits, nuts, cotton and jute	Use has been severely restricted, so food does not appear to be a major pathway of exposure; air may be a pathway because of continued use in termite control (in the United States)
DDT	Pesticide widely used during World War II to protect soldiers and civilians against diseases spread by insects	Fish, dairy products, fat of cattle, hogs, poultry and sheep, eggs, vegetables
Dieldrin	Insecticide used to control insects in soil	Same as for aldrin
Endrin	Foliar (leaf) insecticide used on field crops such as cotton	Current dietary exposure thought to be minimal because of restricted use
Heptachlor	Nonsystemic stomach and contact insecticide used to control insects in soil	Detected in the blood of US and Australian cattle in 1990; current dietary exposure thought to be minimal because of restricted use
HCB	Fungicide used for seed treatment	HCB-treated grain; current dietary exposure thought to be very low because of severely restricted use
Mirex	Stomach insecticide used to control ants, termites and mealy bugs	Meat, fish, wild game, marine bird eggs, sea mammals
Toxaphene	Contact insecticide used primarily on cotton, cereal, grains, fruits, nuts and vegetables and used to control tick and mites in livestock	Dietary exposure thought to be very low because of restricted use; however, there is a local problem with some fish in Lake Superior

Note: PCBs = polychlorinated biphenyls, DDT = dichlorodiphenyltrichloroethane, HCB = hexachlorobenzene.
Source: Contaminant profiles. In: *Health and the environment. The health and environment handbook for health professionals*. Ottawa: Health Canada; 1998. Cat no H46-2/98-2111. Available: www.hc-sc.gc.ca/ehp/ehd/catalogue/bch_pubs/98ehd211/98ehd211.htm

Fig. 5.3 Dietary sources of organic pollutants

aberrant mating behavior, sterility, cancer, and immune system and thyroid dysfunction. Toxicological and laboratory studies confirmed the links between POPs exposure and effects observed in these studies [7].

Unintentionally Produced Chemical Agents

Polycyclic Aromatic Hydrocarbons (PAHs)

Polycyclic aromatic hydrocarbons (PAHs), named also polyaromatic hydrocarbons or polynuclear aromatic hydrocarbons, are hydrocarbons composed from multiple aromatic rings. The simplest of such chemicals are naphthalene, having two aromatic rings, and the three ring compounds anthracene and phenanthrene (Fig. 5.4). This group of chemicals is derived as a byproduct from multiple sources. PAHs are uncharged, non polar molecules found in coal and in tar deposits. They are also produced by the thermal decomposition of organic matter (e.g., in engines and incinerators or when biomass burns in forest fires).

The occurrence of PAHs ambiently, in food, drinking water, tobacco smoke, automobile exhausts, dust, and contaminated air is an increasing concern for the general population given their carcinogenicity and other reported potential health effects including allergy, asthma, cardiovascular, and respiratory diseases [7].

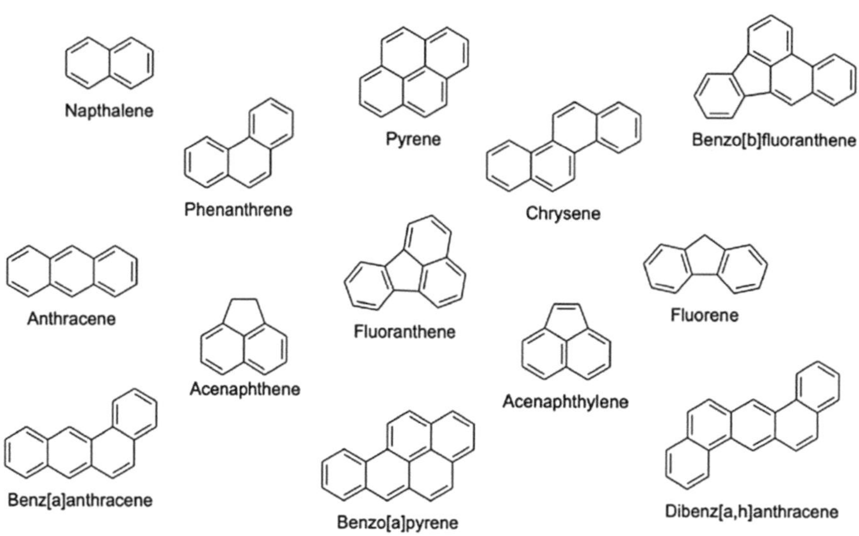

Fig. 5.4 The common PAHS

Tobacco smoking is a source of PAHs. Unburned tobacco, including cigarette tobacco, oral snuff, chewing tobacco, and other smokeless tobacco products, contains fewer PAHs, because most chemicals in smoke are formed during combustion. Nitrosamines, particularly the tobacco-specific nitrosamines 4-(methylnitrosamino)-1-(3-pyridyl)-1-butanone (NNK) and N-nitrosonornicotine (NNN), are by far the most prevalent strong carcinogens in unburned tobacco [7].

A. A significant percentage of polycyclic aromatic hydrocarbons is *Halo-PAHs (Halo-Polycyclic aromatic hydrocarbons)* which contain 2–6 aromatic rings and they have been detected in the environment since 1980s. A variety of chloro-PAHs, bromo-PAHs, and bromo-chloro-PAHs have been identified from different environmental sources, such as municipal incinerator ash, coal combustion, automobile exhausts, and tap water. The identified halo-PAHs containing 4–5 aromatic rings include 7-chloro-BA, 6-chloro chrysene, 6-chloro Benzo[a]pyrene (BaP), 6-bromo-BaP, 1 or 3-bromo-BaP, 1- or 3-chlorobenzo[e]pyrene (1-or-3-Cl-BeP), 7-chlorodibenz [a, h] pyrene, 7-chlorodibenz [a, i] pyrene, and 10-chlorodibenz [a, l] pyrene (Fig. 5.4).

   They are associated with genotoxicity, mutagenicity and carcinogenicity. Due to their widespread presence in the environment and their toxicity, many of these compounds may pose a health risk to humans [9].

   (i) *Chloro-PAHs,* a subgroup of halo-PAHs, are present in urban air samples collected from streets and road tunnels. The detected chloro-PAHs included 9,10-dichlorophenanthrene, 9,10-dichloroanthracene, 3- and 8-chlorofluoranthene, 1- and 4-chloropyrene, 1,3-, 1,6-, and 1,8-dichloropyrene, 7-chloro-BA, and 6-chloro-BaP. Studies have detected 1-chloronaphthalene, 2-chlorofluorene, 9-chlorophenanthrene, 3-chlorofluoranthene, dichlorinated naphthalene, phenanthrene, and fluoranthene in tap water. Since chloro-PAHs were present in tap water, they concluded that chloro-PAHs were formed from reaction of PAHs and chlorine. It was suggested that sunlight irradiation on chlorination of pyrene in soil formed 1-chloropyrene and dichloropyrene, the concentration of which was dependent on irradiation time, source of soil, and water content [9] (Fig. 5.5).

   Benzo[a]pyrene (BaP), a chloro-PAH, has long been the prototype of the study of chemical carcinogenesis by PAHs (Fig. 5.6). Photolysis of BaP in aqueous solution produces initially 6-oxy-BaP radical that upon further oxidation produces 6-hydroxy-BaP and BaP quinines. When BaP is irradiated in benzene, an additional product, a ring opened product (X), is also formed. The oxy-radical of BaP has been proposed to be responsible for DNA damage caused by the combination of UV light and BaP [9].

   The International Agency for Cancer Research (IARC) has classified BaP as a group 1 carcinogen. NTP lists 12 more PAHs in the NTP Twelfth Report on Carcinogens as "reasonably anticipated to be a human carcino-

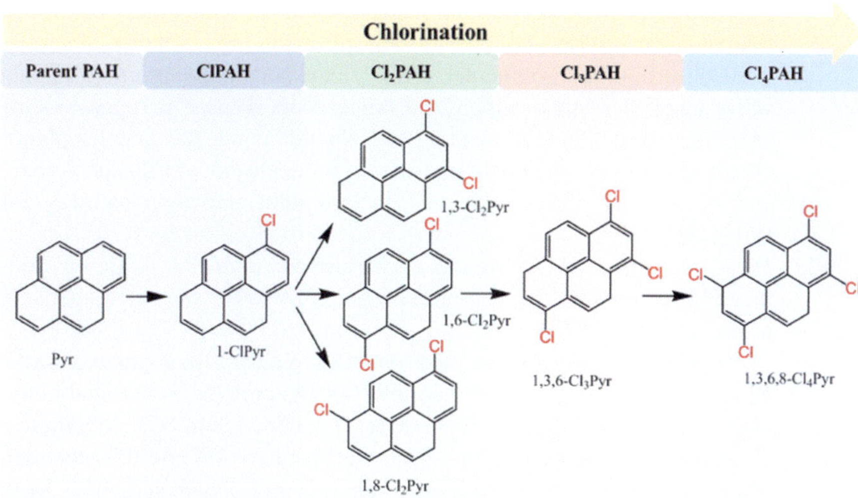

Fig. 5.5 Chloro-PAHs

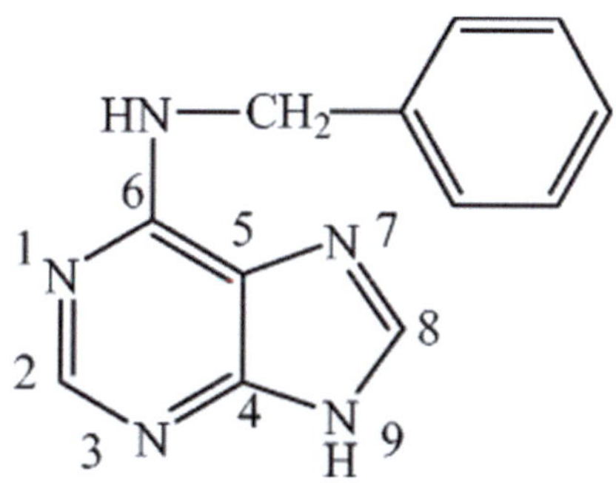

Fig. 5.6 The chemical structure of BaP

gen." These PAHs are benz[a]-pyrine, anthracene, [b]fluoranthene, [j]fluoranthene, dibenz [a, h] anthracene, dibenzo [a, h] pyrene, dibenzo [a, i] pyrene, indeno[1,2,3-cd] pyrene, benzo[k]fluoranthene, dibenzo [a, e] pyrene, dibenzo [a, l] pyrene, and 5-methylchrysene [9].

(ii) Other byproducts of PAHs formed during photoreaction processes include *oxygenated PAHs, PAH quinones, nitro-PAHs, and halogenated PAHs.* Upon light irradiation, PAHs and PAH photoreaction products can absorb light energy to reach increased energy states and they react with molecular oxygen, medium, and coexisting chemicals to produce ROS and other reactive intermediates. These intermediates, including ROS, induce lipid peroxidation and DNA damage including DNA strand breakage and

DNA-adducts with increased oxidation to 8-oxo-2 -deoxyguanosine. These toxicological endpoints are associated with age related diseases. Radioabsorbing environmental PAHs may potentially promote human skin damage, leading to ageing and skin cancers. Thus, it is suggested that risk assessment of environmental PAHs should include the complex photochemical reactions leading to mixtures of products that are also phototoxic and to consider photoreaction as an important activation pathway for PAHs [9].

(iii) *Dichloroacetic acid (DCA) and trichloroacetic acid (TCA)* are environmental contaminants that are peroxisome proliferators and carcinogenic in mouse liver (Fig. 5.7).

Decreased methylation in the promoter regions of the c-jun and c-myc genes and increased levels of their mRNAs and proteins were found in livers of mice exposed to trichloroethylene (TCE), DCA, and TCA. Methionine supplement prevented both the decreased methylation and the increased levels of the mRNAs and proteins of the two proto-oncogenes. This work supported the hypothesis that these carcinogens may act by depleting the availability of SAM, whereas methionine would prevent DNA hypomethylation by maintaining adequate SAM levels [10].

(iv) *Dioxins* are well-known byproducts in the group of chlorinated compounds and 2,3,7,8-tetrachloro dibenzodioxin (TCDD) is the prototypical example of the above class, often referred simply as "dioxin."

Accidental Exposures in History

TCDD is linked to the decrease in the sex ratio in offspring. *The Seveso disaster* occurred in 1976 in Italy and it resulted in the exposure of the population to this dioxin. It was observed that the children born from parents with the highest exposure to TCDD, in the first decade after the exposure, were all female. Many epidemiological studies are underway to study the long-term adverse health effects of this

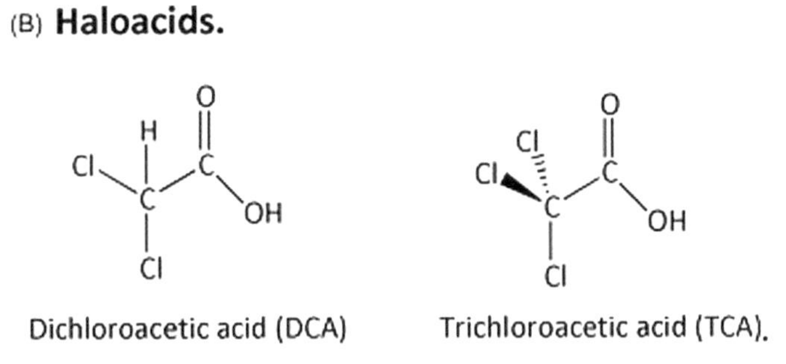

Fig. 5.7 The haloPAHS, DCA and TCA

dioxin including hormonal disruption and cancer. The pre-conception exposure of parents to pesticides is linked to the development of congenital endocrinology malformations of their offspring [4].

Measurement of PAHs levels in the body is based on the measurement of metabolites in human urine, including 1-hydroxypyrene (1-OHP), 1-hydroxypyrene-O-glucuronide, 3-hydroxybenzo[a]pyrene, 7,8,9,10-tetrahydroxy-7,8,9, 10-tetrahydrobenzo[a]pyrene, and other hydroxylated PAHs, that can be used as biomarkers to assess recent exposure to PAHs. Combustion products have a clear effect on the concentrations of the biomarker [4]. Development of biomarkers of exposure to PAHs and related compounds includes detection of protein and DNA adducts, which can be interpreted as indicators of an effective dose [12].

Routine monitoring at workplace exposure to PAHs is performed by polyurethane foam (PUF) in personal samples. Studies showed that, even after short exposure times, PUF can reliably accumulate both gaseous and particle-bound PAHs [12].

Intentionally Produced Products

Polybrominated Diphenyl Ethers (PBDEs)

Another class of POPs includes the PBDEs (Fig. 5.1). These flame retardants have been used in the manufacturing of common household items including furniture, infant products, and electronics. Originally issued in the 1970's, California's Technical Bulletin required that furniture, baby products, and other items resist open flames. PBDEs continue to leach into the environment from older household items. Children living in California have higher serum PBDE levels compared to children living in other areas of the United States as well as children from Mexico and Europe. Exposure to penta-PBDE has been related to lowered fertility, effects on thyroid hormone, and poorer neurodevelopment [13].

PBDEs constitute a class of polyhalogenated aromatic compounds with a basic structure consisting of two phenyl rings linked by an oxygen, with a variable number of bromide molecules from 1 to 10, allowing for 209 different possible congeners, which depend on the number and position of the bromine molecules. Only a few of the 209 different combinations are commercially available, and their toxicity depends on the specific bromine substitution. The congener deca-PBDEs is poorly absorbed and rapidly eliminated. In contrast, the congeners tri- to hexaPBDEs are strongly absorbed, slowly eliminated, and strongly bio-accumulative [1].

PBDEs are widely used as brominated flame retardants, and they are present in a number of consumer goods, plastics, electronics, furniture, and textiles. The EU banned their production in 2004 and 2008. Although PBDEs use has been gradually reduced worldwide, biomonitoring studies indicate that they are still ubiquitous. A number of older products, particularly wastes, such as electronic and electrical items, contain significant amounts of these pollutants that are still being released into the environment, leading to contamination of ecosystems [1].

PBDEs have been found in fish, eggs, human breast milk, and in infants and toddlers, toys probably as a result of contamination by house dust. The exposure routes include diet, ingestion, inhalation of indoor dust and dermal contact. Certain PBDEs and their mixtures have been shown to disrupt the thyroid hormone thyroxine and have been associated with neurodevelopmental deficits in rats and humans. Mixtures of PBDEs have increased cell proliferation in human breast cancer cell lines and they possess carcinogenic effects in humans [14].

Since the production of PBDEs has been banned in many countries, there were substituted by *organophosphate flame retardants (OPFRs)*. OPFRs constituted 20% of the total flame-retardant utilization in Europe in 2006. However, tris(2-ethylhexyl) phosphate (TPHP), tripropyl phosphate (TBP), Tris(1,3-dichloro-2-propyl) phosphate (TDCPP), and tricresyl phosphate (TCP) were identified to antagonize GR pathway. Replacement with brominated flame retardants, such as 2-ethylhexyl 2,3,4,5-tetrabromobenzoate (TBB) and bis(2-ethylhexyl) 2,3,4,5-tetrabromophthalate (TBPH) also exhibit anti-glucocorticoid activities. TPHP, Tritolyl phosphate (TMPP), TDCPP, tris(2,3-dibromopropyl) phosphate (TDBPP) exhibited MR antagonistic properties [14].

Polychlorinated Biphenyl Compounds (PCBs)

Polychlorinated biphenyl compounds (PCBs) are organic chemicals with the structure $C_{12}H_{10-x}Cl_x$ (Fig. 5.1). They are still widely in use, even though their manufacturing has declined drastically since the 1960s, when their toxicity to humans was identified. Some PCBs share a structural similarity and toxic mode of action with dioxins.

PCBs constitute a class of chemically stable aromatic compounds in which 1–10 chlorine molecules are attached to a biphenyl backbone. Theoretically, 209 different chemicals are included in this category. PCBs were widely used in agriculture as pesticides and in a number of industrial applications during the past decades. Their applications included dielectric fluids in capacitors and transformers, hydraulic fluids, building materials, additives and plasticizers in paints, plastics and rubber sealants, copying paper, and adhesives. Due to their widespread use and careless disposal over several decades, PCBs have become widely distributed in the environment. Although bans and restrictions exist at global level, they are still found in the atmosphere, rivers, fish, other wildlife, human breast milk, and other biological samples. The toxic impact of PCBs on human health and on the environment was first recognized in the 1960s, in particular with the Yusho incident in Japan, and later, with the Yu-Cheng accident in Taiwan [1].

Effects on Health

PCBs have been associated with negative effects on cognitive processes, motor development, and reflexes in children. PCBs and brominated flame retardants are indicated to interfere with the function of the thyroid hormone, which is crucial for normal neurodevelopment [8].

Chlorinated Compounds

Chlorinated compounds are used as pepticides and thus they ubiquitously in the environment. Representative members of this group include DDT, mirex, dieldrin, chlordane, heptachlor, endrin, hexachlorobenzene, aldrin, and toxaphene, which are listed in the so-called dirty dozen compounds banned by the Stockholm Convention. These pesticides were extensively used in agriculture and as insecticides for mosquito control between the 1940s and 1970s. Although banned, as a result of their strong persistence in the environment and bioaccumulation, the principal metabolites of DDT are detectable in more than 25% of the general population. The adverse health effects of DDT on animals and human health have been extensively reported in the literature and linked to diabetes, cancer, neurodevelopmental problems in children, miscarriages, and many other health outcomes. Due to agricultural activities, traces of the chemical and its metabolites have been found both in the surface and the groundwater. The presence in the groundwater has been associated with increased risk of cancer among people consuming the contaminated supplies [1].

DDT is a lipophilic compound that bioaccumulates in tissues and it has a long half-life in the human body (Fig. 5.1). Although DDT use was banned in 1972 in the US, DDT use continued in Mexico for malaria control through 2000 and it is currently used for malaria control in a number of countries particularly in Africa. Early life exposure to DDT and its breakdown product dichlorodiphenyldichloroethylene (DDE) have been associated with lowered birthweight and shortened gestation, poorer neurodevelopment in children, and breast cancer in adults [13].

Chlorofluorocarbons (CFCs) and Hydrochlorofluorocarbons (HCFCs)

Chlorofluorocarbons (CFCs) and hydrochlorofluorocarbons (HCFCs) are fully or partly halogenated hydrocarbons that contain only one carbon, hydrogen, chlorine, and Fluorine. They are commonly known as Freon by the DuPont brand. The most common representative is dichlorodifluoromethane (Freon-12). Many CFCs have been widely used as refrigerants, propellants (in aerosol applications), and solvents. (Fig. 5.8).

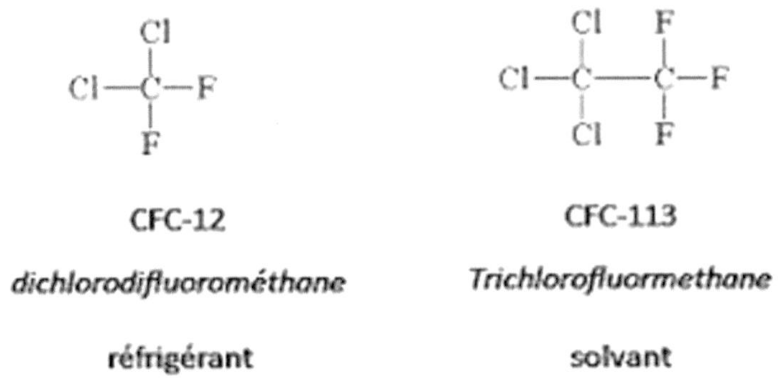

Fig. 5.8 Two chlorofluorocarbons with increase in Cl-, which have different chemical properties

Chlorofluorocarbons (CFCs) are stable in the atmosphere and they may reach the stratosphere. They react with the UV-radiation in the stratosphere to produce chlorine radicals, which interfere with ozone, destruction, and depletion. Due to potential adverse health effects of ozone depletion, chlorofluorocarbon replacements with much lower or absent ozone depleting potential are developed [14].

The toxicology of these compounds that represent HCFCs and HFCs has been studied. New products including HCFC-141b, HFC-134a, HFC-125, HCFC-124, and HCFC-123 show only a low potential for skin and eye irritation. Chronic adverse effects on the liver (HCFC-123) and the testes (HCFC-141b and HCFC-134a), including tumor formation, were observed in long-term inhalation studies in rodents using very high concentrations of these CFC replacements. All CFC replacements are detoxicated in the organism by cytochrome P450 which catalyzes oxidation of C-H bonds. The formed acyl halides are hydrolyzed to give excretable carboxylic acids. Halogenated aldehydes that are formed may be further oxidized to halogenated carboxylic acids or reduced to halogenated alcohols, which are excretory metabolites in urine [14].

Hydrochlorofluorocarbons (HCFCs) are being developed as replacements for CFCs that deplete stratospheric ozone. The lability of HCFCs, which is due to the presence of C-H bonds, it reduces HCFC migration to the stratosphere. HCFCs under development include HCFC-22, HCFC-123, HCFC-132b, HCFC-134a, HCFC-141b, and HCFC-142b. Although most HCFCs appear to show low acute toxicity, some HCFCs are mutagenic in the Ames test [14].

Polychlorinated Naphthalene (PCN)

Polychlorinated naphthalene (PCN) is the product obtained upon treatment of naphthalene with chlorine. The generic chemical formula is $C_{10}H_{8-(m+n)}Cl_{(m+n)}$. Commercial PCNs are mixtures of up to 75 components and byproducts. The material is an oil or a waxy solid, depending on the degree of chlorination.

PCNs include 75 congeners that had been widely used in various industries until the end of 1980s in the form of various mixtures, mainly as capacitor fluids, engine oil additives, cable insulation additives and for wood and paper production. In 2015, these compounds were included in Annex A and C of the Stockholm Convention, which prohibits their production and imposes the need to control their emissions during various industrial processes. The physicochemical properties of these compounds are similar to those of POPs. Currently, the main sources of PCNs in the environment are considered to be emissions from older products, i.e., technical preparations of PCNs, PCBs, and other similar compounds where PCNs occurred as impurities, as well as high-temperature industrial processes, during which PCNs are still formed as unintentional by-products. These processes increase environmental contamination and greater exposure of the general population to PCNs. The exposure occurs primarily by consumption of foods. It has been found that PCNs represent about 5% of the total toxic equivalence value in butter and cheese [11].

Despite the low value of total toxic equivalence in food, their lipophilic properties allow PCNs to accumulate to high levels in the body, especially in the liver and adipose tissue. A significant accumulation of PCNs is detected not only in body's fluids, but also in the placenta and they become dangerous to developing fetuses [11].

Exposure of pregnant females to a mixture of PCNs roughly corresponding to that used in the past (Halowax 1013 and 1014) was found to induce embryotoxic, fetotoxic, and even teratogenic effects. The mechanism of intoxification of the embryo is a very strong CYP1A1 induction in the placenta and in the maternal and fetal liver. Analyzing the results in various fetal organs, the highest levels of HxCN were found in the brains and in the sciatic nerve. The studies demonstrate the high affinity of the analyzed compound to the nervous system and potential induction of neurotoxic effects. Changes in GABA (gaminobutyric acid) neurotransmitter in rat brains confirm the above observations [11].

Sodium Cyanide (HxCN)

Sodium cyanide (HxCN) is an inorganic compound. It is a white, water-soluble compound. In addition, it is a moderately strong base. When treated with acid, it forms the toxic gas hydrogen cyanide. Cyanide has a high affinity for metals, which leads to the high toxicity of this salt. Its main application is in gold mining.

Previous studies show that HxCN (0.11 mg/kg body weight) administration to females during organogenesis is associated with embryotoxic effects, whereas a mixture of diverse congeners of PCNs with a similar composition to the previously used Halowax resulted in not only embryotoxicity, but also teratogenesis. Among all PCNs congeners, hexachloronaphthalene is one of the most bio-accumulative and toxic forms [11].

Perfluorinated Compounds (PFC)

Perfluorinated compounds (PFC) are organofluorine compounds containing only carbon-fluorines. PFCs have properties that result from the presence of fluorocarbons and the functional group.

Drinking water is the primary route of PFCs exposure. While PFCs are persistent in the environment and in the body (half-life in humans is 3–5 years depending on the compound), they are not metabolized and they are not lipophilic. Animal data indicate that PFCs can cause several types of tumors and neonatal death and they may have toxic effects on the immune, liver, and endocrine systems. Data on the human health effects include reported positive associations with cholesterol levels, hepatic enzymes, and adverse reproductive outcomes [12].

Trichloroethylene (TCE)

The chemical compound trichloethylene named IUPAC, is a halocarbon used as an industrial solvent and water pollutant. It is a clear non-flammable liquid with a sweet smell. It should not be confused with the similar 1, 1, 1-trichloroethane, which is commonly known as chloroethene.

This toxicant is linked to autoimmunity. Chronic TCE exposure (mostly occupational, but sometimes environmental) has been linked to a variety of autoimmune diseases including lupus, autoimmune hepatitis, and scleroderma. Signs of immune activation and alterations in lymphocyte subsets have been associated with chronic environmental exposure to drinking water contaminated by TCE or by dermal or inhalational occupational exposure. It was reported that chronic TCE exposure in drinking water during adulthood induces autoimmune hepatitis in autoimmune-prone mice and that the disease development is associated with several changes in $CD4^+$ T cells [15].

Epigenetic Mechanism of Intoxification

TCE altered the DNA methylation profile of the IFN-γ promoter in mouse $CD4^+$ T cells. In a genome-wide DNA methylation and gene expression study, TCE exposure *in vivo* differentially methylated CpGs in regions related to the regulation of T effector cell expansion and differentiation. These results support evidence that epigenetic events may play a role in TCE-induced immunotoxicity and autoimmunity [15] (Fig. 5.9).

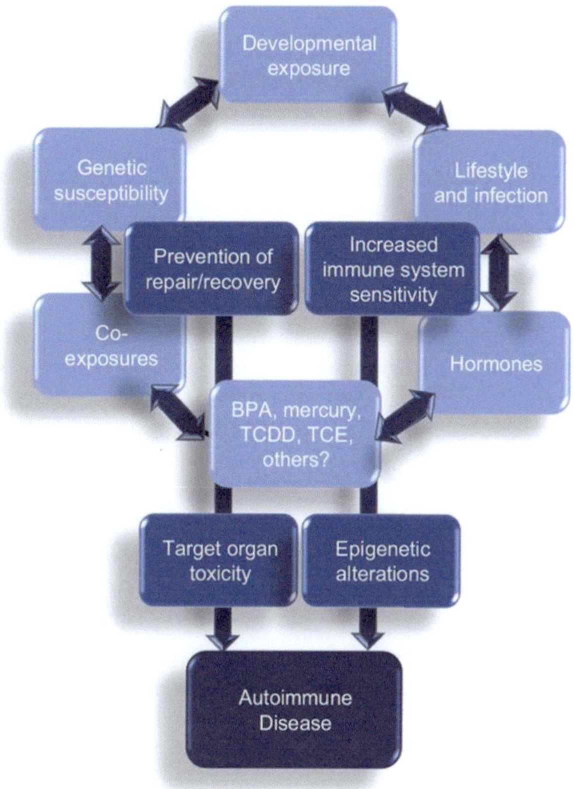

Fig. 5.9 Epigenetic alterations due to trichloroethene [15]

Brominated Flame Retardants (BFR)

Brominated flame retardants (BFR) are now being widely used in different products in our daily life. These products may range from computers, electronics and electronical equipment, textiles, foam furniture, insulating foams, and other building materials [16].

Mechanisms of Intoxification of POPs

The different POPs are toxic by different mechanisms. Since 1970, *childhood cancer* has been rising in Europe and United States by 1% per year and the data show evidence for a tendency for further increase. The causes of childhood cancer are largely unknown. Well known etiopathogenic causes are radiation and exposure to carcinogens, but time of exposure, dosage of exposure, and genetic susceptibility to the development of cancer, have a important role. A number of chemicals are associated to childhood cancer [4].

Because of the strong evidence for them inducing cancer in humans and in animals, PCBs were classified in 2013, by the International Agency for Research on Cancer (IARC), as carcinogenic to humans (Group 1) [1].

In the last few years, several investigations have examined the relation between exposure to environmental POPs and epigenetics and they identified several toxicants that modify the epigenome. Most of the studies conducted so far have been centered on DNA methylation, whereas only a few recent investigations have studied the effects of environmental chemicals on histone modifications and miRNA [10].

Epigenetic research holds substantial potential for developing biological markers to predict which exposures would expose subjects at risk and which individuals will be more susceptible to develop disease. However, a clear relation between epigenetic markers and disease is not yet found. Future prospective investigations are needed to determine whether exposed subjects develop epigenetic alterations over time and, in turn, whether such alterations increase the risk of disease [10].

Mutations in oncogenes due to exposure to POPs have been described. The most interesting example is that 7-chloro-BA which induces a ras (K–ras) oncogene activation, with 95% of the mutations being GGC → CGC at the first base of K ras gene in liver tumors [9].

Apart from the contribution of POPS to childhood cancer, an interesting theme to analyze is contribution to neurodevelopment. Various environmental chemicals, including pesticides, PCBs, and methyl-mercury, are known to be neurotoxic and may alter brain development.

Neurodevelopment is a complex process that begins with the formation of neural tube, the first stage of brain development, followed by a differentiation of the neuronal cells of the central and peripheral nevrous system. During the first two trimesters of fetal development, the basic structure of the brain is formed. Changes in neural connectivity and function occur in the last trimester and the postnatally. Synaptogenesis and myelination, the most prolonged changes, occur in the last trimester and continue postnatally to adulthood. The disruption of these complex processes could lead to neurodevelopmental disorders, such as attention-deficit hyperactivity disorder (ADHD), autism spectrum disorder (ASD), schizophrenia, intellectual disabilities, and neurological disorders, such as Rett syndrome. PCBs are among the most extensively studied environmental contaminants in terms of their effects on neurodevelopment and cognitive functions [4].

Furthermore, PCBs exhibit *endocrine intoxification* mainly due to their endocrine disrupting potential. It is found to increase endocrine receptors' transcriptional activity in breast cancer cell lines and thyroid function. The burden of chemicals in which females are exposed interferes with the hormone development during critical periods of fetal development [2].

Epigenetic Mechanisms of Intoxification

Prenatal exposure to POPs and postnatal exposures may affect DNA methylation. An interesting research was reported regarging CHAMACOS children (Center for Health Assessment of Mothers and Children of Salinas, University of California at Berkeley) [11].

Although there is no conclusive data demonstrating that the POP compounds are dangerous for pregnant women, it has been reported that the CHAMACOS children, whose mothers during pregnancy were consuming fish from the Great Lakes in North America, known to be contaminated with organo-chlorine compounds including PCNs, often manifested delays in psychomotor development and a lower intelligence quotient (IQ). Moreover, a study on the population of women from northern Quebec, who also consumed contaminated fish during pregnancy, demonstrated a correlation between the increased activity of CYP1A1 in the placenta and low birth weight. Moreover, administration of hexachloro-naphthalene (HxCN) (a dose of 1 μg/kg) to pregnant rats during organogenesis is associated with an accelerated initiation of spermatogenesis in the male offspring [11].

Another interesting perspective study examined the DNA methylation of Alu and LINE-1 repetitive elements in relationship to age, sex, and exposure into two classes of POPs, DDT and PBDEs, in Mexican-American children living in California. These children had relatively high DDT/exposure in utero and high childhood exposure to PBDEs. Specifically, DNA methylation of Alu and LINE-1 repeats was lower in the blood of 9-year-old compared to fetal cord blood. Furthermore, girls had lower levels of methylation patterns than boys at both time points. This factor study suggests that host factors, such as age, sex, and prenatal exposure to a specific or to a combination of factors, are potential predictors of repetitive element DNA methylation in children [13].

Global DNA hypomethylation was associated with high serum concentrations of POPs in apparently healthy Greenlandic Inuit. The degree of DNA methylation decreased with increasing concentrations of different POPs such as DDT and the total of all POPs studied. Inuit due to their nutritional habits manifest higher concentrations of POPs than the average general population. The global DNA hypomethylation mechanism may be involved in various chronic diseases, including type 2 diabetes, metabolic syndrome, and coronary heart disease, that have increased in prevalence in this population [2].

However, results regarding the methylation status of exposed individuals to POPs was controversial. In the same population, in a number of 70 Greenlandic Inuits, the exposure to DDT, DDE, organochlorines, and a sum of PCBs showed a significant association to Alu methylation and a similar, but non-significant trend was also found for LINE-1 methylation. However, comparable results were reported in a cross-sectional study of healthy adult Koreans ($N = 86$) with relatively low levels of environmental exposure to organochlorine pesticides and PBDEs. Another interesting study found lower levels of LINE-1 methylation among PCB–exposed individuals who had a maternal 15q11-q13 duplication [13].

A group of POPs, which is quite investigated regarding its mutagenic and tumorigenic properties, is *polycyclic aromatic hydrocarbons (PAHs).* Three metabolic activation pathways of PAHs have been described as responsible for cancer initiation processes: metabolism into by region diolepoxides, radical-cation intermediates, and PAH quinines. Formation of PAH diolepoxide-derived DNA adducts is the principal and most well established metabolic activation pathway leading to cancer initiation *in vivo*. The diolepoxide metabolites, particularly those with the epoxy ring located at the bay region, can react with DNA to form bound DNA adducts. The second pathway involves the formation of PAH radical-cation intermediates, which form easily DNA adducts. For example, they determined that the formation of BaP DNA adducts in mouse skin was mediated by radical cations. The third activation pathway, metabolic formation to PAH quinones, generates ROS that can damage DNA, protein, and cell membranes. All these metabolic pathways lead to the formation of adducts with DNA with subsequent repair disturbances or direct damage of DNA and important proteins (Fig. 5.10) [9].

Nitro-PAHs exhibit their mutagenic effects by other mechanisms. In 1978, the first reports regarding the effects of air pollution and particulate matter to health were published. It was reported that the urban air particulates contained direct acting bacterial mutagens, which they suggested to be nitro-PAHs. Since then, nitro-PAHs have been recognized as a class of widespread mutagenic and tumorigenic environmental contaminants. There are two distinct pathways of nitro-PAH

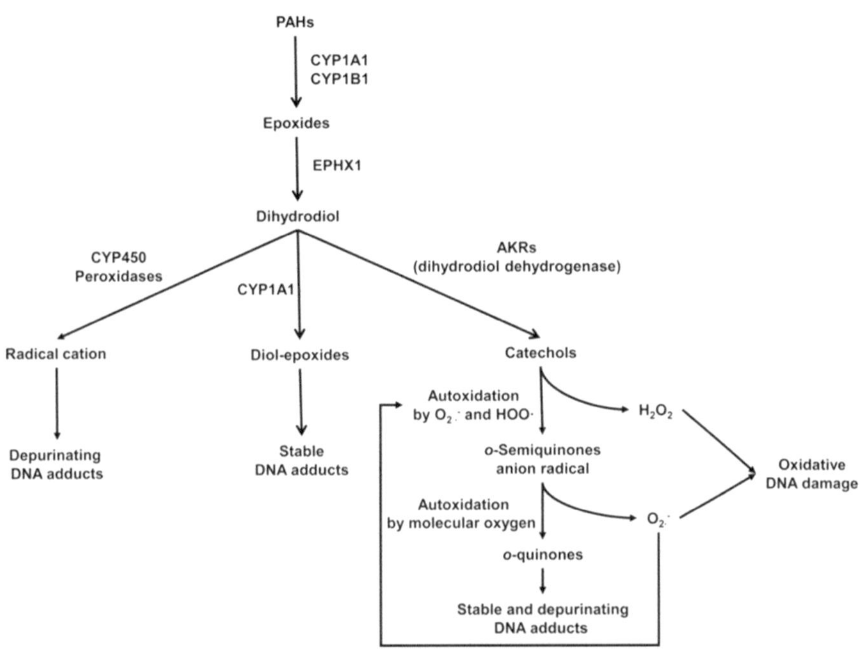

Fig. 5.10 PAHs metabolism into by region diolepoxides, radical-cation intermediates, and PAH quinines

formation in the environment: (i) as a result of incomplete combustion of organic materials and (ii) by atmospheric reaction of PAHs with nitrogen oxides, including NO_2 and NO_3 radicals. Both PAHs and nitrogen oxides are abundant combustion products and environmental contaminants [9].

Nitration of PAHs during the combustion process involves an *ionic mechanism* and the predominant species is the nitric oxide (NO_2^+). For example, the most active position of pyrene is the C-1 position; thus, oxidation of pyrene during the combustion process produces 1-nitropyrene by an ionic mechanism. On the other hand, atmospheric oxidation of PAHs taking place in a gas phase involves a radical reaction mechanism. For example, the oxidation of pyrene in the presence of NOx generated 2-nitropyrene as the sole product and a hydroxyl radical (Fig. 5.11). Oxidation of fluoranthene under similar conditions generated 2-nitrofluoranthene as a predominant product; while 7- and 8-nitrofluoranthene were formed in trace amounts. Thus, during the combustion process, oxidation of fluoranthene produced predominantly 3-nitrofluoranthene, and minor amounts of 7- and 8-nitrofluoranthene. 1-Nitropyrene has been the prototype used for the study of toxicological activities of nitro-PAHs [9] (Fig. 5.11).

Besides the mechanisms by which PAHs induce cancer in experimental animals, the metabolic activation pathways have been determined. The environmental fate of PAHs and the *phototoxicity* exerted by PAHs, as well as their photoreaction products formed in the environment, has been also investigated [9].

Sunlight reaching the earth's surface typically contains 5–6% of UV light. The sunlight reaching the earth surface is 91% visible, 8.7% UVA (320–400 nm), and only 0.3% UVB (280–320 nm). Reaction of PAHs in the environment principally

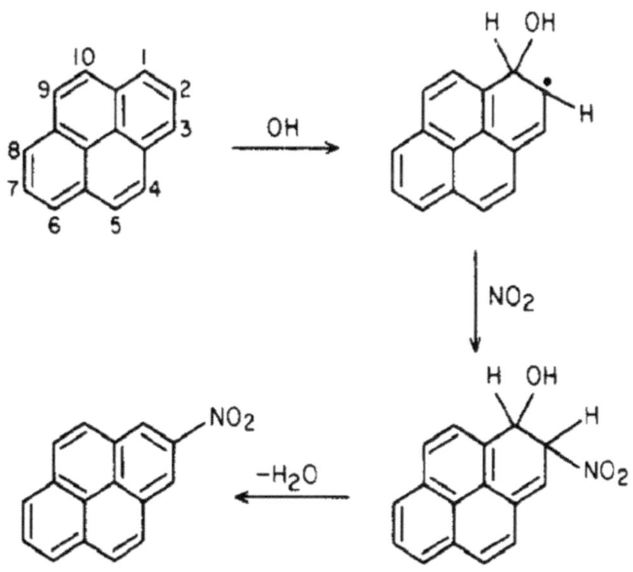

Fig. 5.11 The reaction of pyrene with NO oxidants of pyrene

occurs under the sunlight. As a result of the photoreactions, singlet oxygen, superoxide radical, PAH radical, DNA radical, oxygenated PAH, and other reactive intermediates are formed and they can damage cellular constituents, resulting in acute toxicity [9].

PAHs in the atmosphere could react with different molecules, such as NO_2 and NO_3 radicals or molecular oxygen. The principal PAH photochemical reactions include: (i) oxidation to oxygenated PAHs, which upon further oxidation, generate PAH quinones as end products, (ii) reaction with NO_2 or NO_3 radicals in the atmosphere to form nitro-PAHs, (iii) reaction with halogen-containing materials to produce halogenated PAHs, and (iv) reaction with molecular oxygen to produce ROS. The formation of oxygenated PAHs, PAH quinones, nitro-PAHs, and halogenated PAHs from photoreaction of the environmental PAHs has received much attention [9]. The electron activated molecules as the molecular oxygen induce lipid peroxidation and DNA damage including DNA strand breakage, oxidation to 8-oxo-2 -deoxyguanosine, and DNA-adducts [9].

Therefore, it is more biologically and environmentally relevant to study the photoirradiation of PAHs and their derived photoreaction products under the UVA radiation. PAHs with five or more aromatic rings can absorb visible light (>400 nm). Upon absorbing light energy, PAHs will be excited to an singlet state that can undergo intersystem crossing or electron energy transfer to molecular oxygen, solvents, or biological molecules in the cell to generate reactive species. These reactive species or intermediates can damage cellular constituents, such as cell membranes, nucleic acids, or proteins. Thus, PAHs can be activated by light irradiation to cause photo-induced cytotoxicity. There is evidence that PAH mixtures and individual PAHs are phototoxic to microorganisms, plants, cells, and animals also [9].

More detailed study of the photochemical reactions in PAHs show that there is a selectivity on the region of the reactions, which varies in different PAHs. The formation of photodegradation products is complex due to the ring arrangements. There are different geometric positions in a PAH that are sensitive to photooxidation. Thus, chemical and photochemical reactions of PAHs are highly region selective. To help recognize the chemical reactivity of a region and region-selectivity of a reaction, special terms are conventionally used to designate the selected geometric regions L regions [9].

For example, an L-region is the 9- and 10-positions of anthracene or the 7- and 12-positions of benz[a]anthracene (BA). The L-regions contain carbon positions with the highest electron density and therefore, they most reactive sites of the molecules. Photoirradiation of the L-region in anthracene, BA, BaP, 5-methylbenz[a]anthracene (5-MBA), and 7,12-dimethylbenz[a]anthracene (DMBA), in the presence of molecular oxygen forms, produces endoperoxides as intermediates and the further oxidation produces the quinones as end products [9].

BaP has long been studied as a prototype of chemical carcinogenesis. Photolysis of BaP in aqueous solution yields BaP 1,6-, 3,6-, and 6,12-quinones. When irradiated in benzene, an additional product, a ring opened product is also formed. The series of products due to photoirradiation of BaP include initially 6-oxy-BaP radical that upon further oxidation produces 6-hydroxy-BaP and BaP quinines. The oxy-radical of BaP has been proposed to be responsible for DNA damage caused by the combination of UV light and BaP [9].

UVA and visible light irradiation of BaP and pyrene induced cytotoxicity in a dose dependent manner in human keratinocytes. Part of the induced cytotoxicity is due to the PAH quinones formed from photoirradiation. PAH quinones are redox cycle compounds that can produce ROS and semiquinone radicals. PAHs quinones are also electrophilic and therefore, may covalently bind to cellular nucleophiles, such as proteins and nucleic acids. Therefore, PAH quinones could pose potential toxicity to higher organisms by food chain pathways. Photolysis and enzymatic transformation of BaP produced BaP 1,6-quinone, BaP 3,6-quinone, and BaP 6,12-quinone, exhibited cytotoxicity, and cause DNA damage [13] (Fig. 5.12).

PAHs and PAH mixtures in the environment exhibit photo induced toxicity not only to humans but also to plants, microorganisms, and aquatic organisms. In general, PAHs with a higher number of benzene rings exhibit greater phototoxicity if soluble in aqueous media. PAH mixtures in the environment, such as in marine sediment, are phototoxic to marine vertebrae larvae and juveniles [9].

To summarize, the toxicity of PAHs including mutagenicity, oncogenicity, and genotoxicity are caused by direct DNA damage, DNA adducts, and depletion of detoxification mechanisms of the cell and finally, phototoxicity with subsequent production of genome toxic byproducts. At the molecular level, the causative mechanisms of PAH related health effects are not completely investigated. However, it is suggested that epigenetic mechanisms may be involved. Benzo[a]pyrene (BaP) has

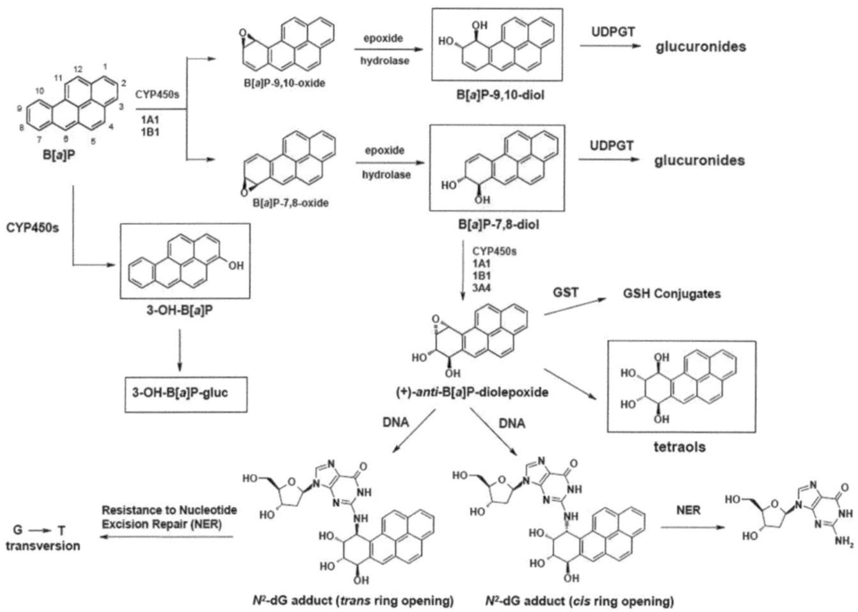

Fig. 5.12 Photoactivation of BaP produces conjugates, such as glucuronides and GSH conjugates, and DNA adducts

been reported to disrupt DNA methylation patterns in experimental models in breast cancer cells. Binding of BaP adducts to DNA decreased methylation by reducing binding and activity of DNMTs. Interestingly, experimental evidence suggests that PAH DNA adducts formation may preferentially target methylated genomic regions and they may interfere with their DNA methylation status [12].

Epigenetic Mechanisms of Intoxification

Important publications related to the DNA methylation status of PAH exposed populations are further described. These studies were conducted in Mexico, Poland, and China. In the Mexican study population ($N = 39$), urine 1-hydroxypyrene was inversely associated with the methylation of LINE-1 and Alu elements. The association, however, was not statistically significant. In the Polish study population ($N = 92$), increasing levels of blood and urine exposure biomarkers were associated with increasing DNA methylation in LINE-1 and Alu elements. Both studies showed consistent directions in the associations between exposure and methylation of the genes *p53* and *IL-6*. The associations, however, were statistically significant only in the Polish study. In the Chinese study ($N = 128$), the $p16^{INK4\alpha}$ promoter methylation status showed a positive correlation with urine 1-hydroxypyrene, which was not consistent with the results from the Polish study. Thus, the methylation effects of PAHs remain controversial [12].

Pesticides

Pesticides are chemicals that are produced in order to control pests. The term includes all the following: herbicides, insecticides, nematicides, molluscicides, piscicides, avicides, rodenticides, bactericide, insect repellent, antimicrobials, and fungicides. The most common of these are herbicides which account for approximately 80% of all pesticide use.

Organophosphate and carbamate insecticides disrupt acetylcholinesterase activity. Although commonly used in the past, many older chemicals have been removed from the market due to their health and environmental effects.

A total of about 890 active ingredients are registered as pesticides in the United States and currently marketed 20,700 pesticide products. Many of these products are described as POPs with the characteristic stability and resistance to biodegradation. The general population is exposed to the residues of pesticides, including physical and biological degradation products by air, and water [2].

Occupational exposure occurring at all stages of pesticide formulation, manufacture, and application involves exposure to complex mixtures of different types of chemicals. The most vulnerable groups are pesticide manufacturing workers, pesticide applicators, floriculturists, and farm workers. Genetic damage associated with pesticides occurs in human populations due to intensive use, misuse, or failure of control measures. Pesticides act selectively against certain organisms without adversely affecting others. Absolute selectivity, however, is difficult to achieve and most pesticides are a toxic risk for humans [2]. Pesticides are the most important method in self-poisoning in the developing world. Three million cases of pesticide poisoning, nearly 220,000 fatal, occur worldwide every year [2].

The ecological risk assessments analysis indicates the high level of risk with the pesticide presence to aquatic environment including organisms. In addition, the study also found that the seasonal application in agriculture, rainfall, and temperature could influence the levels [17].

Effects on Health

Studies have concentrated on the toxicity of these chemicals. Occupationally exposed populations demonstrate increased presence of chromosomal aberrations (CA), sister-chromatid exchanges (SCE), and micronuclei (MN). The International Agency for Cancer Research (IARC) has reviewed the potential carcinogenicity of a wide range of insecticides, fungicides, herbicides, and other compounds. Fifty-six pesticides have been classified as carcinogenic to laboratory animals. Associations with cancer have been reported in human studies for chemicals such as phenoxy acid herbicides, 2,4,5-trichloro-phenoxyacetic acid (2,4,5-T), lindane, methoxychlor, toxaphene, and several organophosphates [14]. Similar studies have been reported for organophosphorus insecticides 6 months after the exposure, malathion and trichlorfon, dimethoate, ethylenebis dithiocarbamate (EDC), chloroorganics, and complex mixtures.

The incidence of congenital malformations due to a parental exposure to pesticides has been investigated. Recent findings suggest that female workers in flower greenhouses may have reduced fertility [2].

Mechanisms of Intoxification

The mechanism of toxicity from pesticides is ROS generation. The CYP2E1 is a cytochrome P450 superfamily involved in the metabolism of many indirect carcinogens. More than 70 different substrates are specifically metabolized by this enzyme. In addition, glutathione *S*-transferases (GSTs) are the most important group of detoxifying enzymes. This family of enzymes presents genetic polymorphisms in human populations responsible for the glutathione conjugation to various reactive species of many chemicals including pesticides. Null genotypes for the *GSTT1* and *GSTM1* genes have been identified to be associated with an increase in cancer risk. *N*-Acetyltransferases (NATs) catalyze reactions of both activation (O-accetylation) and detoxification N-acetylation. Two *NAT* genes are expressed in humans, *NAT1* and *NAT2*. *NAT2* characterizes rapid and slow acetylators with an increased risk for different cancers. Paraoxonases (PONs) are responsible for metabolism of organophosphate based insecticides. Individuals with low *PON1* activity are more susceptible to parathion poisoning than individuals with higher *PON1* activity [2].

Epigenetic Mechanism of Intoxification of POPs

Pesticides have been considered to be potential chemical mutagens: Experimental data revealed that various agrochemical ingredients possess mutagenic properties inducing mutations, chromosomal alterations or DNA damage [2].

Pesticides and/or their metabolites have been recognized as one of the first classes of compounds to act as endocrine disruptors. Dozens of pesticides or pesticide metabolites have been screened for GR disrupting activities, mainly by *in vitro* bioassays. Most of the tested pesticides did not exhibit glucocorticoid activity, but some chemicals were reported to be weak GR antagonists [18].

A recent study found 12 compounds among 34 pesticides and/or metabolites, including five pyrethroids, four organochlorines, ethiofencarb, atrazine, and tolylfluanid, that exhibited apparent GR antagonistic effects and they enhanced glucocorticoid responsive genes. Although banned for decades, the organochlorine pesticides are still commonly detected in the environment and remain a global concern as potential EDCs. Vinzolin, a fungicide, is a GR antagonist [18].

The studies measuring DNA methylation in Alu elements showed consistent statistically significant inverse associations with oxychlordane, p, p′-DDE, and DDT. Increased levels of PCB183, heptachlor epoxide, trans-nonachlordane, and PBDE47 in a study population from Korea ($N = 86$) and PCB (156, 99, and 105), β-HCH, α-chlordane, mirex, sum of PCBs, and total POPs in a study population from Denmark ($N = 70$) were significantly associated with lower DNA methylation in Alu elements. In a population of Japanese women ($N = 399$), serum POPs were inversely associated with the global DNA methylation also. In an elderly population from Sweden ($N = 519$), increasing levels of PCB1 and p, p′-DDE concentrations were significantly associated with increasing global DNA methylation levels. For PFCs, in a study population from the US ($N = 671$), a 12 ng/mL increase in levels was associated with a difference of 20% in average 5-mC levels. Other PFCs did not show statistically significant associations [12].

The well-known example because of its toxicity is DDT. Exposure to dichlorodiphenyl-trichloro-ethane (DDT) induced specific gene hypomethylation of CpG islands in the promoters of *Sst, Gal, Arf1, Ttr, Msx1*, and *Grifin* genes in the hypothalamus of young male rats. Rats treated in utero and postnatally with organochlorine pesticides and PCBs also showed decreased methylation of CpG sites in the promoter of the tumor suppressor gene *p16 (INK4a)* compared to controls. Perfluorooctanoic acid induced gene promoter hypermethylation of *GSTP1* in the human liver. For most POPs, studies evaluating DNA methylation globally showed a trend towards hypomethylation with increasing levels of exposure [12].

Benzene

Benzene is a volatile aromatic hydrocarbon solvent, an industrial chemical in petroleum products and combustion materials and thus a well known air pollutant of the environment. In the past, benzene was used predominantly as a solvent; however, occupational exposure in industrialized countries occurs by the production and use of petroleum derivatives, in particular fuels, in various settings, including factories, refineries, rubber production plants, shoe manufacturing, and printing factories. The general population is exposed to benzene by inhalation; exposure sources include motor vehicle emissions, gasoline evaporative loss at service stations and cigarette smoke. In addition, the consumption of contaminated foods or water can present a secondary route of exposure [19] (Fig. 5.13).

Effects on Health

Chronic exposure to benzene is known to be implicated in various health problems and mainly, in disorders of the hematopoietic system including myelodysplastic syndromes and mainly aplastic anemia and an increased risk of acute myelogenous leukemia [19]. It is established as a carcinogen (International Agency for Research on Cancer 1999) [20].

Epigenetic Mechanism of Intoxification

Exposed to low dose benzene, occupational exposure has demonstrated significant leukocyte DNA hypomethylation of LINE-1 and Alu sequences as well as gene specific hypermethylation of *CDKN2B* gene and hypomethylation of the *MAGE1* gene. Additionally, benzene causes hypomethylation of *RUNX3* gene, hypermethylation of *MSH3* gene, and Sema3C protein. It is also responsible for the downregulation of the expression of *p15* and *p16* modulating the promoter by methylation of

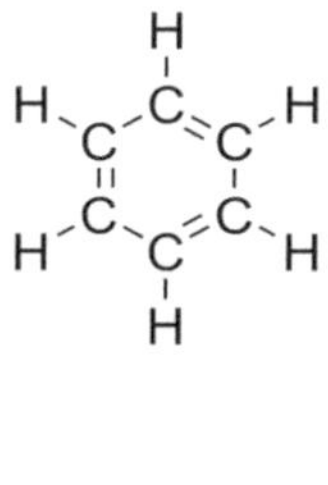

Fig. 5.13 The chemical structure of benzene

LINE-1 and hypomethylation of Alu DNA sequences. It also attributes to hypomethylation of *STAT3*. It is also responsible for the production of multiple miRNAs [19]. Another preliminary study analyzed miRNA expression in seven exposed control matched pairs and identified upregulation of miR-154, miR-487a, miR-493-3p, and miR-668 in exposed subjects [20].

Several studies have examined the possibility that benzene causes leukemia at high concentrations; however, it has since been suggested that exposure to increased levels of products of phototoxic reactions, and the production of catechol, hydroquinone, glucuronide and sulphate conjugates, may be responsible for carcinogenesis. The majority of conjugates is excreted by urine [10, 19]. Catechol and hydroquinone can also be partially transformed into reactive o- and p-benzoquinone, respectively. Benzene is also converted by a ring-opened oxidation product to t-muconic acid, by the precursor t, t-muconaldehyde. In addition, benzene oxide can be a substrate of glutathione-S-transferase, forming mercapturic acid. There is evidence that hydroquinone, phenol, catechol, benzoquinone and muconaldehyde are actually involved in benzene toxicity; whereas the glutathione conjugate of mercapturic acid, and the glucuronide and sulphate conjugates of hydroquinone, catechol, phenol, and benzoquinone are the results of detoxifying processes. These benzene metabolites are different in chemical structure and reactivity; therefore, it is difficult to determine which may contribute the most to producing benzene toxicity or leukemia. t, t-muconaldehyde and 1,4-benzoquinone are the most potent in inhibiting erythropoiesis, followed by hydroquinone and catechol, co-administration of phenol with hydroquinone and catechol, and certain glutathione adducts of hydroquinone [10]. Furthermore, benzene exposure can cause damage to the immunological, neurological, and reproductive systems [19].

Dioxin and Dioxin-Like Chemicals

Dioxin chemicals continue to be of great concern as environmental toxicants (Fig. 5.1). About 30 years ago, the aryl hydrocarbon receptor (AHR) was discovered as a specific binding site for 2,3,7,8-tetrachloro-dibenzo-p-dioxin (DDT). The AHR serves as the archetype for understanding toxicity mediated by chemicals. The fact that toxicity is receptor-mediated has important implications, especially for dose–response relationships [21].

Mechanism of Intoxification

In 1976, an industrial accident contaminated the residential area surrounding Seveso in Italy with very high TCDD levels. The incidence unmasked another side effect of chronic exposure to chemicals and especially to TCDD that was accidentally released in the environment. A significant association between dioxin levels and the frequency of t (14;18) translocation-positive cells and the increased risk for non-Hodgkin's lymphoma (NHL) was found. More than 80–90% of follicular and 20% of diffuse large lymphomas carry the t (14;18) translocation, which is the most frequent chromosomal translocation in human lymphoid malignancies. The t (14;18) translocation results from a reciprocal translocation that juxtaposes the BCL-2 (B-Cell Leukemia/Lymphoma 2) gene on chromosome 18q21 near the immunoglobulin heavy chain (*IgH*) locus at chromosome 14q32. TCDD is a known cancer promoting agent and it has been associated with increased NHL risk in several epidemiology investigations. It was found that the frequency of t (14;18) translocations increased in a dose dependent manner with increasing levels of plasma TCDD, indicating that TCDD exposure may have promoted the expansion of existing t (14;18)-positive clones. TCDD has been shown to affect both cell mediated and antibody immunity and alterations of immune parameters have been documented in the Seveso population [17].

While several environmental pollutants bind the aryl hydrocarbon receptor (AHR), 2,3,7,8-tetrachlorodibenzo-*p*-dioxin or TCDD is the prototypical ligand. Developmental exposure to TCDD and other AHR-binding xenobiotics alter the immune system. In contrast to the primarily immunosuppressive nature of adult TCDD exposure, developmental exposure to TCDD generated immune dysfunction in a different way. Adult autoimmune-prone mice models develop autoimmune glomerulonephritis including increased levels of anti-dsDNA in adulthood. Neonatal exposure to TCDD increased pro-inflammatory cytokines and the autoantibody production for Sjogren's syndrome by signaling events in the neonatal thymus [15].

Experiments in laboratory animals showed genetic differences in AHR gene structure which led to profound differences in responsiveness to dioxin-like chemicals. Humans, however, exhibit relatively few AHR polymorphisms and thus they are not influenced much from AHR polymorphisms. The important result of the studies is that dioxin toxicity is due to AHR-mediated dysregulation of gene expression [21].

Going back to the discovery of the receptor, the following paradox was described. Animal models with functional receptor were prone to cancer, although the main "purpose" of the receptor was binding to toxic and carcinogenic xenobiotic chemicals and thus, it might be tempting to conclude that humans (and other vertebrates) would be better without an AHR. Not having an AHR prevents cancer induction by specific chemicals, presumably because CYP1A1 is not inactivated [21].

New data show that the aryl hydrocarbon receptor (AHR) cross talks with multiple signal transduction pathways and it possesses essential functions in cell proliferation, development, immunity and circadian rhythm. Activation of the receptor by its ligand induces AHR nuclear translocation and heterodimerization. The aryl hydrocarbon receptor nuclear translocator binds with AHR its receptor complexes and AHR response elements (AhREs) in the promoter region of target genes and mainly CYP1A1 [17] (Fig. 5.14).

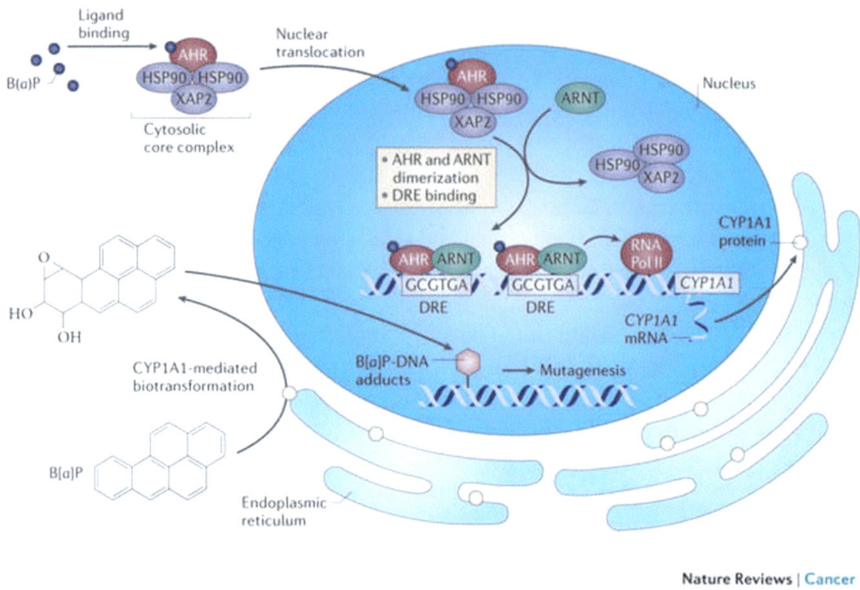

Fig. 5.14 AHR receptor leading to BaP entrance in the cell transformation with following interference in DNA translation

Epigenetic Mechanisms of Intoxification

Epigenetic changes control the transcriptional state of the receptor as well as induction to a silent state of AHR-regulated genes. Moreover, binding of as previously complexes to AhRE motifs recruits multiple transcription cofactors and associated chromatin remodeling proteins that initiate signaling of gene transcription. As previously heterodimers recruit histone acetyltransferase p300 to the target promoter to function as a coactivator that increases the level of histone acetylation and enhances transcription activity. Conversely, HDAC1-DNMT1 complexes, reversibly bound in naive cells at the *CYP1a1* promoter, maintain this gene in a transcriptionally silent heterochromatic state [17] (Fig. 5.14).

Guidelines suggested from WHO and the United Nations Environment Program and advice from experts in the field. The UN suggestions, as a measure for protection from overwhelming chemicals in the environment, are presented [22].

- It is preferable to opt for fresh instead of processed and canned foods. In general, the exposure of foods to contaminants additives, plasticizers, and monomers, can leach and be transferred on food items. Several reports have been published on this subject. The inside of metal cans is normally coated with a thin layer of epoxy resin, which is made from BPA.
- It is preferable to opt for organic food without chemical processing. Exposure to pesticides is linked to many diseases. However, organic foods may also be con-

taminated with pollutants as a consequence of the effects of the entire food chain Most of the times, the exposure is far less than conventional food. The option that the consumption is difficult to be for public use (the organic food market is growing at a compound annual growth rate of around 14.56%, between 2017 and 2024), the prices of organic food products tend to decrease.

- Food in plastic containers should not be heated in a microwave oven. Plastic containers can be replaced by glass or ceramic ones. The migration of the chemicals to the food can be accelerated by increasing temperature. Some plastics, such as polycarbonate, may infiltrate the food. BPA analogues, such as bisphenols have replaced BPA in numerous consumer products. These analogues also exhibit endocrine disrupting activity, cytotoxicity, genotoxicity, reproductive toxicity, and neurotoxicity.
- The consumption of fat dairy or meat products should be reduced. POPs bioaccumulate in the food chain, and they accumulate in fatty tissues. Bioaccumulation leads to biomagnification. POPs are absorbed by lower trophic organisms, such as phytoplankton, that are consumed by zooplankton and finally by fish. They accumulate in fatty tissues of organisms that are then eaten by other organisms, thus magnifying their effect up the food chain. This concentration effect reaches maximum levels in the top predator species, such as humans and other mammals.
- Products such as makeup, perfumes, and skin care products should be free of phthalates, parabens, triclosan, and other chemicals. Products with fragrances generally contain phthalates as carriers. Products that are labelled as "antibacterial" generally contain triclosan. Information about ingredients of cosmetics can be found in databases such as the EWG's Skin Deep Cosmetics Database.
- It is preferable to opt for ecological household cleaning products. Human exposure to BPA, phthalates, and triclosan occurs, besides food, water, and consumer products, by household cleaning products. The use of household cleaning products during pregnancy, at least once per week, was associated with 10–44% greater levels of phthalate metabolites in urine.
- Flame retardant treated furniture should be avoided. PBDEs are widely used as brominated flame retardants in furniture. These have been gradually decreased worldwide since 2004; however, biomonitoring studies indicate that they are still ubiquitous in human blood and breast milk worldwide. A number of older products contain significant amounts of these pollutants that are still being released into the surrounding environment. PBDEs have been found in human breast milk and in infants and toddlers, probably as a result of contamination by house dust.
- Indoor environments should be ventilated regularly. It is estimated that the major source of contamination comes from indoor environments, since we spend about 90% of our life indoors, particularly at home and at the workplace, and also the car. Therefore, the IAQ has a considerable impact on our exposure to contaminants and poses a risk to human health.
- Alternatives to plastic toys are preferred. Children's toys and teething items are generally made of plastics that contain additives which are intended to modify the properties of the polymer. The most common additives or plasticizers, added to increase flexibility, durability, and so on, have been phthalate esters. Children's

normal behavior, such as hand-to mouth and object-to-mouth contacts, increases the exposure through inhalation and ingestion routes. Phthalates have been implicated in various male malformations. Reports on the effects of prenatal exposure to phthalates in humans showed that birth outcomes were associated with high levels of phthalate esters in the urine of the mothers during pregnancy. Although the EU has imposed a limit of 0.1% (W/W) of phthalates in toys, these have been found in much higher concentrations (from 0.1% to 63.34%) [22].

Air Pollution

Air pollution, indoors and outdoors, are the environmental factors with the greatest impact on health in Europe and it is responsible for the largest burden of environmental related diseases. Recent estimates indicate that 20 million Europeans suffer from respiratory problems every day.

Indoor Pollution

Indoor pollution is defined as the presence of physical, chemical, or biological agents in the air of constricted areas, where they are not naturally present in high quantities in the air of this environmental system. If we take into consideration the time people spend indoors in westernized societies, indoor pollution is of significant importance.

The Chemicals, Their Forms, and Distribution in the Environment

The major sources of indoor air pollution are asbestos, biological pollutants, carbon monoxide (CO), lead (Pb), pesticides, radon (Rn), respirable particles, tobacco smoke, and second-hand smoke, as well as volatile organic compounds (VOCs) [20].

Volatile Organic Compounds (VOCs), including formaldehyde, benzene, and tri- and tetra-chloroethylene, are including paints and coatings to underarm deodorant and cleaning fluids, are ambient in the indoor areas. Widely used as components in household products, they are emitted as toxic gases from certain solids and liquids, to be released, while there are used and to some degree, when they are stored [20].

In the 90s in Japan, a total of 76.0% of the patients, presenting with ambiguous general symptoms, were diagnosed as having what is called "*The Sick Building Syndrome*" ("SHS"). Physicians diagnosed 87.6% of those patients as having chemical associated SHS, based on SHS classifications, definition, and diagnostic criteria. Based on the patients' interview, the 50.3% of the patients corresponded to the diagnostic criteria [20].

Detection of the chemical was possible in some cases. The chemical substances that have been detected were formaldehyde, toluene, xylene, ethylbenzene, styrene, paradichlorobenzene, acetaldehyde, and tetradecane. The level of the chemical was measured and compared to acceptable levels in guidelines. Among 287 subjects, the 18 patients (6%) were diagnosed as having multiple chemical sensitivities or single chemical sensitivity [20].

Based on previous studies, it is known that toluene, which is considered one of the major causes of SHS, induces symptoms from the nervous system, e.g., stress and anxiety as well as mucous membrane irritation. Formaldehyde, which is also considered as a major cause of SHS, also induces mucous membrane irritation. [20]. The air of the houses of ten patients also contained paradichlorobenzene and acetaldehyde. It is important for the patients who suffer from chemical intolerance to be careful about daily toiletry necessities that come into contact with the skin, such as insect repellents and deodorants, in addition to being careful about building materials [20].

Formaldehyde (FA) is a high-volume chemical, which is used for disinfection purposes and as a preservative. Also, it is used in the production of resins and binders, which are used in wood-products. Furthermore, formaldehyde is used in the production of plastics, coatings and paints, flooring materials, for textile finishing, and for synthesis of chemicals and it is a component of combustion products. Additionally, it is a major compound derived from ozone initiated reactions with alkenes, e.g., terpenes. Due to its ubiquitous use, formaldehyde is considered a common indoor air pollutant [23].

Due to its importance as an indoor pollutant, in 2010, the World Health Organization established an indoor air quality guideline for short- and long-term exposure to formaldehyde of 0.1 mg/m^3 (0.08 ppm) for all 30-min periods. The majority of the studies showed that indoor air concentrations in Europe and the United States were below 100 µg/m^3 and the median means ranged between 5 and 60 µg/m^3 [23].

Both primary and secondary effects of FA are prevented if the air pollution level is kept within the guideline level. Nasopharyngeal cancer and leukemia were observed inconsistently among studies; new updates of the US National Cancer Institute cohort confirmed that the relative risk was not increased with mean formaldehyde exposures below 1 ppm and peak exposures below 4 ppm [23].

Radon exposure is the best documented environment-related cause of cancer, but it is localized in geographical areas, where radon precursors (uranium) occur naturally in the ground [24].

Radon is the second leading cause of lung cancer and it is recommended that all homes be tested for radon. Radon is an environmental carcinogen caused by the radioactive decay of naturally occurring uranium in the soil. The ionized atoms of radon are incorporated to dust particles and when inhaled expose the lungs to small doses of radiation. Soil composition determines the amount of radon present at home. In zones with high radon levels, radon gas leaks into the home through cracks in the foundation or other open entries [23]. Air pressure inside a building is sometimes lower than pressure in the soil under the foundation. Because of this difference in pressure, a building acts as a vacuum, drawing radon inside from the soil. Typical cracks and openings include joints, where the floor meets the wall, expansion joints in the floor, openings on the floor for pipes and wires, and hollow masonry walls that penetrate the floor. Thus, the circulation of fresh air in the house as well as in the basement floor position of the house contribute to the inhalation of radon from residents [24].

Although there is no known safe level of radon, the EPA has established 4 pCi/L as an acceptable level. Since 1998, continuing efforts by federal and state agencies, including health education campaigns and announcements, encourage radon home testing and they have led to some improvement in radon awareness. Written surveys completed by 692 patients at two primary care clinics in the Minneapolis area revealed that only 24.7% had ever tested their home. Testing rate was higher with greater income and education level and at homes without smoking [22].

Effects on Health

The intoxicated environment at work or at home is called *sick house syndrome (SHS)* and it has been accepted as a disease and a social problem since the 1990's. SHS is a distinctively Japanese concept that generally indicates various health disturbances induced by indoor environmental pollution. Initially, the name of the syndrome was "sick building syndrome." The classification of this entity includes: type 1 (symptoms of chemical intoxication), type 2 (symptoms developed possibly due to chemical exposure), type 3 (symptoms developed not because of chemical exposure, but because of psychological or mental factors), and type 4 (symptoms developed due to allergies or other diseases) [20].

Symptoms are mainly from the central nervous system, the cardiovascular or respiratory system, and also from the skin or mucous membranes, muscle or bones, central nervous system the gastrointestinal tract and possibly from other systems. They include mucous membrane irritation, skin symptoms, headache, and dizziness. All except skin symptoms characteristically improve within a few hours after the individual leaves the "suspected building." The symptoms were, according to their frequency, headache, cough, difficulty in breathing, and itchy eyes [20].

Since the disorder cannot be detected by specific laboratory testing, diagnostic criteria have been developed. The diagnostic criteria include: [20] 1. The cause of the onset of a disease relates to a move to a new house or building, to the reconstruction of a house or building, and the use of new or different daily toiletry necessities, [23] 2. symptoms appear within the particular house/building, [24] when the patient leaves the house or building, symptoms improve or disappear, [22] 3. when indoor environmental pollution is detected, [25] symptoms developed because of psychological or mental factors, and [26] 4. symptoms developed are due to allergies (asthma, dermatitis) or other diseases. When patients met all of the criteria 1–3, they were diagnosed as suffering from SHS [22].

The pathogenesis of SHS has not been clearly elucidated. The mechanism related to indoor chemical substances, intoxication and chemical intolerance, allergies, and/or psychological factors has been suggested [20].

Acute poisoning with formaldehyde causes blood acidity with resulting vision disorders and blindness, hypothermia, and in serious cases coma and death. Indoor air formaldehyde concentrations would not usually distribute beyond the respiratory epithelium and therefore, direct effects are limited. However, systemic effects have been observed in animal models which may be due to secondary effects as airway inflammation and irritation of the eyes and the upper airways [23].

Formaldehyde is metabolized to formate, which is incorporated in tissue components and it is excreted in the urine or oxidized to carbon dioxide. Using a specific method, the formaldeyde concentration in air exhaled was found at levels up to 1.7 ppb; this figure may be higher shortly after smoking a cigarette. Estimated deposition in the upper airways and by product formation was similar in children and adults. However, the concentration detected in most cases was below 0.5 ppb [23].

Lung cancer is the leading cause of cancer death in the United States. The link between smoking and lung cancer is well known and clearly established. Radon gas, however, is the most important known environmental carcinogen, other than tobacco smoke. Approximately 18,600 lung cancer deaths per year can be attributed to residential radon exposure. Case-control studies have shown that cumulative radon exposure (time spent in the house times radon dosimetry) is associated with lung cancer risk. The Environmental Protection Agency (EPA) estimated that 14.4% of lung cancer deaths in the United States are radon related among nonsmokers, while a total of 26% are thought to be radon related. Because tobacco smoke and radon have synergistic effects, smokers who exposed to radon are particularly vulnerable [22].

Epigenetic Mechanisms of Intoxification

Formaldeyde participates in DNA adduct formation. Additionally, exposure response relationship is not nonlinear. Symptom related genetic polymorphisms have not been identified [23]. It is also a genotoxic chemical due to its covalent binding to DNA, causing DNA monoadducts, DNA–DNA crosslinks, and DNA glutathione cross-links that can cause mutations and DNA strand breaks, chromosomal aberration, micronucleus formation, and sister chromatid exchange. In the mucus layer, $CH_2(OH)_2$ diffuses into the epithelial cells which reacts with glutathione proteins, DNA, and RNA. The glutathione adduct is oxidized by the formaldeyde dehydrogenase to the formate adduct [23].

Repair of the FA–DNA mono-adducts include the base excision repair (BER) pathway, and the nucleotide excision repair (NER) pathway, which may repair the intra-strand cross-links [23].

Formaldehyde may play a role in induction of cancer. The detoxification process from formaldehyde is based on the alcohol dehydrogenase 5 (*ADH5*) gene and the *FANCD2* protein involved in the repair of FA–DNA cross-links. The double deletion experimental models have a profound decrease in survival, induced pancytopenia, reduced bone marrow cellularity (including hematopoietic stem and progenitor cells) and colony formation at cultivation of spleen hematopoietic stem cells [23].

Outdoors Pollution

The Chemicals, Their Forms, and Their Distribution in the Environment

Air pollution is responsible for a high burden of environment related diseases in Europe. Recent estimates indicate that 20 million Europeans suffer from respiratory problems due to polluted air. Air pollutants with strongly indicative respiratory health effects are PM (PM2.5, PM10), ozone (O_3), nitrogen oxides (NO, NO_2), and sulfur dioxide (SO_2). In addition, chemicals like polyaromatic hydrocarbons (PAH) and benzene from combustion processes contribute to toxicity and potential health effects [25].

The Environmental Protection Agency (EPA) lists six pollutants representing harmful agents for to public health. These include carbon monoxide (CO), lead (Pb), nitrogen dioxide (NO_2), ozone (O_3), particle pollution, and sulfur dioxide (SO_2). Applied measures were obtained from the EPA for CO, Pb, NO_2, O_3, SO_2 and fine PM being less than 2.5 μm ($PM_{2.5}$) [26].

NO_x compounds are produced by combustion in high temperatures (internal combustion engines, electricity production industries), thunders, and degradation of nitrogen containing products. NO in high concentrations causes irritation of the lung alveoli. It can also substitute iron in heme and thus inhibit oxygen transportation. NO_2 can also cause irritation of the lung epithelium and decrease resistance to infections [26].

Ground-level "*bad*" *ozone* forms when nitrogen oxide gases from vehicle and industrial emissions react with volatile organic compounds (carbon-containing chemicals that evaporate easily into the air, such as paint thinners). In the troposphere, near the Earth's surface, the natural concentration of ozone is about 10 parts per billion (0.00001%). According to the Environmental Protection Agency, exposure to ozone levels of greater than 80 parts per billion for 8 h or longer is unhealthy. Such concentrations occur in cities during periods where the atmosphere is warm and stable [26].

The so-called "bad ozone" in the stratosphere contributes to environmental pollution. Short-term exposure to O_3, one of the most potent single oxidants in the ambient air has been associated with decreased heart rate variability, ventricular arrhythmia, ischemic heart disease, and cardiovascular mortality. O_3 has been shown to produce inflammatory reactions in the respiratory tract as well as systemic inflammation and a procoagulant status. The harmful effects can include throat and lung irritation or aggravation of asthma or emphysema. In addition, there was found a positive association between O_3 levels and fasting total homocysteine (tHcy). However, the results on the association of O_3 with tHcy did not show a consistent pattern [27].

Peroxyacetyl nitrate (PAN) is an oxidant more stable than ozone. Hence, it has capabilities of long range transport that is higher than ozone. It serves as a carrier of oxides of nitrogen (NOx) into rural regions and it causes ozone formation. It is a secondary pollutant present in photochemical smog. It is thermally unstable and decomposes into peroxy ethanoyl radical and nitrogen dioxide gas. It is a lachrymatory substance.

Sulfur dioxide is produced by fossil fuels that contain S and by anaerobic combustion of sulfur compounds performed by microorganisms in the soil and water as well as from volcano activity. It causes irritation of the lungs, increased secretions of the mucous and, in vulnerable subjects, coughing and breathing difficulties. The increased mortality from this chemical is due to its dilution within PM.

Hydrocarbons are produced from the combustion of fuels without catalytic agents, car exhausts, industries, and solvents. Due to degradation of by products in the presence of oxygen, including POPs, the side effects of these pollutants have already been described.

Carbon dioxide is not considered a direct harmful pollutant and it is the final and ubiquitous product of a complete combustion in the presence of oxygen. It is produced by combustion of fossil fuels, combustion of biomass, human activities, and the degradation of plant materials. It is used for photosynthesis and it is absorbed by organisms of the soil. However, it contributes as a solvent to the degradation of ores and it is diluted in the sea and land water. Thus, CO_2 completes a cycle of production use that keeps the concentration in the atmosphere in a relatively stable range [26].

Radiation is the emission or transmission of energy in the form of waves from space. Most ionizing radiation originates from radioactive materials is naturally present in the environment, since most rocks and soil have small concentrations of radioactive materials. Since this radiation is invisible and not directly detectable by human senses, instruments such as Geiger counters are usually required to detect its presence. In some cases, it may lead to secondary emissions of visible light upon its interaction as in the case of Cherenkov radiation and radio-luminescence.

Ultraviolet radiation, of wavelengths from 10 to 125 nm, ionizes air molecules, causing their absorption by air and by ozone (O_3) in particular. Ionizing UV therefore does not penetrate earth's atmosphere to a significant degree. Although present in space, this part of the UV spectrum is not of biological importance, because it does not reach living organisms on earth.

There is a zone radiation of the atmosphere, known as the ozone zone, that not only absorbs 98% of non-ionizing, but also the dangerous UV-C and UV-B radiation. This so-called ozone layer starts at around 20 miles (32Km) and extends upwards. Part of the radiation is within the ultraviolet spectrum, that does not reach the ground, is non-ionizing, but it is biologically hazardous due to the ability to produce single protons and to cause electronic excitation in biological molecules. In contrast, visible light and longer-wavelength electromagnetic radiation, such as infrared, microwaves, and radio waves, consists of photons with too little energy to cause damage and thus this radiation is far less hazardous per unit of energy. About one third of UVA, with a wavelength of 400 nm, penetrates into the epidermis to a depth of 0.1 mm in vivo, whereas about 99% of UVB at 300 nm gets into only 0.03 mm in the epidermis.

Exposures to traffic-related air pollution (TRAP) can be particularly high in transport microenvironments despite the short duration typically spent there. There is a mounting body of evidence that suggests that this is especially true for fine (2.5 μm) and ultrafine (100 nm, UF) particles. Professional drivers, who spend extended periods of time in transport microenvironments due to their job, manifest high exposures [28].

Concentrations of many air pollutants such as 1,3-butadiene, benzene, toluene, ethylbenzene, xylene, nitrogen oxides, carbon monoxide, black carbon, and both fine (2.5 μm, known as PM2.5) and ultrafine (0.1 μm, UF) particles are elevated, albeit to different extents, on and near roadways. These pollutants originate from incomplete combustion of diesel, gasoline and other fuels. Resuspended road dust, mechanical wear and tear of brake pads, and other vehicle components can also contribute to increased particle levels and affect their chemical composition. While several pollutants can cause negative health effects following on or near road exposure, fine and UF particles arguably attract most interest at present, due the increasing evidence that links them with negative respiratory, cardiovascular, and neurological effects [28].

However, considering as pollutants a wider point of view, the *electro- magnetic fields* in the cities, and environmental noise are included. Nowadays, every European citizen is exposed to electromagnetic fields (EMF), which can be characterized in terms of their frequency and amplitude. Neither experimental studies on animals nor epidemiological studies have any conclusive results. On the basis of current scientific knowledge, it is not possible to answer "yes" or "no" to the question of whether the use of mobile phones poses an increased risk of cancer [24].

Environmental noise is probably the environmental factor that affects the largest number of Europeans. The main health risks of noise are annoyance, interference with social behavior as well as speech communication, sleep disturbance and its consequences, cardiovascular effects, hormonal responses, and poorer performance at work or at school [24].

Effects on Health

From the wide range of atmospheric pollutants, PM side effects are of major importance due to the fact that all chemical pollutants could dilute within these particles. PM and especially small particles with a diameter of less than 2.5 μ (PM2.5) are

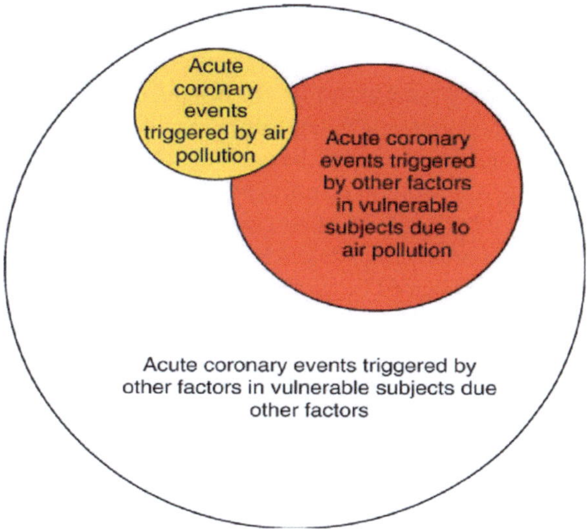

Fig. 5.15 The percentage of acute coronary events related to air pollution [30]

associated with increased mortality, especially from cardiovascular and cardiopulmonary diseases [25].

In 2012, an estimated 3.7 million people worldwide died prematurely due to the effects of ambient air PM pollution. Cohort studies in the United States and Europe have shown associations between exposure to traffic PM and both cardiopulmonary-diseases and cancer [29] (Fig. 5.15).

The association between PM and heart disease was noted in the mid-90s of the last century, when epidemiological evidence showed an association between air pollution and hospital admissions due to cardiovascular disease. Nowadays, epidemiological studies have demonstrated coherent associations between daily changes in concentrations of ambient particles and cardiovascular disease mortality and stroke.

Data showed that there is a 12% increased risk for cardiovascular disease mortality in association with a 10 Ag/m^3 PM2.5 increase in annual averages. This increase was due to ischemic heart disease with an increased risk of 18% per 10 Ag/m^3 PM2.5 and dysrhythmia with an increased risk of 15% per 10 Ag/m^3 PM2.5 [30].

Atherosclerosis is a complex chronic inflammatory process resulting from the interaction between modified lipids, inflammatory cells such as monocytes, T-lymphocytes, and cellular elements of the endothelial walls including smooth muscle cells and endothelial cells. It is characterized by endothelial dysfunction and by subendothelial accumulation of lipids, inflammatory cells, and fibrous tissues. Furthermore, these plaques are a dynamic system and the vulnerable ones are prone to rupture. They are characterized by active inflammation, a thin cap with a lipid core, fissures, endothelial denaturation with superficial platelet aggregation, and an associated vascular stenosis larger than 90% [30].

Mechanism of Intoxification

The principal mechanism for the health effects of particular matter is induced pulmonary and systemic inflammation. Studies show that particles deposit in the lung alveoli and they lead to activation of cytokines by alveolar macrophages and epithelial cells that result in recruitment of inflammatory cells and bone marrow stimulation. Systemic inflammation causes increased plasma viscosity through increased levels of fibrinogen, interleukin 6, and C-reactive protein. Endothelial dysfunction is not only induced by particle exposures, but also from small increases in blood pressure that may occur in association with elevated concentrations of ambient particles. These observations apply in healthy subjects, healthy elderly subjects as well as subjects with underlying cardiovascular disease. Accelerated atherosclerosis and altered cardiac autonomic function may be part of the pathophysiological pathways [30]. Particle injury might be related to the ability of ultra fine particles to effectively promote inflammation and increase blood clotting, which may play a role in cardiovascular mortality [28].

Mild hyperhomocysteinemia is independently associated with an increased risk of cardiovascular disease. Air pollution exposure induces short-term inflammatory changes that may be determined by the levels of homocysteine, which could be a sensitive biomarker [27]. Epidemiologic investigations have demonstrated that high plasma levels of homocysteine is an independent risk factor for vascular disease including coronary artery, cerebrovascular, and peripheral occlusive disease [27]. Ambient or occupational exposure results in the above described health effects not only by enhanced coagulation and inflammatory processes but also by the production of ROS [31]. Although few studies address the potential that exposure to air pollution could alter DNA methylation, environmental particulate exposure has been linked also to the increased risk of lung cancer [24].

Of all biological effects attributed to transport air pollution, those occurring in the respiratory system are probably the most extensively described. Recent work has reported that short-term exposure to fine and ultra fine particles was linked to small decreases in lung function and airway pH to increases in biomarkers of inflammation. Ultra fine particles were better associated with these effects than fine particles, but in both cases, there was often no evidence of a statistically significant effect [28].

Finally, Ultra fine particles are small enough to reach the brain, where they can cause oxidative stress leading to deleterious neurologic effects. One hour of exposure to diesel exhaust has been shown to result in functional changes in the brain; although difficult to attribute to a single pollutant or particle size range. A role for such processes in the causation and progression of neurodegenerative diseases is plausible, although there is a scarcity of data and there is a need for long-term studies from which to draw firmer conclusions. Even less is known about acute neurologic effects following exposures in transport microenvironments [28].

Epigenetic Mechanism of Intoxification

Ultra fine and fine particles can cause oxidative DNA damage in humans and this is a mechanism of carcinogenesis and mutagenesis. Other air pollutants, such as benzene, are also implicated. Although the long term consequences of pollutant mediated oxidative DNA damage in professional drivers is not well documented, the elevated incidence of several types of cancer in this group have been ascribed to TRAP exposure. Moreover, it should be noted that the International Agency for Research on Cancer (IARC) has recently classified diesel exhaust as carcinogenic to humans, and gasoline exhaust as possibly carcinogenic to humans [28].

Analyses of H3K9ac, H3K9me3, H3K27me3, and H3K36me3 levels in blood revealed that only H3K27me3 was significantly associated with exposure to transportation pollution. Specifically, H3K27me3 was 3.3% higher in men than in women, 3.0% higher in office workers than in truck drivers and 3.2% higher in habitual smokers compared to non-smokers. Office workers and truck drivers experienced similar levels of ambient PM_{10} exposures, whereas truck drivers had significantly higher personal levels of $PM_{2.5}$, and elemental component exposures [29].

H3K27me3 is associated with gene inactivation maintained by histone methyltransferases such as zeste 2 (EZH2). Inflammation can induce a decrease in H3K27me3 levels, a pre-mark in normal cells during carcinogenesis. Studies have also demonstrated that air pollutants can modify immune response, which in turn increases susceptibility to pathogens and worsens pre-existing diseases. Specifically, H3K27me3 has been shown to play a role in the immune system's response to cancer. H3K36me3 facilitates the DNA mismatch repair during the cell cycle, and loss of H3K36me3 has been associated with increased risk of death from renal and liver cancers [29].

The discrepancy between findings in office workers and drivers could be explained by the so called "ceiling effect," meaning biological adaptation to traffic-derived pollutant exposures. Thus, the non significant findings in truck drivers may be driven, at least in part, by the fact that truck drivers' exposure level was on average 32% higher. Similarly, exposures to PM_{10} in men were on average 17% and 35% higher than those in women. Thus, a similar disparity in air pollution exposure between men and women, and biological "ceiling" effect on histone H3 modifications, may also explain our observed sex differences in associations between traffic-derived air pollutants and histone modifications [29].

A study, using primary bronchial epithelial cells from non-smokers, has shown dramatic alterations in miRNA expression following treatment with diesel exhaust particles (197 of 313 measured were up or down-regulated more than 1.5-fold), and suggested that the alterations were associated with inflammatory response pathways. Significantly lower peripheral blood NOS_2 promoter methylation was observed in steel plant workers with exposure to PM_{10}. In addition, this same study reported significantly reduced methylation of LINE-1 and Alu repeat elements in blood DNA associated with long term PM_{10} exposure. A recent report examining the effect of ambient particulate pollutants on repeat element methylation, in subjects from the Boston area, found significantly decreased LINE-1 methylation following recent exposure to higher black carbon [24].

The characteristic ability of both mutagenic and non-mutagenic compounds to impact on selective pressure of cellular clones is critical for functional epigenetic alterations. Reduced fidelity of maintenance methyltransferases with aging is one potential explanation for age-related decreases in methylation; while age related increases in methylation could potentially reflect the accumulation of stochastic methylation events over time. Another hypothesized mechanism is a spreading process whereby CpG methylation within the gene moves into the promoter due to the loss of protective boundary elements, which normally protect promoters from CpG-island methylation. Alternatively, for example, in aging, a spreading process subsequent to stochastic methylation alteration (in pathologically normal tissue) without functional consequences for gene expression, may allow the initiation of a spreading process that eventually confers an altered expression phenotype. Yet, these mechanisms do not fully explain the apparent sequence context specificity of these observations, thereby suggesting that the age related alterations of methylation observed may be influenced in part by targeted mechanisms that are yet to be fully explained [24].

Although radiation was thought to function by DNA damage in the form of large deletions and in some cases point mutations, it may also be pathogenic by epigenetic alterations. Silencing of *Cdkn2a* (encoding p15ink4b) was detected in lung tumors of rats after exposure to ^{239}Plu. In human lung adenocarcinoma, *CDKN2A* methylation occurred more often in workers compared to non-worker controls of the Russian MAYAK weapons-grade plutonium plant and the prevalence of this methylation exhibited a dose response with radiation internal exposure dose. Methylation of *CDKN2A* has also been linked to produced by radiation exposures and murine models of radiation-induced lymphoma have also demonstrated hypermethylation of *Cdkn2b* [12].

Alcohol

The increased consumption of alcohol is known to be related to liver injury and cancer. Alcohol is known to interfere with folate absorption in the intestine. Liver releases the folate in order to supply tissues. Folate deficiency can be considered a toxicant with the potential for epigenomic alterations. Moreover, by interference with several steps of methionine metabolism, alcohol leads to inhibition of the classical pathway for the production of *S*-adenosylmethionine (SAM) and to activation of betaine homocysteine-methyltransferase pathway, which acts as a compensatory pathway for the maintenance of SAM levels by folic acid use. Continuous exposure to alcohol could suppress the compensatory pathway and result in liver injury. Changes in methylation enzymes could be the etiology of many cancers related to alcohol overconsumption (International Agency for Research on Cancer 1988) [20].

More specifically, *folate deficiency* interferes with one-carbon metabolism on the establishment and maintenance of DNA methylation and potentially other epigenetic mechanisms. Methylation changes is a common pathway that toxicants induce epigenome-wide effects. One-carbon metabolism is the network of biochemical reactions essential to both DNA synthesis as well as protein and lipid biomolecules synthesis. The reactions involve the transfer of one carbon groups.

Folate (vitamin B_9), a central nutrient in this pathway, donates its methyl group for homocysteine remethylation to methionine. Subsequently, methionine is the methyl donor for all cellular methylation reactions with SAM. Other B vitamins (B_2, B_6, B_{12}) act as enzymatic cofactors in the network. Collectively, B vitamins, homocysteine, and methionine are important contributors to the maintenance of DNA integrity and DNA methylation. Additionally, the one-carbon pathway links nutrient availability with DNA integrity, but it also participates in detoxification processes by the production of glutathione and the glutathione transferase (GST) enzyme family [20] (Fig. 5.16).

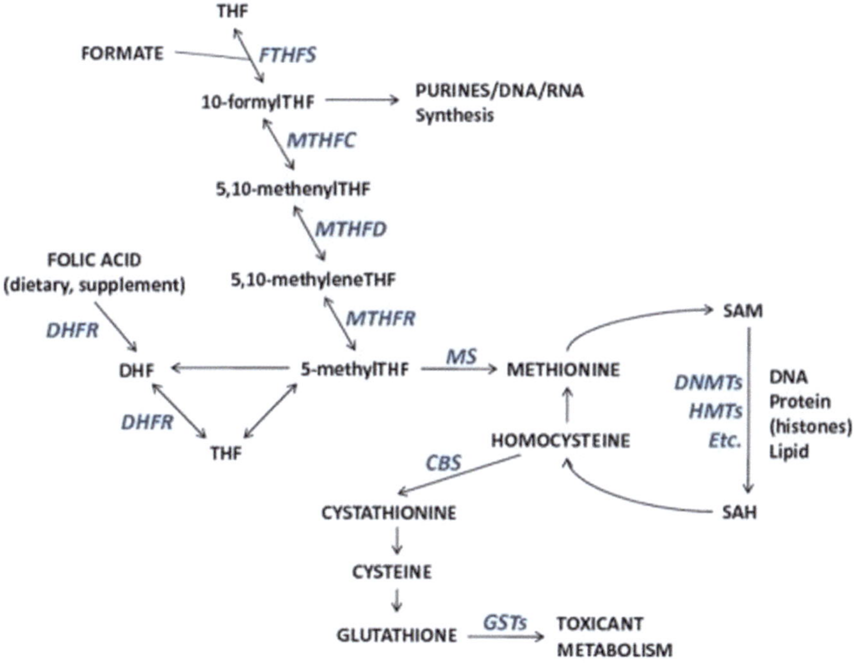

Fig. 5.16 Folic acid as a methyl-donor participates in DNA/RNA synthesis, SAM synthesis, and glutathione production

Effects on Health

Alcohol consumption increases colon cancer risk, although the association may be modified by dietary folate. The epigenetic changes responsible for colon cancer due to alcohol are associated with increased risk of developing LINE-1 hypomethylated and low association for hypermethylated LINE-1 tumors. Thus, subjects with high folate intake are less likely to develop colon cancers.

In addition, head and neck squamous cell carcinoma are also associated with reduced LINE-1 methylation and epigenetic alterations of specific gene regions. A study of breast tumor DNA methylation profiles demonstrated a significant, independent association between alcohol intake and tumor DNA methylation profile. Moreover, ethanol exposure was shown to selectively acetylate histone H3 at lysine 9 (H3K9) in primary culture of rat hepatocytes. Collectively, these studies suggest that a major carcinogenic mechanism of action of alcohol is interference with epigenetic regulation, in part by disruption of one-carbon metabolism [20].

Tobacco Smoking

Tobacco products provide a clear example of cancer caused by a life-style factor carcinogen exposure. There are over one billion smokers and hundreds of millions of smokeless tobacco users worldwide. Tobacco use is by far the most widespread link between exposure to known carcinogens [32].

More than 60 known carcinogens have been detected in cigarette smoke. In general, the stronger carcinogens are PAHs, nitrosamines, and aromatic amines that occur in lower amounts in cigarette smoke (1–200 ng per cigarette) than the weaker carcinogens, such as acetaldehyde (nearly 1 mg per cigarette), catechol, and isoprene. The total amount of carcinogens in cigarette smoke adds up to 1–3 mg per cigarette [32].

Nitrosamines, particularly the tobacco-specific nitrosamines 4-(methylnitrosamino) -1-(3-pyridyl)-1-butanone (NNK) and *N*-nitrosonornicotine (NNN), are the most prevalent strong carcinogens in unburned tobacco. The levels of NNK and NNN in smokeless tobacco products are hundreds to thousands of times higher than those of carcinogenic nitrosamines in any other consumer product designed for ingestion [32]. The synergic effect of other pollutants has been described. Tobacco smoke is often mixed with arsenic due to arsenic based pesticides in the soil. Arsenic has also been detected in the synthetic fluids of electronic cigarettes and the vapors they produce [33].

Effects on Health

The IARC Monograph entitled *Tobacco Smoke and Involuntary Smoking* (published in 2004), concluded that, based on worldwide literature, cigarette smoking increases the risk of all histological types of lung cancer. Additionally, it causes cancer of the oral cavity and the risk is amplified by smokeless tobacco and alcohol consumption. Cigarette smoking is also causally associated with laryngeal, oropharyngeal, and hypopharyngeal cancer and sinonasal or nasopharyngeal cancer. Cigarette smoking is causally associated with cancer of the esophagus, both squamous cell carcinoma and adenocarcinoma. Furthermore, cigarette smoking causes cancer of the stomach, liver, and pancreas, as well as transitional cell carcinomas of the bladder, ureter, and renal pelvis, and renal cell carcinoma. Finally, cigarette smoking is also a cause of squamous cell cervical carcinoma and myeloid leukemia as well as colorectal cancer. Environmental tobacco smoke (ETS) also causes lung cancer. Smokeless tobacco products are established causes of oral cavity cancer [32].

Mechanisms of Intoxification

Carcinogens are the key connection between nicotine addiction and cancer. Nicotine addiction is the major reason for which people smoke. However, nicotine is not carcinogenic by itself. Each cigarette or dip of smokeless tobacco contains a mixture of carcinogens, tumor promoters, and co-carcinogens. Most tobacco carcinogens require metabolic activation to exert their carcinogenic effects. Thus, they are in a competing balance with the host's detoxification pathways [32].

Tobacco carcinogens and their DNA adducts are the mechanism for the development of cancer in smokers [33]. If their formation is inhibited or blocked, so is carcinogenesis. If DNA adducts escape cellular repair mechanisms and persist, they may lead to permanent mutations. Detoxification mechanisms interfere with cells with damaged DNA, which are removed by apoptosis or programmed cell death. If not, a permanent mutation could occur. However, mutations of critical regions of an oncogene or tumor suppressor gene can lead to activation of the oncogene or deactivation of the tumor suppressor gene. Multiple events of this type lead to aberrant cells with loss of normal growth control and ultimately to cancer [32] (Fig. 5.17).

Cancer is a chronic disease and there is a 20–40-year interval from the time of exposure of an individual to a chemical or viral carcinogen until the clinical detection of a tumor. By the time a tumor is apparent, cancer cells have acquired the ability to divide, to invade adjacent cellular architectures, to metastasize and to kill the host. DNA damage can be caused by two categories of carcinogens: those that result from exogenous agents such as chemicals, viruses, and irradiation and those caused by reactive molecules generated by normal cellular processes. Normal

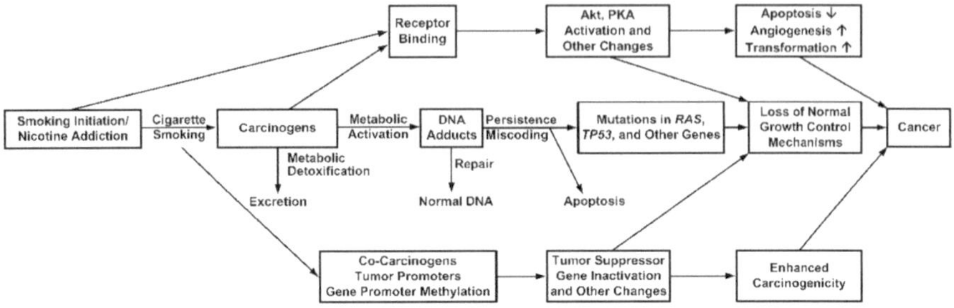

Fig. 5.17 The mechanisms of development of cancer due to smoking

cellular processes that damage DNA include the generation of reactive oxygen and nitrogen species, alkylation, depurination, and cytidine deamination [32].

Besides cancer, several epidemiologic studies have demonstrated the contribution of smoking to increased risk of cardiovascular disease such as myocardial infarction, stroke, and other adverse effects on cardiovascular function. In healthy subjects, tobacco smoking is a common cause of persistent low-level inflammation and smoking also has been associated with elevated levels of homocysteine (tHcy). Studies have demonstrated that high plasma levels of tHcy are an independent risk factor for vascular disease, including coronary artery, cerebrovascular and peripheral occlusive disease. The inflammation induced is accompanied by increased levels of C-reactive protein, fibrinogen, and interleukin-6 [34].

Several mechanisms have been suggested to account for the smoking-related increase in tHcy, including changes in plasma thiol redox status, inactivation of the enzymes of homocysteine remethylation, such as methionine synthase, reduction of nutrients and vitamins and lower levels of plasma folate, vitamin B12, and plasma pyridoxal 5-phosphate. Oxidative stress and endothelial dysfunction, which are enhanced in subjects with hyperhomocysteinemia, have been associated with both cigarette smoking and exposure to air particles [35].

Executive Summary

Persistent Organic Pollutants

- POPs, also known as "forever chemicals," is a general term for carbon-containing compounds that share similar properties.
- Resistance to photolytic and biological and chemical degradation, semi-volatile properties responsible for transportation to kilometers, persistence in the environment, lipophilic which accumulate in the fat of living organisms crossing along the food chain in ecosystems, the food chain in ecosystems, bioaccumulate by the entero-hepatic cycle, biomagnify.

- *The Stockholm Convention (2009, 2012)* on POPs banned 12 substances, "the dirty dozen," which included eight chlorinated pesticides (aldrin, dieldrin, endrin, mirex, chlordane, heptachlor, DDT, and toxaphene), two industrial chemicals (PCBs and HCB), and two by-products (PCDFs and polychlorinated dibenzo-p-dioxin)
- Effects on health: reproductive, developmental, endocrine, immunologic, and carcinogenic effects
- PAHs: byproducts, the major source of production is incomplete combustion of organic materials such as fossil fuels and petroleum, PAHs described: Halo-PAHs.
- PBDEs: flame retardants
- PCBs: pesticides, dielectric fluids in capacitors and transformers, hydraulic fluids, building materials, additives and plasticizers in paints, plastics and rubber sealants, copying paper, and adhesives
- Chlorinated compounds
- PBDEs: flame retardants
- CFCs and hydrochlorofluorocarbons (HCFCs): freon
- Polychlorinated naphthalene (PCN)
- Perfluorinated compounds (PFC)
- Trichloroethylene (TCE): industrial solvent
- Genotoxic
- Mutagenic due to DNA-adducts and oxidative stress as well as photoreaction processes, epigenetic changes in aryl hydrocarbon receptor (AHR)
- Global DNA hypomethylation
- Endocrine disruptors
- WHO and the United Nations Environment Program and advice from experts in the field regarding protection from POPs

Air Pollution

- Indoor pollution: asbestos, biological pollutants, carbon monoxide (CO), formaldehyde, lead (Pb), pesticides, radon (Rn), respirable particles, tobacco smoke and second-hand smoke, and volatile organic compounds (VOCs) such as formaldehyde, benzene, and tri- and tetra-chloroethylene
- Sick house (building) syndrome (SHS): described since 1990's, mucous membrane irritation, skin symptoms, headache, and dizziness. All except skin symptoms characteristically improve within a few hours after the individual leaves the "suspected building." The symptoms were, according to their frequency, headache, cough, difficulty in breathing, and itchy eyes
- Acute poisoning from formaldehyde: blood acidity with resulting vision disorders and blindness, hypothermia and, in serious cases, coma and death
- Formaldehyde: genotoxic, mutagenic due to DNA-adducts
- Outdoors pollution: carbon monoxide (CO), lead (Pb), nitrogen dioxide (NO_2), ozone (O_3), particle pollution, and sulfur dioxide (SO_2)
- Measures of criteria pollutants were obtained from the EPA for CO, Pb, NO_2, O_3, SO_2, and fine PM measuring less than 2.5 μm in aerodynamic diameter ($PM_{2.5}$)

- PM of major importance due to the fact that all chemical pollutants could dilute within these particles. PM and especially small particles with a diameter less than 2.5 μm (PM2.5) are associated with increased mortality, especially from cardiovascular and cardiopulmonary diseases
- Radon
- Mechanism of toxicity: inflammation
- Oxidative DNA damage, higher H3K27me3 associated with gene inactivation, hypomethylation in LINE-1

Alcohol
- Toxification: interferes with folate absorption and methionine metabolism thus DNA metabolism, DNA methylation
- By epigenetic mechanisms predisposes to the development of cancer: colon, head, and neck squamous cell carcinoma, breast cancer

Tobacco
- 60 known carcinogens. The stronger: PAHs, nitrosamines, and aromatic amines that occur in lower amounts in cigarette smoke (1–200 ng per cigarette) Weaker: acetaldehyde (nearly 1 mg per cigarette), catechol, and isoprene
- Cigarette smoking increases the risk of all histological types of lung cancer. Additionally, it causes cancer of the oral cavity, laryngeal, oropharyngeal, and hypopharyngeal cancer and sinonasal or nasopharyngeal cancer, cancer of the esophagus, both squamous cell carcinoma and adenocarcinoma, cancer of the stomach, liver, and pancreas, as well as transitional cell carcinomas of the bladder, ureter, and renal pelvis, and renal cell carcinoma
- Contains a mixture of carcinogens, tumor promoters, and co-carcinogens
- Tobacco carcinogens and their DNA adducts, DNA alterations, mutations of an oncogene or tumor suppressor genes
- Besides cancer, cardiovascular disease

References

Persistent Organic Pollutants

1. Encarnação T, Pais AA, Campos MG, et al. Endocrine disrupting chemicals: impact on human health, wildlife and the environment. Sci Prog. 2019;102(1):3–42.
2. Lee D-H, Jacobs Jr DR, Porta M. Hypothesis: a unifying mechanism for nutrition and chemicals as lifelong modulators of DNA hypomethylation. Environ Health Perspect. 2009;117(12):1799–802.
3. Abelsohn A, Gibson BL, Sanborn MD, et al. Identifying and managing adverse environmental health effects: 5. Persistent organic pollutants. CMAJ. 2002;166(12):1549–54.
4. Heudorf U, Sundermann VM, Angerer J, et al. Phthalates: toxicology and exposure. Int J Hyg Environ Health. 2007;210(5):623–34.
5. Anders MW. Metabolism and toxicity of hydrochlorofluorocarbons: current knowledge and needs for the future. Environ Health Perspect. 1991;96:185–91.

6. Liang X, Yin N, Liang S, et al. Bisphenol A and several derivatives exert neural toxicity in human neuron-like cells by decreasing neurite length. Food Chem Toxicol. 2020;135:111015.
7. Ruiz-Hernandez A, Kuo C-C, Rentero-Garrido P, et al. Environmental chemicals and DNA methylation in adults: a systematic review of the epidemiologic evidence. Clin Epigenetics. 2015;7(1):55.
8. Hans L. Environmental pathology. Kapodistrian University of Athens.
9. Fu PP, Xia Q, Sun X, et al. Phototoxicity and environmental transformation of Polycyclic Aromatic Hydrocarbons (PAHs)—light-induced reactive oxygen species, lipid peroxidation, and DNA damage. J Environ Sci Health C Environ Carcinog Ecotoxicol Rev. 2012;30(1):1–41.
10. Ruiz-Hernandez A, Kuo C-C, Garrido PR, et al. Environmental chemicals and DNA methylation in adults: a systematic review of the epidemiologic evidence. Clin Epigenetics. 2015;7(1):55.
11. Sun H, Brocato J, Costa M. Oral chromium exposure and Toxicity Trichloroethylene (TCE), dichloroacetic acid (DCA), and trichloroacetic acid (TCA). Curr Environ Health Rep. 2015;2(3):295–303.
12. Stragierowicz J, Sitarek K, Grobelski B, et al. The effect of prenatal exposure on disposition of hexachloronaphthalene in female Wistar rats and fetal compartment. Int J Occup Med Environ Health. 2018;31(5):685–95.
13. Strandberg B, Julander A, Sjöström M, et al. Evaluation of polyurethane foam passive air sampler (PUF) as a tool for occupational PAH measurements. Chemosphere. 2018;190:35–42.
14. Huen K, Yousefi P, Bradman A, et al. Effects of age, sex, and persistent organic pollutants on DNA methylation in children. Environ Mol Mutagen. 2014;55(3):209–22.
15. Dekant W. Toxicology of chlorofluorocarbon replacements. Environ Health Perspect. 1996;104(Suppl 1):75–83.
16. Blossom SJ, Gilbert KM. Epigenetic underpinnings of developmental immunotoxicity and autoimmune disease. Curr Opin Toxicol. 2018;10:23–30.
17. Kabir ER, Rahman MS, Rahman I. A review on endocrine disruptors and their possible impacts on human health. Environ Toxicol Pharmacol. 2015;40(1):241–58.
18. Lee D-H, Jacobs Jr DR, Porta M. Hypothesis: a unifying mechanism for nutrition and chemicals as lifelong modulators of DNA Hypomethylation. Environ Health Perspect. 2009;117(12):1799–802.
19. Kurita H, Schnekenburger M, Ovesen JL, et al. The Ah receptor recruits IKKα to its target binding motifs to phosphorylate Serine-10 in Histone H3 required for transcriptional activation. Toxicol Sci. 2014;139(1):121–32.
20. Choudri BS, Charabi Y, Ahmed M. Pesticides and herbicides. Water Environ Res. 2018;90(10):1663–78.
21. Fenga C, Gangemi S, Costa C. Benzene exposure is associated with epigenetic changes (review). Mol Med Rep. 2016;13(4):3401–5.
22. Christensen BC, Marsit CJ. Epigenomics in environmental health. Front Genet. 2011;2:84.
23. Okey AB. An aryl hydrocarbon receptor odyssey to the shores of toxicology: the Deichmann Lecture, International Congress of Toxicology-XI. Toxicol Sci. 2007;98(1):5–38.

Air Pollution

24. Christensen BC, Marsit CJ. Epigenomics in environmental health. Front Genet. 2011;2:84.
25. Nielsen GD, Larsen ST, Wolkoff P. Re-evaluation of the WHO (2010) formaldehyde indoor air quality guideline for cancer risk assessment. Arch Toxicol. 2017;91(1):35–61.
26. Miyajima E, Tsunoda M, Sugiura Y, et al. The diagnosis of sick house syndrome: the contribution of diagnostic criteria and determination of chemicals in an indoor environment. Tokai J Exp Clin Med. 2015;40(2):69–75.
27. Environmental pathology, EU.

28. Hall ES, Connolly N, Jones DE, et al. Integrating public data sets for analysis of maternal airborne environmental exposures and stillbirth, air pollution. AMIA Annu Symp Proc. 2014;2014:599–605.
29. Baccarelli A, Zanobetti A, Martinelli I, et al. Air pollution, smoking, and plasma homocysteine. Environ Health Perspect. 2007;115(2):176–81.
30. Knibbs LD, Morawska L. Traffic-related fine and ultrafine particle exposures of professional drivers and illness. Environ Int. 2012;49:110–4.
31. Zheng Y, Sanchez-Guerra M, Zhang Z, et al. Traffic-derived particulate matter exposure and histone H3 modification: a repeated measures study. Environ Res. 2017;153:112–9.
32. Ozanne SE, Constância M. Mechanisms of disease: the developmental origins of disease and the role of the epigenotype. Nat Clin Pract Endocrinol Metab. 2007;3(7):539–46.
33. Bollati V, Marinelli B, Apostoli P, et al. Exposure to metal-rich particulate matter modifies the expression of candidate MicroRNAs in peripheral blood leukocytes. Environ Health Perspect. 2010;118(6):763–8.
34. Strandberg B, Julander A, Sjöström M, et al. Evaluation of polyurethane foam passive air sampler (PUF) as a tool for occupational PAH measurements. Chemosphere. 2018;190:35–42.

Alcohol, Tobacco Smoking

35. Christensen BC, Marsit CJ. Epigenomics in environmental health. Front Genet. 2011;2:84.
36. Wilson AS, Power BE, Molloy PL. DNA hypomethylation and human diseases. Biochim Biophys Acta. 2007;1775(1):138–62. (1)
37. Wogan GN, Hecht SS, Felton JS, et al. Environmental and chemical carcinogenesis. Semin Cancer Biol. 2004;14:473–86.
38. Coryell M, Roggenbeck BA, Walk ST, et al. The human gut microbiome's influence on arsenic toxicity. Curr Pharmacol Rep. 2019;5(6):491–504.
39. Baccarelli A, Zanobetti A, Martinelli I, et al. Air pollution, smoking, and plasma homocysteine. Environ Health Perspect. 2007;115(2):176–18.

Chapter 6
Endocrine Disruptors

Endocrine Disruptors

The Chemicals, Their Forms, and Their Distribution in the Environment

Endocrine disruptors EDs are environmental pollutants that mimic endogenous hormonal signals [1]. These substances interfere with hormone-dependent functions in the body such as embryonic development, production of sperm, control of the menstrual cycle, the onset of puberty, and cancers in hormone-dependent tissues [2].

Since World War II, more than 140,000 synthetic *chemical compounds* have been produced; about 1000–2000 new compounds are synthesized each year, and approximately 800 chemicals are known, or suspected, to interfere with the endocrine system. Some of them are persistent organic pollutants (POPs) such as polychlorinated biphenyls (PCBs), polybrominated diphenyl ethers (PBDEs), dioxins, furans, and pesticides [3].

These chemicals are widespread and omnipresent. They can be found in Bald eagles in Florida (1952), river otters in England (late 1950s), mink in Michigan (mid-1960s), pheasants, eagles, and ospreys in Sweden (1960s), fish consumers in Japan (1960s), herring gulls in Michigan (1970), western gulls in California (early 1980s), alligators in Florida (1980s), oysters in France (1980s), seals in northern Europe (1988), dolphins in the Mediterranean (early 1990s), and sperm counts in men worldwide (1992); in pharmaceutical residues from humans and farm animals. These are only a few of the areas that are driving public concerns about the impacts of unsafe and unhealthy ecosystems in human health [2].

Dangerous chemical compounds are ambient. They exist in the air we breathe, the food we eat, and the water we drink. They can be found in our everyday lives

A. Salavoura, *Chemical Environmental Pollutants and their Effect on Health*,
https://doi.org/10.1007/978-3-031-80271-3_6

through personal care products, household cleaning products, furniture, and children's toys. Every year, hundreds of new chemicals are produced and released into the market without being tested and they reach our bodies. Permanent exposure to those chemicals is far from being innocent [3].

People are exposed to chemicals by a variety of routes. Some of these chemicals are suggested to have properties of endocrine disruptors. We present few of them:

1. Oral consumption by food and water: Endocrine disruptors include PCBs, dioxins, perfluorinated compounds, DDT, commonly found in industrial waste or pesticide contaminated soil or groundwater. Leaching of chemicals like BPA, phthalates from food and beverage containers, and pesticides residues in food and beverages can also enter the human body by this route
2. Contact with skin and inhalation: Pesticides, commonly used in agriculture, or vector controls (DDT, chlorpyrifos, vinclozolin, pyrethroids). Additionally, it is also quite known that cosmetics, personal care products, antibacterials, sunscreens, medications contain phthalates, triclosan, parabens, and insect repellants and we apply on our skin on a daily basis
3. Intravenous route: Phthalates are commonly used in intravenous tubing enter the human body [3].
4. Biological transfer from the placenta or maternal milk [4].

As endocrine-disrupting compound is defined by the US Environmental Protection Agency (EPA) as "an agent that interferes with the synthesis, secretion, transport, binding, or elimination of natural hormones in the body that are responsible for the maintenance of homeostasis, reproduction, development and/or behavior" [4]. Although, this group of chemicals are quite heterogeneous, they can be classified in two categories:

1. Those that occur naturally (e.g., phytoestrogen: genistein and coumestrol) and
2. Those that are synthesized, such as synthetic chemicals used as industrial solvents or lubricants and their byproducts (e.g., polychlorinated biphenyls (PCBs), polybrominated biphenyls (PBBs), dioxins), plastics (e.g., bisphenol A (BPA)), plasticizers, pesticides (e.g., dichlorodiphenyltrichloroethane (DDT), fungicides (e.g., vinclozolin), and some pharmaceutical agents (e.g., diethylstilbestrol, DES) [4].

They can also be grouped according to their origin:

(i) Natural and artificial hormones (e.g., phytoestrogens, 3-omega fatty acids, contraceptive pills, and thyroid medicines).
(ii) Drugs with hormonal side effects (e.g., naproxen, metoprolol, and clofibrate).
(iii) Industrial and household chemicals (e.g., phthalates, alkylphenol-ethoxylate detergents, fire retardants, plasticizers, solvents, 1,4-dichloro-benzene, and PCBs).
(iv) Side products of industrial and household processes (e.g., polycyclic aromatic hydrocarbons, dioxins, pentachlorobenzene) [4].

In the last decades, the increasing incidence of some medical conditions, such as breast and prostate cancer, types 2 diabetes, obesity, or cardiovascular diseases in the industrialized world, cannot be explained only by genetic factors [3].

Environmental factors could probably provide an explanation for the increasing incidence of these multifactorial diseases. The results of different studies show that several of the chemical products released to the environment could interfere with various metabolic pathways, leading to alterations in development, growth, and reproduction [3].

Effects on Health

The increase in prevalence of many multifactorial diseases are related to the increase of endocrine disruptors [3].

1. *Cancer*: In Great Britain from 1978 to 2007, the overall age of standardized incidence rate of cancer has increased by 25%, an increase of 14% in men and a 32% in women.
2. *Diabetes*: The CDC reports the prevalence of diabetes incidence from 1980 to 2011 increased by 176%.
3. *Obesity*: Childhood obesity has more than doubled in children and quadrupled in adolescents in the past 30 years. In the United States, the prevalence of obesity in children aged 6–11 years increased from 7% in 1980 to nearly 18% in 2012. Similarly, in adolescents, the prevalence increased from 5% to nearly 21% over the same period of time (CDC 2014).
4. *Thyroid disease* is related to PCBs, BPA, TCDD, and PCDF; pentachlorophenol (source of the chemical hexachlorobenzene); HCB, PBDE and TBDEs); phthalates (DEHP, DnOP, DIDP, DNHP and DBP); pesticides: (DDT, HCB, methoxychlor, chlordane and endosulfan).
5. *Nervous system dysfunction*: PCBs, dioxins, DDT and related chlorinated pesticides and metabolites, heavy metals like mercury, and lead.
6. *Female reproductive system*: Precocious puberty, polycystic ovary syndrome, and premature ovarian failure as well as increased growth of endometrium and a higher risk of breast cancer. Puberty: It is observed from over the last century that the age of menarche has been diminished from 16 or 17 years to less than 13. This early onset of puberty may be associated with a greater prevalence of many other disorders, such as insulin resistance, metabolic syndrome, breast and reproductive system cancers. Primary ovarian failure: Studies showed that primary ovarian failure (POF) occurs in about 1% of the female population who are less than 40 years old. Irregularities in menstruation cycles: EDC may interfere with the hormonal regulation of the menstruation cycle causing irregularities, such as long cycles, which may reduce the ability to conceive. Polycystic ovary syndrome (PCOS): The disease is characterized by irregular menstruation cycles and hyperandrogenism and it is associated with a higher prevalence of obesity, insulin resistance, and metabolic abnormalities.

7. *Male reproductive system*: Malformation of reproductive organs: sexual organs, sperm anomalies, hypospadias, and ectopic testes. Sperm quality: A number of studies reported a decline in sperm quality in several countries, such as sperm count, proportion of normal sperm, semen volume, which might be expected to affect fertility.
8. *Fertility*: In men, fertility has decreased in the last decades, at least in some countries, like in Denmark, France, and the United Kingdom.

Mechanism of Intoxification

Endocrine disruptors could interfere with hormone circuits in a number of ways. Regarding the mechanism of action, the classification is shown in the diagram (Fig. 6.1) [4].

1. *EDs inappropriately bind to or activate hormone receptors.*

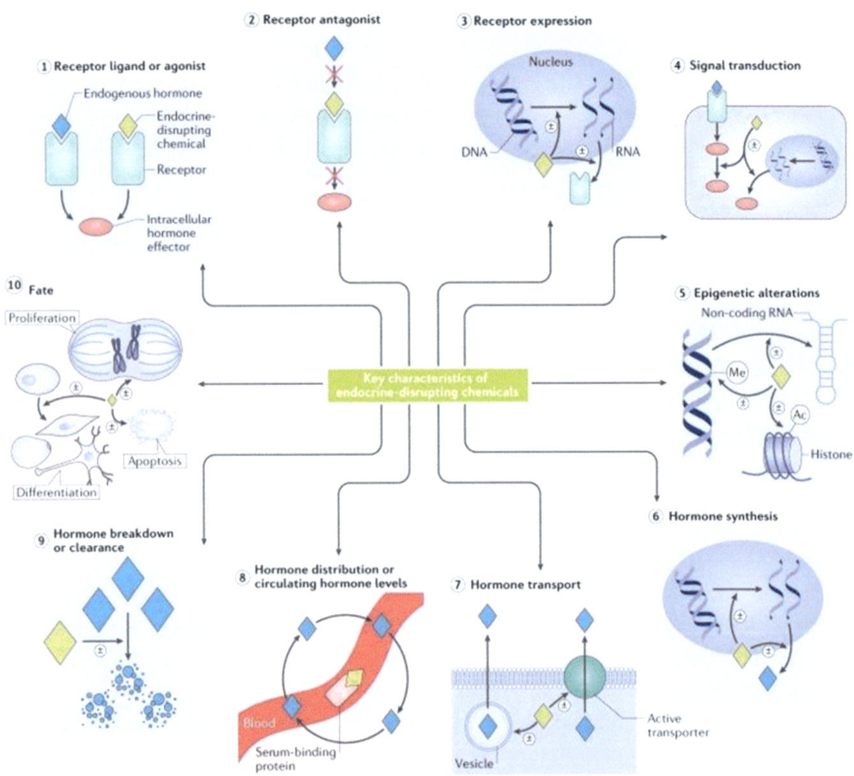

Fig. 6.1 Different mechanisms of action of endocrine disruptors

EDs can mimic the hormonal function after binding to its receptor. There are numerous examples of chemicals that cause adverse effects after binding to nuclear hormone receptors. *Nuclear receptors (NRs)* are a class of proteins which function as transcription factors that regulate gene expression. They are activated by hormones, such as steroid hormones, as well as other lipid-soluble signal molecules, such as retinoic acid, and thyroid hormones. Some of these NRs are considered orphans, as their specific ligand or physiological function remains unknown. So far, 48 nuclear receptors have been detected in humans by the sequencing of the human genome. Some EDs, such as alkylphenols, BPA, dioxins, furans, halogenated hydrocarbons and heavy metals, they can cross the cell membrane and they can bind directly to these receptors either activating them as agonists or inhibiting them as antagonists [48].

EDs that inappropriately activate the estrogen receptors (ERα and ERβ) during development; they increase the risk of infertility in both sexes as well breast and ovarian cancer in women and prostate cancer in men. For example, DDTE binds to ERα and ERβ receptors and it stimulates ER-dependent transcriptional activation and proliferation. The PCB can activate human thyroid hormone receptor-β-mediated transcription. DDT binds to the transmembrane domain and it enhances its stimulation.

2. *EDs can inhibit or block effects of endogenous hormones by acting as receptor antagonists.*

Although antagonism for membrane hormone receptors or intracellular hormone receptors can occur, research has focused on the antagonistic function of nuclear hormone receptors. For example, DDDE, an organochlorine pesticide, inhibits androgen binding to the androgen receptor (AR) and it also inhibits androgen-dependent transactivation of the AR in human and rat prostate cells. Other organochlorine pesticides (such as lindane and dieldrin, which are closely related to the organochlorine insecticide aldrin) also inhibit dihydrotestosterone binding to the AR. As androgens are key regulators of male sex differentiation during fetal development, disruption of androgen action by AR antagonism at a critical developmental window can permanently demasculinize male fetuses and it could lead to malformations of the genital tract.

3. *Alters hormone receptor expression*

As hormone receptors mediate hormone function, their pattern of expression dictates their response to hormone signals. Thus, the availability of binding receptors can determine both the concentration of produced hormones as well as the magnitude of the effect itself. EDs can modulate hormone receptor expression, internalization, and degradation. For example, di(2-ethylhexyl) phthalate decreases the expression of the mineralocorticoid receptor (MR) in the testis of adult mice. Furthermore, BPA alters the expression of estrogen, oxytocin, and vasopressin receptors in brain nuclei, and it reduces degradation of ERβ. The internalization of cell surface receptors is also disrupted by chemicals. For example, DDT prevents the internalization of the TSH receptor.

4. *Alters signal transduction in hormone-responsive cells*

Signal transduction in hormone-responsive cells is mediated by both membrane and intracellular hormone receptors and it could be altered by some EDs. Two classes of receptors are the most extensively studied in the field of endocrinology: the cell surface membrane receptors and nuclear steroid hormone receptors.

An example of cell surface receptor blocking is the BPA, which blocks low glucose induced calcium signaling in the isolated pancreatic glucagon secreting α-cells from adult male mice. Furthermore, in 2018, it was shown that chemicals in ultraviolet filters disrupted calcium signaling in human sperm. Some hormonal signals can use an alternative membrane receptor, for example, the G protein-coupled estrogen receptor GPER.

EDs can also affect signal transduction initiated by nuclear receptors. They interact with coregulatory complexes such as activators and repressors, which are a key part of the molecular machinery determining the downstream response to nuclear hormone receptor activation. The coregulatory factors for the steroid receptor coactivator family are among the most studied. For example, xenoestrogens (such as DES, PCBs, and BPA) induce the recruitment of GR receptors by concomitant binding to ERα and ERβ in a dose-dependent manner. BPA and its analogs also recruit SRC1 to thyroid hormone receptor-β. Another EDs, 4-methylbenzylidene camphor (which is used in ultraviolet filters), also increases the expression of nuclear receptors in female rat hypothalamus.

Furthermore, EDs can attenuate or potentiate hormone action by signal transduction. For instance, the fungicide tolylfluanid impairs insulin action by reducing insulin receptor substrate 1. Methoxy acetic acid potentiates ligand-activated transcription and progesterone receptor-mediated transcription in a manner dependent on MEK1 and MEK2 activity.

5. *Induces epigenetic modifications in hormone-producing or hormone-responsive cells*

Chemicals can exert permanent effects, especially during development and differentiation, by modifying epigenetic processes, including DNA methylation, histone modifications, and non-coding RNA expression. For example, the pesticide methoxychlor increases the expression of the DNA methyltransferase DNMT3B to hypermethylate DNA in the ovary of exposed rats. In addition, di(2-ethylhexyl) phthalate inappropriately demethylates MR DNA in the testis of male mice.

EDs can also change the expression of non-coding RNAs, as is seen in the case of PCBs, altering the developmental trajectories of hypothalamic miRNA expression in a sexually dimorphic manner. BPA and phthalates affect miRNA expression in placental and breast cancer cell lines. A study on non-coding RNAs found that BPA, and DES induced *HOX* antisense intergenic RNA (termed HOTAIR) in human breast cancer cells. In the presence of BPA and DES, the ER-binding region of the HOTAIR promoter was enriched by trimethylation on lysine 4 of histone 3

(H3K4) and by H3K4-specific methyltransferases, which are known to activate transcription. Moreover, BPA exposure enhances H3K4 trimethylation on genes, which are subsequently more sensitive to regulation by testosterone. Estradiol and DES also reduce trimethylation of lysine 27 on histone H3 in the developing uterus of rats.

6. *Alters hormone synthesis*

Hormone synthesis is regulated by both intracellular and distant endocrine feedback mechanisms. For example, thyroid hormone synthesis is controlled by a combination of TSH activation of the receptors of TRC, as well as intracellular cAMP, iodine uptake, and the activation of various enzymes. After transcription and translation, proteins and peptide hormones are packaged into secretory vesicles, where they can be stored. By contrast, steroid hormone synthesis, including the transformation of the prohormone to a hormone, occurs more or less simultaneously with hormone activity.

Some EDs are known to interfere with hormone synthesis. For example, perchlorate can block the uptake of iodine by the thyroid cells, thereby inhibiting thyroid hormone synthesis. Phthalates can reduce testosterone synthesis in the fetal rat testis, resulting in testosterone insufficiency. On the contrary, the herbicide atrazine apparently increases estrogen synthesis in peripubertal male mice, because it increases the activity of aromatase synthesizing estrogen from testosterone. Additionally, several pesticides block the JAK–STAT3 pathway to enhance the promoter usage and expression of CYP19, the gene that encodes the enzyme that converts testosterone to estrogen.

7. *Alters hormone transport across cell membranes*

Due to their lipophilic structure, steroid hormones (including estrogens, androgens, progestins, and adrenal steroids) can move through membranes passively. Other hormones (such as amine, peptide, protein, and thyroid hormones) must be selectively transported across membranes. These passive and selective transport processes can be disrupted by EDs. An example of impaired selective transport, by an exogenous chemical, is low-dose BPA, which reduces calcium entry into mouse pancreatic β-cells to reduce insulin secretion from vesicles. The fungicidal imidazoline modulates ions to enhance insulin secretion. Low-dose dichloro-diphenyl-dichloro-ethylene reduces the passive secretion of corticosterone from rodent adrenal glands.

8. *Alters hormone distribution or circulating levels of hormones*

Hormones typically circulate throughout the body with the blood at low concentrations, often in the parts per billion and trillion range. Depending on its chemistry, a circulating hormone is either transported "freely" (not bound to serum proteins), with or without conjugation (such as glucuronidation or sulfation), or it circulates bound to various proteins. EDs can alter hormone bioavailability by interfering with to the distribution of hormones in hormone-responsive tissues or with the circulation of hormones, which can lead to impaired active hormone delivery by target tissues. For example, BPA causes a concentration dependent decrease in circulating levels of testosterone in male rats and men and the pesticide malathion reduces testosterone levels in serum, testis, and ovaries in several animal models. When diethylstilbestrol is administered intravenously in men subjects, it causes a sixfold reduction of total and 20% of free testosterone. In the same time, a five fold reduction of estradiol is detected with a concurrent sevenfold increase in serum concentration of sex hormone-binding globulin.

9. *Alters hormone metabolism or clearance*

The various hormones (such as protein, peptide, steroid, or thyroid) are inactivated differently. EDs can alter the rates of inactivation, including the metabolic degradation or clearance of hormones, which could alter hormone concentrations and ultimately their activity. For example, after secretion, protein hormones (e.g., gonadotropins) are inactivated by proteases in the blood. On the contrary, steroid and thyroid hormones are metabolized by enzymes that render them hormonally inactive and they increase their water solubility so that they can be cleared rapidly from the bloodstream and excreted. A large number of chemicals could activate glucuronidases, which increase thyroid hormone clearance from the blood. Furthermore, sulfation by estrogen sulfotransferase, which reduces the rate of estrogen clearance from the blood, is inhibited by several PCB metabolites, for example hexachlorobenzene, the major metabolite of fungicides, and several phenol wood preservatives.

10. *Alters the fate of hormone producing or hormone-responsive cells*

Hormones affect tissue structure and organization by affecting the hormone producing tissue survival (e.g., cellular proliferation, migration, or differentiation) and death (i.e., apoptosis) during development and adulthood. In adulthood, many endocrine organs have a stable number of cells (including adrenals and pancreas), whereas other endocrine organs or hormone-responsive tissues depend on growth signals for normal function (such as the testicles to form sperm, the uterine endometrium, and vaginal lining). EDs can alter the number or function of cells in hormone producing or hormone responsive tissues by disrupting or promoting differentiation, proliferation, migration, or cell death [3].

For example, thyroid hormone controls cell proliferation and apoptosis in the developing cerebellum and PCBs can interfere with thyroid hormone signaling to cause abnormal development. Female mice exposed to benzene have increased mammary epithelial cell proliferation, which is observed even weeks after exposure ceases. In cultured human endometrial stromal cells, treatment with the antibacterial agent triclosan increases decidualization. In the brain exposure to a PCB during critical stages of development decreases the numbers of cells expressing ERα in adult female but not male rats. Furthermore, tebuconazole, a common fungicide, impairs embryonic migration, a process essential to placentation [4].

Over the last 30 years, many reports regarding endocrine disruptors have been published. Since the chemicals identified as endocrine disruptor are diverse and they do not seem to share specific similarities in structure and nature, it is quite hard to predict their mechanism of action. However, in a broad-term analysis, it can be seen that some EDs like dioxins, PCBs, PBBs, and pesticides contain halogen substitutions by chlorine and bromine, while others are just heavy metals or metalloids, which may possess with estrogenic activities or act as antiandrogens or thyroid hormone receptor agonists or antagonists [4].

Disturbances of glucocorticoid action have been linked to birth defects, mood and cognitive disorders, cancer, immune disease, allergic disease, metabolic dysfunction, and cardiovascular disease. Mineralocorticoids, which principally act by activating MR, can contribute to the development of hypertension and related cardiovascular pathologies such as myocardial hypertrophy and fibrosis [6].

The importance of GR and MR has been validated by *in vivo* studies. Deletion of GR and MR receptors led to perinatal death in transgenic mice. Partial loss of GR and MR function resulted in altered corticosteroid levels and neuroendocrine abnormalities. GR and MR are involved in a variety of physiological processes that may be disrupted by exposure to environmental contaminants [6]. Bioassays to test GR activity detected high activity levels in studies performed in the Netherlands, the United States, Czech Republic, Switzerland, and China [6].

The phrase "the dose makes the poison" is a traditional toxicological concept and it suggests that larger doses have a greater impact than lower ones. However, in the case of EDs, the dose does not make the poison. Very low doses of chemicals can have the same impact as high doses. At low doses, chemicals can bind to the receptors and they induce a biological response. In contrast, higher doses can saturate the same receptors and they could inhibit the relative pathways and as a result there is not a biological response. The timing of exposure can also be a crucial factor. The toxicity of EDs is a time dependent process, and exposure during critical stages of development may cause long term consequences [2].

From the above, it is obvious that chemicals considered as EDs exert their effect in very different ways which may be much more complicated than originally thought. EDs were originally thought to function primarily by direct interaction with nuclear hormone receptors, which include estrogen receptors (ERs), androgen receptors (ARs), progesterone receptors, thyroid receptors (TRs), and retinoid receptors. Today, research shows that the mechanism of action may be more complicated, than originally recognized [4]. Although evidence is mounting on the adverse effects of EDs on human health, there is still considerable controversy surrounding this area. The complexities of the multiple causes and effects that lie behind the mode of action of EDs are coupled with skepticism and debates due to different interests [2].

The table presents the mechanism of action of the major EDs (Table 6.1).

Table 6.1 Well-known endocrine disruptors and mechanism of action

Bisphenol A	Incrementation of insulin and leptin. Incrementation of TNF-, IL-1. Decreasing adiponectin. Interfering glucose tolerance induction. Incrementation of adipogenesis. Accumulation of lipids in adipocytes and hepatocytes. Increasing oxidative stress. Decreased anti-oxidant capacity
Phthalates	Increased lipid peroxidation. Increased oxidative stress. Interference with insulin receptors. Altered glucose tolerance induction. Reduced glucose oxidation
Dioxins	Abnormal lipids and lipoproteins. Incrementation of TNF-, IL-1, and TGF. Decreased lipoprotein lipase activity. Decreased hepatic vascular endothelial cadherin. Altered glucose tolerance induction. Induction of hepatic expression of CD36. Decreased regulation of insulin receptor-subunit, IRS-1 and Glut-4. Incrementation of peripheral fat mobilization and hepatic FFAs uptake. Decreased hepatic FFAs-oxidation. Increased oxidative stress. Decreased anti-oxidant capacity. Inducing apoptosis
Other POP's	Initiation of steatosis, periportal fibrosis, hepatocellular ballooning, sinusoidal dilations, activated Kupffer cells. Altered glucose tolerance induction. Reduced anti-oxidant capacity. Incrementation of oxidative stress. Incrementation of lipid peroxidation. Introduction of steatosis, inflammation, lipid granulomas, and portal fibrosis. Incrementation of TNF-, IL-10 and interferon. Incrementation of monocyte chemoattractant protein-1 and keratinocyte chemoattractant-1. Reduction of adiponectin. Decreased regulation of hepatic haptoglobin and fatty acid-binding protein. [5]

Phthalates

The Chemicals, Their Forms, and Their Distribution in the Environment

Phthalates are the diesters of phthalic acid (1,2-benzene-dicarboxylic-acid) with an added alcohol moiety. The phthalate products display different properties which depend on the length of the branching side chain. The di-(2-ethylhexyl)-phthalate (DEHP), with eight carbon atoms in the alkyl side chain, is one of the most commonly used phthalates worldwide [2].

After exposure, low-molecular-weight (LMW) phthalates are rapidly metabolized into their respective monoester and hydroxy-diester metabolites and high-molecular-weight (HMW) phthalates are metabolized into monoesters and hydroxy-, oxy-, or carboxy-diester metabolites, which are then excreted by the urine [7] (Fig. 6.2).

DEHP is rapidly metabolized into mono-(2-ethylhexyl) phthalate (MEHP) MEHP is considered an active metabolite as it is 10 times more potent than the parent compound.

Phthalates are a group of multifunctional industrial chemicals and they are used in a wide range of consumer products. They are used as plasticizers to improve flexibility in plastics, such as polyvinyl chloride (PVC), or as additives in the textile industry and as carriers in pesticide formulations. They are added to pharmaceuticals as coatings to time releasing pharmaceuticals, they are also used as industrial solvents and lubricants. Phthalates are frequently added to consumer and personal care products, such as cosmetics, perfume, deodorants, hair sprays, skin cleansers in order to retain color or fragrance, as well as in the manufacturing of medical devices and toys, and moreover in fuels to enhance performance [2].

Due to their high distribution on the market, human exposure to phthalates is widespread and common, particularly since phthalates can be absorbed by the skin, inhaled, ingested, or directly administered [31]. Human exposure to phthalates starts during fetal development and it continues after birth by contact of the newborn with contaminated food sources [1].

The Scientific Committee on Toxicity, Ecotoxicity and the Environment (CSTEE) identified values for four phthalates di-isononyl phthalate (DINP), di-octyl phthalate (DNOP), di-2-ethylhexyl phthalate (DEHP), and di-isodecyl phthalate (DIDP), which concerns with the so called phthalate syndrome [2].

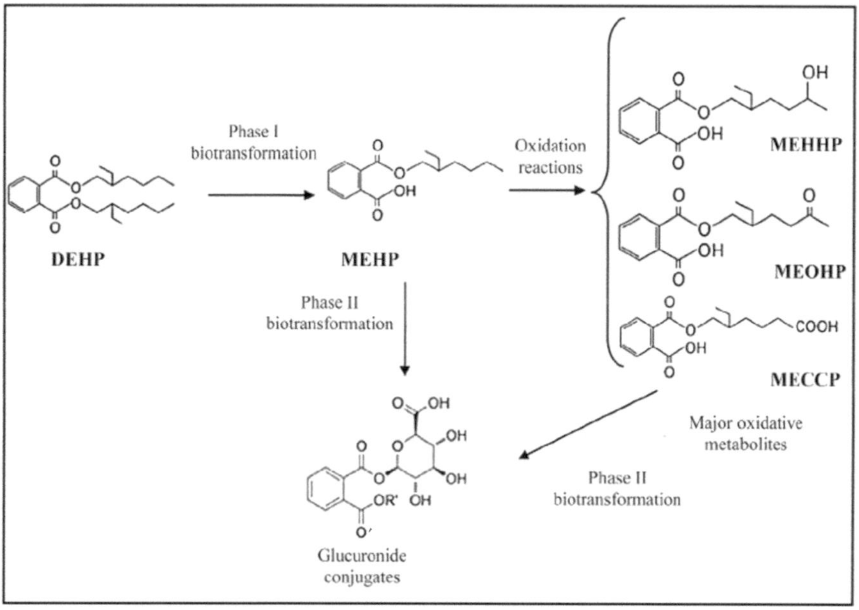

Fig. 6.2 Metabolism of DEHP, the branched chain of phthalates

Recently, biomonitoring studies have demonstrated relatively high urinary levels of phthalate metabolites in the general population of Western countries. A recent European human biomonitoring study reported phthalate exposure levels for children and adults in several countries, with children from Poland experiencing one of the highest levels among the European populations investigated [8]. This exposure is of great concern, as phthalates have been suspected to have anti-androgenic and estrogenic effects, which may alter the reproductive development of children [7]. Between 2012 and 2018, the global consumption of DEHP exceeded 3000 thousand metric tons [2].

Effects on Health

The *phthalate syndrome* is the variety of effects observed in rat experiments and it is characterized by malformations of their reproductive organs, reduced anogenital distances, and nipple retention. Evidence of the human analog of the "phthalate syndrome" has been reported in recent years and it demonstrated the disruption of the androgen signaling pathway. Phthalates have been implicated in various male malformations of external genitalia, such as cryptorchidism and hypospadias. Reports on the effects of prenatal exposure to phthalates in humans showed that birth outcomes and the reduced anogenital distance were associated with high levels of phthalate esters, notably DEHP, in the urine of the mothers during pregnancy. The incidence of these genital malformations is increasing in male newborn babies [2].

In utero, exposure to DEHP decreases fetal testosterone levels by a mechanism that involves the downregulation of CYP11a and CYP17a, which are implicated in androgen biosynthesis. The antiandrogenic effects are not exclusive to DEHP, but also involve other metabolites such as dibutyl phthalate (DBP), butylbenzyl phthalate (BBP), and di-isononyl phthalate (DINP), which also decrease testosterone levels and they disrupt male sexual development. Exposure in adulthood shows that androgen levels decrease, while Leydig cell numbers and steroidogenic enzyme expression levels were normal or moderately increased [8].

Additionally in research, dimethyl-phthalate (DMP) and dibutyl-phthalate (DnBP) showed the highest association with behavioral disturbances, including emotional symptoms and hyperactivity/attention deficit problems and peer relationship problems [8]. These phthalates, namely DMP, DnBP, have also shown statistically significant associations with mathematical skills and the same findings were reported for diethyl-phthalate (DEP) [8]. In a cohort of 404 infants, it was demonstrated that low-molecular-weight phthalate metabolites were positively associated with gestational age and head circumference after adjusting for multiple factors, including creatinine levels [7].

Bioactive phthalates are known to activate the peroxisome proliferator-activated receptor (PPAR) a and c, which are nuclear receptors. The PPARs hetero dimerize with retinoid X receptors (RXR) to bind peroxisome proliferator response molecules and to promote the activation or repression of target genes. PPARa activation is involved in lipid and lipoprotein metabolism, while PPARc regulates adipocyte differentiation, glucose metabolism, and the storage of fatty acids [1].

Bisphenols (BPs)

The Metals, Their Forms, and Their Distribution in the Environment

Bisphenols are a large family of chemicals that are utilized to produce epoxy resins. Bisphenol A (BPA) was first synthesized by Dianin as early as 1891. In the mid-1930s, the estrogenic properties of BPA were discovered during the effort to produce a synthetic estrogen [9].

Commercial production of BPA did not begin before the early 1950s, when the first epoxy resins were developed. To protect foods and drinks from direct contact with metals, epoxy resins are frequently used in the internal coating of food and beverage containers. BPA is a building chemical of polycarbonate plastics, including baby bottles. BPA is also used for some dental sealants, carbonless paper for receipts, digital media (CDs and DVDs), electronic equipment, flooring, tableware, and reusable bottles. A recent study reported that BPA was detectable in dust, air particles, and water, making exposure widespread [9].

Human exposure occurs by food, drinking water, household products, cosmetics, medical equipment, dental materials, and occupational sources. The polymerization of polycarbonate plastics is not complete and as a consequence, the unbound monomer or polymers can easily diluted in liquid environments [2].

BPA, one of the most common produced chemicals worldwide, is a raw material used in many consumer products. Thus, people are continuously exposed to it. However, due to its proven adverse effects on human health, some countries have banned or restricted the use of BPA and research for BPA derivatives have been designed to replace BPA in the manufacturing industry [11]. In 2005, the first study to measure total BPA concentrations in a reference population was reported from the CDC. BPA was detected in 95% of the urine samples analyzed and concentrations up to 5.18 ng/ml (95th percentile) were reported. The National Health and Nutrition Examination Survey (NHANES) examined spot urine samples from 2517 Americans >6 years of age in 2003–2004. BPA was detected in 92.6% of participants, with concentrations ranging from 0.4 to 149 ng/ml (with a mean of 2.6 ng/ml urine) [9].

Furthermore, the mean urinary concentration for BPA among premature infants undergoing intensive therapeutic medical intervention was higher than in the general population (30.3 vs. 2.6 ng/ml urine). Such a high level of exposure may be due Bisphenols (BPs) 276 to the use of medical equipment, including intravenous catheters. While adults have a high capacity to rapidly metabolize and excrete BPA, the fetus and infant have lower hepatic levels of metabolizing enzymes, and thus are at greater risk from exposure to unconjugated BPA. The US Food and Drug Administration (FDA) raised concerns over BPA exposure in fetuses, infants, and young children [9].

The presence of BPA analogs in canned vegetables, fruits, and soft drinks as well as in honey, fish, and mustard is observed by multiple studies. Bisphenol F (BPF) has been found in mustard from white mustard seeds and it is a natural reaction product formed during its preparation. Mustard is one of the most widely used condiments worldwide and it is the main source of BPF in humans in Europe and probably worldwide [10]. In addition to BPA, bisphenol A diglycidyl ether (BADGE) and bisphenol F diglycidyl ether (BFDGE) have been detected in different milk samples from supermarkets and dairy farms while BPF and bisphenol S (BPS)have been detected in dairy products, meat and meat products, vegetables, and cereals [10].

To avoid potential hazards, Canada became the first country to declare BPA to be a toxic compound and required its removal from all infant formula bottles in 2010. BPA was also banned in infant formula bottles by the European Union in 2011 and by the United States in 2012. In France, the use of BPA in any food or beverage packaging was forbidden from January 2015 [9].

In 2015, the production of BPA was estimated at more than 4.5 million tons. Each year, more than 100 tons is released into the atmosphere and it is found in air, drinking water, house dust, food, lakes, sea, and soil. BPA is ubiquitous [2]. Previous studies have shown that BPA can be diluted in liquids from epoxy resins, and other products in contact with foods and drinks. Indeed, a risk assessment study suggested that canned food items contribute to 10–40% of the daily BPA intake. Similarly, the BPA substitutes have also been demonstrated to have detrimental effects on human health [9].

Effects on Health

Although the estrogenic activity of BPA is lower than that of natural estrogens, its prevalence at high concentrations in the environment, including surface, ground, and drinking water, possess a potential risk to human health. In contrast to PCBs, which can persist in the human body for decades, BPA has an average half-life of approximately 6h in humans [2].

BPF and halogenated BPA, used in products of everyday life, are suspected of having toxic effects on fetuses. Numerous studies on the biological effects of BPA have been published, and its potential human health hazards have been extensively analyzed. Previous studies have linked BPA exposure to abnormalities of the reproductive system and to a higher incidence of cardiovascular disease. Along with recent increases in the prevalence of neurobehavioral disorders, evidence has been accumulating that BPA can interfere to nervous system development. Elevated gestational urinary concentrations of BPA have been correlated to adverse behavioral outcomes in children [9].

It seems that genetic predisposition, unhealthy dietary habits, as well as a sedentary lifestyle alone are responsible for the global epidemic of overweightness and obesity. However, there is increasing evidence that a number of chemicals can interfere in hormonal metabolism and regulation of adipocyte function. The hormone imbalance could lead to obesity and the rsponsible chemicals are named 'obesogens'. The mechanism by which obesogens could lead to weight gain or obesity is not clear. It seems that they activate nuclear transcription factors, such as the peroxisome-proliferator-activated-receptors alpha, delta, and gamma (PPARα, PPAR-δ, and PPAR-γ), and steroid hormone receptors that regulate adipocyte proliferation, differentiation, and lipid metabolism, thereby influencing body composition. In addition, changes in the function of transcription factors may regulate gene expression [10].

Mechanism of Intoxification

Molecular studies have revealed a variety of mechanisms by which BPA stimulates cellular responses even at very low doses. BPA was reported to bind to the aryl hydrocarbon receptor, inhibiting the detoxificating metabolism and to the thyroid hormone receptor, inhibiting the transcriptional activity induced by triiodothyronine. A potent antiandrogenic activity of BPA, resulting from its ability to interfere with an agonist binding mechanism to the androgen receptor, was also reported. In addition, BPA has been found to act as a glucocorticoid receptor agonist and it also impairs insulin homeostasis. BPA also activates the pregnane X receptor, a nuclear factor related with metabolism. Thus, the effect of BPA may result from multiple actions on different signaling pathways [9].

Estrogen-related receptor γ (ERRγ) was identified as a BPA target receptor in 2006. BPA binds very strongly to ERRγ. ERRγ is a nuclear receptor with unknown function, but it is thought to play a role in the differentiation and maturation of the fetal brain. When BPA binds to ERRγ, the physiological functions of ERRγ might be disturbed resulting in neurodevelopmental disorders, especially when exposure occurs during critical early periods of nervous system development. The National Toxicology Program of the United States concluded that "there is some concern for effects on the brain, behavior, and prostate gland function in fetuses, infants, and children at current human exposures to BPA" [9].

A number of animal studies reported that BPA exposure during gestational period affects brain development and behavior in later life at relatively low exposures. BPA exposure during organogenesis upregulates *dopamine receptor* function, whereas exposure at other gestational periods does not, suggesting the presence of critical developmental windows of BPA toxicity [9]. Gestational and early life exposure to BPA has been shown to affect the dopamine system by reducing the number of midbrain dopamine neurons in monkeys. However, it exerts its toxicity in males and not females. The mechanism of dysfunction of these neurons is an altered function of tyrosine hydroxylase, which is an important enzyme in the dopamine synthesis. In addition, exposure alters the organization and function of the oxytocin (OT)/vasopressin (AVP) axis. Furthermore, prenatal exposure to BPA affects fetal neocortical development by accelerating neuronal differentiation and migration during the early embryonic stage [9].

BPA exposure is related to behavioral disturbances like aggression, anxiety, cognitive deficits, and learning-memory impairment. The overall outcome is hyperactivity [9].

Many behaviors and neuroendocrine pathways are sexually dimorphic. BPA has sex-specific effects on the expression of behaviors associated with anxiety, activity, and sociality. Exposure to BPA, that disrupts hormone function during critical periods of development, may perturb sex specific or hormonally regulated behaviors. BPA exposure affects exploratory activity and behavior in females but only minimally disrupts the partner preference formation [9].

Like the other EDs was initially examined for effects on sexual dysfunction, malformation, and cancers of reproductive origin [30]. Among the 19 BPA related compounds tested, several, including tetrachlorobisphenol A (TCBPA), bisphenol AF (BPAF), bisphenol B (BPB), 2-(4-hydroxyphenyl)-2-phenylpropane (HPP), 1,1-bis(4-hydroxyphenyl) cyclohexane) (BPCH), 4-hydroxydiphenyl-methane (HDM), and 3,3′-dimethyl-bisphenol A (DMBPA), were found to have high estrogenic activity than BPA [9].

Due the biological effects of BPA, there has been a gradual shift to using bisphenol analogs. Several chemicals that are structurally similar to BPA are presently used in the manufacturing of resins and plastics, as substitutes or replacements for BPA. The biological actions of bisphenol analogs are beginning to be clarified [9].

The production and use of *bisphenol S* (*BPS*, bis-4-hydroxyphenyl sulfone) has recently increased in many countries. BPS, at low-dose doses, is present in food items and it seems to interact predominantly with the cell-membrane affecting signaling and endothelium reticulum leading to potential consequences for cell function. A recent study showed that both BPS and BPA influence hypothalamic development. Given the disrupting effects of BPS and its higher resistance to environmental degradation in comparison to BPA, the proposal to utilize BPS instead of BPA in various products should be viewed with caution [9].

Bisphenol AF (BPAF), a fluorinated derivative of BPA, is used in polycarbonate co-polymers produced in high-temperature for use in electronic materials and polymer applications. BPAF binds to ERα more strongly than BPA and it has a much higher affinity for ERα than ERRγ. BPAF is a strong ligand for both ERα and ERβ, with a threefold greater for the latter. However, BPAF is a full agonist for ERα, but an antagonist for ERβ. Thus, BPAF perturbs physiological processes regulated by estrogen [9].

Bisphenol-A diglycidyl ether (BADGE) is an analog of BPA that is used in manufacturing of coating and resins. BADGE induces adipogenesis in multipotent mesenchymal stromal stem cells at low concentrations. The effects of BADGE appear to be mediated, at least in part, by peroxisome proliferator-activated receptor γ (PPARγ). PPARγ is a member of the nuclear receptor superfamily and it is involved in the pathogenesis of obesity and diabetes. Interestingly, brain PPARγ regulates energy balance and its activation may lead to hyperphagia and an increase in adipose tissue. Thus, BADGE may function as an obesogenic by central nervous system effect [9].

Halogenated BPs are emerging contaminants. Analogs of BPA, where the phenolic moieties are substituted with halogens (Br or Cl), are used as flame retardants. Approximately 200,000 tons of tetrabromobisphenol A (TBBPA) are produced annually and used to protect computer boards and other electric equipment. The presence of TCBPA and other chlorinated analogs in environmental samples has been unequivocally demonstrated. Their presence in human samples, including mother's milk and umbilical cord blood, indicate high prenatal and postnatal exposure [9].

Not all halogenated BPA analogs are ER agonists. In general, the greater the number of bromine atoms, the weaker the ER agonism. TCBPA displays full agonistic activity compared to BPA associated with ERa function, with a 1000 to 10,000-fold lower potency to E2 receptor. While these analogs show marginal estrogenicity in assays, they bind with stronger affinity to PPARγ than BPA. Thus, these halogenated BPA analogs may contribute to onset and disease progression by the receptor PPARγ, similar to BADGE [9].

IUPAC (4-(2-(4-hydroxyphenyl) propan-2-yl) phenol) is an important monomer used for the production of polycarbonates, epoxy resins, polyesters, polysulfones, and polyether ketones and it is also applied as a polymer additive in plasticizers and halogenated flame retardants. It is present in a wide range of applications including baby bottles and linings for metal-based food and beverage cans, ophthalmic lenses, medical equipment, consumer electronics, and electric equipment [2].

Epigenetic Mechanism of Intoxification

It seems that the underlying mechanism of these hormonal changes is epigenetic. BPA causes the enrichment of H3K4 trimethylation and H3K4 specific methyltransferases at the ER binding region of the *HOTAIR* promoter resulting in activation of transcription estrogen responsive cells [39]. Several studies reported BPA can impair the early stages of neurogenesis by decreasing the expression of neural marker genes such as *Sox1, Sox3, Nes, Pax6*, and *Fgf5* during mouse differentiation. Moreover, one metaanalysis study demonstrated that transplacental BPA exposure downregulated the expression of *SOX2* and *PAX6*, potentially affecting childhood neuronal development [11].

Maternal BPA exposure disrupted genomic imprinting in the mouse embryos and placenta by modifying methylation of the loci *Avy* and *CapbIAP* CpG site specific DNA [12]. In one genome wide association study named SMART study population ($N = 35$), only one CpG site in the *TSP50* gene promoter showed increased DNA methylation with elevated BPA exposure [12]. Another region in this gene showed a 26% decrease in average methylation [13].

Parabens

The Chemicals, Their Forms, and Distribution in the Environment

Parabens are alkyl esters of p-hydroxybenzoic acid and they are widely used in cosmetics, pharmaceuticals, personal care products and as food additives to inhibit microbial growth and extend product maintenance. Especially in cosmetics, including foundations, powders, eye shadows, mascara, lip glosses, lipsticks, and nail polishes, and in pharmaceuticals and personal care products such as lotions, sunscreens, cleansers, shampoos, deodorants, hair care products, and toothpaste, which are in frequent contact with skin, lips, eyes, oral mucosa, and hair, have a commercial use [14].

These chemicals have been in use for more than 80 years due to (1) antibacterial and fungicidal properties, (2) high stability at different pH and temperatures, (3) inertness, and (4) low cost, which has resulted in widespread, chronic human exposure [15].

The synthetic parabens include methyl-, ethyl-, propyl-, isobutyl-, butyl-, isopropyl-, and benzyl-, with increasing alkyl chain length, and they show increased antimicrobial activity. Parabens are estrogenic molecules, but exert weaker activity than natural estrogens, which would imply a low risk. The estrogen receptor binding affinities of parabens are 10,000 to 100,000-fold lower than estrogen and they decrease in order as: isobutylparaben > butylparaben > isopropylparaben =. propylparaben > ethylparaben [14].

In a 1984 study, greater than 13,000 cosmetic products were found to contain parabens and by 1987 approximately 7000 kg of parabens was used annually in cosmetics and consumer's clothing. A question that remains unresolved is whether frequent exposure to varying concentrations of these weakly estrogenic compounds poses a threat to humans [14].

In a 2004 survey, the Environmental Working Group reported that women on average used 12 products containing a total of 168 ingredients per day, whereas men used approximately 10 products with a total of 85 ingredients daily (EWG 2004). In addition to adult exposure, children up to 9 years of age are exposed to approximately 61 products daily, suggesting potential endocrine modulating interference (EWG 2007). Collectively, these reports suggest that paraben exposure frequency and concentration vary significantly by individual and exposure is widespread in adults and children of both genders, with women on average receiving the highest exposure [14].

Consumers of these compounds are frequently exposed by the skin and mucosal surfaces. The skin, with one of the largest body surface areas, absorbs approximately 50 to 70% of a topically applied parabens. These products act as endocrine disruptors and they affect the reproductive, immune, and nervous systems [14].

Current regulations on parabens usage differ by country. In 2009, the Danish Environmental Protection Agency (Danish EPA) published a comprehensive report on endocrine disruptor exposure in 2 years old children. The report strongly emphasized that parabens in lotions and sunscreens (especially propyl- and butyl-paraben) should be considered as a possible risk in young children due to potential endocrine disrupting effects, which might manifest in adult life. As a precautionary measure, based on the study from the Danish EPA, in March 2011, Denmark banned the use of propyl- and butyl-paraben in cosmetic products for babies and young children under the age of 3 years. France's National Assembly is currently preparing a similar law to ban the use of endocrine disruptors that include parabens from consumer products [14].

Although the European Union Scientific Committee on Consumer Safety had previously issued statements indicating no cause for concern in relation to the use of parabens in cosmetics, the reports from Denmark led to reevaluate the subject in December 2010. The committee arrived at the conclusion that current data do not provide the essential evidence between dermal absorption of parabens and endocrine toxicity in animal and human studies. In December 2011, however, the EU commission reversed this decision and it is currently preparing a proposal to ban five parabens in cosmetics (isopropyl, isobutyl, pentyl, phenyl, and benzyl esters of 4-hydroxybenzoic acid and their salts), citing "insufficient information to conduct a safety assessment" [14].

Although no ban on paraben use currently exists in the United States, a number of prominent cosmetic manufacturers are starting to move from parabens and instead focus on alternative preservatives or to decrease the number of parabens in their products due to increasing consumer concerns; however, fully satisfactory substitutes have yet to be discovered. The FDA first issued opinion statements in March 2006 and October 2007 about parabens and concluded that "at the present time there is no reason for consumers to be concerned about the use of cosmetics containing parabens" [14].

Effects on Health

Although the estrogenic affinity of parabens to the estrogen 17β-estradiol is low, the frequent topical exposure to formulations containing multiple parabens increase the possibility of high exposure. Recent studies also demonstrate that estrogenic activity varies and it depends to paraben structure especially, the length and branching of the alkyl chain. Due to estrogenic properties, parabens were reviewed by the Endocrine Society and classified as potential EDs. The US Environmental Protection Agency (EPA) in its 1999 report "Pharmaceuticals and Personal Care Products in the Environment: Agents of Subtle Change?" documented in several tests that parabens displayed estrogenic activity [14].

Accumulating data suggests that developmental exposure to parabens during critical gestational and perinatal time windows may produce subtle endocrine changes in later life. Prenatal exposure of pregnant rats to isobutyl-paraben, using slow release maternal silastic capsules (4.36 mg/kg/d), resulted to impaired learning and memory capabilities and to increased anxiety in male offspring, but did not markedly affect the female. Conversely, adult female rats prenatally exposed to isobutyl-paraben showed impaired social recognition. Data suggests that paraben exposure may selectively alter development of the rat brain by gender [14].

Some interesting studies that confirm the EDs function of parabens are included. Treatment of pregnant rat dams with isobutyl-paraben diminished plasma corticosterone levels in dams and enhanced the uterotrophic response to estrogen 17β-estradiol in adult female offspring. Rat fetuses maternally exposed to butylparaben also showed marked depression of ERβ mRNA in their ovaries. Similar methyl and propyl-paraben treatment of neonatal female rats produced impaired postnatal ovarian follicle development [14].

Some studies demonstrated paraben effects on male reproductive organs, testosterone levels, and sperm production. Dosing of rats with butyl- and propyl-parabens at 0.1% in the diet, a daily exposure level below the 10 mg/kg level, adversely influenced male reproductive parameters as evidenced by decreased sperm production, but they were no alteration in the weights of prostate and seminal vesicle was detected [14].

A recent human study in premature babies was performed in neonates weighing approximately 500 g, who were prophylactically administered gentamicin at 2 mg/kg/d and additionally 0.1 μg of parabens from the antibiotic. Although the acceptable daily intake of parabens for adult humans is up to 10 mg/kg/d as recommended in 1974 by the Joint FAO/WHO Expert Committee, the effect of parabens on preterm infants remains unknown [14].

The weak estrogenic capacity of parabens combined with their ability to localize in tissues has raised concerns regarding their potential risk for cancer, especially breast cancer since the number of cases with parabens detected in breast tissues has risen. In 2004, it was suggested that cream or lotion-based cosmetics, containing preservatives such as parabens, used on the underarm, chest, or breast area may increase breast cancer incidence in women. In a recent study, relatively high levels of propyl- (16.8 ng/g tissue) and methyl (16.6 ng/g of tissue) and lower levels of butyl- (5.8 ng/g tissue), ethyl- (3.4 ng/g tissue), and isobutyl-parabens (2.1 ng/g of tissue) were detected in 160 breast cancer tissue biopsies from 40 patients. Interestingly, 7 of these 40 patients had never used underarm cosmetics, suggesting they were exposed to parabens from other sources [14].

Additionally, the FDA accepted the fact that parabens in cosmetic formulations may be responsible for allergic contact dermatitis and skin inflammation in paraben-sensitive individuals or individuals with damaged skin. Moisturizers (creams or lotions) are one of the most commonly used topical products with the highest potential of producing allergic contact dermatitis, irritation, or photocontact dermatitis. Further, analysis of the chemical ingredients in moisturizers showed that parabens were one of the major potential allergens in 170 (62%) of the 276 products [14].

Pesticides

The Chemicals, Thier Forms, and Their Distribution in the Environment

Pesticides are usually designed that they can be invasive to the reproductive and neural systems of the organisms. However, the similarity of the specific function to normal human reproductive system indicates that these chemicals can also affect normal human body. The pesticides that are used commonly include DDT and chlorpyrifos [4]. Organophosphorus pesticides are the most commonly used as insect killer and chlorpyrifos is a example of this class of insecticides. The insecticide is used in both in household and agriculture to control pests such as cockroaches, flies, termites, fire ants, and mosquitoes. However, it is confirmed from studies that the compound is highly toxic with side effects from the nervous system [4].

Effects on Health

DDT has a long history as an endocrine disruptor. The compound was once used randomly as a pesticide in the agriculture, in production of livestock, in household, gardens, but because of its hazardous nature, it was banned few years back. However, the compound is still in use in some countries [4].

DDT can interfere with thyroid, estrogen, androgen, renin-angiotensin, insulin, and neuroendocrine systems which can directly influence the reproductive, cardiovascular, and metabolic systems of human body. All these characteristics suggest that the chemical is one of the most potential candidates of endocrine disruptors [4].

There are a lot of publications that relate the effects of the pesticides on learning and aspects of neurobehavior, especially, DDT and other organochlorines. In addition, there is much animal data on neurodevelopmental and direct neurotoxic effects of all three major classes of insecticides (organochlorines, organophosphates, pyrethroids), which should raise an alarm clock for potential effects to humans. As a consequence of such studies, that have been evaluated in the framework of the EU pesticides legislation, many organochlorines, organophosphates, and pyrethroids have been banned [16].

Thalidomide, α-Phthalimido-Glutarimide

Thalidomide, α-phthalimido-glutarimide, was synthesized in West Germany in 1954. This company, while studying different chemicals in order to produce a low-cost antibiotic, they investigated the potential of this new molecule as a sedative in humans. After tests in different experimental models, they concluded that the medication has no apparent side effects and the company distributed free samples to doctors in West Germany and Switzerland [2].

In 1957, the product demonstrated the appropriate licensing as a sedative; it was commercialized first in Germany and later in other European countries. The drug rapidly became a success and, soon after, was sold in 46 countries. In 1958, the company advised physicians to prescribe thalidomide to pregnant women for morning sickness and nausea, unfortunately without testing the consequences in the fetus. In 1961, reports started to appear, linking thalidomide with a variety of birth defects, including the rare malformation phocomelia. Between 1957 and 1961, thousands of babies were born worldwide with severe congenital malformations (Fig. 6.3). Widukind Lenz, a German pediatrician, found that the administration of the drug between the 20th and 36th day of gestation could lead to the development of malformations in the fetus. The drug is forbidden to be prescribed to pregnant women any more, but it is used with an enormous success in various medical conditions, including erythema nodosum leprosum, immune system disorders, multiple myeloma, cancer, and in many other diseases [2].

Diethylstilbestrol (DES)

Diethylstilbestrol (DES) was used in the past to prevent miscarriages in pregnant women. It was confirmed to be a potential carcinogen for the girls of women, under this treatment during pregnancy, who tended to develop cancers of the urogenital area. In 1971, the compound was then advised by the FDA to stop being prescribed, mainly because of the increased incidence of clear cell carcinoma, a rare vaginal cancer [4].

Diethylstilbestrol was used to treat metastatic prostate cancer, because of its estrogenic effects and it was also prescribed to women during pregnancy to prevent miscarriages and premature labor. In mice models, pre and neonatal DES exposure causes a wide range of gene expression changes. DES treatment results in persistent expression of certain genes, including lactoferrin, epidermal growth factor, and protooncogenes such as *c-fos*, *c-jun*, and *c-myc*. It was confirmed that there is an indirect relationship between DES and methylation changes, showing that estrogen inhibited catechol-O-methyltransferase (*COMT*) gene transcription. COMT is a ubiquitous enzyme catalyzing the transfer of the methyl groups from SAM to one of the hydroxyl groups of catechols in the presence of Mg^{2+}. Inhibition of this enzyme results in inhibition of the methylation process (18).

Fig. 6.3 Birth defects including phocomelia due to thalidomide during gestation

Dibutyltin (DBT)

Dibutyltin (DBT) is the main metabolite of tributyltin, which is widely used to control contamination from boat hulls and fishnets. It is also used as a plastic stabilizer. DBT has had significantly negative impacts on the endocrine system and several neuroendocrine axes. DBT inhibits GR to his ligand and its transcriptional activity by inserting into a site close to the steroid binding pocket of GR. In hepatoma cells that express endogenous GR, DBT also inhibits glucocorticoid-induced expression of phosphoenolpyruvate carboxykinase (PEPCK) and tyrosine aminotransferase (TAT) and it inhibits glucocorticoid mediated of the TNFα NF-kb signaling. Moreover, DBT exposure reduced the glucocorticoid mediated suppression of cytokine production in lipopolysaccharide-stimulated macrophages. These findings suggest that DBT may disturb metabolic functions and the modulation of the immune system by blocking GR activation [6].

The Wide Use of Endocrine Disruptors in Livestock, Agricultural Industry, and Medical Industry
The use of steroid compounds in the livestock industry for the improvement of the reproduction and growth began in the 1940s. Sex steroids like estradiol, progesterone, and testosterone have growth-promoting effects in animals. As a result, these chemicals have been used in agricultural practice for many years, where they are exogenously administered to meat producing animals for the improvement of weight gain. These chemicals further contaminate the soil, water, and air of the environment of the particular area as well as areas in a distance [4]. The influence and presence of sex hormones as one of the environmental pollutants were first experimented by Shore et al. [39].

Waste water from pharmaceutical and other industries can act as a major source of exposure. They can contain hormones, hormones from pharmaceutical products like birth control pills and potential EDs from different detergents, soaps, plastics, food, and personal care products such as fragrances, etc. It is observed from studies that endocrine disruptors like estrogen and androgen, which are ubiquitously present in environment in trace quantities, they could be released from the municipal wastewater of different industries [4].

Additionally, steroid compounds like ethinyl estradiol, mestranol, norethynodrel, and megestrol acetate are used as contraceptives and as treatment of diseases. Investigations regarding the harmful effect of these compounds started at the same time [4].

Many of these compounds might end up in the aquatic environment. The reasons may be incomplete removal of the contaminants during sewage treatment processes, soil run off, or indiscriminate discharges of them into the waterways. Some water treatment systems like flocculation, sedimentation, filtration, and chlorination are not able to eliminate or remove these contaminants [4].

Neuroendocrine disruptors, recognized as environmental pollutants, are found to mimic or antagonize or modulate the synthesis and metabolism of neuropeptides, neurotransmitters, or neurohormones. These reactions result in the several diverse modifications of physiological, behavioral, or hormonal processes of animal body, the modifications might include altered reproduction, development, ability to deal with challenges and many other abnormalities. Endocrine disruptors are known to mainly function by altering hormonal and homeostatic systems of the cell [4].

Cosmetics

The cosmeceutical industry is one of the fastest and profitable growing commercial sectors producing different personal care agents. The global cosmeceuticals market was expected to gain $429.8 billion by 2022, registering a Compound Annual Growth rate of 4.3% during the forecast period 2016–2025. Reports have shown that the average woman and man in the United States use twelve and six cosmetic products, every day, respectively. Among a list of number of 12,000 synthetic and industrial-based chemicals used in cosmetics, fewer than 20% have been recognized safe [17].

The worldwide culture regarding physical appearance works as a mode of communication with other people, both in the workplace, where it designates an acceptable middle line profile to managers and colleagues, and also in personal relations, where it possesses characteristics of the personality and sexuality. A strong correlation exists between physical attractiveness and social status, where a youthful appearance confers economic and social advantages. It is interesting though, that this cultural behavior known to interest mainly women, it is now a demand for men and children also. The legend "You are what you show off" has jeopardized ethical standards [18].

As a consequence, "The Beast of Beauty" and the everlasting youth take their revenge and many people suffer today from the overuse of cosmetical formulations from a very young age due to chronic exposure to toxic ingredients. Some of the ingredients included in these products function as endocrine disruptors imposing a vulnerability towards some of multifactorial diseases of modern society. Is there any reason behind increasing rates of breast cancer in women or infertility in young couples? Given the critical role of estrogen modifying chemicals, e.g., parabens in breast cancer, it is an hypothesis that body exposure to such chemicals in the form of cosmetics along with their possible accumulation could be a potential risk. Thus, the question is still is not answered yet which is beyond our knowledge whether these products are suitable to apply to the skin on a regular basis. Moreover, the inadequate toxicological profile is also questionable [17].

The term "cosmetical" is used to refer to skincare products formulated with biologically active ingredients for daycare of the skin [18]. The term was coined in 1984 by Dr. Albert Kligman of the University of Pennsylvania describing a hybrid category of products in the midway of the spectrum of "cosme-tics and pharmaceuticals." A cosmetical is consensually accepted to exert a "pharmaceutical therapeutic benefit," but not necessarily a "biological therapeutic benefit." For Dr. Kligman, cosmetical represented "a topical preparation that is sold as a cosmetic, but it has performance characteristics that suggest pharmaceutical action." He coined this term around the crucial time of Kligman's experimentation on the anti-aging effects of tretinoin [19].

The term "nutraceutical" was coined in 1989 by Stephen De Felice from "nutrition" and 'pharma-ceutical'. "Nutraceutical is a product prepared from foods, but sold in the form of pills, or powder or other medicinal forms, not usually associated with foods." Nutraceuticals are provided from natural herbs, food industry, and dietary supplements market. The pharmaceutical industry is now trending towards genetically engineered "designer" foods as well. Although nutraceuticals cover most of the therapeutic areas such as anti arthritic, digestive problems, prophylaxis and treatment for cancers, lipid and sugar control, osteoporosis, blood pressure, and depression among others, one can grossly regard a nutraceutical an orally consumed product with nutritional benefits [19].

Despite the prevalent confusion about its definition and role, it would not be an exaggeration to state that almost 30% to 40% of any dermatologist's prescription are cosmeticals. Prescriptions frequently include anti aging creams, treatment of photomelanosis and phototanning as well as treatment of pigmentation related disorders like melasma or freckles, wrinkles reduction as well as anti-inflammatory, neutrachemicals for fat loss, hair growth and hair fall prevention, and maintenance of skin tone and clarity of complexion [19].

The Chemicals, Their Forms, and Their Distribution in the Environment

Chemicals commonly used in the cosmetical industry are: alpha-lipoic acid oral Coenzyme Q10, oral Vitamin B-complex, oral Vitamin C and topical, Vitamin E topical and oral, Hydroquinone topical, Alpha and beta hydroxy acids topical, Polyunsaturated fatty acids oral, Peptides topical, Retinoids oral, Retinaldehyde topical, Retinal esters topical, Retinol topical, Comfrey topical, Fever few topical, Jojoba oil topical, Licorice topical, Pine bark extract topical, Rose topical, Turmeric topical and oral, Milk thistle topical, Lavender topical, Grapeseed oral, Green tea oral, Lycopene oral, Pomegranate oral, Arbutin topical, Kojic acid topical, Soya oral and topical, Aloe vera topical and oral, Chamomile oral, Caffeine oral and topical, Polypodium leucotomos oral, Glutathione oral, parenteral and topical, Phytosterols, oral and topical, Proanthocyanidin oral, Panthenol topical, Ceramides topical, Zinc topical and oral, Licorice plant [glycyrrhiza]—glabridin, Tyrowhite topical, Ellagic acid topical, Sunscreen ingredients, Idebenone, Resveratrol, Hyaluronic acid, Glucosamine, Azelaic acid, Niacinamide, Allium cepa, Allantoin, and Marine protein supplements (MPS) [19]. They are generally classified according to cosmetic appearance and function (Fig. 6.4).

After absorption from the body by dermal penetration, some cosmetic products have reported to act as endocrine disruptors, carcinogens, mutagens, neuro-, and reproductive toxins [50]. Many regulatory authorities across the world are committed to the regulation of commercial activities ensuring the safety, security, and quality control of cosmetic products. Though there are rules and quality control standards to be taken into consideration in cosmetics manufacturing, these regulatory measures are not entirely effective and they are frequently bleached due to financial reasons [17]. Some of the ingredients commonly used in the cosmetical industry are presented.

Formaldehyde and paraformaldehyde releasing compounds are an important class of broadly employed preservatives in cosmetics and personal care products. Formaldehyde, known as oxy-ethylene or formalin (37% concentrated solution of formaldehyde), and paraformaldehyde (polymer derived from formaldehyde) are used in various cosmetic products including liquid soaps, shampoos, shower creams, and lotions. They are used as a disinfectants and preservatives. Many cosmetics use formaldehyde-releasing resins, such as toluene sulfonamide, to provide minute amounts of formaldehyde over long periods of time, ensuring product viability and protecting the product from various microorganisms [53]. Five formaldeyde releasing compounds namely imidazolidinyl urea, diazolidinyl urea, Quaternium-15, 1,3-Bis (hydroxyl methyl)-5,5-dimethyl imidazolidine 2, 4-dione (DMDM) hydantoin, and 2-bromo-2-Nitropropane-1, 3-diol are most commonly used preservatives in the US. Inhalation is considered the prominent cause of formaldehyde acquaintance [17].

Both the imidazolidinyl urea and diazolidinyl urea are well-known formaldehyde releasing compounds and among them, imidazolidinyl urea is the most commonly used. It is used as antimicrobial preservative in several cosmetics and personal care products. Based on clinical trials, imidazolidinyl urea has been identified as a causative agent of dermatitis allergies. The treatment of human fibroblast cells with different concentrations imidazolidinyl urea induced an increased inflammation and it led to a significant loss of cell viability. A 24 h exposure of 0.01–1% imidazolidinyl urea to cultured cells caused moderately cytotoxic effects [17].

Diazolidinyl urea has been used since 1982 as an antimicrobial preservative in manufacturing of various cosmetics and personal care products particularly

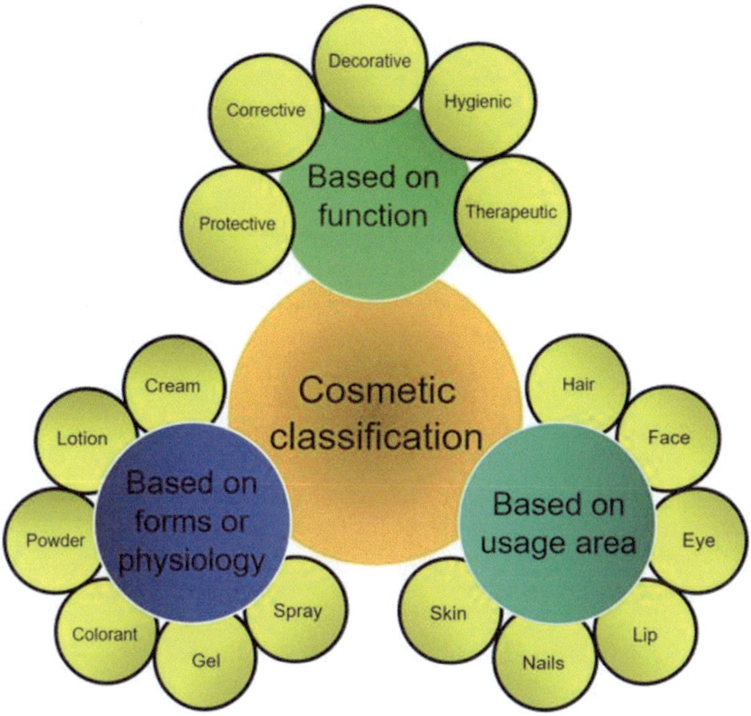

Fig. 6.4 Classification of cosmeticals based on appearance and function

childcare products, skincare products, eye and face makeup and hair, and nails. However, its effect was noted within 3 h at diazolidinyl urea concentrations of 0.1–0.5%. In Europe, the level of diazolidinyl urea in cosmeticals is restricted up to a concentration of 0.5%. Its exposure by dermal applications triggers allergic contact dermatitis. It is also categorized as a carcinogenic and mutagenic agent, because of releasing formaldehyde. In the USA and Europe, diazolidinyl urea is utilized between 5 and 8% of total cosmetics causing a 3.5% dermal irritation rate [17].

Methyl-chloro-isothiazolinone, methyl-isothiazolinone, methyl-chloro-isothiazolinone, and methyl-isothiazolinone (MCI/MI) are widely used as preservatives in personal manufacturing hygiene and cosmetic products due to their strong biocidal effects [17].

Parabens, the esters of para-hydroxybenzoic acid, are used as preservatives in cosmetics including makeup, moisturizers, toothpaste, hair, shaving products, and topical drugs. Parabens are found as additives in processed foods, as well as naturally in blueberries. In combination with other preservatives, parabens inhibit microorganism growth, protecting the user from infection and the product from microbial contamination. Methyl-paraben, ethyl-paraben, benzyl-paraben, butyl-paraben, and propyl-paraben are common congeners used [17]. Low cost, broad activity spectrum, stability over a wide-ranging pH, negligible toxicity, and worldwide regulatory acceptance of paraben, are the major characteristics for their popularity and extensive utilization as preservatives [19]. The use of a combination of preservatives allows manufacturers to use a lesser quantity of paraben, thereby decreasing any associated risks. The three most common parabens found in cosmetics are methyl-paraben (E218), propyl-paraben (E216), and butyl-paraben. Studies have shown that paraben toxicity is largely questioned [20].

In contrast to other cosmetics, methyl-paraben and propyl-paraben are particularly preferred and often used in combination in the cosmetic products. Butyl-paraben is a routinely employed preservative and antimicrobial agent in some cosmetics, baby products, food, and drug manufacture. It is found that the continuous application of these products at the maximum levels may exert a potential risk to human health [17].

Phthalates are derivatives of phthalic acid and they are manufactured as the principal type of plasticizers with worldwide utilization as components of numerous products. Low molecular weight phthalate compounds, namely di-n-butyl phthalate (DnBP) and diethyl-phthalate (DEP), are commonly used in the formulation of cosmetics products, e.g., perfumes, lotions, nail polish, and hair care products. These chemicals are often found in products as plasticizers, solvents, and alcohol denaturants. Commonly, phthalate compounds may be found in the enteric coating of medication (omeprazole, didanosine, mesalamine). The widespread use includes building materials (polyvinyl chloride), adhesives, children's toys, packaging materials, food products, and medical devices [20]. Due to their extensive global consumption (higher than 3 million tons phthalates/year) and environmental ubiquity, human exposure to phthalates is ineventable [17].

Benzalkonium chloride (BAC) is among one of the most common and widely used preservatives in cosmetics. Belonging to the family of cationic surfactants and detergent preservatives, BAC is a quaternary ammonium compound (QAC). It is used as an active ingredient in many cosmetic, personal care, and pharmaceutical products preparations, as well as in some disinfectant solutions owing to its antimicrobial activity. According to the report of Cosmetic Ingredient Review Expert Panel (2008), BAC was utilized in 83 cosmetic products at a concentration of $\leq$0.1%–5% in 1986, and its use is increased to 567 cosmeceuticals (0.46% of total cosmetic products) at concentrations ranging from 0.01% to 0.5% in 2013. Besides being an active ingredient, it remains the most commonly used preservative in topical ophthalmic products and solutions. Although innocent in first place, the use of BAC has been associated to some adverse effects also [17].

Methyl-dibromo-glutaronitrile/phenoxy-ethanol (MDBGN/PE) is used in the manufacture of moisturizing creams as a preservative. Since the 1980s, MDBGN has been used as a biocidal preservative being effective at low concentrations and as a weak sensitizer. Due to these favorable attributes, MDBGN became increasingly popular during the 1990s [17].

Dioxane. The 1,4-dioxane is a commonly found detergent, emulsifier, or solvent in cleansing agents such as shampoo, toothpaste, and mouthwash. Dioxane reduces the harsh nature of compounds, such as sodium sulfate and it enhances foaming capabilities. Although dioxane is not a listed ingredient in cosmetics per se, it is a contaminant from the manufacturing of polyethylene-glycol, poly-ethylene, and polyoxy-ethylene. The FDA does not regulate the content of dioxane in consumer products; however, it does recommend that manufacturers remove dioxane with vacuum stripping [20].

1,4-dioxane (C4H8O2, dioxane). It is an ether with emulsifying, detergent and solvent properties making it a preferred additive in shampoos, mouthwashes, and toothpaste. Accidental significant exposure to this chemical by ingestion can function as a potent carcinogen eliciting cancer of breast, skin, and liver and as potent inducer of irritation in the nasal cavity. It possesses endocrine disruption properties [17, 19].

All in all, the list chemicals with potential or established toxicity profile is noticeable. Trace metals especially in lip cosmetics has become one of the most commonly used cosmetics worldwide. Introduction of trace metals cosmetics, including antimony (Sb), arsenic (As), cadmium (Cd), chromium (Cr), cobalt (Co), copper (Cu), nickel (Ni), manganese (Mn), and lead (Pb) is discussed further [17].

Arsenic (As) used in skin cream and makeup powder, eyeshadows, eye pencils, shampoo, hair gel and conditioner, and showering products could potentially cause skin disease, lung cancer, cardiovascular disease and peripheral neuropathy, and an increased risk of gastrointestinal and urinary tract cancers. It exhibits a high affinity for the skin, and therefore long-term contact results in adverse consequences including variability of skin eruptions and skin cancer [17].

Inhalation of *antimony metal (Sb)*, primarily found in lipsticks, face powder, skin creams, eye pencils, and eyeshadows, could cause respiratory disorders (bronchitis, emphysema, altered pulmonary functions, pneumoconiosis) and adverse gastrointestinal effects, i.e., vomiting, diarrhea, ulcers, and abdominal pain. Sb level in the range of 0.008 to 0.39 μg/g was examined in lipsticks and levels of 0.179–0.46, 0.0134–0.265, and 0.157–0.76 μg/g for makeup powder, skin cream, and eyeshadow, respectively [17].

Owing to its coloring property, *cadmium (Cd)* is used in different cosmetic products, i.e., lipsticks, skin, and hair creams in concentrations ranging from 4.9 to 10.6 μg/g in lip cosmetics and quite low in the other products. The chronic exposure to Cd may cause skin tumors and kidney damage. It can also cause bone brittleness and an increased risk for bone fractures [17].

Mercury in cosmetic products is used in lipsticks, shampoos, hair conditioners and gels, nail polishes, eyeliners, and shadows with varying levels between 0.01 and 12,590 μg/g. Soaps, creams, and contain high concentrations of Hg (between 204 and 4700 μg/g). Cosmetic application may be toxic and cause neurological, nephro-, and gastrointestinal disorders [17].

Cobalt and nickel metals commonly found in lipstick, eyeshadow, face paint, and hair cream are associated with contact dermatitis. Cobalt and its salts are widely used as coloring agents in makeup and light brown hair dyes. Generally, substantial levels of cobalt were detected in face paints, eyeshadows, henna dyes, and hair creams. The highest levels of cobalt, between 122.78 and 253.33 μg/g, were present in the eye shadows of Chinese products. A cobalt concentration ranging from 0.055

to 0.105, 1.05–2.42, and 3.21–5.64 μg/g has been recorded for lipstick, eye shadow, and powdered makeup, respectively [17].

Inorganic lead is found in numerous cosmetics, such as hair dyes (lead acetate) and lipsticks, eye shadow, eye pencil, hair cream. Lead content ranges from 0.25 to 81.50 μg/g as it was detected in Italian, Chinese, and US samples. A high concentration of lead was recorded in shampoos (0.98–1.59 μg/g) and soaps (3.8–4.63 μg/g). Ingestion of large quantities of the lead may interfere with heme and calcium channels synthesis, which are important for nerve transmission [51]. FDA ruled that progressive hair dyes could not contain lead acetate at a concentration exceeded 0.6% w/v and implemented specific regulations for progressive hair dye use [20].

Prevention of Skin Damage

Facial changes associated with aging are skeletal atrophy giving rise to less support to overlying tissues, to the loss of facial fat pad volume leading to the downward migration of esthetic lines that results in folds, deep wrinkles giving a gaunt appearance with protrusion of infraorbital fat pads, and thinning of the lips as well as jaws. A triangular facial shape gives the impression of an aging face, in contrast to an inverted triangle face represents youthfulness. Whilst facial contouring can be addressed by means of surgical and nonsurgical procedures as injections of neurotoxins and synthetic volumizing or stimulatory fillers, topical therapies have been developed to enhance the effects of ablative procedures such as chemical peels, microneedling, and CO_2 ablative laser resurfacing [18]. The onset of action, effectiveness, and tolerance for chemical peels are slow and not immediately observable [18]. Regarding the field of cosmetics, there are products with anti aging properties as well as products which offer sun protection.

Sunscreens can be broadly categorized functionally into absorbers and reflectors. They contain organic chemical compounds (benzo-phenones, avo-benzone, ecamsule, methyl anthranilate, salicylates, cinnamates, etc.) and inorganic metal particles, respectively [18]. Organic compounds absorb solar UV radiation and dissipate it by means of a series of exothermic reactions or high wavelength and low energy irradiation into the surroundings.

Benzophenones (benzophenone-3, oxybenzophenone) are predominant UVB absorbers (peak absorption at 290 nm) with weak absorption of UVA. They are added on products due to the fact that in high concentrations may render the end product sticky and esthetically unpleasant. They have the highest bioavailability and they are readily absorbed into skin and into circulation. They accounted for the highest incidence of photodermatitis in the literature of the last decade.

Avobenzone is a highly effective UVA absorber with UVB coverage (290–400 nm). It has the best safety profile amongst all sunscreen ingredients, but its photoinstability and incompatibility with metal oxide sunscreen ingredients renders it an unpopular choice with the brand names Helioplex TM (Neutrogena, Johnson & Johnson) and ecamsule patented by L'Oreal under the trade name Mexoryl TM.

Methyl anthranilate is safe and effective organic sunscreen ingredient frequently found in formulations that confer photoprotection against UVA. Its stickiness renders it aesthetically unpleasant in higher concentrations.

Titanium dioxide and zinc oxide are metal oxides. In the form of microparticles, these inorganic compounds deflect and they protect the skin beneath from solar UV radiation. Nanoparticles have been developed in an attempt in to overcome esthetic appearance.

Reversal of Photoaging [18]

All-trans retinoic acid is a chemical with hormone function. It moves to the nucleus, where it interacts with the corresponding receptors, retinoic acid receptors (RAR), and retinoid x receptors (RXR) to activate or inhibit the transcription process of elements that modulate keratinization of epidermis, synthesis of collagen, and production of matrix metalloproteinases (MMPs). The antiwrinkle effect is therefore attributed to the combined effect of stimulated pro collagen I synthesis by cell signaling pathways and the inhibition of UV induced damage by MMP. Additionally, it promotes a relatively minor lightening effect by suppressing expression of the enzyme tyrosinase in melanin synthesis, although the mechanism is not completely understood.

Tretinoin is a topical vitamin A application. CEM Griffiths (1992) produced the better concentrations promoting effects on photoaged skin. Based on an abundance of clinical evidence, over-the-counter retinoids are developed from such intermediate forms as retinol and retinyl esters that may produce the same effects and that they may cause less side effects. The basic side effect is irritation of the skin. To notice, these chemicals are unstable upon exposure to sunlight and atmospheric oxygen; therefore, extra caution must be taken with storage.

Antioxidants confer ability of the skin to ameliorate oxidative stress either as topical applications or as ingested compounds. The well known antioxidant is *vitamin C*. Humans lack the ability to synthesize vitamin C due to a mutated gene present in most mammals and therefore are reliant on dietary sources. Only a small proportion of what is absorbed will end up in the skin. Thus, topical vitamin C needs to be in high concentrations (15% L-ascorbic acid) in order to achieve efficacy. It has an *in vivo* half-life of 4 days and the ideal PH for action is pH 3.5. However, either vitamin C or vitamin E alone is ineffective in preventing UV radiation-induced skin damage. Topical 15% vitamin C in combination with topical 1% vitamin E (alpha-tocopherol) can provide synergistic protection against oxidative stress in the skin. Adding ferulic acid (a botanical chemical, a potent phenolic antioxidant found in plants) to vitamins C and E products result in a fourfold increase in the protection against UV induced photodamage.

Vitamin B3 or niacin is interconvertible with niacinamide. It is a precursor molecule to a vast number of coenzymes that play a role in the neutralizing oxidative stress in over 40 cellular biochemical reactions. Studies focus regarding the *in vivo* effects of topical niacinamide (2–5%) using sophisticated specialized skin analysis imaging systems, that capture high-quality, standardized facial images. The antiaging effects consist in improvement of skin barrier and skin tone, reduction of fine wrinkles and reduction of hyperpigmentation. Like most other antioxidants, niacinamide is chemically unstable, when exposed to the atmosphere. Formulating products, in the pH range of 4–7, are produced to avoid hydrolysis which converts niacinamide to nicotinic acid, that is irritant to skin. Nicotinate esters, such as tocopheryl nicotinate and methyl nicotinate have been developed to address skin irritability, but evidence for efficacy is meager.

Alpha hydroxyl acids (AHAs) are among the first to be discovered for skin rejuvenation in two formulations: In high concentrations (e.g., glycolic acid 35% and above), these promote epidermis exfoliation; in low concentrations, they normalize epidermal layers by thinning stratum corneum, whilst promoting thickening of granular layer.

Polyhydoxy (PHA) were subsequently developed that would rectify the irritating effect of alpha hydroxy acids on skin, whilst retaining their skin rejuvenation effects. They function by downregulating the gene responsible for the UV induced production of dystrophic elastic fibers. In addition, antioxidative properties of gluconolactone and lactobionic acid have been demonstrated with the inhibition of oxidative discoloration.

Peptides are long-chain structures rendering the molecules lipophilic, able to absorb water content, and confer hydration to the corneal layer of skin. However, in recent years, scientific interest have evolved into the use of peptides as cellular messengers. Amongst many peptides that have been developed to reduce wrinkles by targeting fibroblasts and stimulating collagenases, one of the best known, palmitoyl peptides has been developed by the French manufacturer Sederma (Sederma, France), made up of palmitoyl-lysine, threonine, lysine, and serine. At a molecular level, this pentapeptide is a fragment of procollagen I making it more lipophilic by palmitoylation. In a double-blind, randomized, placebo-controlled trial in the United States, digital facial imaging offered the best objective evidence of significant improvement in skin texture with reductions in wrinkles at 8 weeks, without a report of skin intolerance.

Topical transforming growth factor (TGF)-beta 1 has been extracted from fetal dermal fibroblasts to develop a skin cream containing a mixture of human growth factors and cytokines. Histology and ultrastructural analysis have demonstrated evidence of skin rejuvenation with thickening of epidermis and reduced solar elastosis and neocollagenesis, which correlates well with the improved appearance of skin texture and wrinkles after 6 months of twice daily topical application (San Francisco, CA, USA). In addition, TGF-beta 1 mediate lightening effects by an increased degradation of important copper-containing enzymes for melanin production as tyrosinase and tyrosinase-related protein 1.

On that note, *stem cells* derived from Swiss Uttwiler Spatlauber apple trees have been extensively studied *in vitro* for their potential effects of skin rejuvenation. Whilst cosmetical products that contain Swiss apple stem cell extract have been developed with significant antiwrinkle effects evident in as little as 2 weeks. Insufficient data are available as evidence of a plausible mechanism of action.

Hydroquinone in low concentrations (2%) had been the standard over-the-counter (OTC) treatment for dyspigmentation for a half century. It binds onto the pigment-producing enzyme tyrosinase to inhibit its activity, thereby inhibiting melanosome production. Due to controversies over the safety of hydroquinone in cosmetical products and an association with malignancies, restricts its constant use.

The need for an alternative skin lightening agent has led to enhanced research and development of some of the currently widely available ingredients in cosmetical industry, such as topical retinoids, vitamins C and B3.

Biological products, like marine extracts, are used as active ingredients in cosmetical. The efficacy and safety depend on the selection of an appropriate extraction process. The end product acquires various physicochemical properties depending on the source material, the type and concentration of the extracting solvent, the ambient pH, and the temperature and pressure conditions of the extraction methodology.

Effects on Health

Depending on their chemical composition, cosmeticals are related to various side effects. A significant number of them are related to skin disorders including contact dermatitis. Although complications due to cosmetical products can be severe like anaphylactic shock, burns or milder as irritant dermatitis, contact allergic dermatitis, etc., are relatively rare. Provided they are used in moderation and as per the prescribing physician's advice with essential precautions followed by the patients, they are quite safe.

The allergenic, carcinogenic, and mutagenic nature of formaldehyde has provoked pronounced public health concern. It is related to a greater risk of developing myeloid leukemia. It could trigger ophthalmic irritation, including conjunctival redness and increased blinking frequency, whereas nose and throat irritations occur beyond its 1.0 ppm concentrations [19].

In 2006, the International Agency for Research on Cancer released a document examining carcinogenic risks of exposure to high concentrations of gaseous formaldehyde, including increased risk of developing nasopharyngeal cancer. In 2009, studies showed that high levels of formaldehyde exposure were correlated with an increased risk in the development of myeloid leukemia. Recently, the amount of formaldehyde in hair smoothing products was deemed dangerous, as formaldehyde released into the air during use was at quantities exceeding the Occupational Safety and Health Administration limit of 0.75 ppm. However, the FDA has ruled that although some hair-smoothing products and fractional photothermolysis solutions exceed the Occupational Safety and Health Administration's limits, the danger of these products vary because of the amount of formaldehyde contained and the intended use [20].

Imidazolidinyl urea and diazolidinyl urea are degraded to formaldehyde products causing different side effects and they are also related to cancer (Fig. 6.5).

Isothiazolinone derivatives are associated with high cytotoxic effects. These substances are characterized as allergens causing hypersensitivity due to its persistent exposure by cosmetic products. Sensitization to derivatives has increased among cleaning professionals, resulting in allergic contact dermatitis. Due to this reason, the Scientific Committee on Consumer Safety of the European Commission

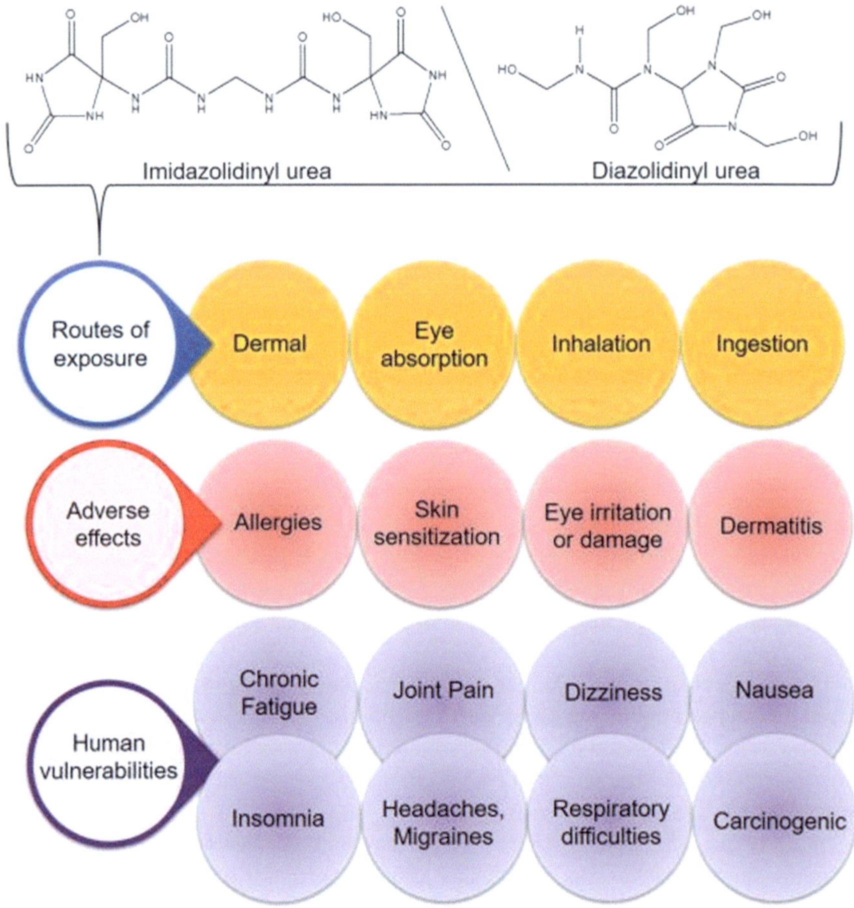

Fig. 6.5 Side effects of imidazolidinyl urea and diazolidinyl urea

prohibited their use in leave on cosmetics, while restricting that in rinse off products. Other less-known isothiazolinone derivatives including benzisothiazolinone and octylisothiazolinone may also trigger skin allergic reactions in humans [17].

The multiple effects of parabens have previously been described. The main toxic effect is the function as endocrine disruptor (Fig. 6.6). Recently, research identified side effects of parabens, mainly on male reproductive health, but also the female reproductive system, with the detection of parabens in the mammary gland and breast cancer tissues and there are reports on genotoxic effects on lymphocytes. Since 2011, the Danish Environmental Protection Agency also banned the use of butyl and propyl-paraben in cosmetic products for children under three years owing to their harmful consequences for newborns and children [19].

Six of the phthalate esters including DEP, DMP, DnBP, BBzP, DOP, and DEHP are currently declared as pollutants. Among these phthalate compounds, diethyl phthalate and dimethyl phthalate are the most common forms. Inhalation, ingestion,

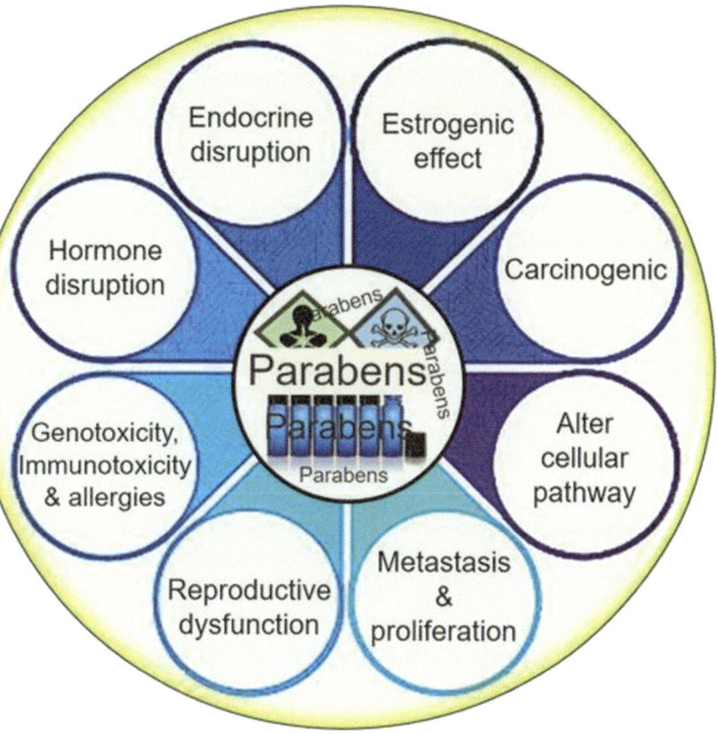

Fig. 6.6 The multiple toxic effects of parabens

and dermal absorption have been demonstrated as the major routes to phthalate exposure following the use of personal care or cosmetics products [17]. Studies show that there is an increased level of phthalates in the indoor atmosphere in urban areas. Due to their volatility, phthalates are found in the air, especially in high temperature climates. Exposure through the diet is possible, with fatty foods, such as dairy and meat containing higher concentrations. It is most likely that the low molecular weight phthalates will be ingested or dermally absorbed, whereas inhalation will occur with volatile phthalates found freely in the air [20].

The FDA has studied the phthalate content in a range of cosmetics. An extended study showed that high concentrations were found in 31 out of 60 adult beauty products (including deodorants, perfumes, lotions, washes, nail polishes, and hair care products) while in 20 products DEP ranged from 80 to 36,006 ug/g. Furthermore, 24 baby products contained diethyl phthalate in amounts ranging from 10 to 274 ug/g. Airborne phthalates were found in polyvinyl chloride flooring and their concentrations may correlate with an increased risk for asthma development [20].

Human evidence of many adverse phthalate-associated public health hazards includes endocrine disruption, reproductive toxicity, carcinogenesis, and other metabolic disorders. DEHP and BBP have been categorized as possible human carcinogens [51]. Several reports have confirmed the relationship between children's exposure to phthalate such as improper sexual development, adverse immune, neuronal, and respiratory disorders, i.e., asthma, and bronchial obstruction [17].

On the other hand, phthalate-associated consequences on adults include early puberty, embryonic loss, abnormal pregnancy duration, pregnancy loss, rise in body weight and obesity, insulin resistance and diabetes, increased risk of breast cancer, preterm birth, and abnormal male's pulmonary functions. Various *in vitro* and *in vivo* experimental studies collaborated that it may also exacerbate hepatic tumorigenesis and hepatotoxicity as well as colon tumorigenesis [17]. DEHP raises melanoma tumor progression [17].

MDBGN is the leading cause of contact dermatitis, representing a 4.5% occurrence in the population. The patch testing data from sixteen different European clinics determined an increasing frequency of MDBGN sensitivity in eczema patients from 0.7% in 1991 to 3.5% in 2000. Considering these data, the Scientific Committee on Cosmetic Products and Non-Food Products of the European Commission banned the application of MDBGN in leave-on cosmetic products and limited to rinse-off products [17].

Metals used in cosmetical possess also toxic function (Fig. 6.7).

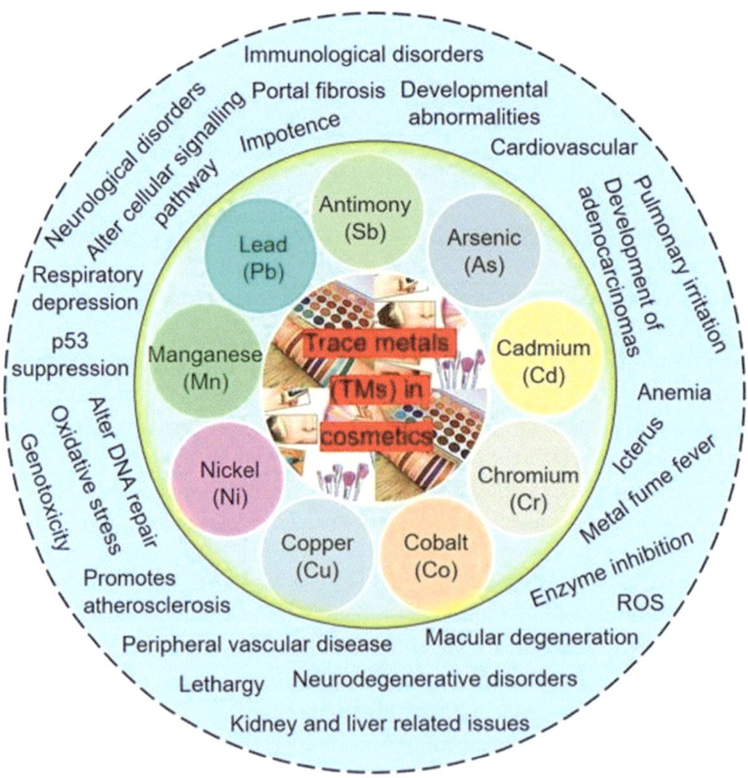

Fig. 6.7 The side effects of metals used in cosmetical industry [17]

The consumer manufacturer enthusiasm revolving around cosmetical has unfortunately led into addition to many chemicals as additives in the commercially available preparations to enhance their properties, qualities, significance, performance, and viability. However, established and potential toxicity of some important chemical ingredients frequently used in cosmetical could be further considered.

Consumer Products

As the variety of chemicals used in consumer products (CPs) has increased, concerns about human health risk have grown accordingly. Even though restrictive guidelines and regulations have taken place to minimize the risks, human exposure to these chemicals and their eco-compatibility have remained a matter of greater scientific concern over the years. A major challenge in understanding the reality of the exposure is the lack of available information on the increasing number of ingredients and additives in the products. Even when ingredients of CP formulations are identified on the product containers, the concentrations of the chemicals are rarely known to consumers. This study is from the few in the literature that measurement of chemical compounds in consumer products has been accurately detected.

In South Korea, 52 deaths and 122 injuries occurred due to the inhalation of aerosolized water containing disinfectants from a humidifier that led to serious lung injuries. Moreover, around 23 million people in the United States of America are currently affected by asthma, which has been suspected as a consequence of chemical exposure by consumption of CPs. According to the National Report on Human Exposure to Chemicals, most of America's population in every age group has detectable levels of phthalate metabolites, bisphenol A, triclosan, and other common EDs in their urine. Furthermore, after usage, chemicals are released directly or indirectly to the environment by wastewater treatment plants resulting in threats to the ecosystem, mainly chemicals that are persistent and cause ecotoxicity.

Among analyzed consumer products, isothiazolinones and phthalates were found in 47% and in 24% of the samples. The highest concentrations were 518 mg/kg for benziso-thiazolone, 7.1 mg/kg for methyl-isothiazolinone, 2.0 mg/kg for diethyl-phthalate, and 21 mg/kg for dimethyl-phthalate. Suspect and non-target analyses yielded six tentatively identified chemicals across the products including benzophenone, ricinine, iodocarb, galaxolidone, triethanolamine, and 2-(2H-Benzotriazol-2-yl)-4, 6-bis (1-methyl-1-phenylethyl) phenol. Our results revealed that selected CPs consistently contain chemicals from multiple classes. Excessive use of these chemicals in daily life can increase the risk for human health and the environment.

Isothiazolinones. Among the analysed products, benzothiazolinone (BIT) was detected in all laundry detergent samples (100%) and frequently found in fabric softener (67%), while methylisothiazolinone (MIT) was detected in body washes (100%) and fabric softener (53%). According to the EU Regulation, BIT is not allowed in leave on and rinse off cosmetics, while MIT and *isothiazolinones* (CMI) can still be used with maximum permissible concentrations of 100 mg/kg and 15 mg/kg, respectively.

Phthalates. Among the eight categories of products analyzed, phthalates were detected in 16 out of 85 products. These were frequently found in lipstick and face cleansers, and rarely detected in dishwasher detergents, fabric softeners, and shampoos. However, the low detection rate in other products suggests that the phthalates might move from plastic packaging materials, as they are the major plasticizer used in polyvinyl chloride (PVC) plastics.

Benzophenone was detected in 33% of the samples and it was mostly found in dishwashing detergents, fabric softener, face cleaner, body washes, and lipsticks. It is mainly used as a sunscreen agent and stabilizer in cosmetics, but also used as a stabilizing agent for plastics and rubber to prevent polymer degradation. Widespread human exposure to benzophenone occurs mainly through the skin and is excreted mostly via urine.

Ricinine is a plant-originated alkaloid used in stabilizers, skin-conditioning agents, emulsion, and surfactants in cosmetics. Due to its relatively high toxicity (LD50 value for mice is 2–3 μg/kg) [21].

Triethanolamine was detected in 28% of samples and was frequently found in laundry detergents (67%), shampoos (42%), and hair dyes (60%). It is used as a surfactant in cleaning formulation, such as soaps and detergents. Fortunately, triethanolamine shows low acute toxicity with LD50 values ranging from 4.19 to 11.26 g/kg; however, it is an irritant to the eyes and skin and it can cause systemic toxicity mainly in the kidneys, liver, red blood cells, and the nervous system.

Iodocarb (IPBC) was detected in 6% of the samples and mostly found in body washes. Due to its good antimicrobial properties, it is frequently used as a preservative in CPs. The maximum concentration of iodocarb allowed in cosmetics is 0.01%. A product exceeding the allowance level should be labeled "irritant" unless otherwise exempted.

Galaxolidone was detected in 32% of the samples and frequently detected in face washes (41%), shampoos (67%), laundry detergents (40%), and fabric softeners (40%). It is indeed a metabolite of galaxolide used as fragrances or odorous components in a variety of CPs. It can concentrate in the blood, fat, and breast milk, and it can affect androgen and progesterone receptors and it also causes stimulation of estrogenic receptors in humans.

Workplace Exposures

Data for 17,524 adults who worked during the 12 months preceding the interview are presented in 2010. The highest prevalence rates of hazardous workplace exposures were typically in agriculture, mining, and construction [22]. The prevalence rate of frequent handling and skin contact with chemicals was highest in mining and construction. After mining and construction, the industry category includes arts, entertainment, and recreation (18.2%) which showed the highest observed prevalence rate for exposure to chemicals at the workplace. Outdoor exposure was most common in agriculture, forestry, fishing, and hunting (85%); construction (73%); and mining (65%). Finally, frequent occupational exposure to VGDF (vapors, gases, dusts, and fumes) was most common among mining (67%), agriculture (53%), and construction workers (51%).

Within manufacturing, workers employed in primary metal (e.g., steel mills and foundries) had high prevalence rates of skin contact with chemicals (48.1%) and VDGF exposure (65.5%). Within other services, workers employed in repair and maintenance, and personal services (e.g., beauty salons) had high prevalence rates of skin contact with chemicals (52.4% and 51.5%, respectively) and VDGF exposure (68.0% and 48.1%, respectively). Industry categories that tended to have the lowest prevalence rate of workers exposed to these work place hazards included several services subsectors: finance and insurance; professional, scientific, and technical services; and education services.

The prevalence rate of frequent occupational skin contacts with chemicals during the 12 months preceding the interview among sample adult workers was 20.6%. The prevalence rate was higher among workers who were male (23.4%) compared to females (17.4%); lowest among workers aged ≥65 years old (11.1%) compared to other ages; highest among workers who had a high school diploma (27.1%) or less education (27.3%) compared to those with higher education levels; and, highest among workers who did not reside in a MSA (26.1%) compared to residing in one. *MSA* is a global organization committed to developing, manufacturing, and supplying safety products that help protect people and facility infrastructures.

The three occupational categories with the highest observed prevalence rate of workers who frequently worked outdoors were farming (90.1%), construction and extraction occupations (79.5%), and protective service occupations (i.e., police, firefighters, and guards, 59.3%). The occupation category with the highest observed prevalence rate of workers with frequent skin contact with chemicals was building and grounds cleaning and maintenance occupations (55.4%).

Working outdoors under hot and humid conditions is a risk factor for heat related illness. In a study of heat-related illness in Washington State in the years 2000 to 2009, industries with the highest risk for heat-related illness were construction (especially roofers and highway, street, and bridge construction), public administration (especially firefighting), and agriculture. We also found other industries and occupations with a high prevalence rate of outdoor work (e.g., mining, utilities, and real estate rental and leasing) which were not reported to be at increased risk for heat-related illness in the Washington State study.

Biomass Management

As petroleum-based resources continue to exhaust, the increased utilization of renewable feedstock has attracted scientific and industrial attention. Biomass is the most abundant renewable resource, of energy and other sources with an annual global production of approximately 1.7×10^{11} metric tons. The main constituents of biomass are cellulose (30–50 wt.%), hemicellulose (20–40 wt.%), and lignin (10–20 wt.%). Conversion of biomass into value added fuels and chemicals is gathering increased attention and many catalysts and exploitation methods have been developed recently (Fig. 6.8).

Catalytic conversion of cellulosic biomass can produce a variety of platform chemicals, such as ethanol, 3-hydroxypropionic acid, succinic acid, furfural,

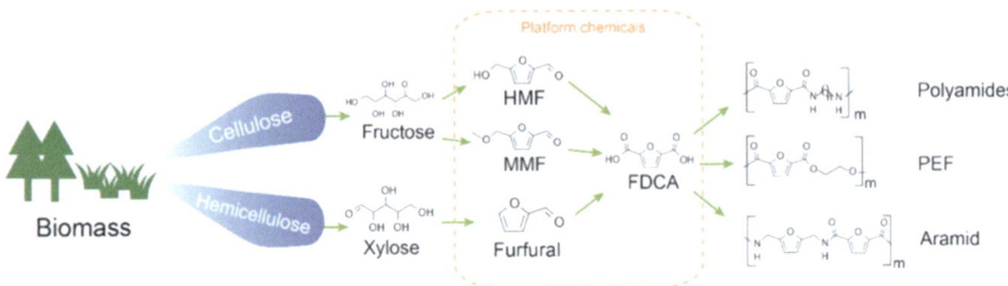

Fig. 6.8 Biomass constituents

levulinic acid, 5- hydroxyl methylfurfural (HMF), 5-methoxy methylfurfural (MMF), and 2,5-furandicarboxylic acid (FDCA) [23].

Among these, *FDCA* is a versatile compound that has been listed as one of the top 12 biobased chemicals by the United States. FDCA is a dicarboxylic acid compound that is widely used as a monomer in the production of a variety of bio-based polymers, such as polyesters, polyamides, and polyurethanes. Since the structure and chemical properties of FDCA are similar to petrol based terephthalic acid, the most exciting application of FDCA is synthesizing renewable poly (ethylene 2,5-furandicarboxylate) (PEF), a substitute for poly (ethylene terephthalate) (PET), which has been widely used in the production of fibers, films, beverage bottles, and food packaging. Chemo-catalysis has become the dominant pathway used for the production of FDCA.

Furan compounds are considered promising building blocks, because of their high chemical versatility and functionality. However, their inherent instability and toxicity to living microorganisms have impeded their practical application and bio-upgrading. Furan aldehydes, particularly furfural and HMF, have attracted increasing industrial research attention. Furfural and HMF can be generated by the triple dehydration of pentoses (xylose) and hexoses (fructose), respectively. Typically, HMF forms spontaneously when sugar is heated, particularly at acidic conditions. Furfural and HMF are found in sugar rich or baked foods, such as coffee, honey, dried fruits, and bread. Humans can consume approximately 30 to 150 mg HMF by their daily diet. They are considered markers of the age of spirits, browning indicators in bread, and biomarkers of excessive storage temperature in juice production. Some studies reported that furfural and HMF are fungal metabolites. In general, aldehydes like furfural and HMF are highly reactive compounds, which can result in the generation of reactive oxygen species. It is worth mentioning that HMF can be useful in the treatment of human diseases, although there is evidence that high doses of furan aldehydes have potential carcinogenic effects. The toxicity of furfural and HMF is one of the most challenging problems in biomass technology, because these compounds inhibit the metabolism and growth of microorganisms by affecting the activities of enzymes involved in glycolysis and the tricarboxylic acid cycle.

Managing Microplastics and Soil Contamination

A main problem of the modern world is management of waste. Possibly the biggest landfill is the ocean. It accumulates various types of waste, called marine litter. This includes metals, glass, ceramics, textiles, paper, and timber. Yet, the largest and probably most harmful fraction of marine litter is plastic debris [24]. *Microplastic and nanoparticles* are widely used in various fields, such as drug carriers, semiconductors, clothing, cosmetics, etc., involved in various aspects of social production and life. A large number of non degradable plastics enter the marine environment and they progressively evolve into microplastic particles (<1 mm) or even smaller. The security of microplastic and nanoparticles appears to be ambiguous [25].

It is estimated that every year between 4.8 and 12.7 million tons of plastic waste enter the ocean from land based sources alone. The high input is linked to plastics durability and lightweight which allows, unlike other types of marine litter, to be easily transported by the wind, waste water, and rivers, when it has been inadequately disposed. This is not only true for macroplastics, but also for microplastics from e.g., car tire wear, clothes washing, and cosmetic products. Marine litter also originates from sea based activities, for example from fishing gear, cargo that are lost, abandoned, or discarded despite international legislation that prohibits the dumping of waste at sea (International Convention for the Prevention of Pollution from Ships, 1988; 1995) [24].

The United Nations Environment Program, UNEP, estimates that up to 18,000 pieces of plastic debris are floating on every square kilometer of ocean. Plastic fragments can be found across the Southwest Pacific in surprisingly high quantities, even in remote and non industrialized places. The durability and slow rate of degradation allow these fragments, constituted by synthetic polymers, to withstand the ocean environment for years to decades [25]. Marine litter is found in all seven ocean compartments: biota, coastlines, at the sea surface and the seafloor as well as in sediments, sea ice, and the water column. The contact with biota has been reported to affect up to 1400 marine species worldwide [24].

Marine contamination caused by plastic debris represents a global problem that has become particularly relevant in recent years, due to the direct impact these pollutants have on the environment, to their potential effects on human health and the economy [25]. The plastic debris follows a geographic distribution and the quantity as well as treatment differs between countries. Many of these are wealthy countries that primarily contribute to the problem. Over 250 thousand tons have already been removed, but there is little or no information available regarding how this waste is treated or used post collection [24].

Marine litter can also lead to a decline in tourism, when washed ashore on beaches, destroy corals at the seafloor, spread invasively to animal species, and they cause accidents in ships. The plastic fraction of marine litter can be particularly harmful, because it is able to travel over long time and distances before breaking down into smaller fragments. Microplastics can then adsorb and release chemicals, which become bioavailable after ingestion [24]. Moreover, it is found that the nano-Fe_3O_4 and polyacrylate styrene components are irreversibly detrimental to the human body, which is manifested as the permanent damage of the respiratory system and even invasion of the lymphatic and epithelial tissues [25]. Ingested and airborne substances are a risk for humans because they can accumulate along the biological food chain [24].

A new and more alarming source of marine contamination has been recently identified consisting of micro and nanosized plastic fragments. Microplastics are difficult to see with the naked eye and they biodegrade in marine environment, representing a problem since they can be ingested by plankton or other marine organisms [26].

Microplastics derive from fragmentation of debris of large dimensions (bags, packaging) or they are directly produced for a application such as abrasives or they are additives for cosmetics (such as microbeads used for skin scrubs). Another source of microplastics is the domestic and industrial washing process of synthetic clothes. The release of microplastics from synthetic clothes is caused by the mechanical and chemical stresses that fabrics undergo during a washing process in a laundry machine. Due to their dimensions, a majority of released microfibers cannot be blocked by wastewater treatment plants [26]. To date, plastic debris is almost ubiquitous in aquatic habitats.

Additional sources of marine contamination are the increasing number of electronic wastes, such as toner particles in the printers produced by nanotechnology methods. Contaminants include micro polyester, polyacrylate styrene copolymer, ethylene propylene, nano-Fe_3O_4, and SiO_2 [58]. In China, the production of toners was about 138.8 thousand tons in 2016 and approximately 10% of toners were wasted. It is worth mentioning that the toners have potential toxicity that cannot be ignored due to their free diffusion and size [25].

On a political level, the United Nations Development Goals demand that the majority of countries worldwide "prevent and significantly reduce marine pollution" by 2025 (United Nations General Assembly, 2015 [25]. The European Union has set an "Action plan for the Circular Economy" implementing a waste hierarchy in which prevention, reuse, recycling, and energy recovery are favored over landfill in this respective order (European Commission, 2015). While this may help to reduce the input of waste in the long run, in the meantime the collection of marine litter is essential to reduce its stock and impact in the ocean. Europe, North America, and Australia undertake major scientific collection efforts, but also Brazil and South Korea are strongly represented. Scientific collection projects tend to focus on the Mediterranean Sea, the Pacific Ocean, and Australia whereas non scientific collection projects are typically conducted in the northern hemisphere. Russia, East and West Africa, and the Western part of South America with the exception of Chile were not identified as locations of marine litter collections (Fig. 6.9) [24].

The scientific collection methods typically focus on the removal of marine litter by hand or by smaller gear whereas non scientific collections involve larger machines such as cranes. The collection methods that were reported within both scientific and non scientific collections were: (1) manual removal of marine litter from the coastline, (2) surface trawling, (3) use of a retention boom meaning a floating barrier that collects buoyant litter at the sea surface, (4) bottom trawling, and (5) diving to pick up marine litter from the seafloor [24].

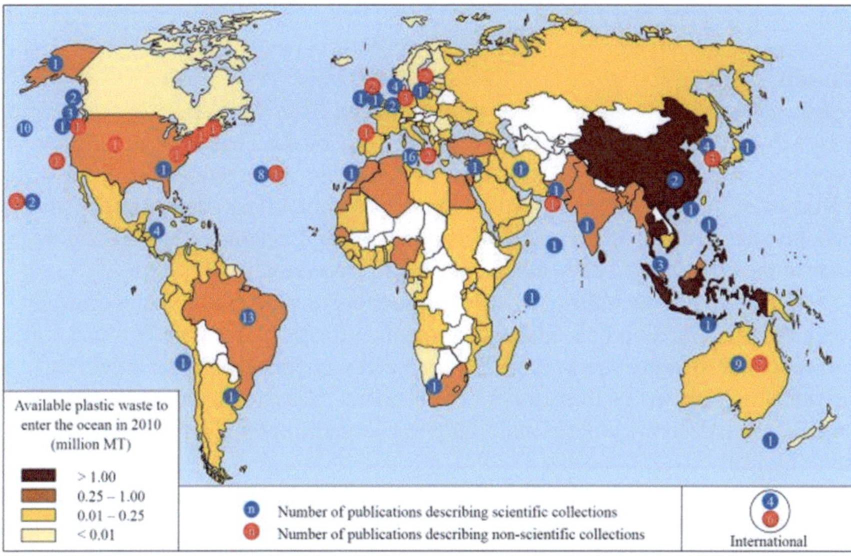

Fig. 6.9 Marine scientific collection of microplastics

The non scientific collections that described the waste treatment of collected marine litter cover all available treatment options: reuse, recycle, energy recovery, and landfill. Prevention of marine litter implies a reduction of waste input into the ocean. Massive collection of marine litter happens in the 10 top countries: China, Indonesia, Philippines, Vietnam, Sri Lanka, Thailand, Egypt, Malaysia, Nigeria, and Bangladesh, but only Malaysia, China, and the Philippines lead collection projects [61]. It is interesting that collections from coastlines comprise 60% plastic, whereas litter from the seafloor contained 52% to 76% plastics [24].

Similarly, soil contamination occurs from microplastics especially due to accumulation in landfills. Another serious problem of soil degradation is metal contamination. Mining and manufacturing industries are the main sources of heavy metals that pollute the soil, ground water, and air. The EPA regulatory limits for different metals are: Ag 0.10, Ba 2.00, Cd 5.00, Cr 0.10, Cu 1.30, Hg 2.00, Ni 0.20, Pb 15.00, Se 50.00, and Zn 0.50 (pm) [27].

A main source of soil contamination that has received publish concern are fertilizers and pesticides. A huge number of fertilizers are added to soils during intensive farming to supply adequate amount of nitrogen, phosphorus, and potassium for crop growth. Use of certain fertilizers that contain phosphorus unconsciously adds toxic elements to the soil including Fe, Hg, and Pb [27]. After application in an environment, these products persist for a period, entering a degradation process, where the biotic factors are considered the most relevant forms [28]. Additionally, several pesticides used widely in horticulture and agriculture practices from the past contained extensive heavy metal concentration. Recently in the UK, 10% of the chemicals were legalized as fungicides and insecticides which contain lead, mercury, manganese, copper, and zinc [27]. Intoxication by such substances remains a threat to human health, being responsible for 3 million poisonings and 200,000 deaths annually [28].

Biosolids are defined as organic solid produced from wastewater treatment processes that can be beneficially recycled. In most of the countries, application of bio-solid materials in agricultural land is a widespread practice that permits the reprocess of biosolids obtained from urban populations. The application of several biosolids like municipal sewage sludge, compost, and manures to land unwillingly leads to the addition of heavy metals in the soil environment [27].

Pollution caused by atmospheric deposition due to air pollution from transportation and industrial activities add on to the contamination of soil and ground water. Pollution due to atmospheric activities has been recognized as a potential threat in the industrial regions of the northern hemisphere. Waste water irrigation has been established to decrease soil pH and to increase organic carbon content and soil conductivity [29].

The methods used for the detoxification of the soil are chemical remediation using high temperatures, chemical dilution using solvents, chemical fixation, biological remediation, phytoremediation, and phyto extraction, all which use microorganisms and plant materials for the cleaning from chemical materials [28].

Biological wastes generated from food and fruit processing industries, municipal markets, and water treatment facilities are a major cause of concern. Conventional means of managing these wastes such as transportation, treatment, and disposal, are proving uneconomical. The need is to develop green and sustainable technologies to circumvent this ever growing and persistent problem. The potential of diverse microbes to metabolize complex organic rich biowastes into a variety of bioactive compounds with diverse biotechnological applications has been largely investigated [30].

Global food waste has been estimated to be 1.6 billion tons and loss of around USD 1 trillion. China, United States, Philippines, and India are the largest contributors to industrial processing waste, which generates 32, 15, 6.5, and 1.8 million tons of waste, respectively [30].

The focus has shifted from the conventional practice of using these wastes as animal feed. These high-value components include dietary fibers, nutrients, proteins, peptides, phenolic compounds, polysaccharides, proteins, flavors and phytochemicals. Implementation of processes to treat waste include:

(a) use biowastes as feed for microbes to produce hydrolytic and other enzymes or further process the hydrolyzed waste,
(b) hydrolysate to a multi-stage fermentative process for producing H2,
(c) use volatile fatty acid rich effluent from the H2 production stage for producing CH4 directly or PHA followed by CH4,
(d) use the inorganic matter-rich effluent as a biofertilizer for enhancing crop production, and
(e) use biowastes for rearing insects which form as a biologic material for further treatment of waste [30].

Executive Summary

Endocrine Disruptors

- Endocrine disruptors (ED) are environmental pollutants that mimic endogenous hormonal signals
- Main groups: phthalates, bisphenols, parabens, pesticides, POPs
- Interfere with hormone-dependent functions in the body such as embryonic development, production of sperm, control of the menstrual cycle, the onset of puberty and cancers in hormone-dependent tissues
- Related to increase of breast and prostate cancer, types 2 diabetes, obesity, cardiovascular diseases, neurodegenerative diseases, problems of infertility
- Different pathways of interfering to hormonal circuits

Cosmetics

- The cosmeutical industry is one of the fastest and profitable growing commercial sectors producing different personal care agents
- A variety of chemicals
- Widespread use
- Skin side effects: contact dermatitis, severe like anaphylactic shock, burns or milder and transient like irritant dermatitis
- Formaldehyde: carcinogenic, myeloid leukemia
- Imidazolidine urea and diazolidine urea degraded to formaldehyde
- Parabens: EDs
- Phthalates: EDs

Consumer Products

- The main exposure pathway takes place in indoor environments
- Used as active ingredients, preservatives, solvent, or additives
- Reproductive inhibition, endocrine disruption, cancer, immune dysfunction, allergies, skin rashes, eye irritation, respiratory problems, and other chronic diseases
- Isothiazolinones and phthalates were found in 47% and in 24% or frequently consumed products

Workplace Exposures

- Hazardous workplace exposures were typically in agriculture, mining, and construction
- Occupational skin contacts with chemicals were detected in 20.6% of cases
- Prevalence rate of workers with frequent skin contact with chemicals was building and grounds cleaning and maintenance occupations (55.4%)

Biomass Management

- Conversion of biomass into value added fuels and chemicals
- Catalytic conversion can produce a variety of platform chemicals, such as ethanol, 3-hydroxypropionic acid, succinic acid, furfural, levulinic acid, 5- hydroxyl methyl furfural (HMF), 5-methoxy methyl furfural (MMF), and 2,5-furandicarboxylic acid (FDCA)

Managing Microplastics and Soil Contamination

- Management of waste: metals, glass, ceramics, textiles, paper, timber, and plastic debris
- Every year between 4.8 and 12.7 million tons of plastic waste enter the ocean.
- Affects up to 1400 marine species worldwide
- Soil contamination that has received public concern are fertilizers and pesticides
- Biological wastes generated from food and fruit processing industries, municipal markets, and water treatment facilities

References

Endocrine Disruptors

1. Christensen BC, Marsit CJ. Epigenomics in environmental health. Front Genet. 2011;2:84.
2. Encarnação T, Pais AA, Campos MG, et al. Endocrine disrupting chemicals: impact on human health, wildlife and the environment. Sci Prog. 2019;102(1):3–42.
3. Kabir ER, Rahman MS, Rahman I. A review on endocrine disruptors and their possible impacts on human health. Environ Toxicol Pharmacol. 2015;40(1):241–58.
4. La Merrill MA, Vandenberg LN, Smith MT, et al. Consensus on the key characteristics of endocrine-disrupting chemicals as a basis for hazard identification. Nat Rev Endocrinol. 2020;16(1):45–57.
5. Zhang J, Yang Y, Liu W, et al. Glucocorticoid and mineralocorticoid receptors and corticosteroid homeostasis are potential targets for endocrine-disrupting chemicals. Environ Int. 2019;133(Pt A):105133.
6. Jankowska A, Polańska K, Koch HM, et al. Phthalate exposure and neurodevelopmental outcomes in early school age children from Poland. Environ Res. 2019;179(Pt B):108829.
7. Martinez-Arguelles DB, Papadopoulos V. Prenatal phthalate exposure: epigenetic changes leading to lifelong impact on steroid formation. Andrology. 2016;4(4):573–84.
8. 12 February 2018, on the use of bisphenol A in varnishes and coatings intended to come into contact with food and amending Regulation (EU) No 10/2011 as regards the use of that substance in plastic food contact materials.
9. Inadera H. Neurological effects of Bisphenol A and its analogues. Int J Med Sci. 2015;12(12):926–36.
10. Andújar N, Gálvez-Ontiveros Y, Zafra-Gómez A, et al. Bisphenol A analogues in food and their hormonal and obesogenic effects: a review. Nutrients. 2019;11(9):2136.
11. Liang X, Yin N, Liang S, et al. Bisphenol A and several derivatives exert neural toxicity in human neuron-like cells by decreasing neurite length. Food Chem Toxicol. 2020;135:111015.
12. Ruiz-Hernandez A, Kuo C-C, Rentero-Garrido P, et al. Environmental chemicals and DNA methylation in adults: a systematic review of the epidemiologic evidence. Clin Epigenetics. 2015;7(1):55.
13. Karpuzoglu E, Holladay SD, Gogal Jr RM. Parabens: potential impact of low-affinity Estrogen receptor binding chemicals on human health. J Toxicol Environ Health B Crit Rev. 2013;16(5):321–35.
14. Polyzos SA, Kountouras J, Deretzi G, et al. The emerging role of endocrine disruptors in pathogenesis of insulin resistance: a concept implicating nonalcoholic fatty liver disease. Curr Mol Med. 2012;12(1):68–82.
15. Hans L. Kapodistrian University of Athens, oral presentation.
16. Sun H, Brocato J, Costa M. Oral chromium exposure and Toxicity Trichloroethylene (TCE), dichloroacetic acid (DCA), and trichloroacetic acid (TCA). Curr Environ Health Rep. 2015;2(3):295–303.

Cosmetics

17. Bilal M, Iqbal HMN. An insight into toxicity and human-health-related adverse consequences of cosmeceuticals — a review. Sci Total Environ. 2019;670:555–68.
18. Lee C-M. Fifty years of research and development of cosmeceuticals: a contemporary review. J Cosmet Dermatol. 2016;15(4):527–39.
19. Pandey A, Jatana GK, Sonthalia S. Cosmeceuticals. In: Pandey A, Jatana GK, Sonthalia S, editors. StatPearls [internet]. Treasure Island: StatPearls Publishing; 2021.
20. Juhász ML, Marmur ES. A review of selected chemical additives in cosmetic products. Dermatol Ther. 2014;27(6):317–22.

Chemicals in Consumer Sources

21. Sardar SW, Choi Y, Park N, et al. Occurrence and concentration of chemical additives in consumer products in Korea. Int J Environ Res Public Health. 2019;16(24):5075.

Workplace Exposures in the US

22. Calvert GM, Luckhaupt SE, Sussell A, et al. The prevalence of selected potentially hazardous workplace exposures in the US: findings from the 2010 National Health Interview Survey. Am J Ind Med. 2013;56(6):635–46.

Biomass Management

23. Yuan H, Liu H, Jieke D, et al. Biocatalytic production of 2,5-furandicarboxylic acid: recent advances and future perspectives. Appl Microbiol Biotechnol. 2020;104(2):527–43.

Managing Microplastics and Soil Contamination

24. Schneider F, Parsons S, Clift S, et al. Collected marine litter. A growing waste challenge. Mar Pollut Bull. 2018;128:162–174 2.
25. Ruan J, Qin B, Huang J. Controlling measures of micro-plastic and nano pollutants: a short review of disposing waste toners. Environ Int. 2018;118:92–963.
26. De Falco F, Gullo MP, Gentile G, et al. Evaluation of microplastic release caused by textile washing processes of synthetic fabrics. Environ Pollut. 2018;236:916–25.
27. Dhaliwal SS, Singh J, Taneja PK, et al. Remediation techniques for removal of heavy metals from the soil contaminated through different sources: a review. Environ Sci Pollut Res Int. 2020;27(2):1319–33.
28. Liu L, Li W, Song W, Guo M. Physical remediation techniques include washing of soil, thermal desorption, and replace or partially replace the contaminated soil with clean soil to reduce the concentration of pollutants. Sci Total Environ. 2018;633:206–19.
29. Cardozo M, de Almeida JSFD, de Samir Cavalcante FA, et al. Biodegradation of organophosphorus compounds predicted by enzymatic process using molecular modelling and observed in soil samples through analytical techniques and microbiological analysis: a comparison. Molecules. 2019;25(1):58.
30. Lee J-K, Patel SKS, Sung BH, et al. Biomolecules from municipal and food industry wastes: an overview. Bioresour Technol. 2020;298:122346.

Chapter 7
Mechanisms of Aging and Cancer

Molecular Models Implicated in Aging and the Development of Cancer

The time dependent accumulation of cellular damage is widely considered the general cause of aging. Although the evolutionary model suggests that aging is a process related to epigenetic modifications from early life, the application of new epigenetic phenomena through life contributes to this process. We could describe as aging the gradual degeneration of organs and tissues due to insults that accumulate with time.

Concomitantly, cellular damage may occasionally provide aberrant advantages to certain cells, which can eventually lead to cancer. Therefore, cancer and aging can be regarded as two different manifestations of the same underlying process, namely the accumulation of cellular damage. In addition, several of the pathologies, such as atherosclerosis and inflammation, involve uncontrolled cellular overgrowth or hyperactivity [1]. At first sight, aging and cancer may seem opposite processes due to similar mechanisms, as aging is characterized by a loss of fitness, while cancer is the consequence of an aberrant gain of cellular fitness. Overall, aging and cancer share common multifactorial origins [1].

Although methylation levels have been thought to be largely stable throughout life, the process of aging is an increasing hypomethylation process. Normal individuals acquire a variability of epigenetic silencing, which is associated with the risk of age related diseases. Given that molecular diversity is a characteristic of neoplastic transformation, it is reasonable to speculate that age related dysregulation of methylation is a fundamental predisposing age related neoplastic transformation process (Fig. 7.1) [2].

A. Salavoura, *Chemical Environmental Pollutants and their Effect on Health*,
https://doi.org/10.1007/978-3-031-80271-3_7

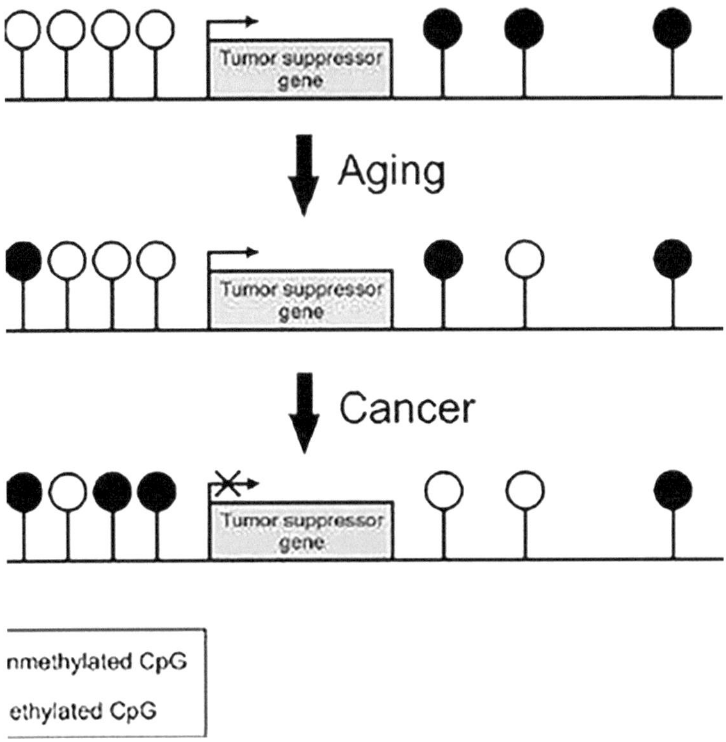

Fig. 7.1 Aging and cancer. (i) Unmethylated normal tissue with expression of a tumor-suppressor gene. (ii) Black areas of hypermethylation due to aging. (iii) Black areas with aberrant hypermethylation suppress the expression of the tumor-suppressor gene

Senescence: The Mechanism of Aging

Senescence and biological aging are the gradual deterioration of functional characteristics. As part of the aging process, senescent cells accumulate in some, but not in all tissues, with a prominent increase in the skin, liver, lung, and spleen. The primary triggers of aging are associated with senescence probably related to the DNA damage, the loss of telomere protective functions, and the depression of the *CDKN2A* locus encoding *p16* and *ARF3* [3].

Senescence is a process that, in the first place, is beneficial for the organism. The central purpose is to initiate a sequence of reactions that eliminate damaged cells and it culminates tissue regeneration by three sequences of events: senescence–clearance–regeneration. However, this beneficial process can be interrupted, particularly in aged tissues by a combination of factors [3].

Senescence is a state of the cell, which is the result of different molecular pathways and it is related to stressors. Multiple stressors and damaging triggers activate signaling cascades that converge the activation of cell cycle inhibitors and tumor suppressor genes. DNA damage agents and telomere loss activate the DNA damage response (DDR), which directly activates *p53* and its downstream transcriptional target *p21*. Many types of senescence are associated with the epigenetic depression of the cyclin-dependent kinase inhibitor 2A (*CDKN2A*) locus (encoding the cell cycle inhibitor *p16* and the *p53* activator (ARF). Reactive oxygen species activate *p16* and *p53* by interfering with the kinases MKK3 (also known as MAPKK3) and MKK6 (also known as MAPKK6), and their downstream kinase effector *p38*. Oncogenic signaling or permanent loss of tumor suppressors activates *p16* and *p53* with the participation of the DDR and ARF pathways. Transforming growth factor-β is a notable component of the senescence-associated secretory phenotype (SASP), which upregulates the cell cycle inhibitors *p21*, *p27,* and *p15* by the SMAD complex. Developmental cues induce senescence by *p21* by activating the *PI3K* and *TGFβ* pathways. Polyploidization and cell fusion upregulate *p21* by the DDR and *p53* and the transcription factor of the early growth response protein 1 (*EGR1*) (Fig. 7.2) [3].

On the other hand, *Clearance* of senescent cells by the immune system is the process of removal of damaged cells in the state of senescence. The removal happens with the contribution of T helper cells and it is also due to the recruitment of inflammatory phagocytic cells, especially macrophages. The process is used in the treatment of multiple cancers such as activation of *p53* in hepatocellular carcinomas and sarcomas or pharmacologic *CDK4* inhibitors (Fig. 7.3) [3].

When this process is ineffective, damaged cells may accumulate and aggravate tissue dysfunction detecting a particular phenotype called the senescence-associated secretory phenotype (SASP). Through independent activities of *p16Ink4a* and

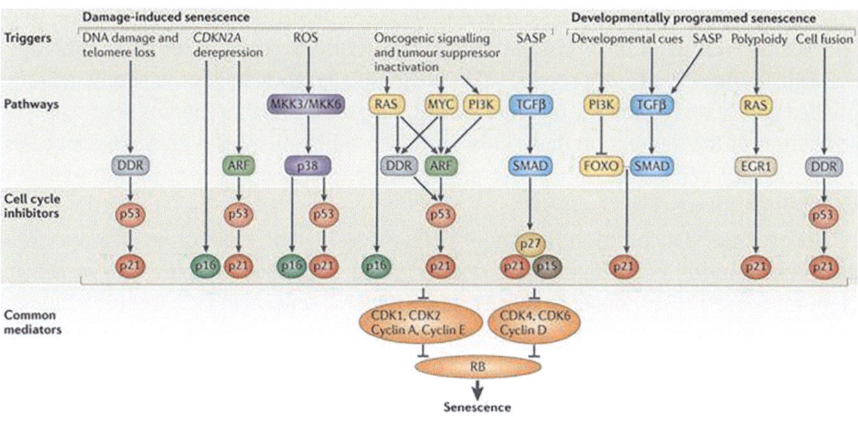

Fig. 7.2 The molecular pathways of senescence depending on the trigger. For example, DNA damage activates the p53 pathway, ROS the p53 and p16 pathway, and oncogenic signals the p16 and p53 pathway. On the contrary, during early development, the developmental cues activate the p21 pathway, polyploidy the p21 pathway, and cell fusion of the p53 pathway. Independently, from the etiology, the same pathways lead to cell senescence

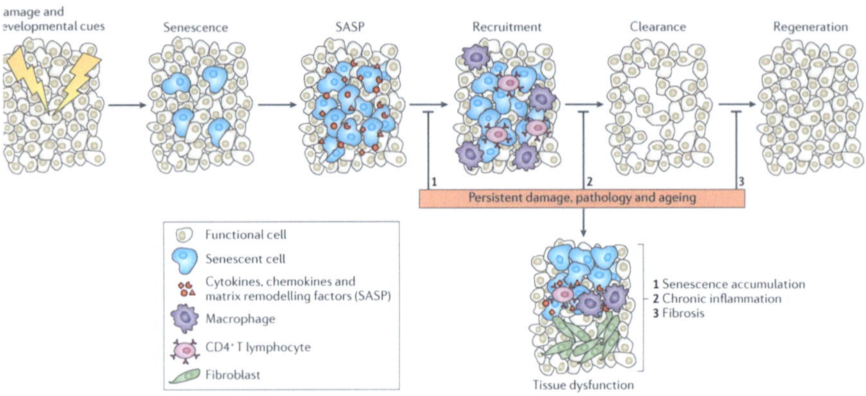

Fig. 7.3 The mechanism of clearance, by activation of the secretory phenotype (SASP)

p21Cip1, irreversible molecular changes could be induced, leading to permanent cell cycle arrest. Senescent cells develop a defined, but heterogeneous, secretory profile termed *senescence-associated secretory phenotype (SASP)*. This comprises a range of different proteins, including several proteins known to play a role in aging and age-related diseases, including matrix metalloproteinases such as MMP3, growth factors, chemokines such as CCL2 and CLL11 [17], and proinflammatory interleukins such as IL1, IL6, and IL8 [5].

When senescence affects not only differentiated cells but also stem and progenitor cells, it limits the regenerative capacity of tissues. The combination of inefficient clearance, excessive SASP, and ineffective regeneration may explain the accumulation of senescence cells during aging and its active contribution to some aging phenotypes as well as in different pathological conditions. Thus, with advancing age, the full sequence of senescence–clearance–regeneration is not entirely accomplished and senescence may become part of the problem rather than its solution. The resolution of the damage in these cases involves a fibrotic scar with senescent cells, inflammatory cells, and fibrotic tissue. Thus, senescence has a different action in adults and elderly (Fig. 7.4) [3].

Senescence is a common finding of cells in aging. The ineffective clearance and poor regeneration performance are due to poor macrophage recruitment or a deficient regenerative response. In these circumstances, senescent cells accumulate and create a stable senescent lesion that aggravates pathology. It has been proposed that the balance between the beneficial and detrimental effects of senescence probably depends on whether cells are transitory or whether they accumulate over time [3]. Consistent with this idea is the observation that interference with senescent cell accumulation in BubR1 progeroid mice delays several of the aging-associated disorders that these animals develop. However, because progeroid syndromes do not mimic the complex degenerative changes of aging completely, the relevance of these findings has remained unclear [4].

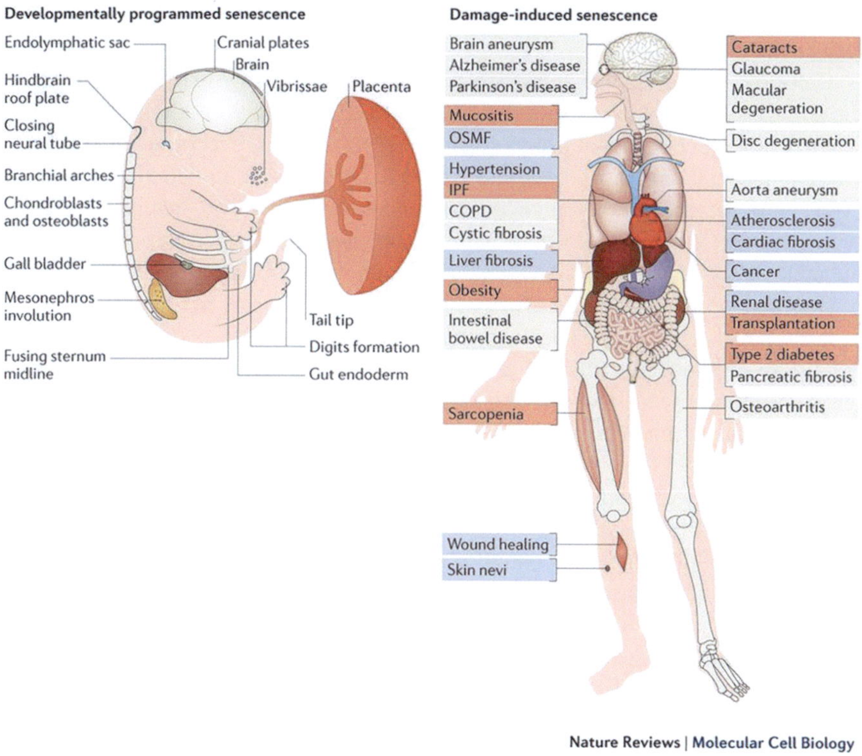

Fig. 7.4 Senescence, a mechanism participating in the developmental process of the embryo as well as in the degeneration of tissues and aging

The *unified model of senescence* proposes that senescence is a component of tissue remodeling both in normal development and pathology. In more detail, cellular senescence coordinates tissue remodeling by three sequential processes: first, a stable proliferative arrest; second, a secretory phenotype (SASP) that recruits immune cells, notably including T helper lymphocytes and macrophages; and, third, the mobilization of nearby progenitor cells that repopulate the tissue [3].

Senescent cells are associated with a spectrum of diseases including metabolic (obesity, diabetes), neurological (AD, PD), muscle, bone, and cartilage related (sarcopenia, osteoarthritis, kyphosis, herniated discs) or tissue dysfunction related (lung emphysema, renal disease, and atherosclerosis). Cellular senescence can be triggered by chemotherapy or ischemia-reperfusion injury during organ transplantation [5].

Cellular senescence has been increasingly linked to the development of vascular pathologies, in particular to atherosclerosis. Plaque formation involves a series of events initiated by endothelial cell dysfunction, vascular permeability, and the monocyte and macrophage mediated release of proinflammatory cytokines. These events induce an initial phase of endothelial proliferation that is subsequently followed by the onset of senescence. SASP amplifies the proatherogenic inflammatory environment [3].

Caloric overload due to nutritional excess and low energy expenditure leads to the storage of energy in adipose tissue. The storage capacity of adipocytes eventually reaches a threshold that triggers a stress response and the recruitment of macrophages. This inflammatory response of the adipose tissue initiates a cascade of events with systemic pathological consequences, including liver steatosis and insulin resistance, which constitute the hallmarks of metabolic syndrome. Interestingly, the adipose tissue presents SASP and upregulation of *p53* and *p21*, both in mouse models and in human patients [3].

Insulin resistance due to obesity is initially compensated by an overproduction of insulin by pancreatic β-cells and an expansion of these cells, but this chronic challenge eventually leads to proliferative exhaustion and loss of β-cell mass. After 12 months on a high fat diet, the atrophic β-cell islets in mice are strongly positive for SASP activity [3]. Experiments on rats show that a chemically induced damage to the rats' pancreas results in fibrosis with areas of abundant senescent pancreatic cells [3].

Loss of muscle function is one of the most prevalent aging associated pathologies. Recent studies have implicated senescence of the muscle stem cells as the underlying cause of aging associated sarcopenia and the loss of muscle regenerative potential. Aged muscles activate *p16* and they are positive for SASP. The upregulation of *p16* and the loss of regenerative potential occurs by a mechanism that involves *p38*, *MAPK*, and *CCN1*. Interestingly, genetic inactivation of *p16* or chemical inhibition of *p38* rejuvenates aged satellite cells and it promotes muscle regeneration upon injury [3].

Senescence state has been implicated in many other diseases. It has been associated with human aneurysms in the brain and the heart. Astrocyte senescence has been proposed as a component of Alzheimer's disease and Parkinson's disease. It is of note that the human brain cortex presents an increase in the number of astrocytes expressing *p16* and *MMP1* (also known as interstitial collagenase); this is exacerbated in age matched patients with Alzheimer's disease.

Other human diseases associated with SASP senescence include glaucoma, fibrosis, intestinal bowel disease, osteoarthritis, and intervertebral disc degeneration. COPD is a lung disease linked to aging and cigarette smoking, characterized by the disruption of the alveolar epithelium and often associated with fibrosis. The alveolar epithelium of patients with COPD contains high levels of *p16* and nuclear *NF-κB* (13).

Senescence as a Mark of the Proinflammatory Profile

Cellular senescence is a well-established cancer defense mechanism presumably by the depletion of stem and progenitor cells and the adverse actions of the senescence associated secretory transformed phenotype, which consists of many proinflammatory cytokines and chemokines, matrix metalloproteinases, and growth factors [4]. Recent studies show that senescent cells have beneficial effects in injury repair [3].

Immunosenescence is also the well-characterized decline in the function of the adaptive immune system, which is accompanied by clonal changes and remodeling of adaptive immunity. Major changes involve thymic degeneration and alterations in T cell subsets, including the exhaustion of CD45RA+ naive T cells and the concomitant clonal expansion of CD45RO+ memory T cells [6].

The age-associated imbalance between adaptive and innate immunity leads to the proinflammatory phenotype with activated innate immunity responses and a decline in adaptive immunity. Franceschi et al. (2000a, b) termed this age-related immune status "inflamm-aging." Antigenic stress, either external or inherent, and oxidative stress increase the demands on the innate immunity system and they lead to the aging type of a low-level but chronic inflammatory phenotype [6].

The master regulator of innate immunity is the NF-kB system, an ancient signaling pathway found in both insects and vertebrates. The NF-kB system links together the pathogenic signals and cellular danger signals and then it organizes the cellular resistance. NF-kB system represents an ancient signaling system, which it has received different molecular forms during evolution. Pathogens have different pathogen-associated molecular patterns, so-called PAMPs. Toll-like receptors (TLRs) represent the conserved host defense receptor system. Both systems are connected with the NF-kb system. However, there is an NF-kB-independent pathway by interferon-regulatory factors, which functions alternatively to the NF-kb system [6].

The regulation of proinflammatory responses during inflamm-aging contains a complex interaction of regulatory mechanisms from the recognition receptor level to the genomic transactivation level of NF-kB target genes. Moreover, inflamm-aging contributes to the appearance of degenerative diseases, and many anti-inflammatory drugs, such as corticosteroids, aspirin, and other nonsteroidal anti-inflammatory drugs can inhibit NF-kB activity and they can retard the progress of degeneration. Dietary restriction is known to extend the life span by inhibiting the inflammatory response also [6].

Archetypal inflammatory pathways such as NF-κB promote an open configuration of chromatin and, therefore, generate epigenetic plasticity for the self repair of damaged tissues. This phenomenon is the so-called transflammation [7].

Moreover, epigenetic regulations contribute to the phenomenon of aging. DNA methylation and protein acetylation are the typical epigenetic regulation mechanisms affecting genes. It is known that DNA methylation can inhibit or totally silence the gene expression. Interestingly, both aging and inflammation are associated with DNA methylation epigenetic regulation changes. There is a general hypothesis that epigenetic changes during aging might direct the aging process. It is known that the aging process is associated both with hypo- and hypermethylation of genes and that these responses are tissue specific. However, it is not known if these changes accelerate aging or vice versa [6].

The so-called *Yamanaka cocktail* reported by the Lopez-Otin group revealed that chronic hyperactivation of the proinflammatory NF-κB signaling constitutes a critical impediment to nuclear reprogramming in both normal and accelerated aging. In addition, hyperactivation of the histone H3 methyltransferase was found to contribute to the loss of epigenetic plasticity following chronic NF-κB hyperactivation [7].

Cells with loss of plasticity are less prone to respond to danger signals and activate the NF-kb pathway leading to chronic inflammation, but at the same time, they are better protected from malignant transformation. In this regard, the cyclic and transient expression of reprogramming factors *in vivo* has recently been shown to increase lifespan in a murine model of premature aging by remodeling of the chromatin. Conversely, cells with more plastic chromatin will be more adaptable to face cell intrinsic or microenvironmental challenges, but they might also provide the molecular power on a tissue's susceptibility to undergo aging associated degeneration or cancer-associated malignant transformation [7].

Reprogramming aged somatic cells into mesenchymal cells reverts some hallmarks of aging, including telomere length, mitochondrial fitness, abnormal nuclear morphologies, loss of heterochromatin markers, and gene expression to a youthful embryonic state. It is suggested that the *in vitro* expression of the four Yamanaka reprogramming factors, *Oct4, Sox2*, *Klf4*, and *c-Myc*, reverts age associated features in mouse and human cells, without the loss of cellular identity [8]. Thus, the hallmarks of aging according to the epigenetic model include 10 different molecular characteristics (Fig. 7.5) [1].

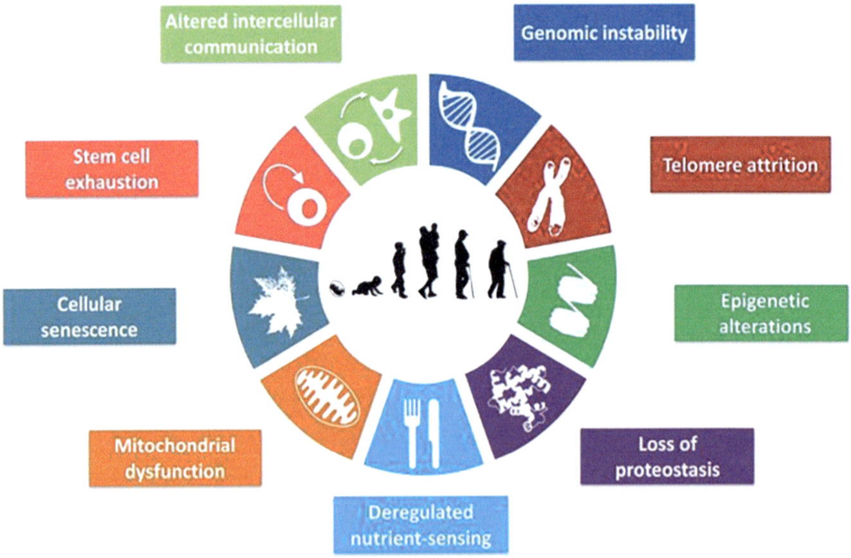

Fig. 7.5 The hallmarks of aging: genomic instability, telomere attrition, epigenetic alterations, loss of proteostasis, deregulated nutrient sensing, mitochondrial dysfunction, cellular senescence, stem exhaustion, altered intercellular communication

1. *Genomic instability*

   Endogenous or exogenous agents can stimulate a variety of DNA damage. Excessive DNA damage or insufficient DNA repair favors the aging process. Chromosomal and point mutations contribute to genomic instability, while the complex network of DNA repair mechanisms contributes to a stable state. The genomic stability system includes specific mechanisms for maintaining the appropriate length and function of telomeres and for ensuring the integrity of mitochondrial DNA (mtDNA) [1].

Nuclear DNA Damage Somatic mutations accumulate within cells from aged humans. Other forms of DNA damage, such as chromosomal aneuploidies and copy-number variations, have also been found associated with aging. Increased clonal mosaicism for large chromosomal anomalies has been also reported. All these forms of DNA alterations may affect essential genes and transcriptional pathways, resulting in dysfunctional cells that, if not eliminated by apoptosis or senescence, may jeopardize tissue and organism homeostasis. This is especially relevant when DNA damage impacts stem cells. A recent report listed more than 50 oncogenic or mitogenic alterations that are able to induce senescence [1].

Mitochondrial DNA Mutations and deletions in aged mtDNA may also contribute to aging. mtDNA has been considered a major target for aging-associated somatic mutations due to the oxidative microenvironment of the mitochondria, the lack of protective histones in the mtDNA, and the limited efficiency of the mtDNA repair mechanisms. However, the causal implication of mtDNA mutations in aging has been controversial, because of the variability of mitochondrial genes, which allows the co-existence of mutant and wild-type genomes within the same cell, a phenomenon that is referred to as heteroplasmy [1].

Nuclear Architecture In addition to genomic damage affecting nuclear DNA or mtDNA, defects in the nuclear lamina, known as laminopathies, can also cause genome instability. Nuclear lamins constitute the major components of the nuclear lamina and they participate in genome maintenance by providing a scaffold for tethering chromatin. Mutations in lamins are related to accelerated aging syndromes such as the Hutchinson-Gilford and associated progeria syndromes [1].

2. *Telomere attrition.* Telomeres, the end parts of chromosomes, lose length in each cell division and thus they are particularly susceptible to age-related erosion. Telomere exhaustion explains the limited proliferative capacity of some types of *in vitro* cultured cells, the so called replicative senescence or Hayflick limit [1].
3. *Epigenetic alterations.* Global hypomethylation and hypermethylation of various tumor suppressor genes and polycomb target genes are hallmarks of aging. Alterations in the acetylation or other modifications of histones, in addition to aberrant DNA methylation patterns as well as of other chromatin-associated proteins, can induce epigenetic changes that contribute to the aging process. Increased histone H4K16 acetylation, H4K20 trimethylation, or H3K4 trimeth-

ylation, as well as decreased H3K9 methylation or H3K27 trimethylation, constitute age-associated epigenetic marks (Fig. 7.6) [1].

Moreover, aging is associated with an increase in transcriptional noise and an aberrant production and maturation of many mRNAs. These aging-associated transcriptional signatures also affect noncoding RNAs, including a class of miRNAs (gero-miRs) that is associated with the aging process and that influences lifespan by targeting components of longevity networks or by regulating stem cell behavior [1].

4. *Loss of proteostasis*. Aging and some aging related diseases are linked to impaired protein homeostasis or proteostasis. Proteostasis involves mechanisms for the stabilization and correct folding of proteins, most prominently the heat-shock family of proteins as well as mechanisms for the degradation of proteins by the proteasome or the lysosome [1].

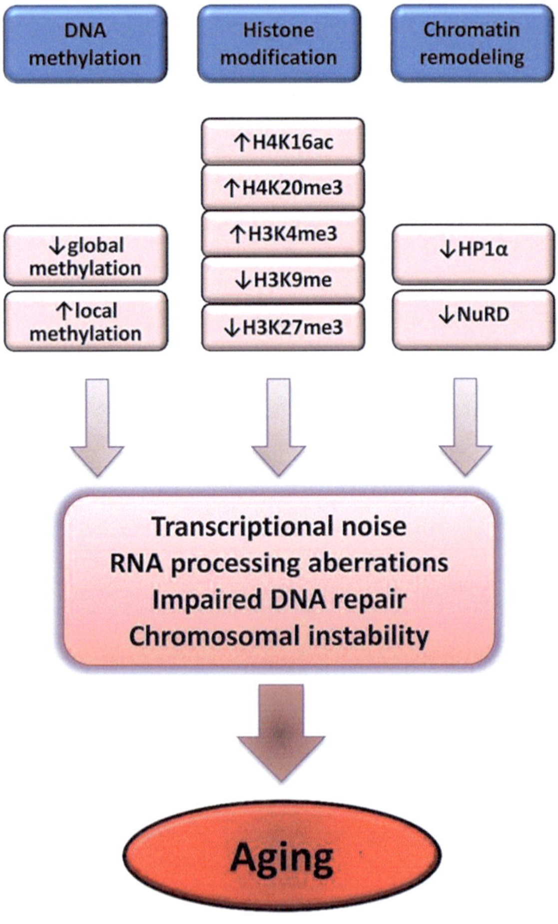

Fig. 7.6 The epigenetic modifications of aging

5. *Deregulated nutrient sensing*. The somatotrophic axis in mammals comprises the growth hormone (GH), produced by the anterior pituitary, and its secondary mediator, the insulin-like growth factor (*IGF-1*), produced in response to GH by many cell types, most notably hepatocytes. The intracellular signaling pathway of *IGF-1* is the same as that elicited by insulin, which informs cells of the presence of glucose [1].

   Paradoxically, *GH* and *IGF-1* levels decline during normal aging, as well as in mouse models of premature aging. Apart from growth and insulin regulation pathways, *mTOR*, for the sensing of high amino acid concentrations, another nutrient-sensing pathway, shows impaired function. The *mTOR* kinase is part of two multiprotein complexes, mTORC1 and mTORC2, that regulate essentially all aspects of anabolic metabolism. Finally, AMPK protein (5' AMP-activated protein kinase) which senses low energy states by detecting high AMP levels and sirtuins, is also impaired. These molecules sense low energy states by detecting high NAD^+ levels. The other two nutrient sensors, AMPK, and sirtuins, act in the opposite direction to IIS pathway, an hormonally regulated cell-signaling pathway, and mTOR, they signal nutrient scarcity and catabolism instead of nutrient abundance and anabolism. Accordingly, their upregulation favors healthy aging [1].
6. *Mitochondrial dysfunction*. Stress signals and defective mitochondrial function generate ROS that, below a certain threshold, induce survival signals to restore cellular homeostasis, but at higher levels can contribute to aging. Similarly, mild mitochondrial damage can induce a response (mitohormesis) that triggers adaptive compensatory processes. Dysfunctional mitochondria can contribute to aging independently of ROS. Dysfunction could happen by a number of mechanisms, for example, mitochondrial deficiencies may affect apoptotic signaling by reduced response to stress and inflammation reactions. Also, mitochondrial dysfunction may directly impact cellular signaling and intraorganellar crosstalk, by affecting mitochondria associated membranes, that constitute an interface between the outer mitochondrial membrane and the endoplasmic reticulum [1].
7. *Cellular senescence*. Cellular senescence can be defined as a stable arrest of the cell cycle coupled with stereotyped phenotypic changes. Senescence observed by Hayflick is caused by telomere shortening and nontelomeric DNA damage and depression of the *INK4/ARF* locus, both of which progressively occur with chronological aging [1].
8. *Stem cell exhaustion*. Consequences of the exhaustion of hematopoietic stem cells (HSCs), mesenchymal stem cells (MSCs), satellite cells, and intestinal epithelial stem cells (IESCs) are a cause of aging.
9. *Altered intercellular communication*. Apart from autonomous intercellular communication, other systems responsible for the communication of different systems in the body, such as endocrine, neuroendocrine, and neuronal communication, are dysregulated. Thus, neurohormonal signaling tends to be deregulated in aging as inflammatory reactions increase, immunosurveillance against pathogens and premalignant cells declines, and the composition of the peri- and extra-

cellular environment changes, thereby affecting the mechanical and functional properties of all tissues [1].

Melatonin (N-acetyl-5-methoxytryptamine) is an endogenously produced indoleamine synthesized from the essential amino acid tryptophan and it is secreted by the pineal gland in a circadian manner, but also by extra pineal organs (i.e., gastrointestinal tract, immune system cells, retina, spleen, liver, kidney, and heart) [9].

Melatonin acts by several mechanisms: directly scavenging free radicals and indirectly upregulating antioxidant enzymes and downregulating prooxidant enzymes. Aging is associated with a significant reduction in endogenous melatonin secretion, resulting in an augmentation of oxidative stress and other metabolic changes. Melatonin receptors have been identified in the human cardiovascular system, mainly localized in the ventricular wall, coronary arteries, aorta, and peripheral arteries, where they exert a protective effect from serious events, such as ischemic heart disease, acute myocardial infarction etc [9].

Melatonin protects the brain by activating protective immune responses and as a defense against ROS damage. It has beneficial effects against mitochondrial dysfunction, considered a major causative agent in neurodegenerative diseases, such as Alzheimer disease (AD), Huntington's disease (HD), and Parkinson's disease (PD). Lipid peroxidation and ROS generation in neurons lead to neuropsychiatric and neurodegenerative disorders. The strong antioxidant activity of melatonin due to the diverse pleiotropic actions has been proposed as an effective neuroprotective agent in diverse neuropsychiatric disorders [9].

Individuals living in highly industrialized countries look for melatonin-supplemented drugs and pills as a way to improve sleep and rest, as the main advantages of the pill include quicker times to fall asleep and quicker jet-lag reductions. It aids in the adjustment and maintenance of a regular circadian rhythm. On the other hand, some of the users complain to experiencing headaches, dizziness, and sleepiness during daylight hours, but these may be placebo effects [9].

However, there are many other potential health effects attributed to melatonin's application. For example, melatonin supplementation is helpful in inflammatory markers, hypertension, oxidative stress, and metabolic syndrome. A positive correlation was found between melatonin-rich foods and clinical metabolic indicators.

Models Related to the Development of Cancer

The immortality of cancer cells could be regarded as pathological self renewal, as it has already been described. The molecular mechanisms of the development of cancer underpin the generation of pluripotent stem cells with deregulation properties and aberrant reprogramming. The similarities between different tumors also suggest that mutations in genes implicated in DNA methylation dynamics might represent a hallmark of cancers with a stem cell origin, and they highlight an alternative view of cancer that may implicated in the development of new treatment stategies. [10]. Tumor progression can be explained by different models depending on the type of the tumor and the organ.

1. The *clonal genetic model of cancer* developed 30 years before and it suggests that cancer could arise from a single cell. According to this view, a series of genetic alterations is responsible for progressive clonal selection and tumor cell heterogeneity. The process gives rise to tumor cells, which have proliferation, invasion, metastasis, and drug-resistance properties. The clonal genetic model has been supported by the discovery of oncogenes and recessive tumor suppressor genes in tumor cytology. The number of the genes that have been reported so far suggests the view that cancer is a heterogeneous group of diseases with diverse etiology and pathogenesis (Fig. 7.7) [24].

Despite its strengths, this classical view has significant limitations. First, progression related genetic changes are inconsistent. In colorectal cancer, for example, there are not detected any mutations related to tumor progression apart from a gatekeeper mutation. However, a series of sequential genetic changes account for each pathological stage of colorectal cancer progression, although the specific order and which specific mutations are implicated after the gatekeeper mutation might not be important in tumor progression. Despite determined efforts by many laboratories, so far, no recurrent mutations have been identified that are responsible for the invasion or metastasis of that tumor. Other studies have not linked cancer progress with specific gatekeeper gene mutations [24].

2. *The epigenetic dysregulation model* suggests that in some tumors there are no specific mutations to the stem cell that initiate the process of the transformation of the normal cell, but a series of epigenetic alterations. The precise regulation of the DNA methylation pattern is involved in the control of cel-

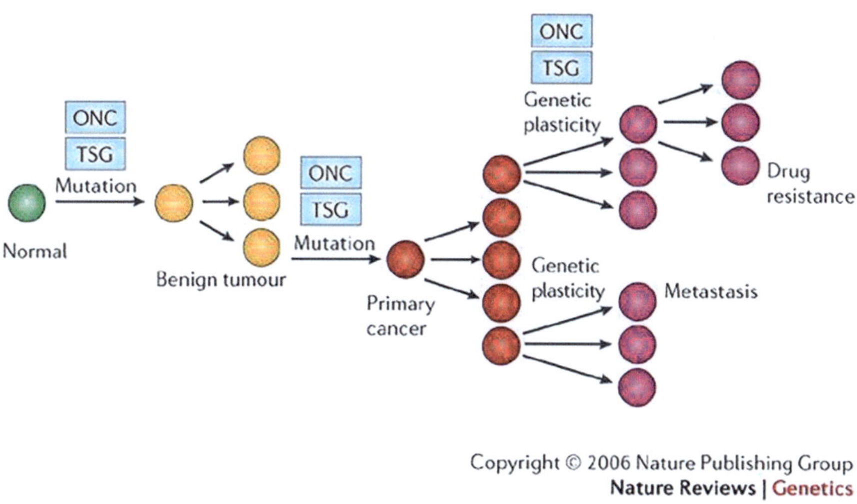

Fig. 7.7 The clonal genetic model for the development of cancer. Continuous mutations and epigenetic mechanisms that offer increased tissue plasticity result in tumor progression, drug resistance, and metastasis

lular plasticity. It is not surprising, therefore, that aberrant DNA methylation patterns are associated with cancer and they constitute a substantial contribution to the altered program of gene expression that is characteristic of malignancy [10].

Although normal genes vary in the degree of methylation, all tumors examined so far, both benign and malignant, have shown a global reduction of DNA methylation. In addition to global hypomethylation, promoters of individual genes show increased DNA methylation levels [10].

Promoter CpG island methylation has emerged as a common molecular defect in cancer cells. Evidence suggests that, for some genes, altered methylation provides a similar selective advantage as mutations or deletions of genes and aberrant chromosomes, an argument that has led to the suggestion that hypermethylation should be considered as one of the two steps in the Knudson's model of tumor suppressor gene inactivation [10].

Important evidence of epigenetic variation can be found in the cancers themselves. Aberrant methylation in neoplastic cells varies between different tissues and it also varies considerably between different individual patients of a given tumor type. One view of this variability is that, as with gene mutations, aberrant methylation occurs randomly and stochastically [10]. Functionally, these genes include tumor suppressor genes, DNA repair genes, as well as potential oncogenes such as cyclooxygenase 2 (*COX2*) and telomerase reverse transcriptase (*TERT*) [10].

The above model applies to many tumor-suppressor genes, including RB1 in retinoblastoma, p16 in melanoma, VHL in renal-cell carcinoma, and APC and Wntsignaling genes in colorectal cancer. Moreover, hypermethylation of specific tumor suppressor genes contributes to the appearance of specific tumors. An increasing number of genes are found to be normally methylated at promoters, but hypomethylated and activated in the corresponding tumors. These include *melanoma antigen family A, 1 and R-RAS* in gastric cancer, *melanoma antigen family A*, 1 (*MAGE1*) in melanoma, *maspin* in gastric cancer, *S100A4* in colon cancer, and various genes in pancreatic cancer. An intriguing recent example of hypomethylation promoting tumor cell growth is the paired box 2 (*PAX2*) gene in endometrial cancer, which is responsible for the responsiveness of the cancer to estrogens and it promotes tumor proliferation. Studies of model organisms support the role of both hypomethylation and hypermethylation in cancer [11].

Apart from global hypomethylation and gene-specific hypermethylation or hypomethylation, it seems likely that aberrant cancer associated DNA methylation that silences genes would occur by an imbalance of the demethylating versus methylating machinery. For example, the inactivation of the TET enzymes, which promote demethylation, or the methyltransferases, such as DNMT3A that promote methylation changes the methylation pattern of a specific region related to tumor progression. Heterozygous mutations in *DNMT3A* are found in 20–30% of cases of acute myeloid leukemia (AML), and they presumably disrupt the normal function of *DNMT3A*, gene which is required to stabilize the different ions and proliferation of pluripotent genes [10].

Chromatin alterations detected in oncogenic tissues include the overproduction of histone methyltransferases that catalyze the methylation of either H3-K4 or H3-K27 residues. Global reduction in monoacetylated H4-K16 and trimethylated H4-K20 is a general feature of cancer cells. Variant forms of histone H2A are involved in DNA repair, X-chromosome inactivation, and other basic cellular processes [11].

According to this model, cancer arises in three steps. The first is an epigenetic alteration of the stem cell within a given tissue, which is mediated by aberrant regulation of tumor progenitor genes (TPG). This alteration can be due to events within the stem cells themselves, the influence of the stromal compartment, or environmental damage and injury. Second is a gatekeeper mutation (GKM), for example, the tumor suppressor gene (TSG) in solid tumors, and rearrangement of an oncogene (ONC) in leukemia and lymphoma. Third is the genetic and epigenetic instability, which leads to tumor evolution. Note that many of the properties of advanced tumors (invasion, metastasis, and drug resistance) are inherent properties of the progenitor cell that give rise to the primary tumor and there is no need for other mutations (highlighting the importance of epigenetic factors in tumor progression) (Fig. 7.8) [11].

Tumor-progenitor genes are defined as the ones that could mediate epigenetic expansion of the progenitor cells and they increase their cancer predisposition, by increasing their capacity for self-renewal and pluripotency, while they block the tendency to limited replicative potential and differentiation. A good candidate for a tumor progenitor gene would be one that acts directly on DNA, with both genetic and epigenetic consequences. One example is *IGF2*, in which a mutation promotes an increase in the progenitor cell compartment, increasing the probability of neoplasia. Another class might include the *APOBEC* (apolipoprotein B mRNA-editing enzyme) family of genes that encode encodes DNA cytosine deaminases. Many B-cell lymphomas are evidently caused by mutations that originated by the activation induced deaminase (AID)-dependent deamination events in tumor suppressor genes [11].

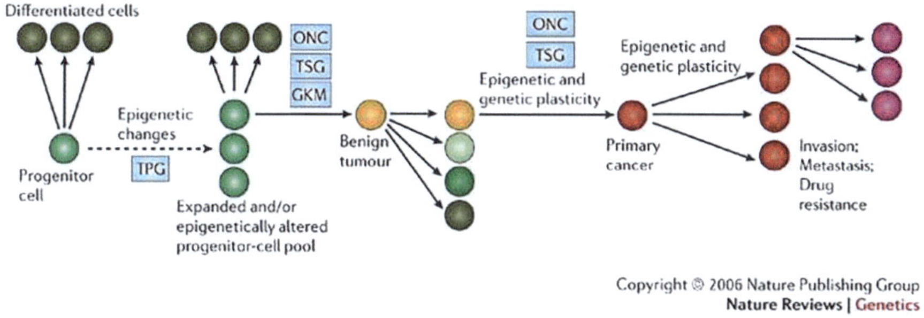

Fig. 7.8 The epigenetic model for the development of cancer that suggests that the first step in tumor initiation is epigenetic alterations followed by genome instability and plasticity

3. In the *reprogramming model of cancer*, both in primary tumors and metastases, *p53* loss and *MYC* promoter senescence are bypassed. The *MYC* protein promotes proliferation and reprograms metabolism. Reprogramming of mouse embryonic fibroblasts to induced pluripotent stem cells by the sequential addition of the transcription factors *OCT4*, Krüppel-like factor 4 (*KLF4*), SRY-box 2 (*SOX2*), and *MYC* requires a mesenchymal to epithelial transition (MET). This model suggests that an alteration in genes responsible for transition states, it could be the initial step of cancer development. To metastasize, primary tumors usually first undergo an epithelial to mesenchymal transition (EMT), during which they lose the expression of adhesion molecules such as E-cadherin and adopt an invasive phenotype, before undergoing a transition that enables successful colonization. Plasticity and reversible phenotype switching between these different states generate phenotypic heterogeneity in solid tumors [10].

In solid tumors, a remarkable degree of heterogeneity is detected, with tumors containing a varying proportion of cells in different stages of differentiation, proliferation, and invasive phenotypes. Although the cancer stem cell hypothesis implies that a hierarchy would be maintained such as in hematological malignancies, different phenotypic states can interconvert. In addition, the tumor hypoxia and the decreased TET activity stabilize epigenetic states that are induced by a loss of differentiation factors to provide long-lived dormant stem cells. The hypoxic cells might have a reduced capacity for altering DNA methylation status by TET-mediated removal of epigenetic modifications. This could represent an advantage under physiological conditions, in which maintenance of stem characteristics would be enhanced in a hypoxic niche [10].

4. As already mentioned, *nonresolved inflammation* is one of the consistent features of the tumor microenvironment. The connection between inflammation and cancer, first perceived in the 19th century, is now accepted as an enabling characteristic of cancer. Current estimates suggest that about 25% of cancers are associated with chronic inflammation sustained by infections (e.g., hepatitis) or inflammatory conditions of diverse origin (e.g., prostatitis). Moreover, tumors that are not epidemiologically related to inflammation are characterized by the presence of inflammatory cells and mediators. Inflammation and cancer are connected by two pathways: extrinsic pathways from conditions that cause nonresolving inflammatory responses and intrinsic pathways driven by oncogenes or tumor suppressor genes that activate the expression of inflammation related programs. Different parameters of the cancer microenvironment and interaction with different cells of the immune system have been described [12].

Cells of the monocyte macrophage lineage exhibit considerable plasticity and diversity. Tissue repair is an important component of the inflammatory response and fibroblasts participate to this process. Tumors have long been considered "wounds that do not heal." Tumor associated fibroblasts are a major component of the cancer microenvironment and an important source of tumor promoting cytokines and growth factors, expressing a proinflammatory gene signature in skin, breast, and pancreatic cancers. A major gene upregulated in these processes is the NF-kb. Fibroblasts are thought to arise from mesenchymal stem cells, but recent data suggest that other bone marrow cells can regulate their activity [12].

Other inflammatory cells that have recently been identified as modifiers of some inflammatory tumor microenvironments are neutrophils and mast cells, which are important not only for the release of cytokines and chemokines, but also for the generation of free radicals and the activation of proteases. Thus, the cellular networks involved differ considerably in different tumors and in tumors originating in different tissues, but a common origin may be detected, the involvement of macrophage lineage [12].

TNF-a, IL-6, and IL-1 are among the most studied inflammatory cytokines in the tumor microenvironment. Many malignant cells constitutively produce small amounts of TNF-a in contrast to normal cells. There is evidence from animal models that malignant cells induce the production of TNF-a that contributes to the growth and spread of tumors of the skin, ovary, pancreas, pleural cavity, and bowel. Whether made by malignant cells or host cells, TNF-a may directly contribute to oncogene activation, DNA damage, and stimulation of mesenchymal cells. However, it is also clear that high doses of TNF-a could have an antitumor activity that justifies its original naming as a tumor necrosis factor [12].

IL-6 is also a tumor-promoting cytokine on both malignant and stromal cells in a range of experimental cancer models. It is a downstream effector of oncogenic *ras* and has been implicated in several human cancers including multiple myeloma and hepatocellular carcinoma. IL-6 can be an autocrine or paracrine growth factor in some malignancies, especially those of hematologic origin, it blocks apoptosis and signals through the STAT3 pathway that is activated in many cancers. The STAT3 gene, promote cell proliferation and angiogenesis, while IL-6 interferes with the tumor microenvironment [12].

Recent research found a significant association between tumor Th2 cell infiltrates and poor prognosis depending on the ratio between Th2/Th1 cells in tumor biopsies, being independently predictive of patient survival. The actions of the cells and cytokines of inflammation in the tumor microenvironment are very context dependent and a new approach to cancer therapy is the attempt to switch the tumor promoting immune suppressive microenvironment to one that kills tumor cells [12].

Executive Summary

Molecular Models Implicated in Ageing Disorders and the Development of Cancer

- Senescence is the gradual deterioration of functional characteristics of the cell
- As part of the aging process, senescent cells accumulate in the skin, liver, lung, and spleen

- The primary triggers of aging-associated senescence probably consist the DNA damage, the loss of telomere protective functions, and the depression of the *CDKN2A* locus encoding *p16* and *ARF3*
- The function of senescence is, in the first place, beneficial in order to clean damaged cells in a series, senescence–clearance–regeneration
- Ineffective clearance leads to senescence-associated secretory phenotype: proinflammatory cytokines and chemokines, matrix metalloproteinases, growth factors, and disease
- Hallmarks of aging: genomic instability, telomere attrition, epigenetic alterations, loss of proteostasis, deregulated nutrient sensing, mitochondrial dysfunction, cellular senescence, stem cell exhaustion, and altered intercellular communication
- Proposed models for the development of cancer: clonal genetic model (a mutation in a single gene accompanied by a series of mutations that offer genomic instability and plasticity), epigenetic dysregulation model (global hypomethylation and specific gene hypermethylation or hypomethylation that activates oncogenes and inhibits tumor-suppressor genes), the reprogramming model (alteration in genes responsible for transition states could be the initial step of cancer development), and the nonresolved inflammation model (cancer is related to chronic inflammation and the secretion of IL-6, IL-8, and TNF-a cytokines from the tumor cell or the microenvironment)

References

1. López-Otín C, Blasco MA, Partridge L, et al. The hallmarks of aging. Cell. 2013;153(6):1194–217.
2. Issa J-P. Epigenetic variation and human disease. J Nutr. 2002;132(8 Suppl):2388S–92S.
3. Muñoz-Espín D, Serrano M. Cellular senescence: from physiology to pathology. Nat Rev Mol Cell Biol. 2014 Jul;15(7):482–96.
4. Baker DJ, Childs BG, Durik M, et al. Naturally occurring p16Ink4a-positive cells shorten healthy lifespan. Nature. 2016;530(7589):184–9.
5. de Keizer PLJ. The fountain of youth by targeting senescent cells? Trends Mol Med. 2017;23(1):6–17.
6. Salminen A, Huuskonen J, Ojala J, et al. Activation of innate immunity system during aging: NF-kB signaling is the molecular culprit of inflamm-aging. Ageing Res Rev. 2008;7(2):83–105.
7. Menendez JA, Alarcón T. Senescence-inflammatory regulation of reparative cellular reprogramming in aging and cancer. Front Cell Dev Biol. 2017;5:49.
8. Mahmoudi S, Brunet A. Bursts of reprogramming: a path to extend lifespan? Cell. 2016;167(7):1672–4.
9. Liguori I, Russo G, Curcio F, et al. Oxidative stress, aging, and diseases. Clin Interv Aging. 2018;13:757–72.
10. Feinberg AP, Ohlsson R, Henikoff S, et al. The epigenetic progenitor origin of human cancer. Nat Rev Genet. 2006;7(1):21–33.
11. Goding CR, Pei D, Xin. Cancer: pathological nuclear reprogramming? Nat Rev Cancer. 2014;14(8):568–73.
12. Balkwill FR, Mantovani A. Cancer-related inflammation: common themes and therapeutic opportunities. Semin Cancer Biol. 2012;22(1):33–40.

Chapter 8
The Food as an Antioxidant System

The Food as a Complex Antioxidant System

The field of nutrition is expanding as a scientific perspective because of the interest of the public and its significance for health, and our knowledge about what is considered a "healthy diet" is based on analytical scientific research and data. It is well known from ancient times that diet is an important factor of a healthy lifestyle. The Greek physician Hippocrates first reported the contribution of the imbalance of "body fluids" as a risk factor in the development of disease. However, our understanding of diet and its effects on health has evolved from simplistic concepts to more detailed and in depth associations. The modern science led to the realization that nutrients, nutritional value, associations of nutritional epidemiology, and molecular characteristics of chemical compounds in food can directly contribute to disease [1].

Indeed, the diet is comprised of a multitude of nutritional and chemical molecules, each capable of regulating a diversity of biological processes, and thus it cannot be used as an approach, similar to the pharmaceutical industry, that is, the "one drug one target" paradigm. Hence, nutrition is a new discipline of science assessing biological networks associated with nutrients [1].

The new discipline field regarding nutrition that has developed is named "*nutritional genomics*," which provides the tools to understand how diseases such as cancer can be induced, up- or down-regulated, by modulating the expression of specific genes with dietary interventions [1]. In other words, it describes the use of genomic tools to understand how a nutritional stimulus affects metabolic pathways and homeostatic control [2]. Moreover, it studies the effect of nutrients on health by altering genome, expression proteome, and metabolome [3].

A. Salavoura, *Chemical Environmental Pollutants and their Effect on Health*,
https://doi.org/10.1007/978-3-031-80271-3_8

Therefore, proteomics aims to characterize all proteins in a biological sample, including their relative abundance, distribution, posttranslational modifications, functions, and interactions with other biological molecules. Metabolomics can be simply defined as the quantitative analysis of all metabolites in a cell system, tissue, or biological fluid. Although these two terms tend to be used interchangeably, the metabolomics discipline is generally associated with drug discovery and refers to a systematic approach to metabolite profiling, whereas the more widely used term "metabolomics" refers more specifically to cell based metabolite profiling [4].

Nutrigenetics, on the other hand, aims to understand how the genetic makeup of an individual coordinates his response to diet and thus as a science considers that underlying genetic polymorphisms and gene variants are associated with differential responses to nutrients and disease states [4].

Therefore, both disciplines are interrelated, and although distinct in their approach, they aim to unravel diet and genome interactions. Nutrigenomics will unravel the optimal diet from within a series of nutritional alternatives, whereas nutrigenetics will provide important information that will assist clinicians in identifying the optimal diet for a given individual or in other words to provide personalized nutrition (Fig. 8.1) [4].

Another aspect to be considered is the concept behind many marketing campaigns that individuals select foods based on health claims. Unfortunately, this appears to be misleading. The primary drivers in food selection are taste, convenience, and price. Where quantitative research has explored the effect of health benefit information on food selection, it has repeatedly been found that only minor changes in food purchase and consumption can be achieved, even with quite intensive campaigns [5].

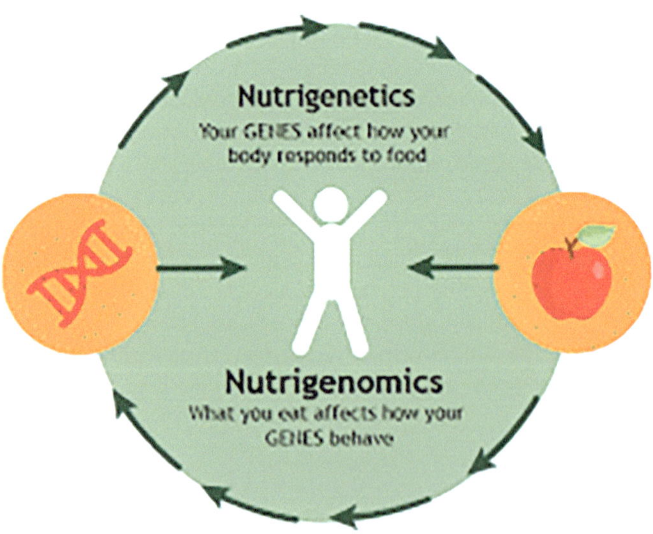

Fig. 8.1 The science of nutrition based on new disciplines including nutrigenomics and genetics

Inflammation and Oxidative Stress as a Cause of Disease

The common result of unhealthy diet preferences is inflammation and the so called "oxidative stress." From an evolutionary point of view, the inflammatory response is inherently protective, eliminating destructive agents and healing damaged tissue. However, chronic inflammation ensues when noxious stimuli persist and the inflammatory response fails to resolve [6].

Chronic inflammation is related to the well known metabolic stress. Excessive nutrient stimulation leads macrophages to undergo a phenotypic switch toward a proinflammatory phenotype. The classic proinflammatory response of macrophages depends on TLRs (Toll-like receptors), activation of nuclear factor κB/c-Jun *N*-terminal kinase, and the production of inflammatory cytokines.

Metabolic syndrome is a chronic inflammatory condition of the adipose tissue. Macrophages account for 30–40% of visceral stromal cells and they display a wide range of activation profiles, from alternative activation in lean adipose tissue to the predominant classic proinflammatory state in adipose tissues [7].

As it has already been described, oxidative stress is a condition in the body that arises when the antioxidant to oxidant balance is altered and redox equilibrium is disturbed, leading to the overwhelming production of free radicals. Redox homeostasis undergoes changes because of a plethora of factors. Conditions contributing to oxidative stress, apart from inflammation, are lack or excess of oxygen, infection, and toxic insults [8].

The brain has a high demand for oxygen and, therefore, its function is impaired by an oxidative enironment due to the production or free radicals. Comparatively, the brain has low antioxidant enzyme ability due to dependence on the antioxidant system of catalase. Secondly, the brain has higher levels of membrane polyunsaturated fatty acids, which are susceptible to free radical mediated attack. All these conditions contribute to a greater susceptibility of the brain tissue to ROS induced oxidative stress that involves elevated toxic radical species formation, membrane lipid degradation by lipid peroxidation, protein oxidation, and oxidative DNA damage (Fig. 8.2) [8].

Oxidative stress is involved in most of the neurodegenerative disorders, such as Alzheimer's disease, Parkinson's disease, and amyotrophic lateral sclerosis; vascular diseases such as atherosclerosis, peripheral vascular disease, ischemia, congestive heart failure, and stroke, as well as early onset of cataracts [9]. Oxidative stress has been implicated in diabetes, renal disorders, asthma, arthritis and other age related diseases [8].

Antioxidants, by definition, either slow or inhibit free radical mediated oxidative stress by different mechanisms. Some antioxidants are endogenous and are produced by the body, while others are exogenous and their main source is food. The various components of the antioxidant defense system are interrelated and they function together to ensure adequate protection of the cell [8] (Fig. 8.3).

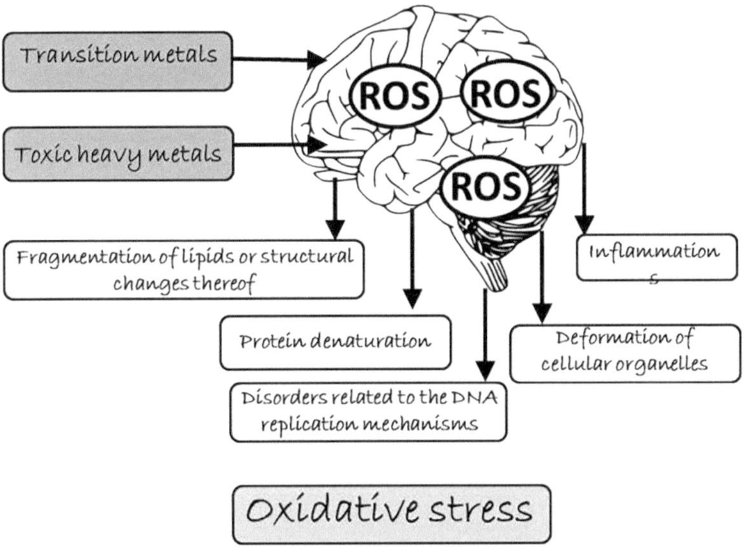

Fig. 8.2 Which factors increase oxidative stress in the brain [9]

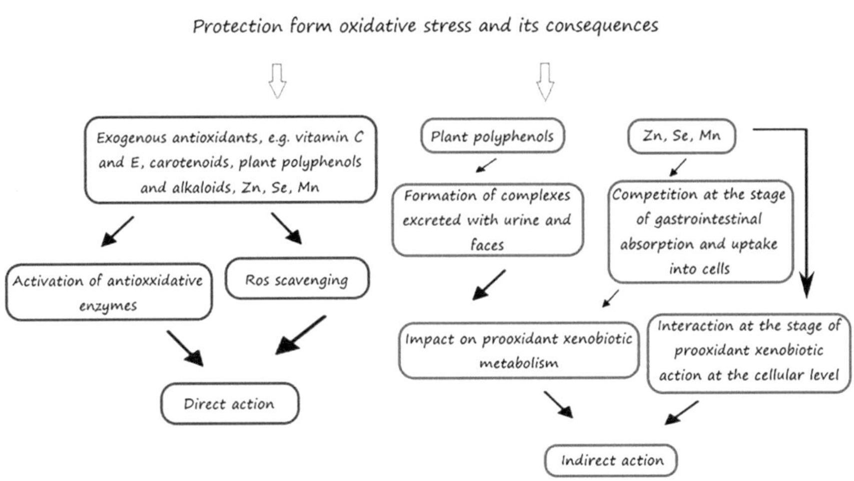

Fig. 8.3 Food provides much to the antioxidant system of the organism

The Microbiome

The human gastrointestinal tract is colonized by an incredibly number of pathogens. Recently, scientific research has focused on the constitution of the microbiome and its association with the development and maintenance of immunity. The bacteria, the so-called microbiota, has proved to be involved in the balance of TH1/TH2 responses responsible for the initiation and elimination of immune responses. Any dysfunction of the system is associated with the development of allergies or autoimmune diseases [9].

These microbes capture energy in the largely anaerobic environment of the colon. In the process of breaking down dietary and host derived substrates, the gut microbiota produces a broad range of metabolic products that accumulate to high levels in the gut. Interestingly, studies reveal that these chemicals impact host biology, either by acting on cells within the gastrointestinal tract or by entering circulation and exerting their effects at distal sites of the body. Given the high level of functional diversity in the gut microbiome and the varied diets that we consume, the repertoire of microbiota derived molecules varies between individuals. Thus, the microbes in our gut and the metabolic end products they produce represent a repertoire that we can potentially control to develop new therapeutics for personalized medicine [9].

Numerous studies have shown that microbial metabolites engage specific host receptors and activate downstream signaling cascades that ultimately change cellular equilibrium. The repertoire of endogenous (host-derived) circulating metabolites is held fairly constant across individuals, reflecting homeostatic control of biochemical pathways for production, recycling, and elimination [9].

It is generally accepted that Bacteroidetes (Bacteroides, Parabacteroides, Alistipes, and Prevotella) and Firmicutes (Eubacterium, Clostridium, and Ruminococcus) comprise roughly 90% of the microbiome of the large intestine, with the balance composed of members of Actinobacteria (Collinsella and Bifidobacterium) and Proteobacteria (Escherichia) Coli [9].

Some of the main sources of food for colonic microbes are resistant starch, nonstarch polysaccharides, host glycans, and dietary protein. The fermentation of starch and nonstarch polysaccharides likely occurs first. Dietary protein and amino acid fermentation occur after starch exhaustion. Diversity is positively correlated with a healthy diet, age, and general good health, and it is negatively correlated with inflammation [10].

Methanogens are anaerobic prokaryotes from the domain archaea that utilize hydrogen to reduce carbon dioxide, acetate, and a variety of methyl compounds into methane. Earlier believed to inhabit only extreme environments, these organisms are now reported to be found in the intestine. Methanogens are involved in hydrogenotrophic metabolism requiring the presence of H_2 to reduce CO_2 to methane, a process termed as "methanogenesis." Alternatively, methanogens can convert acetate into CH_4 (acetoclastic pathway) or methanol and methylamines into CH_4

(methylotrophic pathway). Microbial interactions with the host are known to exert physiological and biochemical effects which can be either beneficial or not [11].

To summarize, apart from the nutritional value of food and healthy eating habits, the chemical reactions in the gut are associated with the development of a healthy microbiome. On the contrary, an unhealthy lifestyle and diet could lead to the loss of the balance of microbiome constitution and to disease.

A Balanced Diet

The human diet consists of macronutrients including carbohydrates, protein, and fat, and micronutrients including vitamins, essential elements, and antioxidant substances.

Carbohydrates or saccharides are biomolecules consisting of carbon, hydrogen, and oxygen atoms. The saccharides are divided into four chemical groups: monosaccharides, disaccharides, oligosaccharides, and polysaccharides. Monosaccharides and disaccharides are referred as sugars.

Carbohydrates of the diet are important for many metabolic pathways. Monosaccharides are the major source of fuel for metabolism, being used both as an energy source (glucose being the most important) and in biosynthesis. When monosaccharides are not necessary, they convert to more storage efficient forms, named polysaccharides. In humans, the stored form of polysaccharides is glycogen in liver. In plants, cellulose and other carbohydrates serve similarly to the storage of polysaccharides. Apart from storage compounds, polysaccharides serve as structural components (e.g., cellulose in plants and chitin in arthropods). The 5-carbon monosaccharide ribose is an important component of coenzymes (e.g., ATP, FAD, and NAD) and the backbone of the genetic molecule, the DNA DNA. Carbohydrates consumed in food provide 3.87 Kcal of energy. The American Institute of Medicine recommends that adults get between 45% and 65% of dietary energy from whole carbohydrates.

Although vegetables and fruits constitute the major source of carbohydrates in an anti inflammatory diet, other carbohydrate rich foods, such as grains, are important. Examples of whole grains include buckwheat, barley, rye, and wild rice. Consumers should choose their whole grains carefully, selecting grains with all their original parts present (bran, germ, and endosperm) and not the refined products that strips out much of the nutrients. Moreover, grains that have been processed by cracking or crushing should be limited due to their high glycemic index. Most of these whole grains are high in fiber, which has been shown independently to reduce inflammatory markers [12].

Although complex carbohydrate foods often provide a lower risk of overweight and obesity compared to high-fat consumption, a high-carbohydrate diet showed only offers transient higher satiety response than high-protein and high-fat diets [13].

Glycemic load is defined as the quantity of carbohydrates ingested, multiplied by the rate at which the carbohydrate enters the bloodstream [13]. Glycemic variability, or high postprandial glucose peak, was recently suggested to be an even more important glucose measurement in the blood than HbA1c or single time point measurements, for example, morning fasting blood glucose. In an attempt to reduce elevations of blood glucose, the pancreas secretes insulin. It is interesting to note that apart from lowering the blood glucose, insulin controls the conversion of linoleic acid into arachidonic acid by activation of the enzyme delta-6 and delta-5 desaturase. Hence, the greater the insulin response to high glycemic load carbohydrates, the more arachidonic acid is produced [12].

High blood glucose potentially generates radical oxygen species by mitochondrial oxidation chain and NADPH oxidase (nicotinamide adenine dinucleotide phosphate). The inflammation induced leads to the expression of TNF-α, IL-6, and plasminogen activator inhibitor-1 [13].

Postprandial hyperglycemia increases oxidative stress, the coagulation process, endothelial dysfunction, and inflammation. Thus, reduced postprandial hyperglycemia is an important preventive and therapeutic goal against vascular diseases, type 2 diabetes mellitus, and inflammation related noncommunicable diseases [13]. The following illustration shows the pyramid of foods relative to their glycemic response (Fig. 8.4).

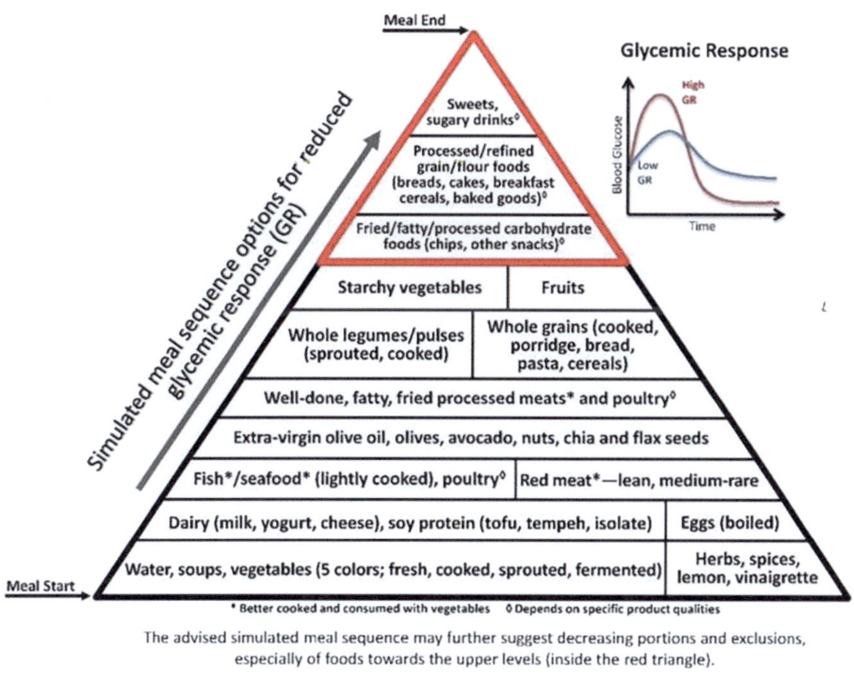

Fig. 8.4 The pyramid of foods depending on their glycemic response

Dietary fiber, apart from increasing satiety, it also lowers blood sugar responses, especially the soluble fiber. It improves glucose metabolism and it reduces related inflammation in individuals with type 2 diabetes. Meals based on fiber rich, such as bread and vegetables, are associated with approximately 40% lower blood glucose and insulin responses, compared with a fast food meal matched for energy and macronutrients [13].

Vinegar (acetic acid), a common ingredient in the traditional Mediterranean diet, slows down the carbohydrate digestive process by inhibiting gastric emptying. Thus, it can prevent blood sugar spikes. Recent studies showed that adding 1–2 tablespoons of vinegar to a meal containing white bread or white rice can lower postprandial blood glucose by 25–35% and it can increase postmeal satiety by more than two fold. Additionally, lactic acid contributes to the reduction of blood glucose [13].

Proteins are large biomolecules, or macromolecules, consisting of one or more long chains of amino acid residues. A linear chain of amino acid residues is called a polypeptide. A protein contains at least one polypeptide. Short polypeptides, containing less than 20–30 residues, and they are rarely considered to be proteins, but they are commonly called oligopeptides. The individual amino acid residues are bonded together by peptide bonds and adjacent amino acid residues. Most microorganisms and plants can biosynthesize all 20 standard amino acids, while humans should obtain some amino acids from their diet. The suggested content of protein in an adult diet is 25–30%.

Protein in an anti-inflammatory diet should be primarily plant based, with some sources of fish and small amounts of lean meat. The amount and the type of fat contained within the protein source are important. Animal protein contains higher levels of w-6 fatty acids (FAs), while plant and fish protein contains higher levels of w-3 FAs, which is considered as anti-inflammatory compounds [12].

A high protein diet demonstrates a long term impact on weight loss, mostly by its high satiating properties than a similar calorically high carbohydrate or high fat diet. The gastrointestinal tract sensors register proteins as satiating and stimulating hormonal signaling that reaches the brain. Amino acids in the gastrointestinal tract induce the release of CCK (cholecystokinin) and anorexigenic substances such as glucagon-like peptide 1 and peptide YY. Thus, prioritizing high-protein foods early in the meal sequence before carbohydrates and fat would be expected to increase their satiating effect more than with the opposite direction order [13]. The following picture depicts the sanitation index of different foods relative to the quantity of protein (Fig. 8.5).

An important aspect of meat consumption includes the cooking method. Meats cooked at high temperatures produce heterocyclic amines and polycyclic aromatic hydrocarbons, and they create advanced glycation end products, which are proinflammatory. Furthermore, the heterocyclic amines and polycyclic aromatic hydrocarbons have been shown to be mutagenic and to cause cancer in animal models. Although a direct link to cancer has not been established in humans, epidemiologic

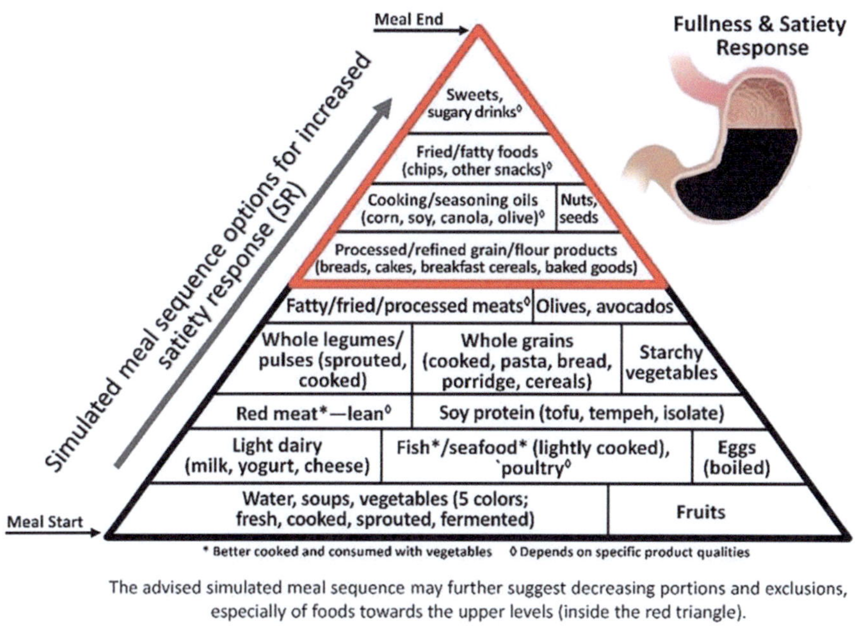

Fig. 8.5 The sanitation index of different meals

studies have found associations with cancer among individuals consuming large amounts of meat cooked well-done or grilled [12].

Although an anti-inflammatory diet permits the intake of some animal protein, plant based proteins should predominate. Soy legumes, such as edamame, tempeh, or tofu, are an excellent source of plant based protein. Finally, mushrooms are a good source of protein, containing polyphenols and other anti-inflammatory phytonutrients [12].

Fats are one of the three main macronutrients. Fat molecules consist of carbon and hydrogen atoms and they are soluble only in organic solvents, while they are insoluble in water. Examples include cholesterol, phospholipids, and triglycerides. Fats and oils are categorized according to the number and bonding of the carbon atoms in the aliphatic chain. Saturated fats have no double bonds between the carbons in the chain. Unsaturated fats have one or more double-bonded carbons in the chain. The nomenclature is based on the nonacid (noncarbonyl) end of the chain. This end is called the omega end or the n-end. Thus, alpha-linolenic acid is called an omega-3 fatty acid, because the third carbon from that end is the first double-bonded carbon in the chain. Some oils and fats have multiple double bonds and they are called polyunsaturated fats. Unsaturated fats can be further divided into cis fats, which are the most common, and trans fats, which are rare.

Fats serve both as energy sources for the body and store energy in excess of what the body needs immediately. Each gram of fat when burned or metabolized releases about nine food calories (37 kJ = 8.8 kcal). Fats upon degradation produce their constituents, glycerol, and fatty acids. Glycerol itself can be converted to glucose by the liver and so become a source of energy.

Saturated and trans fatty acids (SFAs, TFAs) stimulate a proinflammatory state contributing to the metabolic syndrome. These fatty acids can be directly involved in several inflammatory pathways, autoimmunity, allergy, cancer, atherosclerosis, hypertension, and heart hypertrophy, as well as other metabolic and degenerative diseases; and indirectly by alteration in the gut microbiota, they are associated with oxidative stress and inflammation [13].

SFAs are present in small amounts in meat and dairy products, while TFAs are primarily consumed in foods prepared with partially hydrogenated vegetable oils (e.g., bakery goods). The process of hydrogenation converts cis double bonds of unsaturated fats to trans double bonds, thereby producing solid fat. TFA consumption has well established adverse effects on serum lipids, primarily due to the modulation of liver function and the metabolism of lipoprotein. However, evidence also suggests TFAs to be proinflammatory. In clinical trials and observational studies, TFA consumption has been associated with increased markers of systemic inflammation, including tumor necrosis factor (TNF), CRP, and interleukin 6 [12].

Another important dietary contributor to inflammation is the level of *w-3 and w-6 fatty acids* consumed. In general, w-3 FAs are considered anti-inflammatory, while w-6 FAs are proinflammatory. However, both FAs are essential nutrients for the body, and it is the ratio of these FAs that likely determines inflammation levels. Although the precise ratio promoting inflammation is unknown, a ratio of w-6: w-3 FAs >10:1 is believed to be proinflammatory. Several studies have demonstrated a positive impact on various disease states (cardiovascular disease, rheumatoid arthritis, asthma) when the w-6: w-3 FAs ratio increases [12].

The metabolism of w-6 and w-3 FAs clarifies their effect on the inflammatory process. Linoleic acid, a w-6 FA, serves as the precursor for a number of proinflammatory eicosanoids produced from arachidonic acid. As a result, the more w-6 FAs consumed in the diet, the greater the propensity for increased levels of inflammation. On the contrary, w-3 FAs exert a number of anti-inflammatory effects. Opposite to the effect of insulin, w-3 FAs inhibit the delta-6 desaturase enzyme, reducing the production of arachidonic acid. Additionally, when arachidonic acid is produced, w-3 FAs can indirectly limit the generation of proinflammatory eicosanoids by competing with arachidonic acid in the phospholipid bilayer of cell membranes. Finally, w-3 FAs can defuse existing inflammation by contributing to the production of a class of anti-inflammatory eicosanoids known as resolvins, which are believed to regulate the activation and trafficking of polymorphonuclear leukocytes [12].

Fish protein is rich in w-3 FAs especially fresh water fatty fish such as salmon, mackerel, halibut, sardines, and herring. Farm raised fish have lower levels of w-3 FAs. Beef produced from naturally grazing cattle has lower w-6: w-3 ratios as compared with conventional beef. In addition to grass feeding practices, dairy and meats have been shown to contain higher levels of anti-inflammatory w-3 FAs versus their nonorganic counterparts. A diet low in w-6 PUFA-rich oils (e.g., corn, sunflower, and safflower) and high in w-3 PUFA plant sources (e.g., green leafy vegetables, chia, and flax linseed) or w-3 LCPUFA from animal sources can together reduce responsive cellular and plasma concentrations of inflammatory factors [13].

Epidemiological studies have demonstrated that the consumption of LC-PUFA affects many physiological processes including growth, neurological development, lean and fat mass accretion, reproduction, innate and acquired immunity, and the incidence and severity of many chronic and degenerative diseases including cancer, atherosclerosis, stroke, arthritis, diabetes, osteoporosis, and neurodegenerative, inflammatory and skin diseases. Their disparate roles in health can be associated with their ability to serve as ligands or to be processed to ligands for a number of critically important transcription factors, including peroxisome proliferator-activated receptors, hepatic nuclear factor-4, sterol regulatory expressed binding proteins, liver-X-receptor, and nuclear factor, which are able to disseminate the biological activity of LCPUFA throughout the various tissues of the body [1].

Beyond the impact of specific nutrient groups, chronic inflammation is affected by caloric intake. Excess caloric intake, particularly in sedentary individuals, results in increased adiposity. Adipose tissue is a metabolic active tissue that directly contributes to chronic inflammation by the release of proinflammatory cytokines, including TNF-α and interleukin 6. As adipocytes enlarge, further inflammation ensues as a result of infiltration and activation of macrophages that release additional inflammatory cytokines. A number of other proinflammatory alterations occur as a result of excess adipose tissue, including increased insulin resistance and sympathetic nervous system activation [12].

In contrast to excess caloric consumption, calorie restriction with adequate nutrition intake has been shown to exhibit important anti-inflammatory effects. Reducing the number of adipocytes in the body by calorie restriction lowers the level of pro-inflammatory adipokines and cytokines. The same reduction of fat cells also improves insulin sensitivity and it reduces plasma glucose levels, thereby lowering the production of glycation end products. Separate from its effects on adipose tissue, calorie restriction enhances endogenous corticosteroid production, promoting anti-inflammatory effects in the body. Finally, caloric restriction may increase parasympathetic tone, contributing to the suppression of cytokine mediated inflammation [12].

Among caloric macronutrients, a satiating hierarchy depicts protein > carbohydrate > fat. It is known that fat delays gastric emptying [13].

The oxidative stress axis of a diet is currently another parameter of a healthy diet (Fig. 8.6).

The World Health Organization (WHO) makes the following five recommendations with respect to both populations and individuals.

1. Maintain a healthy weight by eating roughly the same number of calories that your body is using.
2. Limit intake of fats. No more than 30% of the total calories should come from fats. Prefer unsaturated fats to saturated fats. Avoid trans fats.
3. Eat at least 400 g of fruits and vegetables per day. Potatoes, sweet potatoes, cassava, and other starchy roots do not count. A healthy diet also contains legumes (e.g., lentils, beans), whole grains, and nuts.
4. Limit the intake of simple sugars to less than 10% of calories (below 5% of calories or 25 g may be even better).
5. Limit salt from all sources and ensure that salt is iodized. Less than 5 g of salt per day can reduce the risk of cardiovascular disease [13].

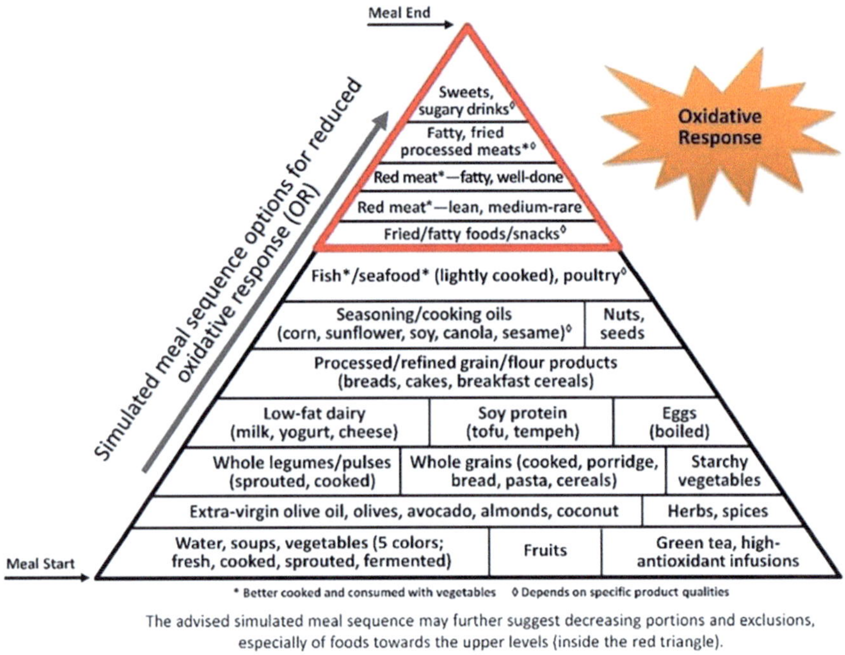

Fig. 8.6 The meal sequence pyramid per oxidative response axis. Foods are ranked according to their potential to enhance reactive oxidation species postprandially. Foods at the bottom of the pyramid have the lowest response because they are the least oxidative and they have the highest quantity of antioxidants, which could neutralize ROS. Thus, they are recommended for starting the meal. The next levels include relatively whole, minimally processed foods. The highest levels include highly processed foods by technologies that generate high amounts of ROS, with little-to-no antioxidant content, and foods high in prooxidative potential like fried foods and high iron content in red meat [13]

Natural Antioxidants

Natural antioxidants may be as effective as pharmaceuticals in regulating oxidative stress and its consequences. They are complex molecule antioxidants from a variety of food sources, such as vitamins, plant pigments, and secondary plant metabolites such as phenolics and complex polyphenols. Multiple natural antioxidants in the food function as part of the cell antioxidant mechanism.

Vitamins

The Chemical Structure, Forms, and Distribution in the Environment

Vitamins are involved in one-carbon metabolism as methyl donors and their levels regulate the levels of methyl groups available for the maintenance of methylation.

Vitamin C (dehydroascorbic acid) was isolated in 1930 and was proposed to treat respiratory infections. Its popularity increased after 1970 when Nobel laureate Linus Pauling concluded that vitamin C would prevent and alleviate the common cold. Vitamin C has been widely sold and used as a preventive and therapeutic agent [14]. However, the public opinion that vitamin C contributes to the elimination of infections has been debated.

First research results have not be proved by subsequent studies. On the contrary, it was reported that regular ingestion of vitamin C had no effect on common cold incidence. The failure of vitamin C supplementation to reduce the incidence of colds in the general population indicates that vitamin C supplementation during an infection is not justified. However, it was proposed that early initiation of vitamin C in acute respiratory virus infections could have potentially some impacts on the duration of the disease. According to a study with an acceptable number of participants, when treatment started within 24 h of the onset of symptoms, the mean duration of vitamin C treated colds was 3.6 days, whereas the duration was 6.9 days with no treatment [14].

However, the effect of vitamin C administration during colds has statistically significant effects in children. Studies show an 18% reduction in the duration of colds by administration of 1 g or 2 g/day, compared to adults (8%), which demonstrates treatment with vitamin C in children [14].

Effects on Health

How vitamin C exhibits its beneficial effects has been investigated. Vitamin C is an efficient *water-soluble antioxidant* and the effects on the immune system can be explained by the protection against oxidative stress generated during infections. Dozens of studies have found that vitamin C may affect phagocytosis and chemotaxis of leukocytes and the replication of viruses by the production of interferon. Phagocytes have a specific transport system by which the oxidized form of vitamin C is imported into the cells, where its levels are replenished. Furthermore, heavy physical stress generates oxidative stress, and the antioxidant role of vitamin C can explain its effects on respiratory symptoms in physically stressed people [14].

Although mean plasma vitamin C concentrations are typically around 50 μmol/L, neutrophils accumulate intracellular vitamin C concentrations against a concentration gradient which is thought to indicate an important role for the vitamin in these cells. Thus, the depleted vitamin C status of neutrophils observed during infectious episodes could potentially compromise their function [15].

Vitamin C, as a potent free radical scavenger together with glutathione, is biochemically intertwined. Glutathione (GSH) actively recycles oxidized dehydroascorbate back to ascorbate and GSH deficiency is accompanied by low tissue levels of ascorbate. It has also been predicted that increasing vitamin C levels diminishes GSH pools. Based on this evidence, vitamin C, if used in safe and tolerable doses, appears to attenuate GSH depletion under stress conditions and to promote an increased antioxidant status, while high doses may have opposite results (Fig. 8.7) [16].

In general, vitamin C is considered safe in doses up to several grams per day. Although there has been speculation about the potential harm of large doses, it has been reported to be unfounded. Doses of 0.01 g/day protect against scurvy, and 100 g intravenously within a few hours do not have any side effects. Doses considered safe reach 30 g/day [14].

It was described how *glutathione* reduces oxidative stress in order to maintain redox balance, it enhances metabolic detoxification, and it regulates the immune

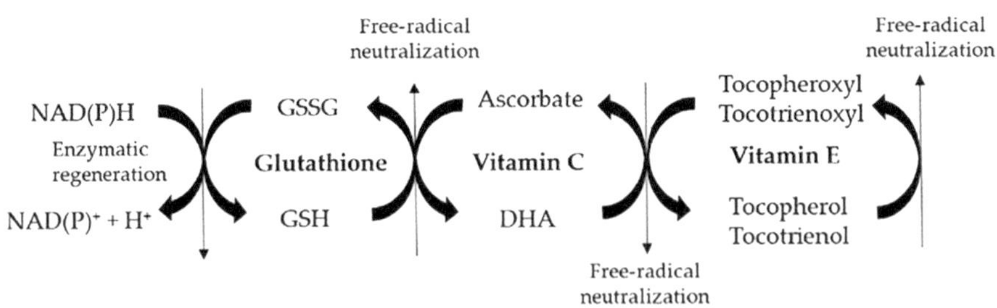

Fig. 8.7 The cycle of ascorbate-dehydroascorbate (DHA) is interlinked with glutathione-glutathione disulfide (GSSG) redox cycle and it neutralizes free radicals such as H+, while it is interrelated with other antioxidant systems also

system. As a result, it is possible that supporting the body's endogenous levels of glutathione would be important for maintaining health and compromising disease and causal relationships between low glutathione and disease risk have been described. Although glutathione is an endogenous antioxidant, to some extent, glutathione levels may be supported by dietary or supplemental nutrients [17]. Foods that have some evidence to improve glutathione status are protein sources, brassica vegetables, polyphenol rich fruits and vegetables, herbs and spices, green tea, and omega 3 fatty acid rich foods such as fish [17].

Riboflavin, vitamin B2, is a necessary coenzyme for the activity of glutathione reductase, which converts the oxidized glutathione into its reduced form. While there are not enough studies to confirm that riboflavin deficiency negatively impacts glutathione levels, there is an indication that homocysteine production and methylation processes are eliminated by the deficiency of the enzyme. Additionally, *pantothenic acid* or *vitamin B5* may also help support glutathione synthesis by its role in ATP production. *B12 deficiency* is associated with lower glutathione levels also [17].

Pyridoxine (vitamin B_6) and its biologically active form, pyridoxal 5′-phosphate, are coenzymes in the conversion of homocysteine to cysteine, which supports GSH biosynthesis. It has an indirect role in GSH homeostasis by sequestering free radical species and by increase of GSH concentrations. Homocysteinemia and chromium toxicity are high-stress environments, where vitamin B_6 is especially critical for maintaining GSH pools. Oral vitamin B_6 supplementation increased renal and hepatic GSH levels in rats with chromium-induced toxicity, and it increases hepatic GSH levels in rats, despite the fact that homocysteine levels remain high [16].

Combined supplementation of vitamin B_6 (200 mg/day) and folic acid (15 mg/day) for 1 month increased the whole blood GSH levels in both hemodialysis patients and healthy patients. Similarly, Parkinson's disease patients supplemented with folic acid and vitamins B_{12} and B_6 had improved GSH status [16].

Other vitamins could be provided as supplementation alternatives in order to improve the activity of antioxidant mechanisms. Fat-soluble *vitamins (A, D, E, and K)* have been evaluated for their potential effects on GSH homeostasis. These vitamins exhibit antioxidant properties and there is evidence that the status of vitamin A, vitamin D, and vitamin K may affect GSH levels. However, excessive supplementation has been associated with adverse effects [16].

Vitamin E (Tocopherol) is a lipid-soluble antioxidant that protects the cell membrane against oxidation and it supports other antioxidant reactions such as the GSH status. Studies suggest that vitamin E supplementation may prevent GSH oxidation and it could ultimately increase its levels in tissues. Based on clinical studies, it appears that vitamin E, when taken in safe doses, has the potential to increase red blood cells and brain GSH concentrations Furthermore, tocopherol has been shown to reduce the risk of CVD, inhibiting the oxidation of LDL lipids [16, 18].

Folate is known to contribute to the production of S-adenosylmethionine (SAM), a methyl donor of cytosines for the process of DNA methylation. Folate deficiency is related to DNA hypomethylation. *Biotin (vitamin 6 or 7 or vitamin H)* is attached to histones and it contributes to the repair of DNA damage. *Tryptophan and niacin (vitamin B3)* are essential amino acids that are converted to nicotinamide adenine dinucleotide, a substrate for poly (ADP-ribosylation) of histones and other DNA binding proteins, which are necessary for DNA repair and apoptosis. Niacin supplementation may reduce the frequency of DNA damage in human lymphocytes. Nutrients other than folate also play roles in one-carbon metabolism: choline and methionine are the major dietary sources of methyl groups, whereas vitamins B6 and B12, riboflavin, and zinc are coenzymes and cofactors, respectively, for various steps in one-carbon metabolism, as it has already described [19]. In this respect, it is of interest that one cup of broccoli contains 25% of the daily requirements for folate, as well as significant amounts of B vitamins [20].

Plant Pigments: Carotenoids

The Chemical Structure, Forms, and Distribution in the Environment

The term carotenoids includes nearly 600 chemicals that modulate a broad spectrum of functions. They consist of a family of organic pigmented compounds that are produced by fungi, several bacteria, algae and plants. In plants, carotenoids contribute to the photosynthetic machinery and they protect them from photo-damage. They occur in all plants capable of photosynthesis, a process that converts CO_2 into O_2 in the presence of sunlight. Generally, carotenoids absorb wavelengths between 400 and 550 nm, and hence the compounds are present in red, orange, or yellow color fruits and vegetables. They are abundantly found, for example, the orange and yellow vegetables and fruits are rich in β-carotene and α-carotene, respectively. Approximately 50 carotenoids are found in a diet. Carotenoids are classified into two groups, according to their chemical constituents: *Xanthophylls*, which are the oxygenated derivatives and they include astaxanthin and a-cryptoxanthin, and *Carotenes*, which are hydrocarbon carotenoids and they include lycopene, lutein and a, b-carotene [21].

Some common household cooking methods, for example, boiling, steaming, and microwave cooking, do not markedly change the extent of the carotenoid content in food. However, there are several factors that affect carotenoid absorption, bioavailability, breakdown, transport, and storage. For example, the dietary intake of fat (in the form of salad dressing, cooking oil, for instance, olive oil or vegetable oil) at the same meal with carotenoid consumption, cooked vegetables or raw vegetable has been found to effectively influence the absorption of some carotenoids. The bioavailability of carotenoids is reduced when the same meal contains a number of them due to competition of different carotenoids in absorption. In addition, dietary fiber from plant sources, for example, guar gum and pectin, was found to decrease carotenoid absorption and the localization of carotenoids within the chromoplasts and chloroplasts of plants may reduce the bioavailability [21].

Effects on Health

The antioxidative activity of carotenoids depends on their ability to trap radicals. The number of conjugated double bonds in these molecules corresponds to their effectiveness in ameliorating free radicals [22].

Astaxanthin (3,30-dihydroxy-β, β-carotene-4,40-dione) is a red pigment belonging to the subclass of xanthophyllin carotenoids. Astaxanthin is composed of a long, nonpolar, conjugated molecule connecting polar ion rings located on both ends of the chain. Due to the extended conjugation, astaxanthin reacts with many free radicals and the presence of polar ion rings containing hydroxylic and carbonyl groups provides the molecule with higher antioxidative capacity compared to other carotenoids. Its properties are due to the presence of numerous unconjugated multiple bonds. Astaxanthin can neutralize singlet oxygen and free radicals, thus, preventing lipid peroxidation. It is found mainly in microalgae, mushrooms, and complex plants, as well as in salmon, trout, and crustacean cells. It has been demonstrated that the most important effects of astaxanthin's activity include the prevention of neuron damage and the inhibition of cell membrane peroxidation [19].

A-cryptoxanthin, lycopene, and lutein are found in orange fruits, tomatoes, tomato products, and dark green vegetables, respectively. The unsaturated nature of these carotenoids makes them prone to oxidation influencing the color of the source. Other factors like pH, light, and temperature can also affect the color and nutritional value of foods [21].

Lutein suppresses the NF-κB activation pathway. The anti-inflammatory and antioxidant properties of lutein (1–25 μM/24 h) reduce the risk of coronary artery and cardiovascular disease in the elderly population. Egg yolk is a highly bioavailable source of zeaxanthin and lutein. Lutein consumption (one soft-boiled egg per day for 5 weeks) was shown to reduce the oxidized low density lipoprotein and it is reversibly related to the development of atherosclerosis. High plasma lutein levels were found to protect the myocardium from ischemia injury by decreasing oxidative stress and apoptosis [21].

Lycopene, a carotenoid that is found in tomatoes, is an important antioxidant source. Meta analysis studies have demonstrated an inverse relationship between lycopene intake and cardiovascular disease risk. This favorable effect could be attributed to the decreased inflammatory response and low cholesterol level as well as to reduced oxidation of biomolecules [18]. It is estimated that the effectiveness and antioxidant activity of lycopene is 10 times higher than that of vitamin E [23].

Adequate dietary intake of carotenes prevents from cardiovascular disease, atherosclerosis, and cancer. Several studies have found that consumption of carotenoid rich fruits and vegetables can prevent cancers such as prostate and cervical cancer [21].

Carotenoids, especially lutein, have been reported to decrease the risk of certain cancers such as colon, prostate, and lung. However, a previous study revealed that smokers who were supplemented with β-carotene experienced increased lung cancer incidence. These findings were not associated with the nicotine level of cigarettes smoked, suggesting that smokers should probably avoid β-carotene supplementation [21].

Carotenoids have been demonstrated as an eye sight protecting agent. Some carotenoids are categorized as provitamin A and they constitute molecules with an unsubstituted β-ion ring (γ-carotene, α-carotene, β-carotene, and β-cryptoxanthin) which can be converted into the retina in photosensitive agents. Zeaxanthin and lutein, are the macular pigments found in the human retina. Macular pigments can absorb short wavelengths and high energy blue light and subsequently protect the retina from photochemical damage. This pigment can protect against UV induced peroxidation and it neutralizes ROS [21].

Much information indicates that carotenoids may limit neuronal damage from free radicals. The data from the 2011–2014 National Health and Nutrition Examination Survey involving 2796 participants aged ≥60 years demonstrated that lutein and zeaxanthin supplementation (2.02 mg/day) may prevent cognitive decline. Carotenoids delay neurodegenerative disease progression by several pathways, for example, they suppress proinflammatory cytokines, they trigger Aβ peptide production, and they reduce oxidative stress. β-carotene binds to neuronal receptors associated to Alzheimer disease, such as histone deacetylase and P53 kinase receptors (Fig. 8.8) [21].

In a well balanced diet, the intake of carotenoids, such as lutein, is sufficient to promote health. However, supplementation is needed in cases of chronic disease or

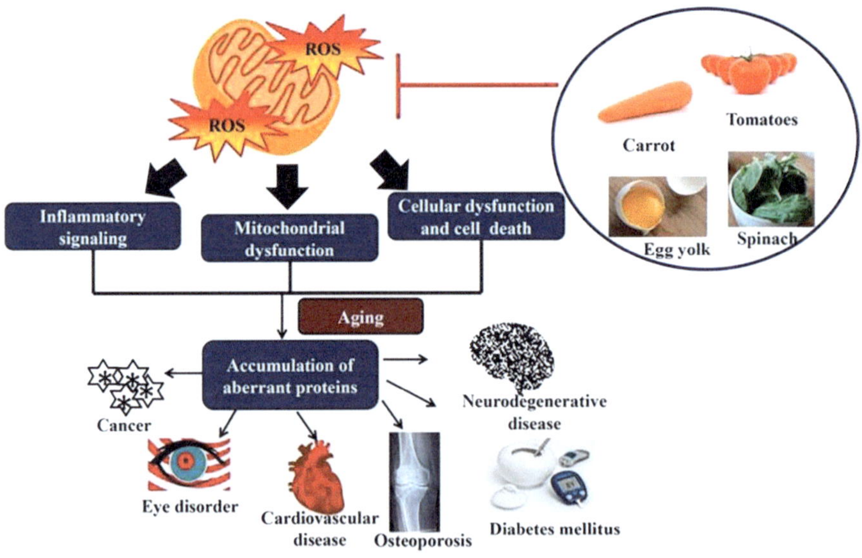

Fig. 8.8 The multiple interference of carotenes as an antioxidant mechanism in health

inadequate absorption. Several studies conducted both *in vitro* and in animal models have revealed that the use of lutein is safe without teratogenic and mutagenic outcomes. Despite the fact that no toxic effect was observed, the established a upper safety limit for daily lutein consumption is 2 mg/kg [21].

Polyphenols

The Chemical Structure, Forms, and Distribution in the Environment

The majority of the phytochemical polyphenols are present in plants as polymers or as glycosylated molecules, for example, conjugated to sugars, wherein the sugar moiety is termed glycone and the polyphenol is called the aglycone. Most of the polyphenols in dietary plant sources occur as complex molecules linked to the sugar residues (*O*-glycosides), but some occur as *C*-glycosides. Polyphenol *O*-glycosides undergo hydrolysis in the lumen of the intestine and the sugar residue is released upon the action of β-glucosidase, setting the aglycone free. On the other hand, the *C*-glycosides are resistant to intestinal enzymatic hydrolysis. It is interesting to note that the polyphenols, after the uptake by the intestinal cells, undergo metabolic biotransformation similar to the xenobiotics [22].

Two major pathways of the biosynthesis of the polyphenols have been identified in plants: (i) the shikimate pathway and (ii) the polyketide or acetate pathway. Structurally, the plant polyphenol has a phenolic or a benzene ring with a hydroxyl group attached, and also certain substituents such as ester and glycosides present as functional groups [22]. Polyphenols in the human body act as antioxidant and anti inflammatory agents as well as preoxidants that promote oxidative stress. Another quality of polyphenols is their chelating properties, the ability to form complexes with trace metals. For example, they attenuate iron caused DNA damage by keeping iron in the +3 oxidative state (Fig. 8.9) [18].

Simple phenolic and polyphenolic secondary plant metabolites are known to play a role in the plant as protectants against oxidative stress and ultraviolet radiation, wound healing, defense against microbes and fungi, plant competitors, and they offer disease resistance. However, the mechanisms of such protection and their physiological role in plants are not completely understood [22].

Polyphenols are divided into different classes, according to their chemical structure and mainly the number of aromatic rings, which include phenolic acids, stilbenes, flavonoids, lignans, and curcuminoids [24]. The most abundant polyphenols are flavonoids, and their division into subclasses depends on the degree of oxidation of the pyran ring: flavonols, flavones, isoflavones, flavanones, anthocyanins, flavanols, catechins, chalcones, aurones, and others (Fig. 8.10) [25].

Humans and animals cannot synthesize polyphenols and, therefore, humans consume polyphenols with plant foods as part of their diet [25]. Polyphenols are the

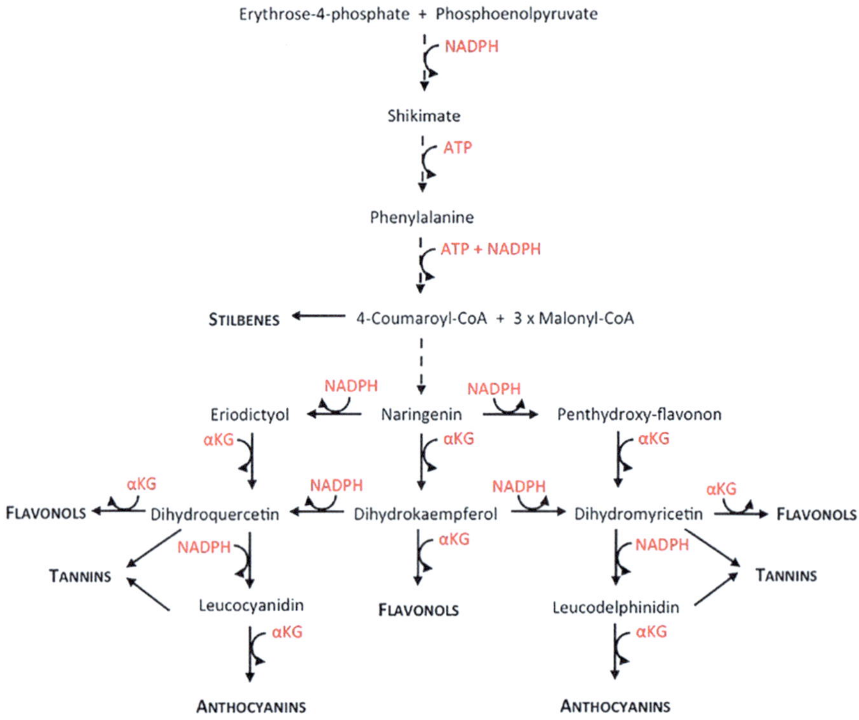

Fig. 8.9 The production of different polyphenols with the shikimate pathway

most abundant antioxidants in our diet and they are commonly present in fruits, vegetables, cereals, olives, dry legumes, licorice, chocolate, and beverages, such as tea, coffee, and wine [24]. However, the change from fresh prepared foods to processed foods has reduced the consumption of plant based natural chemicals, such as polyphenols [26].

Although several plant foods contain complex mixtures of polyphenols, numerous factors, such as climate (sun exposure, precipitation), agronomy, storage, and maturity at the time of harvest, can affect their content. Furthermore, simply peeling fruits or vegetables can significantly reduce the polyphenolic content, since these substances are often present in high concentrations in the external parts of the plant product [24].

Phenolic acids. The great interest in phenolic acids is associated with their use in food technology due to their potential in food preservation [24].

Flavonoids give a colorful appearance to many plants. Their production in the plant is correlated with stressful conditions, such as ultraviolet radiation, lack of water, or soil with increased chemicals [19]. They are secondary plant metabolites contained in fruits and vegetables. Some of them are specific to particular foods, such as flavanones in citrus fruits, isoflavones in soy, and phloridzin in apples. On the other hand, other polyphenols, such as quercetin, are found in a plethora of

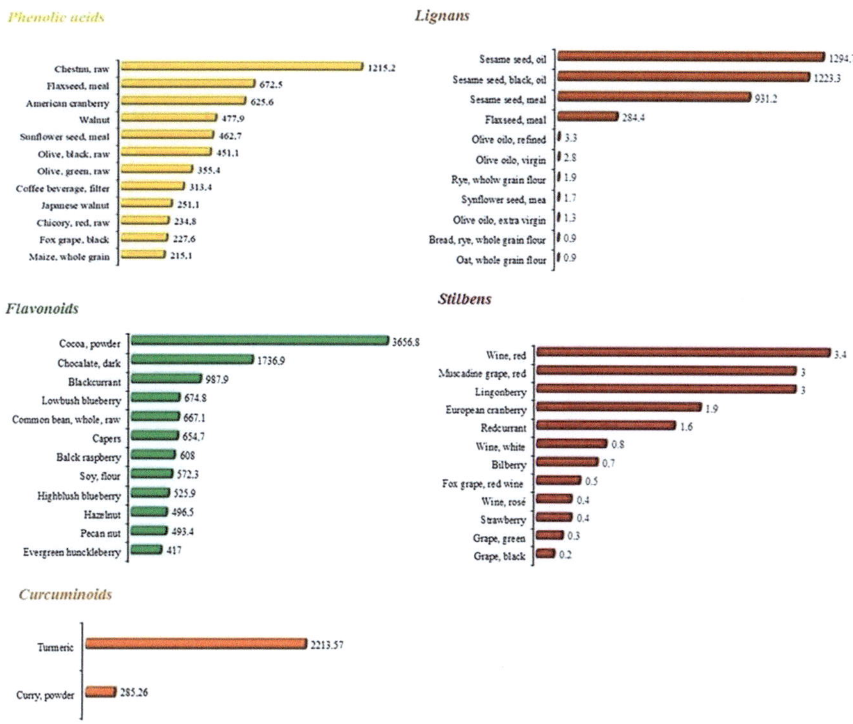

Fig. 8.10 The classes of polyphenols

vegetable products [24]. Onion, broccoli, apple, cherries, blackberries, and tea contain flavonols; oranges contain flavanones; grapes and cherries contain anthocyanins; while legumes and soya are known for isoflavon polyphenols. All these have been shown to exhibit different biological effects, including antibacterial, antiviral, analgesic, antiallergic, hepatoprotective, cytostatic, apoptotic, estrogenic, and anti-estrogenic functions [18].

The flavonols *quercetin* and *epigallocatechin-3-gallate* have gained scientific interest. Quercetin primarily enters into the diet as quercetin-3-glucoside. Subsequently, it is hydrolyzed in the small intestine and it is rapidly absorbed. Foods rich in quercetin include principally apples, berries, and grapes, as well as red onions, broccoli, black tea, green tea, pepper, red wine, tomatoes, and some fruit juices. The way of cooking food also has a noticeable effect on quercetin content. For example, onions lose about 75% of their initial quercetin content after boiling for 15 min, 65% after microwaving, and 30% after frying. Recent works describe the role of this molecule against chronic diseases, such as type-2 diabetes by stimulating insulin secretion (Fig. 8.11) [24].

Lignans are present in a wide variety of plant foods, including seeds (flax, pumpkin, sunflower, poppy, sesame), whole grains (rye, oats, barley), bran (wheat, oat, rye), beans, fruit (berries in particular), and vegetables. Among edible plant

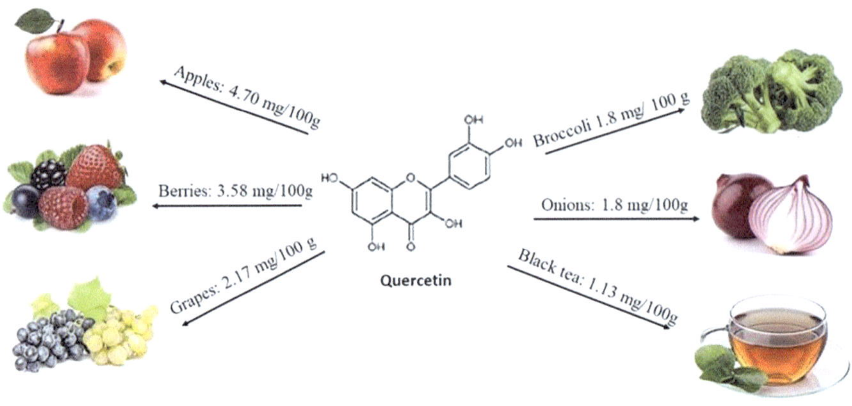

Fig. 8.11 Quercetin, a flavonoid polyphenol, appears in many vegetables, fruits, and beverages

components, the most concentrated lignan sources are sesame and flax seeds. Sesame seeds exhibit the highest concentration of sesaminol followed by cashew nuts. Among vegetables, the *Brassica* family contains pinoresinol, while spinach, white potatoes, and mushrooms contain low amounts of lignin [24].

Stilbenes are present in the human diet and the main representative is resveratrol. This compound was first isolated from the roots of *Veratrum grandiflorum* in 1940 and later from the roots of *Polygonum cuspidate.* It is produced by plants in response to infection by pathogens or to a variety of stress conditions. This polyphenol has been detected in more than 70 plant species, including grapes, berries, and peanuts. Several studies highlighted that this compound has a wide range of biological activities (Fig. 8.12) [24].

Curcumin is a polyphenolic ingredient widely studied among curcuminoids, which is responsible for the yellow color of turmeric, the most popular and common Asian Indian curry spice called the golden spice obtained from the rhizomes of the plant *Curucuma longa,* a member of the ginger family, Zingiberaceae. Recently, curcumin has gained attention from nutritionists, biomedical scientists, pharmacologists, drug discovery scientists, clinicians, and, above all, the common people worldwide as a preventive molecule or therapeutic agent for several disorders and diseases (Fig. 8.13) [22].

The most common applications of curcumin are in dietary supplements, in cosmetics and as flavoring for foods, such as turmeric flavored beverages, turmeric plants, and curry powder, in South and Southeast Asia, as a food additive for orange yellow coloring in prepared foods [24].

Many fruits contain phenolic compounds and they are therefore characterized by strong antioxidative properties. Studies have been conducted in rats poisoned with prooxidative metals and they have received fruit extract or juices for detoxification purposes. It was demonstrated that the use of extracts or juices of pomegranate, strawberries, watermelon, and *Garcinia Indica* fruit, which are popular in Hindu medicine, led to an improvement in cerebral antioxidative potential. The

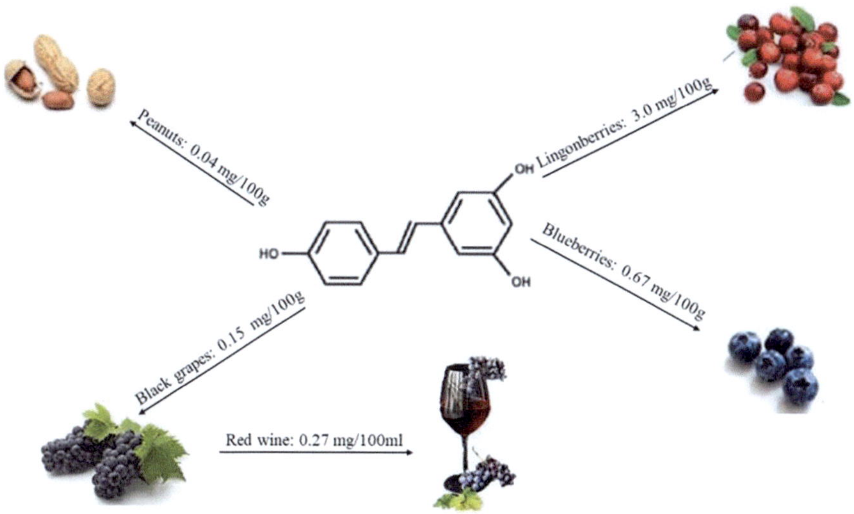

Fig. 8.12 Stilbenes appears in tree nuts, wine, and berries

Fig. 8.13 The plant *Curucuma longa,* a member of the ginger family, Zingiberaceae

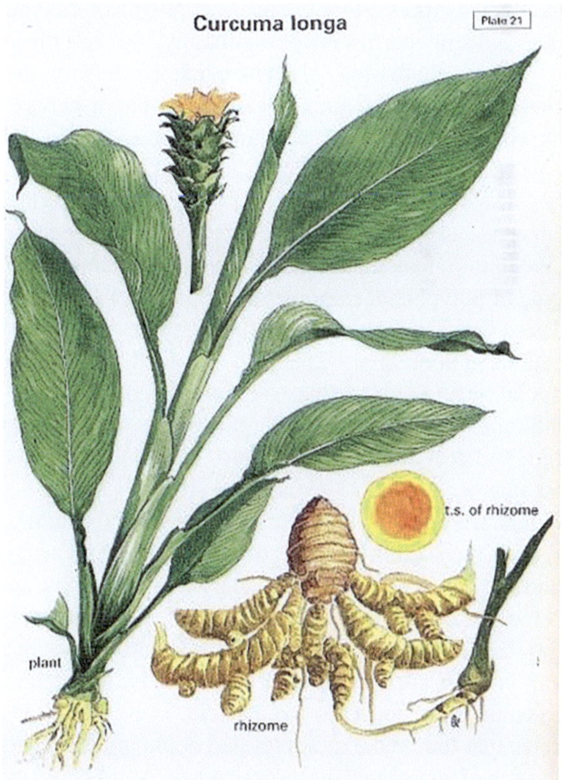

antioxidative properties of these fruits result from the presence of large amounts of phenolic compounds, primarily phenolic acids and catechins, epicatechins, flavan-3-ols, kaempferol, quercetin, and luteolin. Apart from the phenolic compounds, these fruits also contain a number of other substances with known antioxidative properties such as vitamins C and E, β-carotene, selenium, and anthocyanins. Moreover, red flesh watermelons also contain lycopene, as well as saponins and tannins [19]. However, the properties of other fruits and vegetables are described in more detail.

Berries are a term used to include fruits such as strawberries, raspberries, blueberries, black currants, blackberries, among others. Berries are characterized as being low in calories and having a high content of vitamins such as vitamins A, B complex, C, and E; essential minerals such as calcium, magnesium, and selenium as well as fat (especially unsaturated) and dietary fiber such as pectin (soluble fiber) [23].

Berries have the highest antioxidant capacity because of their high content of antioxidant compounds such as polyphenols (procyanidins, quercetin, phenolic acids, and mainly anthocyanins), carotenoids, and vitamin C. It is of note that the amount of antioxidant components varies between berry species and it is also dependent on climatology, the grade of ripening, postharvest treatments, and processing, which is the cause that each berry has a very different nutritional profile. It has been estimated that the average polyphenol content of black currants, raspberries, and strawberries is approximately 300–100 mg/100 g [23].

There is increasing scientific evidence showing that berry intake is closely associated with the prevention of cardiovascular diseases by several mechanisms such as improving lipid profile, inflammation, endothelial function, hypertension, diabetes type II, and coagulation. For example, the consumption of 1–2 portions of strawberries, raspberries, and blueberries daily is associated with a lower risk for cardiovascular disease. This protective effect seems to be attributed to polyphenols. Several epidemiological studies have indicated diverse health benefits of anthocyanins, a type of polyphenol responsible for the color of berries (from red to purple to blue). These benefits in the prevention of atherosclerosis or other cardiovascular complications include improvement in endothelial function due to the activation of endothelial nitric oxide synthase signaling, reduction of oxidative stress, modulation of inflammation response, and improvement of the lipid profile. Blueberries, strawberries, and acai are the berries most frequently studied due to their preventive effects on atherosclerosis [23].

The fruit most widely consumed in several countries are *apples* in different forms such as fresh fruit, juice, or dried. Apples contain several macronutrients and non-nutrient compounds such as dietary fiber, minerals, and vitamins (vitamins C and E, some provitamin A carotenes, lutein, folic acid, potassium, and magnesium) and they are also an important source of natural chemicals. Apples contain important antioxidants such as phloridzin, phlorizin, quercetin, catechins, procyanidins, epicatechin, rutin, and chlorogenic acid. There is growing scientific evidence that indicates that the intake of apples and apple products (extracts and juices) may reduce the risk of age related diseases (cancer, cardiovascular diseases, asthma, Alzheimer's disease, diabetes mellitus II, weight management, bone health, pulmonary function, and gastrointestinal protection) by several mechanisms, due to their antioxidant, anti inflammatory, antiproliferative, and cell signaling properties. In addition, apples and apple products decrease lipid oxidation and improve cholesterol levels [23].

Grapes (*Vitis vinifera*) are a fruit commonly found in the Mediterranean diet and they are also one of the most popular and extensively cultivated fruits worldwide. It is calculated that about 60 species of grapes can be found today. The natural chemicals in grapes are concentrated mainly in the skin, seeds, and juice. Although resveratrol (trans-3,4′,5-trihydroxystilbene) is the most known polyphenol detected in grapes, other phytochemicals such as carotenoids, flavonoids, and others can also exert important health effects [23].

Resveratrol is a polyphenolic phytoalexin produced by many plants for its antifungal properties. A similar function has pterostilbene, a dimethyl ether analog of resveratrol produced by the bark and wood of *Pterocarpus marsupium family Leguminoseae*, which is a large tree that commonly grows in the central, western, and southern parts of India and in Sri Lanka [27].

Resveratrol is found in many fruits, nuts, flowers, seeds, and bark of many plants, such as raspberries, mulberries, plums, peanuts, berries, cranberries, some pines, and roots and stalks of giant knotweed and Japanese knotweed, but resveratrol's most abundant natural sources are grapes, which are used to make wine. The content of resveratrol in wine is dependent on the length of time the grape skins are present during the fermentation process. Therefore, the concentration of resveratrol is significantly higher in red wine than in white wine and grape juice. The content in wine ranges from 0.2 to as much as 40 mg/L. Resveratrol inhibits transcription factor NF-kB, activator protein-1, and cAMP response [27]. In addition to heart related effects, resveratrol is able to regulate metabolism and it is useful in the treatment of diabetes, neurodegenerative diseases, and cancer. In humans, a dose of 450 mg per day is considered safe, while overuse could be harmful [24].

In the last years, resveratrol, are gaining importance as having a potentially important component of health, especially cardiovascular health. The popularity of resveratrol comes from the *French Paradox* concept, which originated in the 1980s by French epidemiologists that figured that although French people had unhealthy diets, they did not suffer from cardiovascular diseases. The protective cardiovascular effects were mainly attributed to relatively high wine consumption [23].

In 1992, Renaud and co-workers published an interesting study, pointing out what is known as the French paradox. The low susceptibility of French people to cardiovascular diseases was linked to the use of red wine. Red wine and obviously grapes are a source of resveratrol, a powerful antioxidant with benefits with anti-inflammatory effects [24].

Moderate intake of alcohol, up to 1 drink per day for women and 2 drinks per day for men, is a component of both the Mediterranean and Okinawan diets. Heavy drinking, in comparison, is associated with higher levels of inflammatory markers such as CRP. However, the importance of moderation should be highlighted given the detrimental impact of heavy drinking on health and well being, including increased cardiovascular disease and rates of certain cancers [28].

Olives, the fruit of the olive tree (*Olea europaea*), are botanically a stone fruit. *Olea europaea* is a small tree that belongs to the family Oleaceae and it is endemic to warm temperate regions of the world. The tree is usually scattered throughout the coastal regions of the eastern Mediterranean Basin, neighboring coastal areas of northern Africa, southeast Europe, western Asia, and northern Iran. Although *Olea europaea* is now cultured in many parts of the world, the Mediterranean area still serves as the most fructuous area, accounting for approximately 98% of the world's cultivation [23].

Different varieties of olives can be found on almost all the continents. In the Mediterranean region, olives are especially important as they are ingredients of Mediterranean cuisine, being the principal source of fat. Natural raw olives, however, are rarely consumed, because specific processes are needed to avoid the high bitterness of the fruit. Additionally, several processes can be used to extract oil from olives, with mechanical processes being better than chemical processes [23].

The Olea europaea tree has a long history of therapeutic and nutritional values. Over the centuries, extracts from olive leaf were known for their therapeutic effects. The ancient Egyptians used the leaves to preserve pharaohs. The health benefits of olive oil are mainly due to the existence of a high content of monounsaturated fatty acids (MUFAs) and functional bioactives such as tocopherols, carotenoids, phospholipids, and phenolics, with several biological activities. The most relevant bioactive compounds found in olive oil are phenolic compounds, benzoic and cinnamic acids, polyphenols, flavonoids, and secoiridoids (42). Research shows that the olive extracts significantly induced the expression of miRNA 153, miRNA 145, and miRNA 137. Decreasing the target gene expression of these miRNAs in glioblastoma stem cells, are related to anticancer activity [29].

The tomato fruit belongs to the *Solanaceae family* and it is one of the most economically important and most frequently produced plants in terms of food production. Tomatoes contain a great number of natural chemicals additionally to a wide range of vitamins, minerals, and antioxidants. There is scientific evidence on the beneficial effects of tomato fruit on chronic age related diseases, including cardiovascular diseases, hypertension, diabetes, obesity, and cancer [23].

Tomatoes present a high content of water (93–95%) and contain 3% carbohydrates, 1.2% proteins, and 1% of total lipids. Their vitamin content is important, especially vitamins A, E (mainly α-tocopherol), C, and folate, as well as minerals such as potassium and iron, among others. Absorption of iron is favored by the presence of vitamin C known to improve the bioavailability of iron. Thus, tomato juice can provide around 10–20% of the recommended daily needs for iron. Other acids also present in tomatoes include succinic, oxalic, and tartaric acid, but these only constitute a small fraction compared to citric acid. In addition, tomatoes are an excellent source of bioactive compounds such as phenolic compounds (phenolic acids and flavonoids) and carotenoids (lycopene, β-carotene, lutein, and zeaxanthin). It is of note, however, that the variety of tomatoes, the grade of ripening at harvest, the irrigation system, light, temperature, air composition, and other factors can affect the chemical composition of tomatoes [23].

Phenolic compounds plus vitamin C provide the highest total antioxidant activity of ripe tomatoes (83%), while the lipophilic fraction (carotenoids, vitamin E, and lipophilic phenols) represents only 17% of the total. Lycopene is the main carotenoid found in tomatoes, followed by β-carotene [23].

Health beneficial constituents of *whole grains* include phytochemicals, bioactive carbohydrate fractions, peptides, and other phytonutrients. They contain high amounts of phytochemicals, especially phenolic compounds. A series of meta-analyses have equally reported an association between increasing intake of tomatoes and the reduced risk of noncommunicable diseases such as cardiovascular diseases, coronary heart diseases, stroke, metabolic syndrome, and cancers, as well as a positive effect on gut microbiota [23].

In addition to the above, it is well known that they are good sources of dietary fiber, vitamins, minerals, and many non nutrient substances that are beneficial for our health, such as flavonoids, plant sterols, and other antioxidants. In fact, the WHO and other health organizations recommend an increased intake of up to 400 g/day to improve overall health (Fig. 8.14) [23].

An archetypical example of high-quality antioxidants is broccoli. *Brassicaceous vegetables* are a well-known family of cruciferous vegetables. The genus and species Brassica oleracea include cabbage, Brussels sprouts, and cauliflower apart from broccoli.

Broccoli contains provitamin A (β-carotene), vitamin C, and vitamin E and it is a good source of phytochemicals due to the high content of carotenoids (lutein, zeaxanthin, β-carotene), phenolic compounds (mainly flavonoids) sulfur glycosides, and minerals (calcium, magnesium, phosphorus, selenium, potassium, and sodium). Several studies have reported the potential anticarcinogenic and antioxidant properties of broccoli rich diets. In fact, higher consumption of glucosinolates, tocopherols, and carotenoids is associated with a lower risk of cancer, cardiovascular disease, and poor eye health, respectively [23].

Glucosinolates are major active anticarcinogenic compounds. The glucosinolate molecule is composed by a glycone moiety, common to all members of this group, and a variable aglycone side chain. Many of these glucosinolates are hydrolyzed to produce isothiocyanates with characteristic side chains. Sulforaphane (1-isothiocyanate-4-methylsulfinylbutane) is the most studied molecule obtained by the consumption of broccoli. Other phenolics are also found in broccoli, including flavonoids, caffeic acid derivatives as chlorogenic acid, and sinapic acid derivatives as sinapic acid, which vary in concentration across different cultures [30].

Broccoli also contains antioxidant vitamins including tocopherols and carotenoids (lutein and zeaxanthin). In general, brassicaceous seeds and sprouts are good sources of vitamins C, E, K, B1, B2, and folate. It also contains minerals such as sodium, potassium, calcium, magnesium, chlorine, potassium, and sulfur, and trace elements such as iron, zinc, copper, manganese, and selenium. Broccoli has the ability to accumulate high levels of selenium with the majority of the seleno amino acids in the form of selenium-methyl-selenocysteine [30].

Legumes are another group of foods with high nutritional value. They are naturally low in fat, practically free of saturated fat and cholesterol, and they have a low

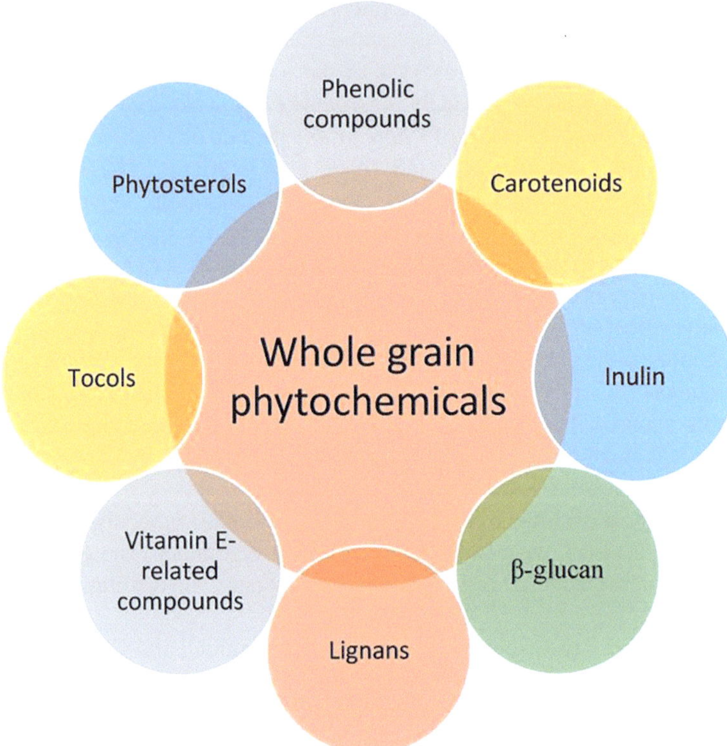

Fig. 8.14 Whole grain phytochemicals

glycemic index. Legumes also provide a significant proportion of fiber and B vitamins, iron, zinc, and other important minerals. The legume food group was found to be an important dietary predictor of longer survival. Considering the uric acid generation potential of the legume's food group, they possess a moderate risk due to a high purine load [31].

Garlic, onions, shallots, leek, and chive belong to the *Liliaceae family*. Garlic has been widely used in many cultures for both culinary and medicinal purposes. At present, garlic is one of the herbal medicines most frequently investigated and used as a complementary therapy [23].

Garlic is characterized by its high content of sulfur compounds such as diallyl sulfate, allicin or S allyl cysteine sulfoxide and allicin. The main bioactive compounds present in garlic is allicin (up to 1.8% in a garlic bulb). The main compounds responsible for the flavor and odor of garlic are mostly sulfur-containing, nonvolatile amino acids (thio-sulfinates), such as S allyl cysteine sulfoxide. Moreover, several *in vitro* studies have demonstrated its antibacterial, antiviral, and antifungal effects. Several clinical trials have also reported anticancer properties. The dose which offers immune enhancement is suggested to be 1.8–10 g/day [23].

Curcumin demonstrates antioxidative and anti-inflammatory activities. It has the ability to bind ROS, to influence the concentration of glutathione, and to activate antioxidative enzymes. Curcumin shows a high affinity to fats and thus it gains the ability to cross the blood-brain barrier. Studies conducted on laboratory animals confirm the positive effects of curcumin on cerebral antioxidative status. In mice poisoned with prooxidative lead acetate, it was observed that curcumin had positive effects on the level of neurotransmitters in the hippocampus, where it increased the levels of serotonin and decreased the levels of dopamine. In the brain, curcumin facilitates the reduction of β-amyloid aggregation. Despite its low bioavailability, experimental studies confirm the high antioxidative properties of curcumin and thus it is is provided as an additional treatment in neurodegenerative diseases [19].

Novel animal and human studies indicate that curcumin can influence the function of different immune cells, such as T lymphocyte subsets, macrophages, dendritic cells, B lymphocytes, and natural killer cells [32]. The immunomodulatory activity of curcumin may involve direct targeting of TLRs. The activation of TLR4 is attributed to the regulation of various transcription factors mainly nuclear factor (NF-κB), activator protein 1, signal transducer, and STAT pathway as well as their downstream signaling pathways [33]. Although several studies have revealed multiple mechanisms for the anticancer action of curcumin, epigenetics in different disease models has been identified [22].

Curcumin has been identified as a promising antitumor compound capable of suppressing tumor initiation, promotion, invasion, and metastasis. The anticarcinogenic effects are attributed to an increase in the activation of macrophages and NK cells. Curcumin can enhance NK cell cytotoxicity, thereby playing a crucial modulatory role in the remission of tumor progression. However, due to its dual role both as an proinflammatory as well as antinflammatory agent depending on its concentration, its function could be related with side effects due to overproduction of cytokines. This paradox may be modulated by changing the different concentrations of curcumin to achieve the optimal therapeutic dose, which can induce tumor cell apoptosis and can enhance antitumor immunity [33].

All in all, different meta-analyses show inverse associations between the intake of fruits, vegetables, whole grains, and nuts and the risk of coronary artery disease, stroke, cardiovascular disease, cancer, and other diseases. The strongest reductions in risk was observed at an intake of 800 g/d for fruits and vegetables, 225 g/d for whole grains, and 15–20 g/d for nuts. Whole-grain and nut consumption was also inversely associated with mortality from respiratory disease, infections, and diabetes [32].

The association between the consumption of different fruits and the prevention of specific cancers is interesting. For example, an inverse association is observed between the intake of citrus fruits and bladder cancer risk, between citrus fruits, cruciferous vegetables, green leafy vegetables, and lung cancer risk, bananas and spinach with colon cancer, apples, pears, peaches, nectarines, apricots, strawberries, lettuce, and estrogen receptor-negative breast cancer, a slight inverse association between lettuce and advanced and fatal prostate cancer. It is also reported an inverse association between the intake of whole grains and colorectal cancer risk. On the contrary, a positive association was observed between corn and advanced and fatal prostate cancer, and between the intake of strawberries, Brussels sprouts, green pepper, and tomatoes with pancreatic cancer [32].

Mechanism of Action

Although polyphenols are not considered an essential nutrient, mounting evidence suggests that they are an important contributor to optimal health, partly by the reduction of inflammation. The anti-inflammatory effect of polyphenols was initially attributed to their ability to neutralize free radicals due to the availability of the free hydroxyl groups of conjugated double bonds. Polyphenolic compounds may also increase endogenous antioxidant defense by the regulation of Nrf2, an important cellular redox transcription factor involved in detoxification processes. Polyphenols activate gene transcription factors that inhibit the activation of NF-kB, an important regulator responsible for the inflammation pathways [26].

Polyphenols are able to interact with cell signaling pathways, modulating gene expression in two different ways: (i) influencing the activity of transcription factors and (ii) epigenetically, modulating miRNAs [24].

Curcumin is a dietary polyphenol from turmeric with numerous pharmacological activities [32]. Curcumin has a dual function since it can express properties as a prooxidant in generating ROS and as an antioxidant in cellular systems. The oxidative stress induced by curcumin has been considered as the probable mechanism of action of the compound to act as an anticancer agent. The biphasic action of curcumin depends upon its concentration that is used in experiments, for example, at higher concentrations ($\sim$50 μM), curcumin acts as a prooxidant and at lower doses ($\sim$10 μM), the same ingredient acts as an antioxidant [22].

An important mechanism of gene regulation, related to the repair of oxidative damage, is the reversible *acetylation of histones*. The addition of acetyl groups to histones promotes gene expression by creating an open chromatin conformation. In contrast, removal of acetyl groups by histone deacetylases (HDACs) leads to a closed DNA conformation that effectively prevents transcription [20].

A study on human hepatoma cells reveals that curcumin treatment lowers histone acetylation (hypoacetylation) by inhibiting the HAT (histone acetyltransferase) activity and without an effect on the HDAC activity, which is linked with the enhancement of ROS production in cells. Especially, it has been demonstrated in studies with animal models that curcumin protects against cardiac hypertrophy and heart failure involving epigenetic regulation by HAT. One of the established mechanisms of epigenetic regulations of gene expression is maintaining a balance between acetylation and deacetylation by a controlled regulation of the activities of HATs and HDACs [22].

It is suggested that polyphenols and curcumin have been described as therapeutic agents for cardiovascular diseases. HDAC, especially HDAC2, has also been recognized to play an important role in inflammatory gene expression and inflammatory lung pathologies, including, asthma, and airway diseases, since histone acetylation upregulates the lung inflammatory genes and could cause lung inflammation [22].

Broccoli and broccoli extracts have gained interest due to their beneficial effects on the prevention of cancer. They act by multiple mechanisms and mainly by anti inflammatory pathways, effects on signal transduction, epigenetic effects, and modulation of the colonic microflora. Known antioxidant systems in broccoli are vitamin C and sulforaphane by modulation of Nrf2 related to mediated stress response and the glutathione metabolism. In addition, it seems that bioactive ingredients in broccoli alter the expression of CYP45 enzymes that conjugate carcinogens. Several interacting signaling pathways, including ERK, JNK, and MAPK signaling, may be involved in the apoptosis induced by sulforaphane and related glucosinolates, as well as flavonoids contained in broccoli. These pathways sense extracellular stresses, leading not only to the induction of Nrf2-mediated transcription but also to the transmission of apoptotic and cell-cycle arrest signals, thus promoting genomic stability [18].

Recent reviews consider folate, B vitamins, and also selenium as regulators of the process. Various mechanisms including epigenetic effects and actions on telomerase activity and mitochondria contribute to the stabilization of DNA. Glucosinolates such as indole-3-carbinol (I3C) produced by the cleavage of the glucosinolate glucobrassicin that are beneficial. However, it has been reported that some of them could also develop into potent mutagens upon activation by myrosinase [23].

Soya

The Chemical Structure, Forms, and Distribution in the Environment

Soy, Glycine max (L.), is a plant of Asian origin belonging to the *Fabaceae family*. In Asian regions, soy is often used for several traditional cuisine dishes. High protein content, together with lower carbohydrate content, characterizes soy as a unique vegetable protein source compared to other legumes. There are several soy foods, traditional Asian foods, and new-generation products, such as soy milk, edamame, tofu, soy cheese, miso, soy sauce (tamari, shoyu), tempeh, natto, sufu, yuba, soy flour, soy protein isolated (SPI), and meat analogs. Soybeans can be blended and heated to extract soy milk, which can also be treated with magnesium chloride or calcium sulfate to form curd and produce tofu. In addition, different fermentation treatments are useful to get natto, tempeh, soy sauce, and sufu [34].

Soy is gaining more and more attention because of the health benefits associated with its consumption. The major health benefit from the consumption of soy is the high content of isoflavones, a member of flavonoids. Beyond isoflavones and soy protein, soybeans are a good source of unsaturated fatty acids, B vitamins, fiber, iron, calcium, zinc, and other bioactive compounds that make soy a good candidate of a high nutritional food. Soy fiber content is primarily pectic polysaccharides. Soybean oligosaccharides have been proposed as prebiotic or sugar alternatives [34].

Isoflavones belong to a structural class of nonsteroidal phytochemicals called phytoestrogens that possess a chemical structure and function similar to animal endogenous estrogens (57). The main sources of isoflavones are legumes from the family *Fabaceae,* namely, soybean (*Glycine max*) as a source of daidzein, genistein, and glycitein, and red clover (*Trifolium pratense)* as a source of formononetin and biochanin A (58). Other known isoflavones are lignans and coumestans [34]. Isoflavones are contained in different legumes, apart from soy, such as kidney beans, navy beans, red clover, and Japanese arrowroot called kudzu, but only soybeans represent a relevant source. Isoflavone content in soy foods varies among brands and preparations [34].

Phytoestrogens are naturally occurring nonsteroidal phenolic plant compounds and they can be divided into two main groups: flavonoids and nonflavonoids. Flavonoids include isoflavones, coumestans, and prenylflavonoids, and nonflavonoids include lignans (22). Genistein (7,4′-dihydroxy-6-methoxy-isoflavone), daidzein (7,4′-dihydroxy-isoflavone), glycitein (7,4′-dihydroxy-6-methoxy-isoflavone), biochanin A (5,7-dihydroxy-4′-methoxy-isoflavone), and formononetin (7-hydroxy-4′-methoxy-isoflavone) belong to isoflavone phytoestrogens [34] (Table 8.1).

The word "phytoestrogen" comes from the Greek term for plant ("*phyto-*") and from the term "*estrogen,*" which is a hormone that influences female fertility in vertebrates. Phytoestrogens are compounds found in plants. However, the fertilizing functions of these molecules were first reported in the 1940s in connection with the "clover disease" which appeared in Australia in sheep grazing subterranean clover. The "disease" manifested the symptoms of diverse reproduction disorders [34].

Equol (7-Hydroxy-3-(4′-hydroxyphenyl)-chroman), a daidzein metabolite, is sometimes considered a phytoestrogen, but it is actually a plant product. It is a metabolic product of intestinal bacteria (22). Only a percentage of 50–60% of Asian

Table 8.1 Different isoflavones

Soy isoflavone	R1	R2	R3
Genistein (aglycone)	OH	H	OH
Daidzein (agrlicone)	H	H	OH
Glycitein (aglicone)	H	OCH_3	OH
Genistin	OH	H	Glucoside ($C_6H_{12}O_6$)
Daidzin	H	H	Glucoside ($C_6H_{12}O_6$)
Glycitin	H	OCH_3	Glucoside ($C_6H_{12}O_6$)
Acetylgenistin	OH	H	Acetyl glucoside ($C_6H_{12}O_6$ - CO - CH_3)
Acetyldaidzin	H	H	Acetyl glucoside ($C_6H_{12}O_6$ - CO - CH_3)
Acetylglycitin	H	OCH_3	Acetyl glucoside ($C_6H_{12}O_6$ - CO - CH_3)
Malonylgenistin	OH	H	Malonyl glucoside ($C_6H_{12}O_6$ - CO - CH_2 - COOH)
Malonyldaidzin	H	H	Malonyl glucoside ($C_6H_{12}O_6$ - CO - CH_2 - COOH)
Malonyldlycitin	H	OCH_3	Malonyl glucoside ($C_6H_{12}O_6$ - CO - CH_2 - COOH)

people and 30% Western individuals among vegetarians can produce equol, a metabolite of daidzein detected in urine after the consumption of soy foods. This suggests an inter-individual capacity in the metabolization of isoflavones that may result in varying effects among individuals (Fig. 8.15) [34].

Equol formation occurs in the distal intestine and colon indicating that most soy isoflavone metabolism takes place in the human gut by a heterogenic pool of different strains of microorganisms that convert the glycoside form of isoflavones to the aglycon form, equol. Also, intake of other dietary components may influence equol production: a diet with high total carbohydrates and low saturated fatty acid has been associated with equol production, while antibiotic treatments had a negative impact on urine equol excretion (Fig. 8.16) [34].

Apart from flavonoids and specifically genistein, soybeans also contain a wide range of natural chemicals such as phytic acid (1.0–2.2%), sterols (0.23–0.46%), and saponins (0.17–6.16%). Few studies have focused on nonisoflavone natural chemicals, such as saponins but with the growing use of soy and soy foods, a more detailed identification and quantification of other soy ingredients apart from isoflavones will follow [34].

Natural chemicals including phytic acid and saponins display a wide range of important bioactivities including antioxidative, antiviral, anticancer, hepatoprotective, and cardiovascular protective effects. Some soy products may contain higher concentrations of other natural chemicals than isoflavones [34].

Soybeans are rich in *protein*, but also in *unsaturated fatty acids*, especially linoleic acid, a ω6 polyunsaturated fatty acid. The substitution of foods rich in saturated fatty acid for soy products shows an improvement in cholesterol concentration and a reduced risk for coronary heart disease (CHD). Polyunsaturated fatty acids could contribute to the protective effects of soy consumption by the influence on inflammatory parameters [34].

Lignans of soybean are considered another group of natural chemiclas, found also in pulses, cereals, and seeds. Intestinal bacteria convert plant lignans to enterolignans, known as enterodiol and enterolactone, which are efficiently absorbed. Lignans may exert anticancer activity mediated by interaction with estrogen receptors; however, antiestrogenic and estrogenic activity of enterolignans in humans is still debatable [34].

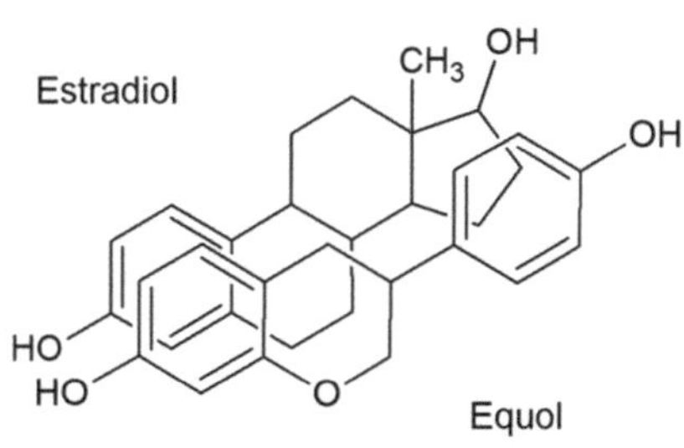

Fig. 8.15 Similarities of the chemical structure of equol and estradiol [34]

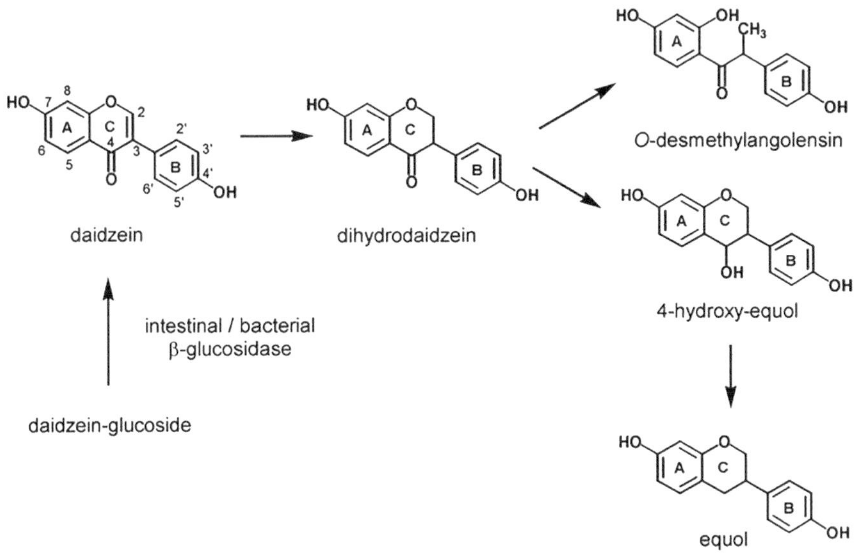

Fig. 8.16 The production of equol from daidzein isoflavone [25]

Additionally, soy could be a good source of *calcium and iron* with major interest in vegetarian nutrition. Soybeans are good sources of calcium as defined with fractional calcium absorption of 0.414 and 0.310 from low and high-phytate soybeans, respectively, compared to 0.377 from cow milk. Absorption of iron from soya products has better bioability due to the presence of a more bioavailable plant ferritin in legumes [34].

Effects on Health

Isoflavones and soy proteins are soy nutrients that aroused the most research interest due to the swift to vegetarian diets nowadays. FDA authorized a health claim about the reduction of coronary heart disease linked to soy protein consumption of at least 6.25 g per portion with a total of 25 g per day that aroused the interest of the food industry for soy. FDA relied on a meta analysis of the effects of soy protein on serum lipid profile in 38 clinical trials that evidenced a relationship between soy consumption and total cholesterol, LDL, and triglyceride blood levels [34].

In 38 controlled clinical trials, the intake of an average of 47 g per day of soy protein showed a 9.3% reduction in total cholesterol, 12.9% in LDL, and 10.5% in triglycerides, but the overall effect was lower than what was initially proposed.

Meta analysis also suggests a modest effect in lowering blood pressure by soy proteins. A recent meta-analysis of randomized controlled trials of the last 10 years showed a significant reduction of serum LDL, triglycerides, and total cholesterol concentrations, especially in hypercholesterolemic patients. A stronger effect emerged with whole soy and soy foods compared to soy extracts and there was no effect for isoflavone supplementation suggesting that other factors than isoflavones may contribute to the preventive outcome of soy consumption [34]. Regarding diabetes, some studies showed a marginal to significant association with risk reduction [34].

In epidemiological studies, Asian traditional diets rich in phytoestrogens were linked to lower risks of coronary heart disease and there is evidence of possible benefits in hormone dependent prostate, colon, breast, and ovarian cancers, as well as menopausal symptoms, osteoporosis, obesity, cognitive dysfunction, and overall risk reduction for noncommunicable diseases. However, the debate on the hormonal interference by phytoestrogens after research in experimental models and infants fed on soy based formula raised concerns. Moreover, dietary soy intake has been associated with an increased risk of thyroid disorders, bladder cancer incidence, dementia, and breast cell proliferation. Soy use during pregnancy was shown to alter the epigenome in offspring and it can have health implications of in utero stimuli [34].

The hormonal benefits are related mainly to genistein. This enzyme inhibits human DNA topoisomerase II enzyme activity *in vitro*, so in high amounts could have an anticancer activity. In breast cancer cell lines, phytoestrogens modulated chromatin transcription by epigenetic modifications such as methylation and acetylation of histones. The protective role from cancer is suggested from epidemiological studies showing reduced breast cancer risk in Asian women exposed to soy since infancy. A systematic review and meta analysis on association between isoflavones and endometrial cancer showed that dietary soy isoflavones from soybeans were associated with the reduction of the risk of endometrial cancer [34].

As mentioned above, intestinal activation of soy phytochemicals by microbiota could be crucial for the absorption and efficacy of soy bioactive compounds. Additionally, soy itself can shape gut microbiota toward a beneficial constitution. Influences on gut microbiota could be partially linked to the beneficial effects of soy on metabolic syndrome [34].

Soy-based formula is commonly used in Western countries for allergy to cow milk proteins or as an alternative to vegetarian choice. Even though soy-based formula was often used for infants with *atopic disease*, 10–14% of subjects with a diagnosis of cow milk allergy were also positive for the soy allergy, because soybean glycinin protein of 30 kDa cross-reacts with casein. Enterocolitis and other clinical manifestations of soy allergy overlap to those of cow milk allergy [34].

Mechanism of Action

Soybean isoflavones have been identified as natural chemicals therapeutic flavonoids for the treatment of colorectal cancer, where as the isoflavones suppress metastasis of the tissue through epigenetic modulation of DNA methylation and histone modifications. In particular, this study highlights genistein, one of the soybeans' isoflavones, as an effective epigenetic modulator of colorectal cancer metastasis, and dietary genistein may be beneficial for colorectal cancer [22].

The well-investigated property of isoflavones is their ability to interact with estrogen receptors (ERs) due to structural similarity to 17β-estradiol. Isoflavones are low-potency estrogen molecules compared to estradiol with an affinity for ERα and ERβ 1/1000 times and 1/3 times, respectively (57). This property to mimic estrogen activity and act as phytoestrogens has gained high scientific interest (Fig. 8.17) [22].

It has been suggested that soy isoflavones may function as selective tissue estrogenic activity regulators (STEAR) and selective estrogen receptor modulators (SERM) with different mechanisms than direct interaction with receptors. However, phytoestrogens implement not only estrogenic activity but also antiestrogenic or no activity depending upon the tissue. ERα and ERβ display distinct expression patterns in males and females, thus phytoestrogens do not exert their activity as classical estrogen agonists [34].

Antioxidative effects of flavonoids may protect from cancer development in which free radicals play a well known role. *In vivo* and *in vitro* studies suggested that the beneficial effects of soy on the prevention of some cancers are not only based on the antioxidant properties but mainly on estrogenic, antiestrogenic, and not hormonally dependent mechanisms of action. Soy peptides, such as conglycinin, released from the degradation of flavonoids in the gut may have a preventive role in cancer, hypertension, hypercholesterolemia, obesity, and oxidative stress [34].

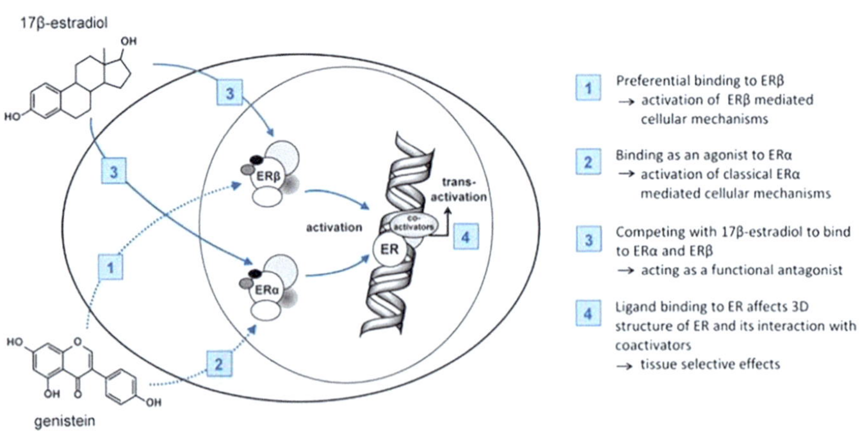

Fig. 8.17 The interaction of genistein with estrogen receptors (22)

The antioxidant effects of soy are mainly related to genistein. The enzyme also enhances the activities of antioxidant enzymes, such as catalase, glutathione peroxidase, and superoxide dismutase. The antioxidant capacity of isoflavones has been shown to be comparable to vitamin E. However, isoflavones undergo extensive metabolism in the liver and gut, so antioxidant capacity could be affected [34].

Genistein, through epigenetic modulations, including chromatin remodeling and DNA methylation, leads to the activation of tumor suppressor genes and suppression of the survival of cancer cells. Soybean isoflavones and specifically genistein have been recorded as being beneficial to colorectal cancer. It acts as an epigenetic modulator causing chromatin remodeling and DNA methylation that leads to tumor suppressor genes [18].

Beverages

In addition to the macronutrients that are included in an anti-inflammatory diet, beverages can enhance anti-inflammatory processes. *Drinking tea* instead of coffee or sugary beverages can add an anti-inflammatory benefit. Tea is made by infusing the dried leaves of the plant *Camellia sinensis* in near-boiling water, which releases antioxidants and polyphenols that contribute to reducing systemic inflammation. Green, black, and white teas contain these beneficial phytonutrients, with green and white having the highest levels [16].

Green tea as a beverage is consumed worldwide as a health-promoting drink due to the presence of phytochemicals, which are considered to act as anticancer agents, retain a healthy heart, and prevent cataracts. Epicatechin-3-gallate is the predominant polyphenol present in green tea and it has been shown to possess anticancer, antitumor, and anti-inflammatory actions [15].

Epigallocatechin-3-gallate (EGCG) is the most studied molecule of the flavanol class. It is a catechin conjugated with gallic acid and it is abundant in *green tea and cocoa*-based products, and pome fruits. Cocoa has the highest content of catechins, followed by prune juice and broad bean pod. Moreover, açaí oil, obtained from the fruit of the açaí palm, is abundant in epicatechin and catechin [17]. In a study on human breast cancer cells, *EGCG* and the novel prodrug of EGCG (pro-EGCG) have been shown to inhibit cell proliferation and the transcription of the human telomerase reverse transcriptase (*hTERT*), the catalytic subunit of telomerase, that is crucial in sustaining the telomere chain length and tumor formation, the latter being by the epigenetic regulation of the estrogen receptor [17].

Cocoa and its main derived products, such as chocolate, have been extensively used for years in many cultures. It is obtained from raw cocoa beans from *Theobroma* cacao fruits of the cocoa tree. In order to obtain cocoa, several processes are applied to the raw seeds, including fermentation, drying, roasting, winnowing, alkalization, and cinching [18].

Many herbs and spices have an impact on the inflammation similar to the other components of an anti-inflammatory diet and they can enhance the overall anti-inflammatory effect of food. *Ginger and turmeric* are the two herbs with the largest amount of data supporting their impact on inflammation. They inhibit IL-2, TNF-α, and IL-8, which are proinflammatory cytokines; they also inhibit leukotriene and prostaglandin synthesis. Additionally, garlic, cayenne, and oregano have anti-inflammatory properties [21].

Herbal extracts have long been considered antiseptics, antidepressants, anti-inflammatories, expectorants, immune system fortifiers, and potent anticancer agents. Around 80% of people in the world use a certain form of herbal treatment. From 1984 to 1994, 60% of the drugs confirmed by the FDA were isolated from natural sources such as plants. Herbal medicines are among alternative and complementary medicines most commonly. Among the 121 drugs prescribed for cancer treatment, 90 are from herbal sources. A study showed that out of the 65 new medicines recorded for cancer treatment during 1981–2002, 48 were derived from natural agents. Like many therapies, people may use herbal medicine to feel healthier [19].

Caffeine, an alkaloid, it is well known that besides increasing alertness, it has some implications in the reduction of ROS. Caffeine, present in coffee, blocks the receptors of adenosine A1 and A2 and it contributes to increased activity of the central nervous system. The blocking of adenosine A1 receptors causes an increase in the release of neurotransmitters such as acetylcholine, noradrenaline, and dopamine. The inactivation of adenosine A2 receptors leads to increased dopamine activity and to a stronger bonding to the D2 receptor, thus protecting the dopaminergic receptors against toxic factors. The influence of caffeine on dopamine activity may be significant in age neurodegenerative disorders. A study conducted demonstrated that administration of caffeine (30 mg/kg^{-1} per day) mitigated neurodegenerative lesions in the brains of mice poisoned with Cd and it also improved their cognitive functions and memory. Additionally, the cited authors studied the cerebral antioxidative status and degree of neuron apoptosis *in vitro* and *in vivo* on BV-2 cell lines. The antioxidative influence of alkaloids on the brains of rats has been confirmed in multiple studies [23].

Biogenic amines (BAs) found in humans have endogenous, dietary, and microbial origins, and they mediate a wide variety of biological activities. For example, histamine, an endogenously produced bioamine from tyrosine, is a potent vasodilator released by degranulating mast cells activated by allergens binding immunoglobulin E receptors. Tyramine is a bioamine found in cheese that is normally metabolized by monoamine oxidases (MAOs). In patients taking an MAO inhibitor (MAOi) for depression or Parkinson's disease, the ingestion of large amounts of tyramine can trigger a life-threatening hypertensive crisis. Several gut-resident bacteria can produce tyramine as a decarboxylation product of tyrosine, and strains producing and consuming tyramine have been routinely cultured from human feces. The effect of tyramine on the human body is unknown, but it possibly contributes to hypertension [24].

Melatonin is a tryptophan-derived substituted indoleamine (*N*-acetyl-5-methoxytryptamine) widely found in living, evolutionarily distant organisms. This molecule was isolated in 1958 by Lerner et al. from a cow pineal gland. For decades, melatonin was considered an animal neurohormone with an impact in circadian rhythm regulation, seasonal reproductive cycles, and mammalian immune system modulation, but currently, these studies are outdated. In the 1990s, an essential and novel function was discovered and attributed to this molecule, a highly potent hydroxyl radical scavenger, a finding that was considered the starting point of extensive studies on melatonin antioxidant protection, which turned out to involve numerous mechanisms beyond radical scavenging [30].

Melatonin's antioxidant properties mainly consist of (1) direct scavenging of reactive oxygen (ROS) and reactive nitrogen (RNS) species, (2) acceleration of antioxidant enzyme activity, (3) protection against oxidative damage, (4) synergistic effects with other antioxidants, and (5) improvement of the electron transport efficiency in mitochondrial respiratory chain, limiting free radicals' overproduction through electron leakage reduction. Moreover, not only melatonin but also its derivatives act as natural electron donors, highly efficient against oxidative stress (Fig. 8.18) [30].

With concerns to melatonin biosynthesis in humans, the exogenous amino acid tryptophan, through the action of tryptophan hydroxylase and aromatic acid decarboxylase enzymes, is converted to the neurotransmitter, serotonin. In the subsequent step, serotonin is converted into melatonin by the influence of arylalkylamine N-acetyltransferase and hydroxy indole-*O*-methyltransferase enzymes (Fig. 8.19) [30].

One of the earliest known functions of melatonin is its impact on circadian rhythms' regulation. Light is the environmental factor known to affect melatonin secretion. Melatonin acts directly on the suprachiasmatic nucleus) (SCN) and it modulates clock function. Circadian signals from the SCN are transmitted to the pineal gland by a multisynaptic pathway, which involves the neural projections from the SCN to the paraventricular nuclei (PVN), the intermediolateral cell column of the spinal cord (IML), superior cervical ganglion (SCG), and finally to the pineal gland (SCN → PVN → IML → SCG → pineal gland). The pineal gland is located in the midline of the brain [30].

Circadian clocks have developed to adapt biological functions to specific times within the day or night. This clock controls diurnal sleeping and waking cycles, body temperature, and hormone release. SCN cells receive neural cues from the retina and they send the information obtained on photoperiodic status to the pineal gland, which then synthesizes them and it releases melatonin, distributing the time signal to the rest of the body [30].

Melatonin receptors are widely distributed in the body. They exist in cardiovascular, immune, and endocrine systems, reproductive and gestational tissues, and even in the skin and gastrointestinal tract. GPR50, a melatonin-related receptor, has also been identified and it is detected in various brain structures and peripheral tissues in humans [30].

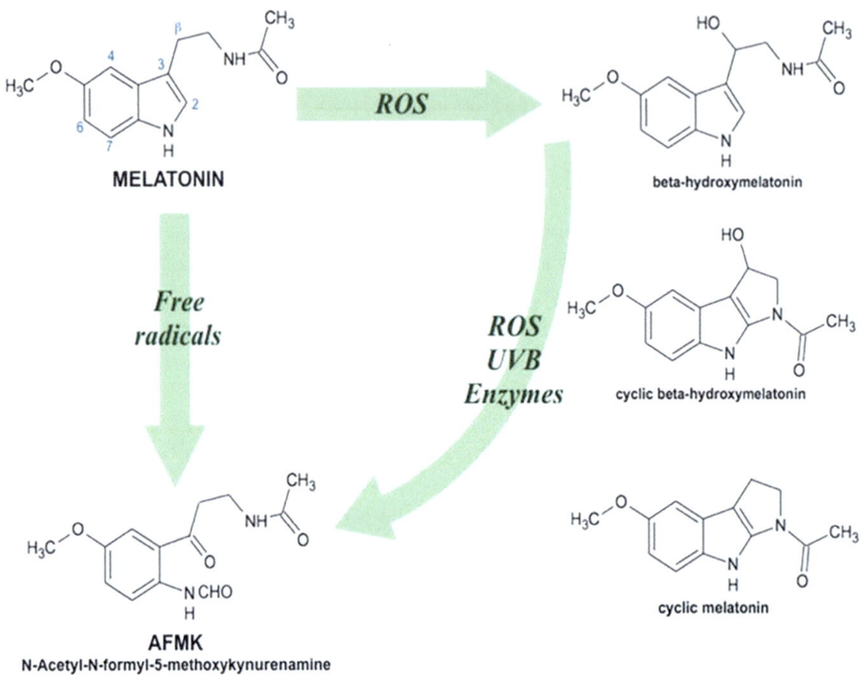

Fig. 8.18 The unknown role of melatonin

Melatonin has beneficial effects against mitochondrial dysfunction, considered a major causative agent in neurodegenerative diseases, such as Alzheimer's, Huntington's disease, and Parkinson's disease. Some studies indicate that melatonin in the range of 50–100 mg/day is effective in reducing oxidative mitochondrial dysfunction in experimental models.

Individuals living in highly industrialized countries look for melatonin-supplemented drugs and pills as a way to improve sleep and rest, as the main advantages include quicker times to fall asleep and quicker "jet-lag" reductions. On the other hand, some of the users also referred to experiencing headaches, dizziness, and sleepiness during daylight hours (overdose symptoms), but these may have placebo effects [30].

Genetically Modified Products

Genetically modified foods (GM foods), also known as genetically engineered foods (GE foods), or bioengineered foods, are foods produced from organisms that have had changes introduced into their DNA using the methods of genetic engineering. Genetic engineering techniques allow the introduction of new traits and a greater

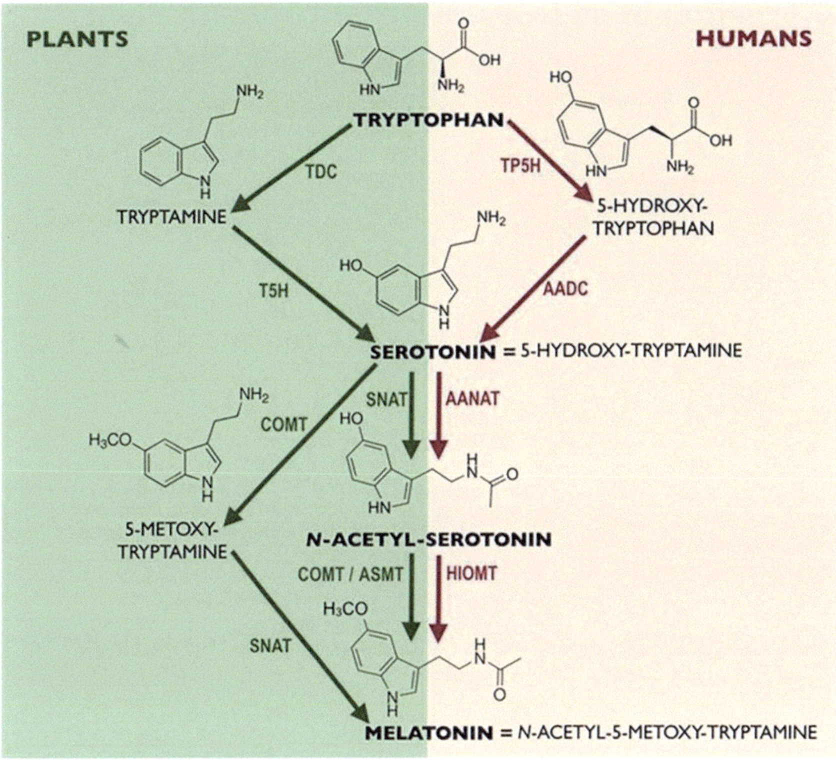

Fig. 8.19 The production of the hormone melatonin

control over traits when compared to previous methods, such as selective breeding and mutation breeding (Fig. 8.20).

Commercial sales of genetically modified foods began in 1994 when Calgene first marketed its unsuccessful Flavr Savr delayed-ripening tomato. Most food modifications have primarily focused on cash crops in high demand by farmers such as soybean, corn, canola, and cotton. Genetically modified crops have been engineered for resistance to pathogens and herbicides and better nutrient profiles.

The Directive 2001/18/EC on the deliberate release and placing on the market of GMOs states that "Member States and the Commission shall ensure that GMOs, which contain genes expressing resistance to antibiotics in use for medical or veterinary treatment, are taken into particular consideration, when carrying out an environmental risk assessment, with a view to identifying and phasing out antibiotic resistance markers in GMOs, which may have adverse effects on human health and the environment" [35].

Concerns regarding GMOs are related to allergenicity, horizontal gene transfer, and nutritional value. As far as allergenicity is discussed, there are no case reports

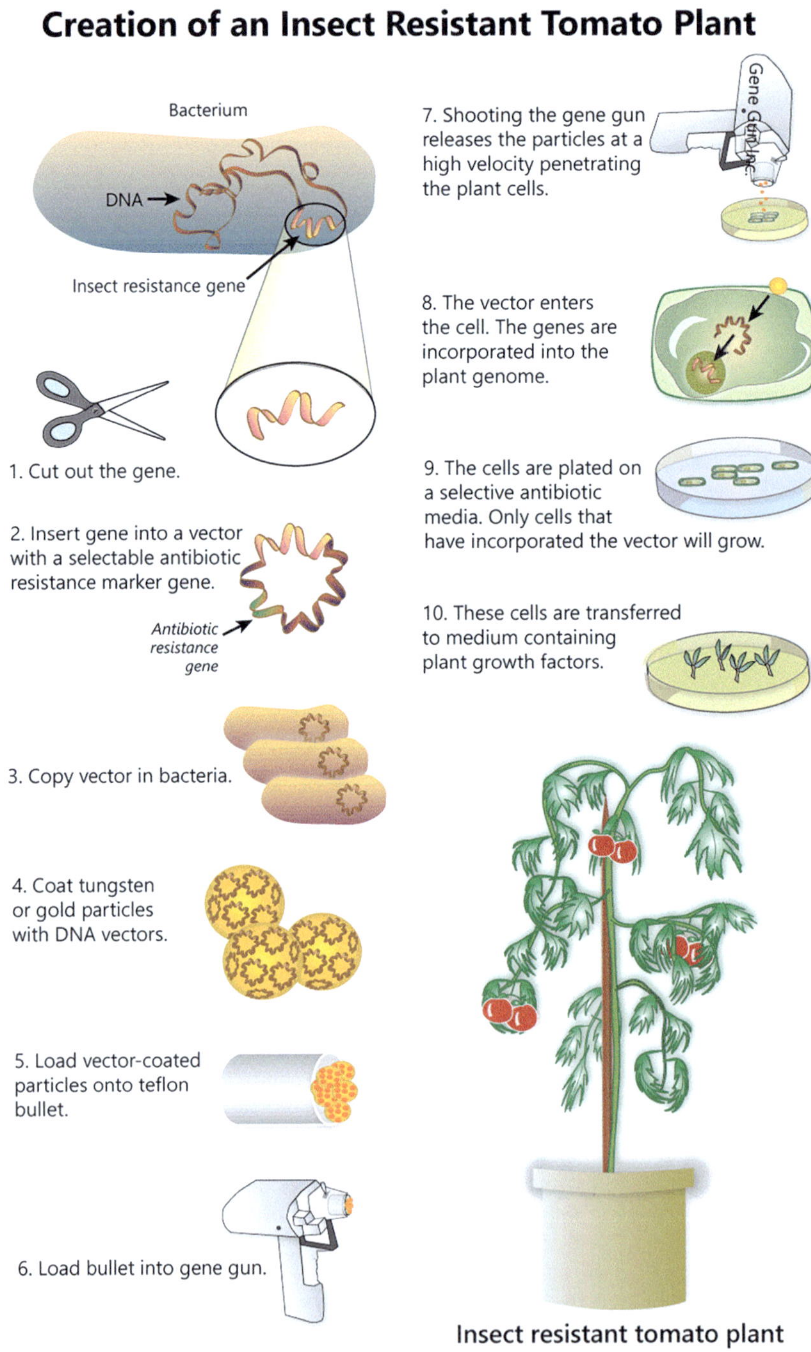

Fig. 8.20 A genetically modified tomato

of allergic reactions or immunotoxin effects resulting from GM feed consumption as compared with non GM feed. The compositional analyses of GM crop varieties, known to be allergenic, have been assessed for market approval by the European Union in the last few years (i.e., GM soybean). So far, there have been no indications that the crops may comprise increased levels of allergens. Due to the fact that these products are widely used in livestock, there were some concerns about the allergenicity of soya products in calves and pigs [35].

Changes in the structure (e.g., posttranslational modifications) of the expressed trait protein may result in unintended effects on the immune response. The recombinant form of the bean a-amylase inhibitor expressed in a GM pea may result in increased antigenicity and adjuvanticity. Similarly, the GM maize could have these properties. These concerns have not been proved [35].

The likelihood of horizontal gene transfer of GMO related DNA to different species was the main concern for these cultures. Horizontal gene transfer refers to the transfer of genetic material to neighboring living cells or organisms independently from reproduction. It is a process taking place extensively among prokaryotes and viruses, playing a key role in the evolution of genomes. It could happen in soil, water, gastrointestinal tracts of humans and animals, as well as in some foods (e.g., milk). It can occur naturally or be induced *in vitro* by chemical or physical conditions. For genetic engineering, these risks are commonly addressed by legislation [35].

Although it is difficult to provide a realistic estimate of DNA intake for farm animal diets, it has been calculated that the DNA content of most food crops is lower than 0.02%. For a 600-kg dairy cow, fed with GM maize as an ingredient, transgene DNA consumption would amount to 2.6 mg per day, which represents 0.00042% of the total dietary DNA intake. Generally, based on studies performed with animals fed by genetically modified soy, maize, and cottonseed, recombinant plant DNA has never been detected in milk, eggs, or blood, even when DNA fragments of multicopy chloroplast genes could be detected [35].

From GM plants, the only genes that are expected to be successfully transferred to bacteria include antibiotic resistance genes employed in the transformation process. Potential recipient bacteria for antibiotic resistance genes include bacteria that infect plants or reside on plant surfaces, soil bacteria, endosymbionts, and mammalian gastrointestinal bacteria that consume plant products. Antibiotic resistance genes are widespread in microbial populations. The transcriptional genes within GMOs could confer a selective advantage to recipient cells and they could probably pose a risk for the micro and macro environment of the cultivation [35].

The debate on the advantage of using GM crops for their capacity to control mycotoxin content is still open and it needs to be further supported with evidence of the potential benefits of the mitigation effects to be presented to the stakeholder community. The topic is of particular interest as the impact of mycotoxins on farm animals is an important issue and mycotoxin reduction is crucial for animal health, productivity, and farm economy. All GM maize events (mainly Bt maize) authorized for import, processing, and cultivation and commercially available in the EU market, were produced for the management of insect infestation, but not for fungal or mycotoxin control. However, insect-resistant GM maize crops show evidence of reduction of fumonisin contamination as an advantageous side effect with potential unproven positive consequences of the derived feed on animal health. For other Fusarium toxins and aflatoxins, however, the evidence of possible control remains inconclusive [35].

To date, most GM crops on the market are first-generation crops, which involve enhanced agronomic traits such as insect resistance (Bt) and herbicide tolerance (Ht) to improve crop production. First-generation GM crops are considered substantially equivalent to conventional, non-GM lines. The DNA insert that leads to the synthesis of a gene product in these plants does not interfere with the overall metabolism of the plant cell. With regard to their nutritive value, the majority of animal feeding studies have not found significant differences in production and health parameters of animals that have consumed first-generation crops in comparison to conventional, non-GM crops [35].

Executive Summary

The Food as a Complex Antioxidant System

- Personalized nutrition: nutrigenomics and nutrigenetics
- Oxidative stress and chronic inflammation are the consequences of unhealthy eating habits. Chronic inflammation is related to metabolic syndrome
- The antioxidant properties of foods
- The microbiome and its implication in metabolism
- Carbohydrates. A healthy diet (carbohydrates 55–75% and only 10% sugars)
- Glycemic load is defined as the quantity of carbohydrate ingested, multiplied by the rate at which that carbohydrate enters the bloodstream
- High blood glucose consequences: ROS by mitochondrial oxidation and nicotinamide adenine dinucleotide phosphate (NADPH) oxidase, increased expression of TNF-α, IL-6, and plasminogen activator inhibitor (PAI)-1
- Protein in an anti-inflammatory diet should be primarily plant-based, with some sources of fish and small amounts of lean natural meats
- Saturated and trans-fatty acids (SFA, t-FA) stimulate a proinflammatory state contributing to insulin resistance
- w-3 FAs are considered anti-inflammatory, while w-6 FAs are proinflammatory. A ratio of w-6: w-3 FAs >10:1 is believed to be proinflammatory
- Chronic inflammation is affected by high caloric intake in contrast to calorie restriction

Natural chemicals

- Natural chemicals antioxidants are simple molecule antioxidants from a variety of food sources
- Vitamins: one carbon metabolism, methyl donors, and their levels regulate the levels of methyl groups available for maintenance of methylation

- The preventive role of vitamin C for the common cold is not documented
- Carotenoids: a family of naturally occurring organic pigmented compounds
- Adequate dietary intake of carotenes prevents cardiovascular disease, atherosclerosis, and cancer
- Carotenoids have been demonstrated as an eye-sight protecting agent (provitamin A) and limit neuronal damage
- Polyphenols are divided into different classes, according to their chemical structure and mainly the number of aromatic rings, which include phenolic acids, stilbenes, flavonoids, lignans, and curcuminoids
- Humans and animals cannot synthesize polyphenols; therefore, humans consume polyphenols with plant foods as part of their diets
- Polyphenols are the most abundant antioxidants in our diet and they are commonly present in fruits, vegetables, cereals, olives, dry legumes, licorice, chocolate, and beverages, such as tea, coffee, and wine
- The shift away from freshly prepared foods to processed foods has reduced the consumption of plant-based natural chemicals, such as polyphenols
- Anti-inflammatory effect (neutralizing free radicals, regulation of Nrf2, an important cellular redox transcription factor involved in detoxification processes)
- Curcumin has been identified as a promising antitumor compound capable of suppressing tumor initiation, promotion, invasion, and metastasis
- The anticarcinogenic effects of curcumin are attributed to an increase in the activation of macrophages and NK cells
- Curcumin treatment lowers histone acetylation (hypoacetylation) by inhibiting the HAT (Histone acetyltransferase) activity without an effect on the HDAC activity
- The major health benefit from the consumption of soy is the high content of isoflavones a member of flavonoids
- Isoflavones belong to a structural class of nonsteroidal natural chemicals called phytoestrogens that possess a chemical structure and functions similar to animal endogenous estrogens
- Asian traditional diets rich in phytoestrogens were linked to lower risks of coronary heart disease and a possible benefit in hormone-dependent prostate, colon, breast, and ovarian cancers, menopausal symptoms, osteoporosis, obesity, cognitive dysfunction, and overall risk reduction for noncommunicable diseases

Genetically Modified Products

- Genetically modified foods (GM foods), also known as genetically engineered foods (GE foods), or bioengineered foods, are foods produced from organisms that have had changes introduced into their DNA using the methods of genetic engineering. Genetic engineering techniques allow the introduction of new traits as well as greater control over traits when compared to previous methods, such as selective breeding and mutation breeding
- GMOs that contain genes expressing resistance to antibiotics in use for medical or veterinary treatment are taken into particular consideration when carrying out an environmental risk assessment

- GM crop varieties known to be allergenic have been assessed in the market
- The likelihood of horizontal gene transfer of GMO-related DNA to different species was the main concern for these cultures

References

The Food as a Complex Antioxidant System

1. Mutch DM, Wahli W, Williamson G. Nutrigenomics and nutrigenetics: the emerging faces of nutrition. FASEB J. 2005;19(12):1602–16.
2. Liguori I, Russo G, Curcio F, et al. Oxidative stress, aging, and diseases. Clin Interv Aging. 2018;13:757–72.
3. Ozanne SE, Constância M. Mechanisms of disease: the developmental origins of disease and the role of the epigenotype. Nat Clin Pract Endocrinol Metab. 2007;3(7):539–46.
4. Ferguson LR, Schlothauer RC. The potential role of nutritional genomics tools in validating high health foods for cancer control: Broccoli as example. Mol Nutr Food Res. 2012;56(1):126–46.
5. Feinberg AP, Ohlsson R, Henikoff S, et al. The epigenetic progenitor origin of human cancer. Nat Rev Genet. 2006;7(1):21–33.
6. Ricker MA, Haas WC. Anti-inflammatory diet in clinical practice: a review. Nutr Clin Pract. 2017;32(3):318–25.
7. Cui J, Zhou B, Ross SA, et al. Nutrition, microRNAs, and human health. Adv Nutr. 2017;8(1):105–12.
8. Kulkarni AC, Kuppusamy P, Parinandi N. Trinity of normoxia, hypoxia, and hyperoxia in disease and therapy. Antioxid Redox Signal. 2007;9(10):1717–30.
9. Van Treuren W, Dodd D. Microbial contribution to the human metabolome: implications for health and disease. Annu Rev Pathol. 2020;15:345–69.
10. Oommen AM, Griffin JB, Sarath G, et al. Nutrition and antioxidant system roles for nutrients in epigenetic events. J Nutr Biochem. 2005;16(2):74–7.
11. Chaudhary PP, Conway PL, Schlundt J. Methanogens in humans: potentially beneficial or harmful for health. Appl Microbiol Biotechnol. 2018;102(7):3095–104.
12. Winiarska-Mieczan A, Baranowska-Wójcik E, Kwiecień M, et al. The role of dietary antioxidants in the pathogenesis of neurodegenerative diseases and their impact on cerebral oxidoreductive balance. Nutrients. 2020;12(2):435.
13. Shapira N. The metabolic concept of meal sequence vs. satiety: glycemic and oxidative responses with reference to inflammation risk, protective principles and Mediterranean diet. Nutrients. 2019;11(10):2373.

Phytochemicals

14. Hemilä H, Chalker E. Vitamin C for preventing and treating the common cold (review). Cochrane Database Syst Rev. 2013;2013(1):CD000980.
15. Liugan M, Carr AC. Vitamin C and neutrophil function: findings from randomized controlled trials. Nutrients. 2019;11(9):2102.
16. Gould RL, Pazdro R. Impact of supplementary amino acids, micronutrients, and overall diet on glutathione homeostasis. Nutrients. 2019;11(5):1056.

17. Minich DM, Brown BI. A review of dietary (Phyto)nutrients for glutathione support. Nutrients. 2019;11(9):2073.
18. Feinberg AP, Ohlsson R, Henikoff S, et al. The epigenetic progenitor origin of human cancer. Nat Rev Genet. 2006;7(1):21–33.
19. Kulkarni AC, Kuppusamy P, Parinandi N. Oxygen, the Lead actor in the pathophysiologic drama: enactment of the trinity of normoxia, hypoxia, and hyperoxia in disease and therapy. Antioxid Redox Signal. 2007;9(10):1717–30.
20. Adebo OA, Medina-Meza IG. Impact of fermentation on the phenolic compounds and antioxidant activity of whole cereal grains: a mini review. Molecules. 2020;25(4):927.
21. Cione E, La Torre C, Cannataro R, et al. Quercetin, Epigallocatechin Gallate, Curcumin, and Resveratrol: from dietary sources to human MicroRNA modulation. Molecules. 2019;25(1):63.
22. Malireddy S, Kotha SR, Secor JD, et al. Phytochemical antioxidants modulate mammalian cellular epigenome: implications in health and disease. Antioxid Redox Signal. 2012;17(2):327–39.
23. Tan BL, Norhaizan ME. Carotenoids: how effective are they to prevent age-related diseases? Molecules. 2019;24(9):1801.
24. Pabich M, Materska M. Biological effect of soy Isoflavones in the prevention of civilization diseases. Nutrients. 2019;11(7):1660.
25. Winiarska-Mieczan A, Baranowska-Wójcik E, Kwiecień M, et al. The role of dietary antioxidants in the pathogenesis of neurodegenerative diseases and their impact on cerebral oxidoreductive balance. Nutrients. 2020;12(2):435.
26. Shahzad F, Anderson D, Najafzadeh M. The antiviral, anti-inflammatory effects of natural medicinal herbs and mushrooms and SARS-CoV-2. Infect Nutr. 2020;12(9):2573.
27. Ricker MA, Haas WC. Anti-inflammatory diet in clinical practice: a review. Nutr Clin Pract. 2017;32(3):318–25.
28. Efferth T. From ancient herb to modern drug: Artemisia annua and artemisinin for cancer therapy. Semin Cancer Biol. 2017;46:65–83.
29. Ferguson LR, Schlothauer RC. The potential role of nutritional genomics tools in validating high health foods for cancer control: Broccoli as example. Mol Nutr Food Res. 2012;56(1):126–46.
30. Jakše B, Jakše B, Pajek M, Pajek J. Uric acid and plant-based nutrition. Nutrients. 2019;11(8):1736.
31. Aune D. Plant foods, antioxidant biomarkers, and the risk of cardiovascular disease, cancer, and mortality: a review of the evidence. Adv Nutr. 2019;10(Suppl4):S404–21.
32. Rizzo G, Baroni L. Soy foods and their role in vegetarian diets. Nutrients. 2018;10(1):43.
33. Abdollahi E, Momtazi AA, Johnston TP, et al. Therapeutic effects of curcumin in inflammatory and immune-mediated diseases: a nature-made jack-of-all-trades? J Cell Physiol. 2018;233(2):830–48.
34. Krizova L, Dadakova K, Kasparovska J, et al. Isoflavones. Molecules. 2019;24(6):1076.

Genetically Modified Products

35. Case studies on genetically modified organisms (GMOs): potential risk scenarios and associated health indicators.

Chapter 9
Reprogramming

Reprogramming

DNA methylation is a major epigenetic modification of the genome that regulates crucial aspects of its function. Genomic methylation patterns in somatic differentiated cells are generally stable and they are transmitted with accuracy during cell divisions. However, in mammals there are at least two developmental periods in germ cells and in preimplantation embryos in which methylation patterns are reprogrammed in a genome-wide pattern, generating cells with broad developmental potential. Reprogramming is likely to have a crucial role in establishing totipotency in normal development, while it is incomplete in cloned animals. The role of reprogramming in stem cell differentiation is also envisaged. Thus, the first erasure of previous generation methylation patterns happens during the gestational period and a short time afterward, as it has already been described (Fig. 9.1) [1].

The use of somatic cells for cloning fertilization is inefficient, and fewer than 6% of cloned mice embryos are efficient. The process is also error prone, and the cloned mammals of different species exhibit a variety of abnormal phenotypes, including placental hyperplasia, large fetus, immune disorders, and shortened life span. Tissue-specific gene expression profiles in cloned mice are also extremely different from those generated by natural reproduction. Interestingly, it is suggested epigenetic alterations and incomplete epigenetic erasure as a possible cause of the abnormalities in gene expression profile and phenotype [2].

In normal conditions, the chromosomal DNA of germ cells is the focus for reprogramming. Somatic cells acquired epigenetic modifications during developmental and postnatal life. In normal conditions, epigenetic modifications during life are not inherited by the next generation, which is called the Weismann Barrier. Environmental epigenetic modifications on chromosomes in germline cells and somatic cells must be different (Fig. 9.2) [2].

A. Salavoura, *Chemical Environmental Pollutants and their Effect on Health*,
https://doi.org/10.1007/978-3-031-80271-3_9

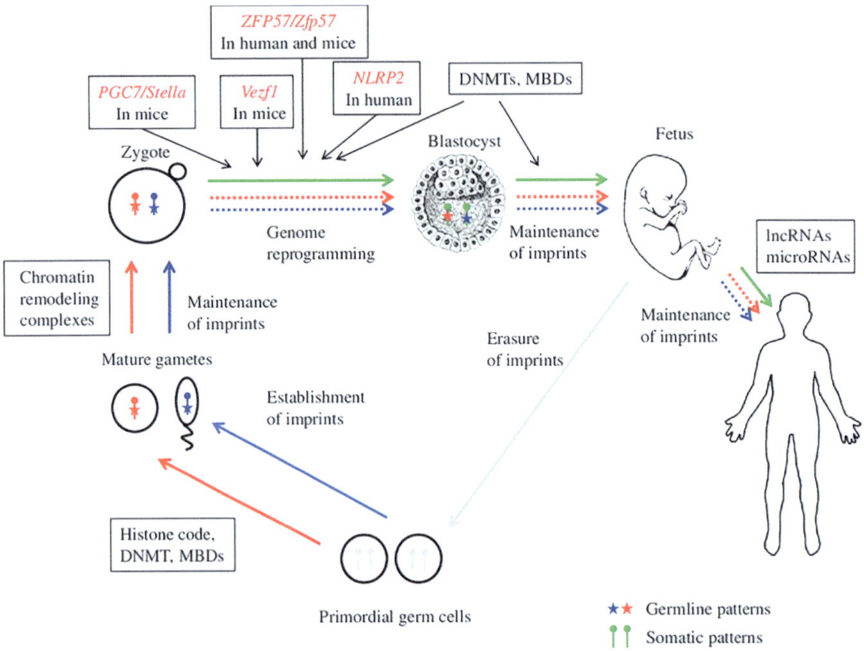

Fig. 9.1 There are two stages of gross erasure of imprints: germ lines and preimplantation of the embryo [1]

Reprogramming could be a result of the response to environmental insults throughout life, but it is transient. Permanent alterations lead to chronic inflammation and the development of age-related diseases including cancer. Local stress and inflammation responses to extrinsic and intrinsic insults might arise first to respond to tissue stress related danger signals (PAMPs and DAMPs) or to chronic tissue malfunction (DNA damage, senescence), later evolving into more systemic responses, such as the expression of the senescent phenotype. The convergence to an inflammatory phenotype is called inflammaging and it is defined as the low, but chronic levels of inflammation associated with many pathological conditions (Fig. 9.3) [2].

Nuclear reprogramming may constitute a previously unrecognized route by which human tissue responds to injury, stress, and disease. These phenomena induce transient epigenetic plasticity followed by cell differentiation and replacement of diseased cells. This process may lead to either acutely resolved tissue repair or, alternatively, to undesirable, chronically unresolved tissue damage. Aging and cancer might thus be viewed as the consequence of an unresolved stimulation of beneficial, nonreparative *in vivo* reprogramming-driven epigenetic plasticity in response to chronic, but reversible, senescence-inflammatory signaling [2].

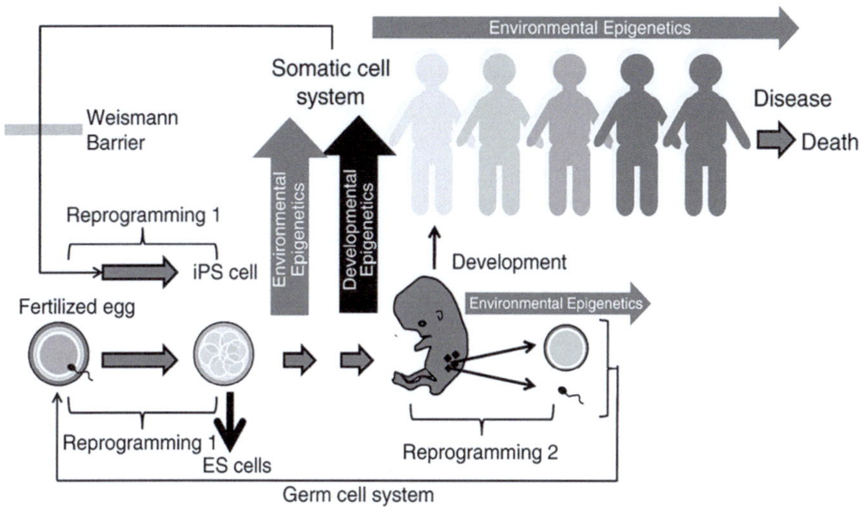

Fig. 9.2 The Weismann Barrier in reprogramming [2]

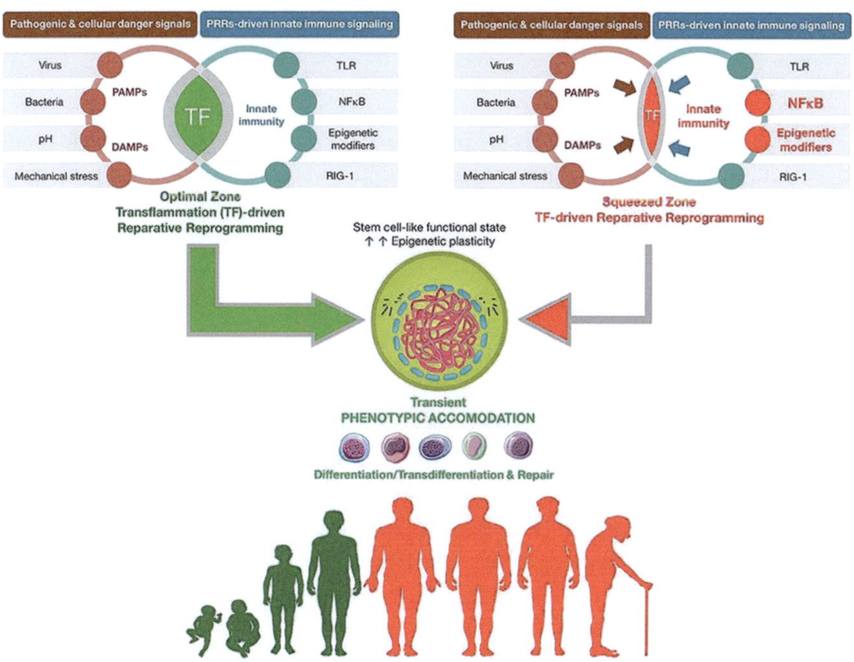

Fig. 9.3 Reprogramming as a result of inflammation and individualized variation [2]

Shinya Yamanaka won the Nobel Prize in Physiology or Medicine in November 2012 for his discovery that somatic cells could be induced to form pluripotent stem cells by overexpression of four transcriptional factors (*Oct4*, *Sox2*, *KLF4* and *cMyc*, or *OSKM*). Yamanaka and others subsequently revealed that these transcriptional factors activate a network of genes required for pluripotency. Subsequently, it was reported that the Yamanaka method to generate pluripotent stem cells was dependent on the activation of innate immunity [3].

In a previously published study, data revealed that the retroviral vectors carrying the Yamanaka factors were more than vehicles. Indeed, the retroviral vectors, by activating the innate immune signaling (TLR3) were integral to the reprogramming effort. Specifically, when researchers exposed human fibroblasts to cell permeant peptides of the Yamanaka factors, they were not able to obtain efficient generation of pluripotent stem cells. Furthermore, the activation of genes in the downstream core pluripotency network was delayed and depressed [3].

Damage-associated molecular patterns (DAMPs) or pathogen-associated molecular patterns (PAMPs) can be sensed by an array of receptors. Unpublished evidence shows that the activation of a wide range of these pattern recognition receptors (PRRs) causes global changes in epigenetic modifiers, which increase the probability for an open chromatin state, a process termed "transflammation." In this state of epigenetic plasticity, the cell is poised to respond to microenvironmental cues that may change its phenotype. This is an unrecognized response of innate immune activation, which results in greater epigenetic plasticity. This response may permit an adaptive form of phenotypic fluidity so as to have a better response to environmental stresses [3].

Because T and B lymphocytes are professional immune cells, they may be prone to epigenetic plasticity. Indeed, it has previously shown that such professional immune cells are easier to reprogram to pluripotency than hematopoietic stem cells and that this effect is in part dependent upon NFkB expression [3].

Epigenetic Changes in Behavioral, Neurodevelopmental, Psychiatric, and Neurodegenerative Diseases

Gene–environment interaction research in psychiatry is a newly developing field and part of neuroscience. It is widely accepted that mental disorders have a significant environmental contribution as a causal factor, but research should be multidisciplinary in order to overcome the genetic predisposition. The complementary disciplines of epidemiology, experimental neuroscience, and neuroimaging techniques fuel gene–environment hypotheses and it is a method for investigating underlying neural mechanisms [4].

Human behavior is a complex characteristic to analyze. How the organism interacts with the environment to produce the variation of behaviors and reactions to stimuli has been a focus of psychology since its initiation as a scientific discipline.

There is now no question that cumulative experiences throughout life history interact with genetic predisposition to shape the individual's behavior. Recent evidence suggests that events in past generations may also influence how an individual responds to events in their own life history by epigenetic mechanisms [5].

"Genetic changes are stable and rarely reversed, whereas epigenetic changes are often reversed." The early stages of life from conception to early childhood are the time of maximal neuronal plasticity. Although the individual's capacity to respond to environmental changes or insults is based on heritable phenotypic variation, it is during this early period that hormones and genotype predispose an individual's responses to future experiences as well as the susceptibility to develop disorders. An important interface between the environment (either internal or external) and the genotype is accomplished by epigenetic modifications (Fig. 9.4). Due to the fact that epigenetic marks are heritable, it seems that the experiences of previous generations can modify regulatory factors affecting gene expression. However, behavior is substantially influenced not only as an affector, but also as an effector system [5].

The genetic predisposition cannot be neglected and it is expressed from the perspective of the interaction of inheritance the so-called nature or instincts. Initial studies emphasized that such behaviors were products of natural selection, the result of genes acting in the brain to generate behaviors that were beneficial for survival, but innate. Thus, various levels of CNS biological organization support the development and display of these species typical behaviors [5].

The comparative approach in psychology emphasizes the dynamic, interactive process between intrinsic imprinted behaviors characteristic of the species with environmental and social interaction. The final behavioral phenotype consists of the

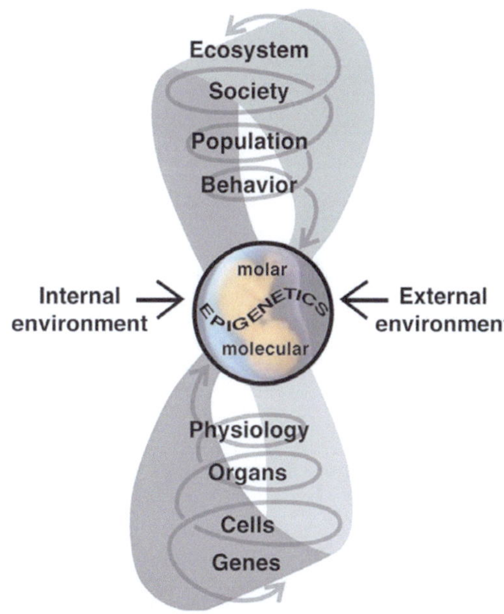

Fig. 9.4 Environment, genes, and personal associations related to experience synthesize a complex network that modifies behavior

foundation for psychobiology, a vibrant field of psychiatry. In this context, the way the individual is grown influences his behavior as an adult and it is likely that the activity of the neural circuits that underlies these behaviors must also be affected. However, this organizational principle is challenged by epigenetic modifications related to the individual's own life history (context-dependent) or inherited from previous generations (gene-dependent) (Fig. 9.5) [5].

The best example of a context-dependent epigenetic modification on behavior is the experiments of Meaney and colleagues. In a long series of elegant studies, this group has demonstrated that the nature and amount of care a pup receives from the mother modulates the offspring reaction to stress later in life largely by effects on the glucocorticoid receptor (GR) in the hippocampus. This maternal effect can cross generations, but its heritability depends upon the pup's experience in the first week of life. Recently this group has documented that being reared by a high quality mother results in the expression of the transcription factor A (*NGFI-A*), a nerve growth factor inducible protein, that binds to the first exon of the GR gene, resulting in increased expression of GR. High-quality maternal care during this critical period demethylates *NGFI-A* and it acetylates histones. Cross-fostering can reverse these molecular and behavioral changes, by the infusion of methionine, a histone deacetylase inhibitor into the hippocampus, which could can reverse these events [5].

Another well-known experimental observation is the determination of a transgenerational epigenetic imprint on mate preference behavior in rats (Fig. 9.6). The left panel shows that follow-up of the behavior in three generations is detected after gestational exposure to vinclozolin, a common-use fungicide with endocrine-disrupting (EDC) properties. The right panel illustrates the testing apparatus for mate preference. Two groups of animals were tested. The control group, which was not exposed to vinclozolin, was the F3 generation of a lineage. Third-generation females from the ED lineage and the control lineage were tested with males from

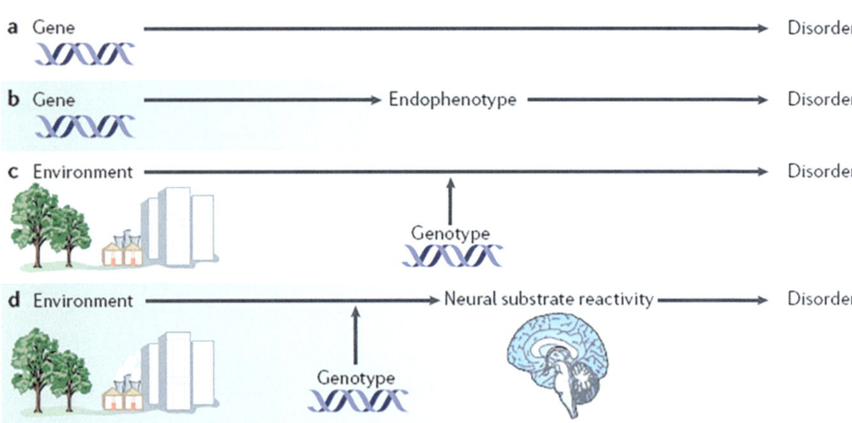

Fig. 9.5 The complexity of the structure of behaviors. Genes, internal processes, the influence of the environment, and the accessibility of neuronal circuits built a disorder

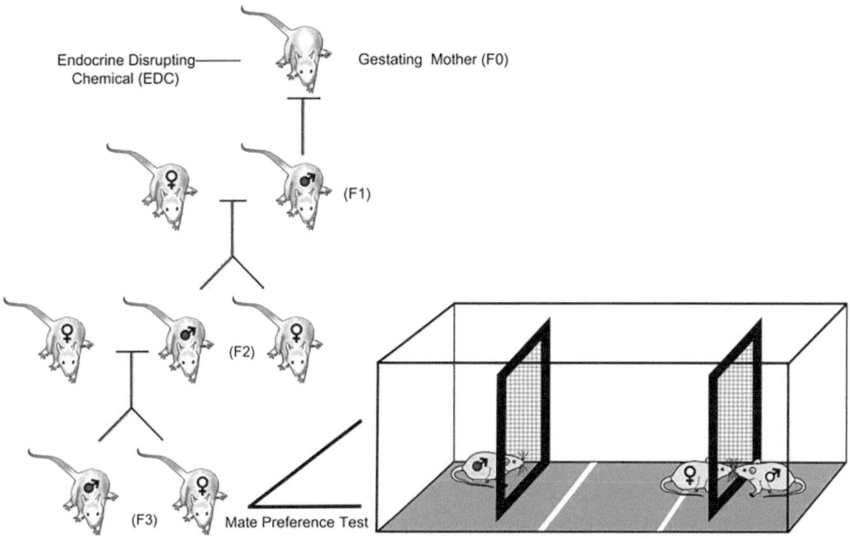

Fig. 9.6 The mate preference of a third-generation rats, one exposed and one not exposed to vinclozolin

both lineages in simultaneous mate preference tests; males from the ED lineage (indicated by red-filled male symbols) and the control lineage (black-white symbol) were similarly tested with females of both stimulus types. The trials are conducted under dim red light during the nocturnal phase. The female could move freely in their chamber but separated from the stimulus males by a wire mesh. This enabled the animals to communicate by olfactory, pheromonal, or behavioral cues, but physical interaction was limited to touching across the wire mesh [5].

The results were sex-specific. The females discriminate and they prefer males who do not have a history of exposure, while males do not exhibit such a preference. In social encounters with rodents and other animals, time is spent engaged in mutual facial investigation. This appears to be a critical aspect of the assessment process and it may underlie our finding that males investigate females equally, while females spend a greater amount of time investigating males from the control lineage. It is known that pheromones from the vomeronasal organ and urine are involved in mate recognition in rodents. Methylation analysis revealed that the Major Urinary Protein 4 (MUP4) is one of the candidates imprinted genes induced in the vinclozolin generation males [5].

In more detail, neurodevelopmental pathologies, such as mental retardation and autism; neurofunctional disorders such as schizophrenia and bipolar disorder; and neurodegenerative disorders such as Alzheimer's disease, Parkinson's disease, and Huntington's disease are now recognized to have epigenetic perturbation as a causative factor [8]. Neurodevelopmental disorders are a group of conditions characterized by impairments of social skills or intelligence with onset in the developmental period. According to the Diagnostic and Statistical Manual of Mental Disorders, Fifth Edition (DSM-V), they include intellectual disability (Intellectual Developmental Disorder), communication disorders, autism spectrum disorder (ASD), attention-deficit hyperactivity disorder (ADHD), specific learning disorders, and motor disorders [9].

Childhood Neurodevelopmental Diseases

To start with, mental retardation has a genetic and nongenetic component genetic factors are confirmed in less than 10% of cases [8].

Autism or autism spectrum disorder (ASD) is a childhood neurodevelopmental disorder that is behaviorally defined and psychiatrically diagnosed based on a spectrum of qualitative impairments in social interaction and communication including restricted and stereotyped patterns of behavior, interests, and activities. Despite controversies regarding its etiology, an ASD diagnosis remains under the diagnostic criteria of a purely psychiatric disorder [10].

In recent years, many studies indicated that children with an autism spectrum disorder diagnosis have brain pathology suggestive of ongoing neuroinflammation in different regions of their brain mainly the cerebral cortex and cerebellum. Evidence of neuroinflammation in ASD includes microglial activation and a proinflammatory profile of cytokines [10].

It is interesting to notice that the prevalence of the disease increases rapidly in developed countries, affecting as many as 1/80 of children in the USA [8]. In 2011–2013, the estimated prevalence of ASD and other neurodevelopmental disabilities in children aged 3–17 years in the USA were 2.24% (1 in 45) and 3.57% (1 in 28), respectively. Alarmingly, the Autism and Developmental Disabilities Monitoring (ADDM) Network reported that surveils 8-year-old children showed that the prevalence of ASD has increased from 0.66% to 1.46% over 10 years from 2002 to 2012. Similarly, the prevalence of ADHD in children from 3 to 17 years also showed a 25.64% increase from 7.8% to 9.8% from 2003 to 2015. The reasons for the increasing trend in ASD and ADHD are still controversial; explanations include changes in diagnostic criteria, reporting methods, or other factors such as environment, culture, and social-economic status that may affect the prevalence of these neurodevelopmental disorders. A previous study found indeed that the changes in diagnostic criteria alone account for only 33% of diagnosed cases and a combination of changes in diagnostic criteria and reporting methods account for 60% of the increase in prevalence of ASD [9].

Genetic factors predispose probably to the sensitivity of the individual to different environmental insults. High risk for developing autism has been associated with mutations leading to decreased phosphatase and tensin homolog (*PTEN*) and tuberous sclerosis protein 1 and 2 (*TSC1/2*). These proteins are upstream inhibitors of the mammalian target of rapamycin (*mTOR*) which leads to microglia and mast cell proliferation. Recent papers reported that offspring of maternal immune activation in mice led to increased IL-6 and IL-17, and contributed to ASD-related behaviors [11].

While initial studies suggested a strong genetic heritability of autism, recent studies with larger sample sizes have demonstrated a lesser influence, including a study of over 14,000 children with autism in Sweden that demonstrated a heritability of 50%, supporting an equally strong role for environmental risk factors [11]. In the largest twin study to date, it was found that environment accounted for 55% of the variance in autism risk among twins. Genetic factors or critical time periods may influence how xenobiotics or noninherited conditions alter brain connectivity and determine whether the exposure results in autism as opposed to other deficits. Large gene-environment studies are one approach to capture the complexity of this disorder [6].

The list of suspected etiological agents for ASD begins with known neurotoxins and neurodevelopmental toxins but also extends to compounds that operate by mechanisms outside the central nervous system (CNS), such as immune dysregulation, altered lipid metabolism, and mitochondrial dysfunction, among others. Other environmental factors acting during this period include infections, medications, and pregnancy complications. Evidence regarding a protective association of maternal nutrition, particularly folic acid, is strong [6]. Maternal smoking, plastic-derived chemicals (bisphenol A), or others in the close environment such as BPA, DEHP and DBP, dioxin, persistent organic pollutants, chemical metals, and hydrocarbons are among a number of chemicals described in the literature (Fig. 9.7) [6].

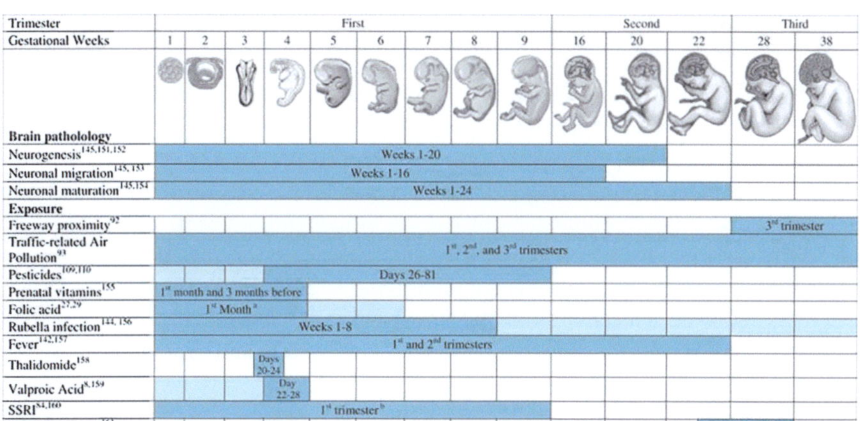

Fig. 9.7 The time during gestation that different chemical exposures interfere with neuronal development

Psychiatric Disorders

The first approach assumed direct linear relations between genes and behavior. The goal of this approach has been to correlate psychiatric disorders with individual variations in DNA sequence. This has been attempted using both linkage analysis and association analysis, with regard to many psychiatric conditions such as depression, schizophrenia, and addiction. Although a few genes have replicated evidence of associations with these diseases, the overall progress has been slow. Because of inconsistent findings, many scientists have despaired in their efforts to find a straightforward association between genotype and diagnosis [4].

Another approach to psychiatric genetics is to incorporate information about the environment. This gene–environment interaction approach differs fundamentally from the main gene-effect approaches, with regard to the assumptions about the causes of psychiatric disorders [4].

Gene–environment interaction studies in psychiatry are new, but some of the initial findings are intriguing. An example is the hypothesis that a functional polymorphism in the promoter region of the gene encoding the neurotransmitter-metabolizing enzyme monoamine oxidase A *(MAOA)* would moderate the effect of child maltreatment in the cycle of violence. Results showed that maltreated children, whose genotype conferred low levels of MAOA expression, more often developed conduct disorder, antisocial personality, and adult violent crime than children with a high-activity MAOA genotype. In another study, it was proposed that a functional polymorphism in the promoter region of the serotonin transporter (5-HTT) gene would moderate the influence of stressful life events on depression. Individuals with one or two copies of the 5-HTT "short" allele exhibited more depressive symptoms, diagnosable depression, and suicidality following stressful life events than individuals with two copies of the "long" allele (Fig. 9.8) [4].

The characterization of subjects' genetic vulnerability as opposed to their resilience on environmental stress needs to move beyond single genetic polymorphisms. New approaches will use information about biological pathways to identify gene systems and study sets of genetic polymorphisms that are active in the pathophysiology of each disorder [4].

There are a lot of important questions to be answered. Why do some people who are exposed to an environmental stress develop mental disorders, while others do not? Why do some disorders excessively afflict one sex or one age group? How can two people experiencing the same environmental stress later develop very different disorders? How does an environmental stress, especially one that is psycho-social in its nature, "get" under the skin to alter the nervous system and generate mental disorders? [12].

There is an increasingly health interactive view of how the micro and macro environment interacts with genetic risk to shape vulnerability for psychosis. A comprehensive review of the incidence of schizophrenia corrects the once-fashionable hypothesis that since there is no geographic variation in the incidence of schizophrenia, the cause must be genetic. However, data demonstrated widespread

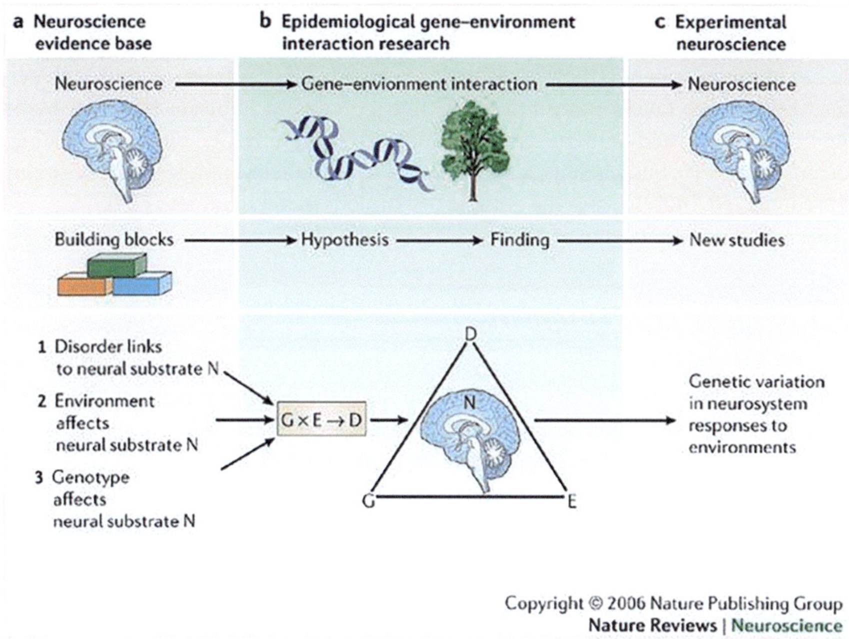

Fig. 9.8 The building blocks in the etiology of behavioral disorders [4]

geographic, temporal, ethnic, and other demographic variations in the incidence of schizophrenia, giving a new impulse to interactive research and models of disease causation. There is good evidence for environmental factors that contribute interactively with genetic factors to the population psychosis morbidity force: cannabis, urbanicity, minority status, early trauma, and prenatal and early childhood exposures. For example, confirmation that cannabis use is not only associated with schizophrenia, but the use tends to manifest expressions of the psychosis phenotype, particularly isolated psychotic symptoms, has been reported with a prevalence of 5–15%. The proxy environmental risk factor of growing up in an urban environment explains around 30% of all schizophrenia incidence, making it the most important of all environmental factors [13].

Inflammatory responses play an important role in the pathophysiology of numerous mental illnesses including depression. In animal studies, toll-like receptors (TLR) are representative innate immune receptors that mediate inflammatory responses. TLR2- and TLR4-deficient mice failed to show social defeat stress-induced behaviors, indicating that the innate immune system plays a key role in mental disease evolvement and it points out the importance of anti-inflammatory treatment in mitigating depression. In fact, previously, increased levels of typical inflammatory cytokines and chemokines, such as interleukin (IL)-6 and C-reactive protein, are frequently observed in the serum and/or plasma of depressed patients. To be precise, during depression, the brain's immune cell microglial cells are activated by switching on the sensitivity of TLR2 and TLR4 receptors, inducing afterward inflammatory responses and leading to depressive symptoms. Moreover, the blood–brain barrier (BBB) of depressed patients is weaker, leading to high permeability; which favors more monocyte lineage cells to infiltrate from the blood into brain tissues, and to brain structures including the prefrontal cortex (PFC), hippocampus (HIPP), and nucleus accumbens (NAc). All these brain regions are highly interconnected, highlighting the complexity of depression treatment [14].

Neurodegenerative Disorders

Alzheimer's disease (AD) stems from all multifactor neurodegenerative disorders; it is a primary degenerative disease of the brain caused by the deposition of proteins with pathological structures, which causes neuronal death and loss of inter-neuron connections. Characteristic histopathological symptoms include the accumulation of amyloid plaques in the cerebral cortex and pathological Tau proteins in neurons and neuroglia [15].

A major factor leading to the inflammation of CNS is the increase of ROS induced by environmental pollutants as highlighted in the previous sessions. Phagocytosis and excretion of multiple inflammatory factors lead to generalized inflammation. Hyperactivation of microglia and astrocytes induces neuron apoptosis and damages the blood–brain barrier which is crucial to the integrity and correct functioning of the nervous system. This facilitates the formation of insoluble β-amyloid (Aβ), which shows neurotoxic properties [15].

Mitochondrial damage is considered the main factor determining the pathomechanism of Parkinson's disease. The disease is caused by the gradual atrophy of dopaminergic neurons in the substantia nigra, which leads to dopamine deficiency in the striatum. Dopamine is an important neurotransmitter synthesized and released by dopaminergic neurons of the central nervous system. The pathomorphological characteristics of PD reveal the intraneuronal presence of Lewy's bodies, which leads to neuron death caused by the mechanism of apoptosis. A number of factors have to be taken into account in the context of PD, including both genetic (the first identified mutation was of the α-synuclein gene PARK1 in chromosome) and environmental, in particular the effects of free radicals and oxidative stress [15].

Huntington's disease (HD) is caused by a mutation entailing the presence of an increased number of copies of three CAG nucleotides in the *IT15* gene. As a consequence, the abnormal mHtt protein is formed (a mutated version of the Huntingtin protein), which leads to irreversible neuronal damage, mainly in the basal ganglia. Huntington's disease is inherited by an autosomal dominant mutation and it is characterized by a high degree of penetrance. Experimental data have been published that indicate the key role of mitochondrial dysfunction in the pathogenesis of HD. As observed in those studies, a key factor influencing the toxic effects of mHtt is the direct or indirect interaction between this protein and mitochondria, most likely triggering changes in mitochondrial membranes, which leads to increased production of ROS and the dysfunction of the respiratory chain (50). An increasing number of studies currently suggest that dyshomeostasis of transition metals may constitute a part of HD pathogenesis. In particular, iron (Fe) and copper (Cu) play the roles of the pathology's mediators [15].

Multiple sclerosis (SM, Sclerosis multiplex) is a chronic, inflammatory degenerative disease of the central nervous system. It is characterized by multifocal and time-scattered emergence of inflammatory demyelinated lesions causing damage to and loss of axons. The pathogenesis of the disease is complex and not yet fully characterized. Leukocyte rolling from blood to the central nervous system results in the activation of microglia that release proinflammatory cytokines leading to extensive damage, primarily of the deep white matter in the region of optic nerves, corpus callosum, and periventricular matter, as well as in subtentorial regions and in the vicinity of the spinal cord, particularly in its cervical section. The inflammatory process is the driving force of the demyelination [15]. Overall, most degenerative diseases have a common denominator, oxidative stress.

Conclusion

As described, the epigenome is the 'fingerprint' of the environment on our genes and it is unique for every human. It is fixed during "critical windows" of development, but it is modified during the whole lifespan. The epigenetic mechanisms are analyzed in detail because they can explain the mechanism of action of chemicals in the generation and the etiology of disease. These parameters affect not only the individual but the following generations since these modifications can be transmitted to the following generations, emphasizing the importance of healthy lifestyle and environment.

The various chemicals are abundant in everyday life including the workplace and house utilities. They overflow the soil and the food, the water we use and drink, and the air we breathe. Depending on the geographical area of habitat, different metals accumulate in the food chain and they could have disastrous health outcomes. Products of human activity, especially the combustion of fossil fuels used in amazing rhythm for transportation have become a major issue for big cities. In addition, thousands of chemicals, such as plastics, are produced and we do not really know their consequences to health.

People are concerned about the rise of diseases such as neurodevelopmental and neurodegenerative diseases as well as the peaking of cancer, but they probably do not know how they can participate in the limitation of these problems. From a scientific point of view, research has offered a lot of interesting results regarding the implication of these chemicals on health and the mechanisms of the generation of disease, which rocketed the production of commercially available new treatments.

Although this study aspired to give information regarding the nature of each chemical, the impact on health, and the mechanism of disease due to exposure to the specific chemical, it is an effort to concentrate and sensitize scientists on the importance of moving toward more environmentally friendly practices.

It is obvious from all the above that a clean environment is essential for human health and well-being. Environmental awareness is the process of understanding the fragility of the environment and the importance of its protection. Promoting environmental awareness is an easy way to become an environmental steward and participate in creating a brighter future for the next generations. When people "go green," they are practicing *environmental awareness*. The term means exactly what you expect to being aware of the natural environment and making choices that benefit rather than harm. In recent years, environmental awareness has gained increased attention. It is fair to say it has become a bit of a trend: from environmentally friendly tiny homes to organic food, environmental awareness is a hot topic. However, it is nothing new and it ought to be more than just a passing fad, because of its importance.

Here are several cause-and-effect problems that harm our environment: *Oil exploitation*. This issue is one that causes a great deal of environmental destruction. Our dependence on fossil fuels is a global addiction that affects every aspect of the world. Oil destroys the atmosphere, soil, and marine life. Combustion of fossil fuels adds to the increased atmospheric CO_2, which in turn causes the progression of global warming and ocean acidification.

Deforestation. Millions of acres of forest are cut down for industrial benefits, such as large-scale farming, oil mining, and the production of paper goods. Deforestation causes wildlife and biodiversity extinction. The International Union for Conservation of Nature (IUCN) has a Red List of environmentally threatened species with up-to-date information. Oftentimes, the cause for their threatened existence is listed as loss of habitat as it is for many Amazonian species. *Soil contamination and erosion*. Soil contamination and degradation are two important considerations that affect soil formation and quality. Soil contamination can be defined as the critical increase of chemicals and other toxic materials in the natural soil environment, while performing industrial, domestic, or agricultural activities, or by improper disposal of wastes.

Production of plastic goods. Currently, our society creates a great deal of waste, and much of that waste is comprised of plastic. According to the Environmental Protection Agency (EPA) in 2010 alone 31 million tons of plastic waste was created. This waste ends up all over the globe in both land and water, a good example is the Great Pacific Garbage Patch. Not only is plastic waste an issue, but the production of plastic is also dependent on fossil fuel combustion. According to the U.S. Energy Information Administration (EIA) in 2010, 191 million barrels of liquid petroleum gases (LPG) and natural gas liquids (NGL) were used in the USA alone to produce plastic goods.

Chemicals and other harmful substances used to process ores can enter waterways and the natural environment, when not managed appropriately.

The production of nonenergy minerals and metals is highly concentrated, with 10 countries accounting for almost 70% of the global nonenergy mineral, metals, and mineral fuels production in 2016.

On the other hand, *air pollution* is a major problem to be solved in big cities due to internal immigration. About 26% of the incoming solar energy is reflected to space by the atmosphere and clouds, and 19% is absorbed. The warmth of the earth's surface and atmosphere is essential for the creation and sustainability of life.

However, the increase in radioactive forcing from human activity contributes to further warming of the planet. The so-called *greenhouse gases* that are produced include water vapor (36–70%), carbon dioxide (9–26%), and other emissions like sulfides, CO, NO_x, methane, and others. Carbon dioxide (CO_2) and other greenhouse gases act like a blanket, absorbing radiation and preventing it from escaping into outer space. The net effect is the gradual heating of the Earth's atmosphere and surface, a process known as global warming.

Climate change is the term for the alternation of the Earth's weather patterns. The greenhouse effect is a very large (if not the largest) contributor to this problem. Climate change is increasing the frequency and intensity of extreme weather events such as heat waves, droughts, floods, and tropical cyclones; aggravating water management problems; reducing agricultural production and food security; increasing health risks; damaging critical infrastructure; and interrupting the provision of basic services such water and sanitation, energy and transport. Potential risks to health include deaths from thermal extremes and weather disasters, vector-borne diseases, a higher incidence of food-related and waterborne infections, photochemical air pollutants, and conflict over depleted natural resources. Oceans have warmed, the amounts of snow and ice have diminished and sea level has risen. Within the last hundred years, the global average sea level rose by 20 cm approximately, as oceans expanded. The Arctic's Sea ice extent has shrunk in every decade since the 80s.

Building materials choices have their greatest impact on *indoor air quality*. Selecting the right finishes, flooring, ductwork, and insulation ensures that living spaces do not expose residents to dangerous chemicals or allergens. *Construction projects* around the world are having a significant impact on our environment, on a local and global scale. Each stage of the construction process has a measurable impact. Regardless of the location, size, or purpose of the construction project, environmental impacts are inevitable; they run the risk of affecting nature, the built environment, and the surrounding neighbors. Impacts can range from noise to pollution, wildlife to flora and fauna.

In recent decades, as a result of increased environmental awareness and stringent environmental regulations, environmental issues in supply chain design and management have received considerable attention from researchers and decision-makers in the biomass-based production sector. Environmental sustainability is critical in the design and management of a supply chain network, by simultaneously handling land and natural resource utilization, environmental management, waste management, and pollution prevention practices.

Executive Summary

Reprogramming

- Massive demethylation and imprinting erasure at germ lines and in preimplantation period
- Reprogramming is a more efficient programming to environmental insults and it is also detected in mesenchymal somatic cells
- The Weismann Barrier: somatic cells cannot acquire epigenetic modification to decedent lineages
- Reprogramming of somatic cells is transient, otherwise it leads to chronic inflammation and disease (transflammation)
- The Yamanaka cocktail activates a network of genes required for pluripotency due to activation of innate immunity by TLR receptors

Epigenetic Changes in Behavioral, Psychiatric Neurodevelopmental, and Neurogenerative Diseases

- Gene–environment interaction research in psychiatry is a newly developing field and part of neuroscience
- Neurodevelopmental pathologies such as mental retardation and autism, neurofunctional disorders such as schizophrenia and bipolar disorder, and neurodegenerative disorders such as Alzheimer's disease, Parkinson's disease, and Huntington's disease are related to epigenetic alterations
- Genetic moderation implies that differences between individuals, originating in the DNA sequence, bring about differences between individuals in their resilience or vulnerability to environmental stimuli and the synthesis of these characteristics underlies the background for the development of many pathological conditions of the mind

References

Reprogramming

1. Reik W, Dean W, Walter J. Epigenetic reprogramming in mammalian development. Science. 2001;293(5532):1089–93.
2. Sakurada K. Environmental epigenetic modifications and reprogramming-recalcitrant genes. Stem Cell Res. 2010;4(3):157–64.
3. Cooke JP, Sayed N, Lee J, et al. Innate immunity and epigenetic plasticity in cellular reprogramming. Curr Opin Genet Dev. 2014;28:89–91.

Epigenetic Changes in Behavioral, Psychiatric Neurodevelopmental, and Neurodegenerative Diseases

4. Caspi A, Moffitt TE. Gene–environment interactions in psychiatry: joining forces with neuroscience. Nat Rev Neurosci. 2006;7(7):583–90.
5. Crews D. Epigenetics and its implications for behavioral neuroendocrinology. Front Neuroendocrinol. 2008;29(3):344–57.
6. Lyall K, Schmidt RJ, Hertz-Picciotto I. Maternal lifestyle and environmental risk factors for autism spectrum disorders. Int J Epidemiol. 2014;43(2):443–64.
7. Kalkbrenner AE, Schmidt RJ, Penlesky AC. Environmental chemical exposures and autism spectrum disorders: a review of the epidemiological evidence. Curr Probl Pediatr Adolesc Health Care. 2014;44(10):277–318.
8. Zahir FR, Brown CJ. Epigenetic impacts on neurodevelopment: pathophysiological mechanisms and genetic modes of action. Pediatr Res. 2011;69(5 Pt 2):92R–100R.
9. Tran NQV, Miyake K. Neurodevelopmental disorders and environmental toxicants: epigenetics as an underlying mechanism. Int J Genom. 2017;2017:7526592.
10. Kern JK, Geier DA, Sykes LK, et al. Relevance of neuroinflammation and encephalitis in autism. Front Cell Neurosci. 2015;9:519.
11. Theoharides TC, Asadi S, Patel AB. Focal brain inflammation and autism. J Neuroinflammation. 2013;10:46.
12. Ehlers S, Kaufmann SH. Infection, inflammation, and chronic diseases: consequences of a modern lifestyle. Trends Immunol. 2010;31(5):184–90.
13. van Os J, Krabbendam L, Myin-Germeys I, et al. The schizophrenia envirome. Curr Opin Psychiatry. 2005;18(2):141–5.
14. Huang Q, Liu H, Suzuki K, et al. Linking what we eat to our mood: a review of diet, dietary antioxidants, and depression. Antioxidants (Basel). 2019;8(9):376.
15. Winiarska-Mieczan A, Baranowska-Wójcik E, Kwiecień M, et al. The role of dietary antioxidants in the pathogenesis of neurodegenerative diseases and their impact on cerebral oxidoreductive balance. Nutrients. 2020;12(2):435.

Index

A

Antioxidant mechanism, 44, 341, 346

B

Bisphenols, 42, 243, 264, 272, 275–279, 299, 307, 385

C

Cancer development, 82, 326, 328, 364
Carotenoids, 344–347, 352–355, 373

E

Effects on health, 48–52, 58–59, 66, 73–78, 86–89, 105–107, 115–119, 129–131, 139, 144–145, 155–156, 159–162, 168–169, 177–180, 183–184, 186–187, 191–192, 197–199, 218–219, 225, 237, 239, 246–247, 250–251, 256, 257, 259, 265–266, 274–277, 281–283, 295–306, 329, 342–347, 362–363
Endocrine disruptors (ED), 97, 106, 119, 202, 218, 238, 259, 263–307
Epigenetic changes, 9, 20, 24, 26, 27, 34, 52, 59, 67, 98, 99, 110, 112, 135, 138, 140–141, 150, 182, 185, 202, 203, 218, 242, 256, 259, 279, 317, 319, 380–389, 392
Epigenetic mechanisms, 4, 5, 8, 9, 13, 15, 28, 59–61, 66–68, 80–83, 98–100, 110–113, 124–127, 135–138, 148–153, 167, 182, 185, 195, 205, 228–229, 231–236, 238–240, 242–244, 247–248, 253–255, 260, 279, 323, 381, 389
Epigenome, 1–28, 60, 63, 67, 79, 152, 363, 389
Evolutionary models of multifactorial diseases, 24–28, 311

F

Food as an antioxidant mechanism, 341

G

Gene-environmental interactions in behavior and psychiatric disease, 380, 386, 392
Glutathione, 44–50, 52, 59–61, 63, 64, 78, 83, 95, 96, 100, 134, 140, 147, 172, 175, 177, 187, 188, 200, 201, 205, 237, 240, 247, 255, 288, 342, 343, 357, 359, 365

H

Heterochromatin, 7, 10–12, 18, 19, 29, 67, 318
Hypomethylation and disease, 31–35

M

Mechanism of disease, 390
Mechanisms of aging, 312–316
Mechanisms of development of disease, 31
Metals, 37, 63, 227, 265, 347, 385
Models related to the development of cancer, 31, 323, 327
Multifactorial diseases, 1–3, 27, 28, 31, 117, 265, 287

A. Salavoura, *Chemical Environmental Pollutants and their Effect on Health*,
https://doi.org/10.1007/978-3-031-80271-3

O
Oxidative stress, 37–52, 59, 60, 64, 67, 68, 80, 91, 92, 94–96, 106, 107, 109, 110, 112, 119–122, 131–138, 140, 141, 146–148, 163, 173, 184, 186–188, 193, 194, 200–203, 206, 252, 258, 259, 272, 293, 317, 322, 331–333, 335, 340–342, 345–347, 352, 358, 364, 367, 372, 388, 389

P
Parabens, 243, 264, 280–282, 287, 289, 290, 296, 297, 307
Persistent organic pollutants, 215–219, 227, 258, 385
Pesticides, 26, 42, 49, 70, 71, 102, 129, 144, 202, 217, 223–225, 230, 231, 236–239, 242, 244, 256, 259, 263–265, 267–271, 273, 283, 305, 307, 308
Phthalates, 243, 244, 264, 265, 267–269, 272–275, 290, 296–300, 307
Polyphenols, 337, 341, 347–359, 365, 373
Products of everyday use, 276

R
Reactive oxygen species (ROS), 37–48, 52, 60, 78–80, 91, 95, 96, 98, 99, 107, 110, 118, 131–135, 146, 147, 163, 172, 173, 176, 179, 180, 184, 187, 194, 195, 201, 202, 204–206, 221, 232, 234, 235, 237, 252, 254, 302, 313, 321, 322, 340, 346, 357, 358, 366, 367, 372, 388, 389
Reprogramming at critical lifetime windows, 377
Reprogramming of somatic cells, 318, 377

S
Senescence, 19, 113, 312–328, 378
Source of exposure, 70, 85, 114, 115, 142, 286

T
Toxicity of metals, 49, 52

V
Vitamins, 41, 44, 83, 91, 93, 102, 128, 153–156, 183, 193, 201, 204, 255, 258, 287, 293, 295, 334, 341–345, 352, 354–356, 359, 365, 372, 373

Y
The Yamanaka cocktail of ageing, 317

MIX
Papier aus verantwortungsvollen Quellen
Paper from responsible sources
FSC® C105338

If you have any concerns about our products,
you can contact us on
ProductSafety@springernature.com

In case Publisher is established outside the EU,
the EU authorized representative is:
Springer Nature Customer Service Center GmbH
Europaplatz 3, 69115 Heidelberg, Germany

Printed by Libri Plureos GmbH
in Hamburg, Germany